APPLIED BIOMEDICAL ENGINEERING MECHANICS

APPLIED BIOMEDICAL ENGINEERING MECHANICS

Dhanjoo N. Ghista

CRC Press
Taylor & Francis Group
Boca Raton London New York

CRC Press is an imprint of the
Taylor & Francis Group, an **informa** business

CRC Press
Taylor & Francis Group
6000 Broken Sound Parkway NW, Suite 300
Boca Raton, FL 33487-2742

© 2009 by Taylor & Francis Group, LLC
CRC Press is an imprint of Taylor & Francis Group, an Informa business

Library of Congress Cataloging-in-Publication Data

Ghista, Dhanjoo N.
  Applied biomedical engineering mechanics / Dhanjoo N. Ghista.
     p. ; cm.
  Includes bibliographical references and index.
  ISBN 978-0-8247-5831-8 (alk. paper)
  1. Biomechanics. 2. Human mechanics. I. Title.
  [DNLM: 1. Biomechanics. 2. Cardiovascular Physiology. 3. Diagnostic Techniques and Procedures. 4. Movement--physiology. 5. Musculoskeletal Physiology. 6. Sports--physiology. WE 103 G422a 2008]
  QP303.G54 2008
  612.7′6--dc22   2008003601

Visit the Taylor & Francis Web site at
http://www.taylorandfrancis.com

and the CRC Press Web site at
http://www.crcpress.com

In Loving Memory of

My Dearest

Father and Mother,

Grandparents and Great-Aunt,

Maternal and Paternal Uncles and Aunts

Contents

Section III Glucose–Insulin Regulation (in Diabetes) Engineering Mechanics

Section IV Orthopedic Engineering Mechanics

Section V Fitness and Sports Engineering Mechanics

Preface

There has been a long-standing need for a comprehensive book on bio-medical engineering that covers the analyses and applications of biomedical and physiological systems as well as human fitness and sports. In addressing this need, this book can serve as a definitive textbook as well as a major reference source.

The book's contents include the following:

1. An introductory chapter that develops the foundation of how physiological systems and their assessment can be described by means of governing differential equations, whose parameters can be combined into nondimensional physiological diagnostic indices
2. A section on cardiological engineering mechanics, including three chapters dealing with cardiac mechanics, left-ventricular contractility indices, and vascular mechanics
3. A section on pulmonary engineering mechanics, including three chapters on lung ventilation and disease diagnosis, lung gas-transfer mechanics and determination of O_2 and CO_2 diffusion coefficients, and lung-ventilatory indices for extubation of chronic obstructive lung disease patients from mechanical ventilatory support
4. A section on glucose–insulin regulation (in diabetes) engineering mechanics, which has three chapters covering glucose–insulin regulatory analysis, responses of glucose and insulin to glucose tolerance tests, and indices for differential diagnosis of diabetic patients and those at risk from becoming diabetic
5. A section on orthopedic engineering mechanics, involving three chapters dealing with the analyses and design of internal bone fracture-fixation plates as well as the design analyses of human spinal vertebral body and intervertebral disc as optimally designed human body structures
6. A section on fitness and sports engineering mechanics, on (i) heart-rate variation during and after exercising on treadmill, optimal walking, and jogging modes requiring minimal work expenditure, and analyses of hip joints to determine their stiffness and damping coefficients, and (ii) analyses of sports events

(namely soccer, baseball, basketball, and gymnastics) delineating the analytical basis for intricacies of their techniques and perform-ance (such as the basis of curving soccer kicks, baseball throws and batting technique, and high-performance of Yurchenko layout vault)

In addressing this comprehensive range of topics, the book covers a wide spectrum of engineering mechanics disciplines of solid and fluid mechanics, dynamics and vibrations, gas diffusion and transfer, and control systems. The book can therefore be ideally employed as a textbook for a biomedical engineering course at the senior undergraduate level or at the graduate level.

In the first section on cardiological engineering mechanics, Chapter 2 describes left ventricular (LV) mechanics. This chapter deals with (1) deter-mination of the pressure drop across a stenotic valve, (2) determination of the constitutive properties of mitral and aortic valves from their static and dynamic analyses, (3) determination of the intra-LV blood flow velocity and pressure distributions (in normal and myocardial infarcted cases) before and after nitroglycerin administration as a means of deciding if coronary bypass surgery would benefit patients with myocardial infarcts, and (4) LV passive and active elastances, as measures of LV pressure dynamics response to LV volume changes and of LV contractility.

Chapter 3 deals with left ventricular contractility indices. Here, we first determine the wall stress (σ) in an ellipsoidal model of the LV, normalize it (to σ^*) with respect to LV pressure, and employ this noninvasive computa-tional index $(d\sigma^*/dt)_{max}$ as a contractility index. It is found that this index has a good correlation with the traditional LV contractility index $(dp/dt)_{max}$, which requires determination of LV pressure by cardiac catheterization. It is also found that more ellipsoidally shaped LVs have higher values of this contractility index $(d\sigma^*/dt)_{max}$. The second part of this chapter deals with the formulation of the sarcomere model of a myocardial-wall fiber, and the expression of its contractile element characteristics (of force vs. shortening velocity) in terms of the monitorable data (of LV pressure, volume, wall thickness, and myocardial volume). Again, this index bears a good correl-ation with $(dp/dt)_{max}$.

Chapter 4 is on vascular biomechanics. It deals with (1) noninvasive determination of aortic pressure (as well as aortic stiffness and peripheral resistance) in terms of LV volume (ejected into the aorta) versus time data, and auscultatory diastolic and systolic pressures; (2) determination of aortic constitutive property (of E vs. σ) from measurement of pulse wave velocity, aortic dimensions, and auscultatory diastolic pressure; (3) arterial bed per-ipheral resistance (as the ratio of mean arterial pressure and flow rate) and arterial impedance (as the ratio of arterial pulse pressure and flow rate); (4) the phenomenon of wave reflection at aortic (arterial) bifurcations; and (5) the composition and amplitude variation of the composite wave in aorta, with an interesting postulation that in an ideal situation the heart is located at the site of minimum value of the composite wave pressure amplitude so that it has minimal after-load.

The second section of the book is on pulmonary engineering mechanics. In this section, Chapter 5 deals with lung ventilation modeling (with application to lung disease determination) based on the differential equation of lung volume (V) response to lung driving pressure (P_N) (in an intubated patient), in terms of lung compliance (C) and resistance to airflow (R). These parameters, R and C, are then combined into nondimensional ventilatory performance indices. The lung volume response expression can also be fitted to the lung volume data in terms of R and C as well as the product of pressure and compliance terms. The corresponding ventilatory performance index would not require intubation of the patient. A two-lobe lung model is also developed, and its total lung volume expression is determined. By fitting this expression to the monitored lung volume data, we can evaluate R, C, and ventilatory indices of left and right lung separately.

Chapter 6 deals with lung gas-transfer performance analysis. We first deal with inspired and expired air composition analysis, based on mass balances. We then derive the expressions for O_2 and CO_2 diffusion coefficients in terms of O_2 consumption rate and CO_2 production rate (from the inspired–expired air composition analysis), alveolar air O_2 and CO_2 partial pressures, and O_2 and CO_2 concentrations in the venous blood.

Chapter 7 deals with evaluation of the lung status of mechanically ventilated intubated patients, and the index for deciding that they are ready to be weaned off the mechanical ventilator. This index is expressed in terms of lung capacitance (C) and flowrate resistance (R), as well as monitored tidal volume (TV), breathing frequency (RF), and peak inspiratory pressure (P_k). The values of R and C are obtained by again modeling the lung ventilatory volume response to its driving pressure, and evaluating them in terms of the monitored values of lung volume (specifically, tidal volume) as well as inspiratory peak pressure and pause pressure (when the lung volume is maximum).

In the third section on glucose–insulin regulation (in diabetes) engineering mechanics, Chapter 8 first deals with the basics of blood glucose–insulin regulatory mechanics. This entails second-order differential equations modeling and solutions of glucose and insulin blood-concentration responses to glucose injection for three different types of glucose inputs: glucose input as a step function, glucose input as an impulse function, and glucose input as a rectangular pulse. The solutions of the governing equations of glucose and insulin blood concentrations to these three inputs are expressed in terms of the intrinsic parameters (α, β, γ, and δ) that relate the time rates of change of blood glucose and insulin concentrations to blood glucose and insulin concentrations as well as glucose input function, in the form of two first-order differential equations. These two first-order differential equations are combined into second-order differential equations of glucose and insulin responses to glucose inputs. These second-order differential equations have parameters of attenuation (or damping) constant (A) and system natural frequency (ω_n), which are in turn expressed in terms of the intrinsic parameters (α, β, γ, and δ). The relations between these parameters A and ω_n (and hence between α, β, γ, and δ), in turn, enable us to designate the system response as underdamped (for normal subjects), overdamped (for diabetic subjects), and critically damped (for subjects at risk of becoming diabetic).

Chapter 9 entails analytical simulation of the oral glucose tolerance test involving the governing equation of blood glucose concentration (y) response to glucose ingestion. This governing equation is a second-order damped-oscillatory differential equation in glucose concentration (y) response to an impulse glucose-input function. This model equation parameters are the system's damping constant (A), natural frequency (ω_n), and damped frequency (ω_d). The equation is solved for (1) underdamped response pertaining to the data of "glucose concentration versus time" for a normal subject and (2) overdamped response to simulate the "glucose concentration versus time" data of a diabetic subject. The equation solutions are fitted to the data, and the model parameters are determined analytically.

The values of the model parameters are distinctly different for normal and diabetic subjects. The purpose of this analytical simulation of the glucose-tolerance test (by means of the solutions of the governing differential equation) is to provide an analytical method to characterize the glucose concentration (y) versus time data in terms of the values of the model parameters. This is deemed to more reliably represent the entire data (of y vs. t), instead of merely employing discrete values (y) of the data to differentially diagnose diabetic subjects and borderline diabetic subjects (whose data is represented by a critically damped solution of the governing equation) from normal subjects.

Chapter 10 constitutes solutions of the governing differential equations of glucose concentration (y) and insulin concentration (x) to impulse glucose input function (to simulate glucose tolerance test data). These solutions' expressions (representing underdamped, overdamped, and critically damped responses) of the system differential equations model are fitted to the "y versus t" and "x versus t" data. Depending on the value of the regression correlation coefficient, a particular response function (i.e., underdamped or overdamped or critically damped function) is selected to best fit the data. It is found that the data of some subjects, who were clinically classified as normal or diabetic, are better fitted by means of a critically damped solution; this then placed these subjects in the category of being borderline diabetic or at risk of becoming diabetic. Next, the model parameters are combined together in the form of indices characterizing glucose and insulin concentration data. These two indices are further combined into one index, which is evaluated for all the patients studied. It is found the values of this index fall in distinct ranges for normal subjects, diabetic subjects, and borderline diabetic subjects. Hence, it can be concluded that this index value can be reliably employed for the differential diagnosis of diabetes.

We now move to Section IV on orthopedic engineering mechanics. Therein, Chapter 11 deals with osteosynthesis of fixation of fractured bone by means of a bone plate and screws, when the bone–plate assembly is subjected to axial loading and bending loading. Both analytical and finite-element solutions are carried out. The issue that is explored at length is to design the placement of the screws (with respect to the fractured site) and the stiffness grading of the plate, such that at the fracture site the bone callus is not subjected to tensile stresses and that away from the fracture site the bone has minimal stress shielding. A novelty explained in this chapter is our deploying a helical plate for fixing bones with helical cracks. A detailed finite-element stress analysis is carried out to demonstrate how a helical plate (and its screws) can be employed to provide maximal stiffness to the fractured bone–plate assembly. The advantage of the helical plate is that the screws fixing the plate to the bone are in different planes, and thereby provide optimal stiffness to the fractured bone–plate assembly under varied loading conditions.

Chapter 12 is on the analysis of the spinal vertebral body (VB), modeled as a hyperboloid cortical-bone shell, subjected to axial, bending, and torsional loadings. It is shown that under all of these loading states, the forces are transmitted across the VB (from the top to the bottom of the VB) as axial forces through the generators of the hyperboloid shell. In other words, the vertebral body is shown to be so intrinsically shaped and designed such that it has only axial forces transmitted through it. This makes it bear heavy loads with minimal weight (represented by a thin cortical shell thickness). We can employ this intrinsically optimal VB design concept to propose the design of an anterior fixator made up of two rings (fixed to the upper and lower end-plates) and connected by straight generators to form a hyperboloid shaped structure (resembling a cane stool), into which the fractured vertebral body fragments can also be deposited to eventually form a solid fixator.

We next go on to analyze the intrinsic design of an intervertebral disc. The disc is modeled as a thick-walled isotropic cylinder, filled with nucleus-pulposus (NP) fluid material. When this disc model is axially loaded, the NP fluid is also pressurized, in addition to the disc wall being stressed by the axial loading. The pressurized NP then exerts radial pressure on the cylindrical disc wall, and subjects it to further radial and circumferential stresses. Now the disc wall material has a stress-dependent elastic modulus (typical of anatomical structures). Hence when the disc wall is further stressed by the radial pressure exerted on it by the NP fluid, its elastic modulus value is enhanced, and the resulting radial displacements do not increase in the same proportion as the increasing axial load. In other words, the disc is able to contain the radial displacements under increasing axial loading, without bulging radially. This is the feature of its optimal intrinsic design. Now, when a disc has radial cracks (due to being excessively loaded), the NP fluid seeps out of the cracks onto the surrounding nerve roots, and causes back pain. The orthopedic solution for such a herniated disc is to denucleate the herniated disc. However, based on our analysis, if the NP fluid is removed, then the disc wall is no longer radially stressed and its modulus is not correspondingly increased. In fact, we have shown that a denucleated disc undergoes greater deformation compared to a normal disc for the same level of loading. Hence, a better solution would be to place a jell sac in place of the NP fluid, which can simulate the role of the NP fluid.

We now come to the final section on fitness and sports engineering mechanics. Here, Chapter 14 describes the biomechanics of a fitness index, optimal jogging modes, and assessment of the hip joint pathology. The first part of the chapter deals with the formulation of a cardiac fitness index composed of the parameters of a first-order differential equation modeling of heart-rate response during and after treadmill exercising. This index is shown to clearly differentiate fit subjects from unfit subjects. We next analyze human jogging by stipulating that for an optimal jogging mode, a subject would involve minimal muscle actuation if he or she were to have the stride frequency of the free-swinging leg simulated as a double-compound pendulum. The stride frequencies are derived in terms of the limb segment's masses, lengths, and locations of the center of mass of upper and lower limb segments, and the mass moment of inertia at the centers of masses of the upper and lower limbs. The lower of the two computed stride frequencies is employed to stipulate the optimal jogging leg frequency. This optimal jogging mode is especially recommended for subjects undergoing cardiac rehabilitation. Next, we want to ensure that jogging is not causing hip problems. For this purpose, we model the swinging leg by means of a second-order differential equation of free damped oscillatory motion of the swing angle (θ) of the simple compound leg-pendulum model, in terms of a viscous damping constant (b), and a joint-stiffness parameter (k). The solution of the governing differential equation is obtained (for the case of small damping) as a "θ versus t" damped-oscillatory response. From the measured amplitudes of the extreme values of θ, we then evaluate the parameters b and k, to characterize the joint pathology.

Chapter 15 analyzes how spin can impart lateral acceleration and force (due to the magnus force effect) to a soccer ball (while kicking it), and make the ball swerve. The resulting planes of the ball trajectory (normal to the ground) are computed to simulate some real data derived from videos of world-cup soccer matches. It can be seen how (1) a right footer taking a corner kick from the left corner can make the ball swerve toward the goal (to deceptively beat the goal keeper) by imparting an anticlockwise spin to the ball and (2) how a left footer, taking a corner kick from the left corner, can make the ball swerve away from the goal (to facilitate heading into the goal) by imparting a natural clockwise spin to the ball.

Chapter 16 describes the mechanics of pitching a baseball, of ball–bat interactions, of batting, and of an optimal bat, replete with theory and applications in the form of simulations. The first part of the chapter demonstrates (1) how a pitcher imparts spin-induced lateral deflection to the baseball to pitch a curve ball and (2) how spin-induced drop and lift can be imparted to the baseball. In the second part of the chapter, on bat–ball collisions, the primary emphasis is on where the ball should strike the bat, such that maximum energy is transferred by the bat to the ball or maximum ball speed is imparted to the ball. This "sweet spot of the bat" can be defined as the center of percussion, or the node of the fundamental bat-bending vibrational mode. The third part of the chapter, on the mechanics of the bat, deals with the ideal bat weight for a batter. Now based on the conservation of momentum equation for bat–ball collision, a player can transfer maximum momentum to the ball either (1) by using a lighter bat and swinging his or her arm more vigorously or (2) by using a heavier bat and leaning his or her body more into the ball when striking it.

The final Chapter 17 is on the dynamic analysis of gymnastics' Yurchenko layout vaulting. The Yurchenko layout vault, pioneered by Natalia Yurchenko in the 1982 World Cup Gymnastics competition, comprises a forward running approach, followed by a cartwheel half-turn to orient the body such that the back faces the vaulting horse at the point of takeoff from the springboard. The gymnast then takes off from the springboard using a back-flip action to impact the horse, and finally completes a one-and-half somersault rotation with the body fully extended (or laid out) before landing. The chapter discusses the optimal technique for this Yurhcenko layout vault (for a given gymnast), defined by a decrease in horse impact time, and the position and alignment of the body segments at the end of the postflight. For this purpose, a five-segment rigid-linked model is developed, which consists of the hand, whole arm, upper and lower trunk, and the whole leg. In this model, each segment has a center of mass (CM), the segments are linked by hinges, gravitational forces are exerted at the segment's CMs, the ground reaction forces on the segments are considered to act at the centers of pressure, and the effect of the segment's muscles is to produce moments at the joints. The governing equations of motion are formulated for the segments. For input anthropometric, kinematics, and ground reaction-force data, we can obtain solutions for the muscle moments at the joints and for the joint reaction forces. The optimization procedure determines the set of joint torques and kinematics required to produce this optimal technique (as indicated above), in terms of vault duration and loading angle.

As can be noted, the book covers the detailed analyses of (1) cardiological, pulmonary, glucose–insulin regulation systems, to address their medical applications in terms of disease assessment, (2) the most effective orthopedic osteosynthesis designs as well as of spinal vertebral body and intervertebral disc that make them intrinsically optimally designed structures, and (3) of sports events and simulations, to provide insights into the techniques required for high performance of these sports events. The book is tailored to serve as a textbook for a one- or two-semester biomedical engineering mechanics course. However, it can also be effectively employed by clinicians for assessment of physiological systems, by anatomists to obtain insights into optimal anatomical designs in nature, and by sportsmen and sports coaches to optimize performances.

Dhanjoo N. Ghista
Singapore
Email: d.ghista@gmail.com

Acknowledgments

I am very grateful to my colleague Assistant Professor Dr. Sridhar Idapalapati for his immense help in making the book manuscript ready for production. I am also thankful to Satttu Sreenu Babu for his help in wordprocessing
the chapters and in formatting them for production.

Author

Professor Dhanjoo Ghista is a pioneer and a world authority in biomedical engineering. He has developed biomedical engineering programs and departments at universities in the United States, Canada, India, United Arab Emirates, and Singapore. He has published several books on many aspects of biomedical engineering, including physiological mechanics, human body mechanics and dynamics, cardiac mechanics, cardiovascular engineering and physics, orthopedic mechanics, osteoarthro mechanics, spinal injury medical engineering, biomechanics of medical devices, cardiac perfusion and pumping engineering, and biomedical and life physics. His academic involvements span engineering, medical, and even social sciences (wherein he has authored a definitive book on socio-economic democracy and the world government, to provide the basis of sustainable communities and sustainable peace).

In an academic career spanning over 40 years, he has taught many courses, including physiological engineering, clinical engineering, biomechanics, and cardiovascular engineering. He has published over 400 papers in peer-reviewed journals and conference proceedings. He has guided several PhD students and postdoctoral fellows, many of whom have gone on to hold responsible positions in universities and research institutes. He has been editor of *Automedica* and *Renaissance Universal*, and has served on the editorial boards of *Mechanics in Medicine and Biology* and *Biomedical Engineering Online*. He has been reviewer of grant proposals for several national agencies, including the International Spinal Trust, Medical Research Council (Canada), Ontario Research Foundation, and Agency for Science and Technology (Singapore).

Author

Dhanjoo N. Ghista (Formerly) School of Mechanical and Aerospace Engineering, Nanyang Technological University, Singapore. E-mail:d.ghista@gmail.com

Contributors

A. Terry Bahill Department of Systems and Industrial Engineering, University of Arizona, Tucson, Arizona, U.S.A.

David G. Baldwin (Formerly) Major League Baseball, Yachats, Oregon, U.S.A.

Sridhar Idapalapati School of Mechanical and Aerospace Engineering, Nanyang Technological University, Singapore

Leslie Jennings Department of Mathematics and Statistics, University of Western Australia, Perth, Australia

Michael Koh School of Sports, Health and Leisure, Republic Polytechnic, Singapore

Ramakrishna Kotlanka (Formerly) School of Mechanical and Aerospace Engineering, Nanyang Technological University, Singapore

Geok Hian Lim School of Mechanical and Aerospace Engineering, Nanyang Technological University, Singapore

Xiang Liu (Formerly) School of Mechanical and Aerospace Engineering, Nanyang Technological University, Singapore

Kah Meng Loh   (Formerly) Division of Engineering, Science & Technology, University of New South Wales Asia, Singapore

Jor Huat Ong   School of Mechanical and Aerospace Engineering, Nanyang Technological University, Singapore

Rohit Pasam   Department of Bioengineering, University of Illinois at Chicago, Chicago, Illinois, U.S.A.

Meena Sankaranarayanan   Signal Processing Department, Institute for Infocom Research, Singapore

Dittakavi Sarma   Department of Biomedical Engineering, Osmania University, Hyderabad, India

Liang Zhong   Department of Cardiology, National Heart Center, Singapore

1

Biomechanics in Medical Diagnosis in the Form of Nondimensional Physiological Indices

Dhanjoo N. Ghista

CONTENTS

1.1 Scope

This introductory chapter delineates the theme and scope of the book. Herein, we are providing a novel concept of physiological systems analysis, in terms of nondimensional physiological indices (NDPIs), for quantifying patient health and disease status as well as patient improvement. We have developed NDPIs for several physiological phenomena and systems, and indicated as to how they can be employed diagnostically. NDPIs have been formulated and evaluated for (1) left-ventricular pumping performance, (2) cardiac fitness and conditioning, (3) lung ventilatory function, (4) oral glucose tolerance test, (5) arteriosclerosis, (6) atherosclerosis and peripheral resistance, (7) mitral valve property, and (8) osteoporosis. This chapter is based on my paper, Ref. [1].*

1.2 Introduction

The concept of nondimensional physiological number index is quite new, and has been adopted from engineering, where nondimensional numbers (made up of several parameters) are employed to characterize a regime or strata disturbance phenomena. For example, in a cardiovascular fluid-flow regime, the Reynold's number

$$N_{\mathrm{re}} = \rho V D / \mu \tag{1.1}$$

is employed to characterize the conditions when N_{re} exceeds a certain critical value, at which laminar flow changes to turbulent flow, which can occur in the ascending aorta when either the aortic valve is stenotic (giving rise to murmurs) or in the case of anemia (decreased blood viscosity).

Similarly, we can construct other such physiological numbers to characterize disturbance from physiological homeostasis. In physiological medicine, the use of nondimensional indices or numbers can provide a generalized approach by which unification or integration of a number of isolated but related events into one NDPI can help to characterize an abnormal state associated with a particular physiological system. The evaluation of the distribution of the values of such NDPI(s), in a big patient population, can then enable us to designate normal and disordered ranges of NDPI, with a critical value of NDPI separating these two ranges, as illustrated in Figure 1.1. In this way, NDPI(s) can help us to formulate physiological health indices (PHIs), not only to facilitate differential diagnosis of patients but also to assess the severity of the disease or disorder. Herein, we have formulated several such new NDPIs [1].

---

* With permission from the publisher World Scientific Publishing Co. (Singapore).

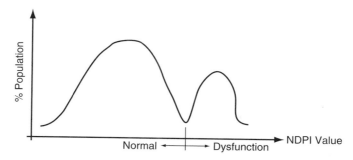

FIGURE 1.1

Integration of a number of isolated but related events into one nondimensional physiological index (NDPI) can help to characterize an abnormal state of a particular physiological system.

In this chapter, we have gone one step further and also applied this concept of nondimensional indices to characterize the cost-effective performance of a hospital unit, namely, the intensive care unit. Finally, we indicate how a hospital operating budget can be distributed among its various departments such that each department operates cost effectively.

1.3 Formulation of NDPI(s) for Some Physiological Systems

1.3.1 Index for Efficiency of Left-Ventricular Pumping

We formulate the expression for left-ventricular (LV) pumping efficiency (ε) as follows:

$$\varepsilon = \frac{\text{Useful LV power output to eject into the aorta blood (or ejection power output)}}{\text{(LV power to develop the intra-LV pressure during isovolumic contraction)} + \text{(ejection power output)}}$$

$$= \frac{\dfrac{(P_{sy} - P_{ed})V_s}{T_{isv}}}{\dfrac{1}{2}\dfrac{(P_{sy} - P_{ed})V_{ed}}{T_{isv}} + \dfrac{(P_{sy} - P_{ed})V_s}{T_s}} \tag{1.2}$$

$$\varepsilon = \frac{2[(P_{sy}/P_{ed}) - 1](V_s/V_{ed})(T_e/T_{isv})}{[(P_{sy}/P_{ed}) - 1](T_e/T_{isv}) + 2[(P_{sy}/P_{ed}) - 1](V_s/V_{ed})} \tag{1.3}$$

where (as illustrated in Figure 1.2) (1) P_{ed} and P_{sy} are the end-diastolic and maximum-systolic pressures, (2) V_s and V_{ed} are the stroke volume and end-diastolic volume, and (3) T_{isv} and T_e are the time periods of isovolumic contraction and ejection.

All of the quantities in Equation 1.3 can be measured noninvasively, and hence ε can be evaluated noninvasively. Let us obtain some idea of the order of magnitude of ε. Let us assume that, in a normal case,

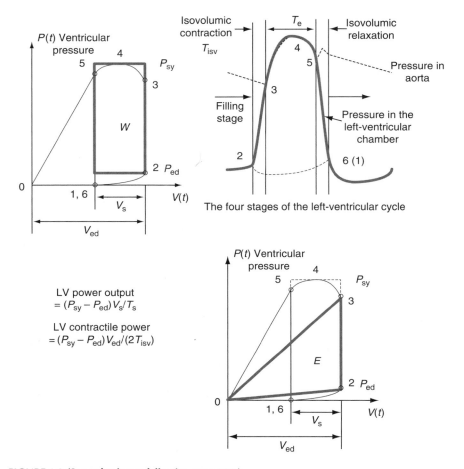

FIGURE 1.2 (See color insert following page 266.)
Left-ventricular work (W), energy input (E), power output and input, and efficiency (ε). (Adopted from Ghista, D.N., *J. Mech. Med. Biol.*, 4, 401, 2004.)

$$\frac{V_s}{V_{ed}} = \text{ejection fraction} = 0.6, \quad \frac{T_{isv}}{T_e} = \frac{0.06\ \text{s}}{0.36\ \text{s}} = \frac{1}{6}, \text{ and } \frac{P_{sy}}{P_{ed}} = 5$$

Hence, ε is of the order of 15% for a normal person (i.e., with no cardiac disease). In other words, the useful power is 15% of the power required to develop the requisite contractility (to in turn raise the intra-LV pressure) in a normal person, and could well go down to even 10% or below in the case of a failing heart.

1.3.2 Assessing Cardiac Fitness and Conditioning by Means of a Treadmill Test

The cardiac-fitness model [2] consists of a first-order differential equation system model describing the heart rate (HR) response (y) to exertion (walking and jogging on a treadmill) monitored in terms of the work exertion (WE) as measured by normalized $\dot{V}O_2(x)$, where x and y are defined as follows:

$$x = \frac{\dot{V}O_2(t) - \dot{V}O_2(\text{rest})}{\dot{V}O_2(\text{rest})}, \quad y = \frac{HR(t) - HR(\text{rest})}{HR(\text{rest})} \tag{1.4}$$

where $\dot{V}O_2$ and HR represent the oxygen-consumption rate and heart rate, respectively. It is to be noted that both x and y are nondimensional quantities.

The subject is asked to exercise on the treadmill for a period of time, t_e (minutes). During this period, the $\dot{V}O_2$ and $HR(t)$ (and hence x and y) are monitored. Now we develop a model to simulate the $HR(t)$ response to $\dot{V}O_2(t)$ or exertion, (1) during exercise (i.e., for $t < t_e$) and (2) thereafter for $HR(t)$ decay after the termination of exercise (i.e., for $t > t_e$).

For a person, the model equation is represented by

$$\frac{dy}{dt} + k_1 y = C_0 x(t) \tag{1.5}$$

where the variation of the normalized $\dot{V}O_2$ (or WE) is given by

$$x(t) = D(1 - e^{-k_2 t}) \quad \text{for} \quad t < t_e \tag{1.6}$$

$$x(t) = x(t_e)e^{-k_3 t} \quad \text{for} \quad t > t_e \tag{1.7}$$

and $t_e = $ time when exercise terminates.

The y solutions to Equations 1.5 through 1.7 are represented by

$$y = \left(\frac{C_0 D}{k_1}\right)(1 - e^{-k_1 t}) + \left(\frac{C_0 D}{(k_1 - k_2)}\right)(e^{-k_1 t} - e^{-k_2 t}) \quad \text{for} \quad t \le t_e \tag{1.8}$$

$$y = \frac{C_0 x(t_e)}{(k_1 - k_3)}\left(e^{-k_3(t - t_e)} - e^{-k_1(t - t_e)}\right) + y(t_e)e^{-k_1(t - t_e)} \quad \text{for the recovery period } t \ge t_e \tag{1.9}$$

where k_1, k_2, k_3, C_0, and D are the model parameters that can serve as cardiac-fitness parameters. These responses are depicted in Figure 1.3.

1.3.2.1　Monitored and Computed Results for a Typical Normal Subject

For a sample patient, the monitored $\dot{V}O_2(x)$ for $t < t_e$ is represented analytically (by means of Equation 1.6) as $x = 2.6547(1 - e^{-0.25t})$, for $t \le t_e$. Hence, k_2 and D in Equation 1.6 are 0.25 m^{-1} and 2.6547, respectively. The values of C_0 and k_1 are obtained by making Equation 1.8 simulate the $(y-t)$ data in Figure 1.3.

The solution for y (Equation 1.8) during the stress test (for $t \le t_e$) is then obtained as

$$y = 0.797(1 - e^{-0.85t}) + 1.129(e^{-0.85t} - e^{-0.25t}), \quad \text{for} \quad t \le t_e \,(=9\,\text{min}) \tag{1.10}$$

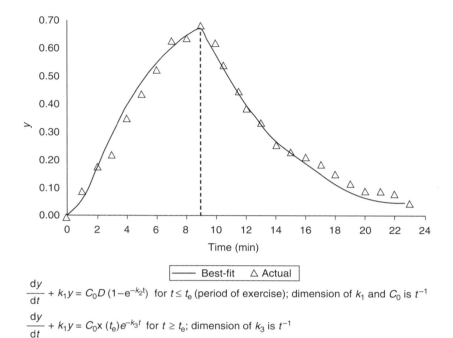

$$\frac{dy}{dt} + k_1 y = C_0 D \left(1 - e^{-k_2 t}\right) \text{ for } t \le t_e \text{ (period of exercise); dimension of } k_1 \text{ and } C_0 \text{ is } t^{-1}$$

$$\frac{dy}{dt} + k_1 y = C_0 \text{x} \left(t_e\right) e^{-k_3 t} \text{ for } t \ge t_e; \text{ dimension of } k_3 \text{ is } t^{-1}$$

FIGURE 1.3

Graph of y (the computed HR response) versus t. (Adopted from Lim, G.H., Ghista, D.N., Koo, T.Y., Tan, J.C.C., Eng, P.C.T., and Loo, C.M., *Int. J. Comp. Appl. Technol.* (*Biomed. Eng. Comp.*), 21, 38, 2003.)

i.e., the value of the parameter k_1 in Equation 1.7 is 0.85 m^{-1}.

Likewise, the solution for HR(t), as given by Equation 1.9, for y after the stress test (during the recovery period) is obtained by making Equation 1.9 simulate the $(y - t)$ data (for $t \ge t_e$) in Figure 1.3

$$y = 1.088 \left(e^{-0.28t} - e^{-0.85t}\right) + 0.68 e^{-0.85t} \quad \text{for} \quad t \ge t_e \quad (1.11)$$

for the value of the parameter $k_3 (= 0.28 \text{ m}^{-1})$ in Equation 1.9.

Now, the parameters D, k_1, k_2, and k_3 can be combined into a single cardiac-fitness index (CFI):

$$\text{CFI} = \frac{k_1 k_3 t_e}{k_2 D} \quad (1.12)$$

For this patient, the value of CFI is 3.23. We need to evaluate CFI for a big spectrum of patients, and then compute its distribution curve to determine the efficiency of this index, in order to yield distinct separation of CFI ranges for healthy subjects and cardiac patients. This CFI can then also be employed to assess improvement in cardiac fitness following cardiac rehabilitation regime.

1.3.3 Nondimensional Lung Ventilatory Performance Index

When a person breathes, the lung volume (V) can be taken to be the response to the lung inflation pressure. We formulate a lung ventilatory performance index (LVPI) as follows:

$$LVPI = RC \ (BR) \tag{1.13}$$

where BR is the breathing rate, and R and C are resistance to airflow and lung compliance in the following differential equation of lung-volume (V) ventilatory-response to lung inflation pressure (P_N, represented by the right-hand side of the following equation):

$$R\frac{dV}{dt} + \frac{V}{C} = P_N = P_1 \cos \omega t + P_2 \sin \omega t \tag{1.14}$$

wherein $P_N = (P_m - P_p) - P_e$ (end-expiratory pressure), as illustrated in Figure 1.4.

This model can be employed to monitor the lung-volume response to lung inflation pressure in mechanical ventilation of patients with chronic obstructive lung disease [3] as well as in normal patients by devising and placing a pressure transducer in the inflation duct of a spirometer [4]. The pleural pressure (P_p) is measured by placing a baloon catheter transducer through the nose into the esophagus, and adopting the esophageal tube pressure to be the pressure in the pleural space surrounding it.

In Figure 1.4, V is the lung volume in liters (L), the right-hand side terms constitute the driving pressure in cm H_2O (= pressure at the mouth–pleural pressure), R is the resistance to airflow (in cm H_2O s L^{-1}), C is the lung compliance (in L/cm H_2O), and P_1 and P_2 are the magnitudes of cosine and sine terms of the driving (oscillatory) net pressure P_N = (mouth-pressure minus pleural pressure) minus end-expiratory pressure.

For a typical P_L cyclic pressure profile (Figure 1.4), given by

$$P_1 = -1.84 \text{ cm } H_2O, \quad P_2 = 3.16 \text{ cm } H_2O, \quad \omega = 0.5 \ \pi \text{ rad s}^{-1} \tag{1.15}$$

the solution to Equation 1.14, to satisfy the condition of $\dot{V}(t=0) = 0$, is given by

$$V = P_1 C \frac{(\cos \omega t + \omega \tau \sin \omega t)}{1 + \omega^2 \tau^2} + P_2 C \frac{(\sin \omega t - \omega \tau \cos \omega t)}{1 + \omega^2 \tau^2}$$
$$+ e^{\frac{-t}{\tau}} \left(\frac{\omega C \tau}{1 + \omega^2 \tau^2} \right)(P_2 + P_1 w \tau) \tag{1.16}$$

where $\tau = RC$. By fitting this lung-volume solution to the clinically monitored lung-volume data (by parameter-identification method), we can evaluate the parameters: $R = 1.24$ (cm H_2O) s L^{-1}, $C = 0.21$ L (cm H_2O)$^{-1}$. Now let us evaluate the nondimensional LVPI given by Equation 1.13

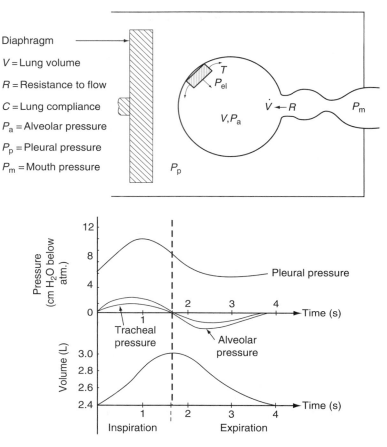

FIGURE 1.4
Lung pressures and volume as functions for normal breathing. (Note that the pressure extremes occur before the volume extremes.) (Adopted from Ghista, D.N., *J. Mech. Med. Biol.*, 4, 401, 2004.)

$$\text{LVPI} = RC \text{ (BR per minute)}$$

where BR, the breathing rate $= 15/\text{min}$ or $0.25/\text{s}$ for the data provided in Figure 1.4. For our case study, the value of LVPI is 3.9.

Let us now see how lung disease will influence R, C, and hence LVPI. For instance in emphysema, the destruction of lung tissue will produce a more compliant lung and hence a larger value of $C = 0.5$ L (cm $H_2O)^{-1}$, yielding a value of LVPI of about 10. In asthma, there is increased airway resistance (due to contraction of the smooth muscles around the airways) to say $R = 5$ (cm H_2O) s L^{-1}. The breathing rate can also go up to BR $= 20/\text{min}$. Hence, the value of LVPI can go up to 20. In the case of lung congestion due to mitral valve disease, it would be important to determine LVPI, so as to serve as an indicator for determining cardiac condition (in end-stage heart disease). By determining the distribution of a big patient population, we can determine the LVPI ranges for normal and disease states, and can hence employ this model to diagnose lung disease states.

A comprehensive analysis of lung ventilation performance (with application to lung disease characterization) is provided in Chapter 5.

1.3.4 Nondimensional Diabetes Index with Respect to Oral Glucose Tolerance Testing

Clinically, patients are diagnosed to be diabetic based on their blood-glucose concentrations at half-hour intervals during an oral glucose tolerance test (OGTT). On the other hand, the blood-glucose concentration responses during OGTT can be modeled as responses to oral ingestion of glucose, modeled as an impulse function $G\delta(t)$. For OGTT simulation (entailing digestive and blood-pool chambers), the differential equation, governing blood-glucose response (y) to oral ingestion of glucose, can be represented as follows [5]:

$$y'' + 2Ay' + w_n^2 y = G\delta(t), \quad y \text{ in g/L}, \quad G \text{ in g(L)}^{-1}(h)^{-1}$$

or

$$y'' + \lambda T_d y' + \lambda y = G\delta(t) \tag{1.17}$$

where

$\omega_n(=\lambda^{1/2})$ is the natural oscillation frequency of the system

A is the attenuation or damping constant of the system (in h^{-1})

$\omega(\omega_n^2 - A^2)^{1/2}$ is the angular frequency (in rad/h) of damped oscillation of the system

$\lambda(=2A/T_d = \omega_n^2)$ is the (proportional-control) parameter (in h^{-2}), representing pancreatic-insulin response proportional to the blood-glucose concentration (y)

T_d is the (derivative-time) parameter (in h), representing regulation of glucose concentration proportional to rate-of-change of glucose concentration (y')

The input to this system is taken to be the impulse function due to the orally ingested glucose bolus [G], while the output of the model is the blood-glucose concentration response $y(t)$.

For an impulse glucose input, a normal patient's blood-glucose concentration data are depicted in Figure 1.5 by open circles. Based on the nature of these data, we can simulate them by means of the solution of the oral glucose regulatory (second-order system) model (Equation 1.17), as an underdamped glucose concentration response curve, given by

$$y(t) = (G/\omega)e^{-At} \sin \omega t \tag{1.18}$$

where A is the attenuation constant, $\omega = (\omega_n^2 - A^2)^{1/2}$ is the damped frequency of the system, ω_n (the natural frequency of the system) $= \lambda^{1/2}$, and $\lambda = 2A/T_d$.

The model parameters λ (or ω_n), T_d (or A and T_d), and G are obtained by matching Equation 1.18 to the monitored glucose concentration $y(t)$ data (represented by the open circles). The computed values of parameters

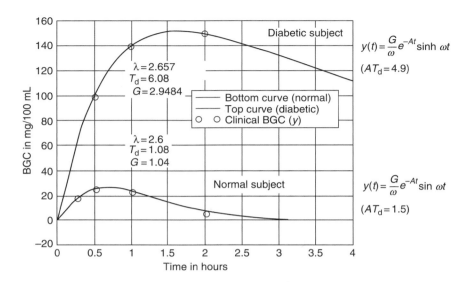

FIGURE 1.5
OGTT response curve: $A = 1.4$ h^{-1} for the normal subject (i.e., higher damping coefficient value); $A = 0.808$ h^{-1} for the diabetic patient (i.e., lower damping coefficient). (Adopted from Dittakavi, S.S., and Ghista, D.N., *J. Mech. Med. Biol.*, 1, 193, 2001.)

are $\lambda = 2.6$ h^{-2}, $T_d = 1.08$ h. This computed response is represented in Figure 1.5 by the bottom curve, fitting the open-circles clinical data.

For a potential diabetic subject, the blood-glucose concentration data are depicted by closed circles in Figure 1.5. In order to model-simulate these data, we adopt the solution of model Equation 1.17, as an overdamped glucose concentration response function:

$$y(t) = (G/\omega)e^{-At} \sinh \omega t \tag{1.19}$$

This function is made to match the clinical data depicted by closed circles, and the values of λ and T_d are computed to be 0.27 h^{-2} and 6.08 h, respectively. The top curve in Figure 1.5 represents the blood-glucose response curve for this potentially diabetic subject. The values of the fitted parameters (T_d, λ, A, and G) for both normal and diabetic patients are indicated in the figure, to provide a measure of difference in the parameter values.

Now, we come to the interesting part of this model, by formulating the nondimensional diabetes index (DBI) as

$$\text{DBI} = AT_d = \frac{2A^2}{\lambda} = \frac{2A^2}{\omega_n^2} \tag{1.20}$$

The value of DBI in Figure 1.5 for the normal and the diabetic subjects is 1.5 and 4.9, respectively. We have further found (in our initial clinical tests) that DBI for normal subjects is <1.6, while the DBI for diabetic patients is >4.5. This is a testimony of the efficacy of the model, and especially for the nondimensional DBI.

A comprehensive analysis of oral glucose tolerance test (OGTT) is provided in Chapter 9.

1.3.5 Characterization of Aortic Stiffness or Arteriosclerosis

The blood flow in the artery is pulsatile and the pulse-wave velocity (PWV) is given by $PWV = (Eh/2a\rho)^{1/2}$, where E is the elastic modulus of the artery, a is the arterial (cylindrical tube) radius, h is the arterial wall thickness, and ρ is the blood density. For a circular cylindrical arterial tube of radius a and wall thickness h, we can express the arterial wall stress σ and elastic modulus E, as follows:

$$\sigma = \frac{Pa}{h} = \frac{130Pa}{h}\,N/m^2, \quad E = \frac{2(PWV)^2 a\rho}{h}, \quad E = E_0 + m\sigma \qquad (1.21)$$

in terms of (1) the aortic dimensions a and h, (2) the auscultatory (or automatedly) measurable diastolic pressure (P), and (3) the PWV monitored by determining both the arterial diameters (echocardiographically) at two sites Δx apart and the time Δt for the pressure pulse wave to travel the distance Δx. Table 1.1 depicts the computed values of σ and E at four independent times for a typical subject. All the quantities in Equation 1.21 can be monitored noninvasively, in order to compute the values of E and σ.

The values of E and σ in columns 5 and 6 of Table 1.1 can be represented as

$$E(N/m^2) = 4.2\sigma + 0.5 \times 10^5 (N/m^2) = m\sigma + C \qquad (1.22)$$

where m is the slope of E–σ graph (assumed to be a straight line) and C is the y intercept of the line.

We will now define the arteriosclerotic nondimensional index

$$ART - NDI = mE_0/(\text{mean diastolic pressure}) \qquad (1.23)$$

For the above patient, the value of the ART – NDI is

$$ART - NDI = \frac{(4.2)(0.5 \times 10^5 \, N/m^2)}{(87 \times 137 \, N/m^2)} = 17.6 \qquad (1.24)$$

and will be much higher for arteriosclerotic patients, which we can determine by conducting clinical test applications of this analysis.

TABLE 1.1

Computation of E and σ from Measurements of PWV, a and h

P (mmHg)	PWV (m/s)	a (mm)	h (mm)	$E\left[\dfrac{N}{m^2}\right]$	$\sigma\left[\dfrac{N}{m^2}\right]$
80	5.3	4.1	1.10	2.13×10^5	3.38×10^4
85	5.4	4.5	1.00	2.6×10^5	4.97×10^4
90	5.42	5.0	0.90	3.01×10^5	5.97×10^4
95	5.5	5.0	0.90	3.38×10^5	6.68×10^4

1.3.6 Noninvasive Determination of Aortic Elasticity (*m*), Peripheral Resistance (*R*), and the Aortic NDI

Figure 1.6 depicts schematically the outflow tract of the left ventricle (LV) into the ascending aorta and a blood-control volume V in the ascending aorta. We can represent the aortic pressure (P) response to outflow rate $I(t)$ from the LV into the aorta (with reference to the blood-control volume, V), as derived in Figure 1.6, by [6]

$$(\mathrm{d}P/\mathrm{d}t) + \lambda P = mI(t) \tag{1.25}$$

where $m =$ volume elasticity of aorta (in Pa/m^3), $\lambda = (m/R)$ in s^{-1}, and the LV outflow rate is given by

$$I(t) = (A)\sin(\pi/t_\mathrm{s})t + (A/2)\sin(2\pi/t_\mathrm{s})t \quad \text{for} \quad 0 < t < t_\mathrm{s}\,\text{(systole)}$$
$$= 0 \quad \text{for} \quad t > t_\mathrm{s}\,\text{(diastole)} \tag{1.26}$$

If $t_\mathrm{s} = 0.35$ s and the stroke volume (SV) is known (from, say, echocardiography), then we have (from Equation 1.26)

$$\int_0^{t_\mathrm{s}} \left[(A)\sin(\pi/t_\mathrm{s})t + (A/2)\sin(2\pi/t_\mathrm{s})t \right] = \mathrm{SV} \tag{1.27}$$

where $A = \pi\,(\mathrm{SV})/2t_\mathrm{s}$.

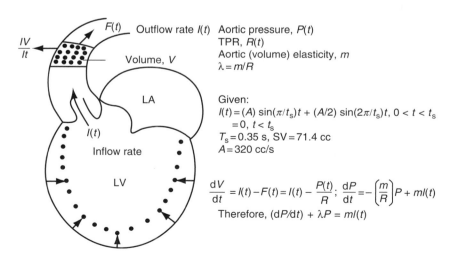

FIGURE 1.6
To derive the equation for aortic-pressure response to the stroke-volume or LV ouput rate $I(t)$. (Adopted from Ghista, D.N., *J. Mech. Med. Biol.*, 4, 401, 2004.)

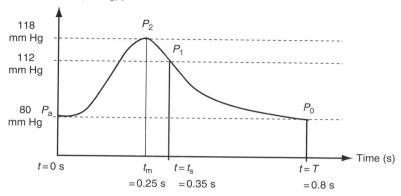

FIGURE 1.7
Computed aortic pressure profile. (Adopted from Ghista, D.N., *J. Mech. Med. Biol.*, 4, 401, 2004.)
P_2 = auscultatory systolic pressure; $P_0 = P_a$ = auscultatory diastolic pressure; $(0 < t < t_s)$ = aortic systolic phase (during which blood is ejected into the aorta from the left ventricle); $(t_s < t < T)$ = aortic diastolic phase.

$$\text{Thus, if } SV = 71.4\,\text{cc, then } A = 320 \text{ cc/s} \tag{1.28}$$

The solution of Equation 1.25 for the aortic diastolic and systolic periods is obtained as follows:
    Aortic diastolic pressure expression (Figure 1.7):

$$P_d(t) = P_1 e^{-\lambda(t-t_s)}, \quad P_1 = \text{pressure at the start of diastole}$$
$$= P_0 \text{ (at } t = T) = \text{pressure at the end of systolic phase}$$
$$\text{or auscultatory diastolic pressure} \tag{1.29}$$

$\therefore P_d(t) = P_0 e^{-\lambda(T-t)}$, where $T = 0.8$ s
    Aortic systolic pressure expression (Figure 1.7):

$$P_s(t) = \left(P_0 + \frac{mA\omega}{\lambda^2 + \omega^2} + \frac{mA\omega}{\lambda^2 + 4\omega^2}\right)e^{-\lambda t} + mA\left(\frac{\lambda \sin \omega t - \omega \cos \omega t}{\lambda^2 + \omega^2}\right)$$
$$+ \frac{mA}{2}\left(\frac{\lambda \sin 2\omega t - 2\omega \cos 2\omega t}{\lambda^2 + 4\omega^2}\right), \quad \omega = \frac{\pi}{t_s} \tag{1.30}$$

We now (1) incorporate into Equations 1.29 and 1.30 the auscultatory data on P_0 (= 80 mm Hg) and P_2 (= 118 mm Hg) with $T = 0.8$ s, as well as (2) invoke continuity in diastolic and systolic pressure expressions to (3) put down the following three equations (in three unknowns: m, λ, and t_m at which $P_s = P_2$):

$$P_d(\text{at } t_s = 0.35 \text{ s}) = P_s(\text{at } t_s = 0.35 \text{ s}) \tag{1.31}$$

$$dP_s/dt = 0, \text{ at } t = t_m = 0.25 \text{ s} \tag{1.32}$$

$$P_s(t = t_m) = P_2(=118 \text{ mm Hg}) \tag{1.33}$$

to compute: $\lambda = 0.66 \text{ s}^{-1}$, $m = 0.78$ mm Hg cm^{-3}, $R(=m/\lambda) = 1.18$ mm Hg cm^{-3}s, for $t_m = 0.25$ s and $T = 0.8$ s. By substituting the values of these parameters into Equations 1.29 and 1.30, we can determine the aortic pressure profile, as illustrated in Figure 1.7.

We now formulate an index:

$$\text{Aortic number} = \lambda T = mT/R \tag{1.34}$$

where $\lambda = m/R$ in the governing differential equation (Equation 1.25).

For the given auscultatory data, and for the above computed parameters: $m = 103 \times 10^6$ Pa m^{-3}, $R = 156 \times 10^6$ Pa m^{-3}s, and $\lambda = 0.66$ s^{-1},

we obtain the Aortic number $= \lambda T = (0.66 \text{ s}^{-1})(0.8 \text{ s}) = 0.52 \tag{1.35}$

1.3.7 Mitral Valve Property Characterization (to Provide Interventional Guidelines)

Determining the in vivo constitutive property of the mitral valve (MV) (for a quantifiable estimate of its calcific degeneration) constitutes another example combining "clinical-data monitoring and processing" with "modeling-for-clinical diagnosis." Herein, we combine (1) the vibrational analysis of MV, along with the use of echocardiography (to determine the MV geometry) and spectral phonocardiography (of the first heart sound [FHS] associated with MV vibration) to determine the second peak frequency (f_2) of the FHS spectrum and (2) the static analysis of the semicircular MV leaflet model (held along its circular boundary), as illustrated in Figure 1.8, to (3) obtain the following expressions given by Equations 1.36 and 1.37 for stress (σ) and modulus (E) of the MV leaflet membrane [7]:

$$\text{Stress } (\sigma) \text{ in the mitral valve leaflet membrane} = \frac{\pi^2 f_2^2 a^2 \rho}{(K_{mn}/2)^2} = \frac{\pi^2 f_2^2 a^2 \rho}{(K_{11}/2)^2} \tag{1.36}$$

where
  a is the radius of the semicircular leaflet
  ρ is the leaflet membrane density per unit area
  K_{mn} is the mth zero of the nth order Bessel function $J_n(k)$
  m (number of nodal circles) $= 1$
  n (number of nodal diameters) $= 1$
  $K_{11} = 3.832$

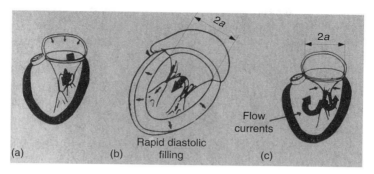

FIGURE 1.8

Functional mechanics of the mitral valve: (a) Mitral valve opening at start of left-ventricular diastole. (b) As the filling left ventricle distends, traction is applied through the chordae tendineae to the valve cusps pulling them together. (c) Start of ventricular systole seals the valves together by the high internal pressure and the flow pattern in the ventricular chamber. It is at this point in time that the MV starts vibrating. (Modified from Ghista, D.N., *J. Mech. Med. Biol.*, 4, 401, 2004.)

$$\text{Modulus } (E) \text{ of the mitral valve leaflet membrane} = \frac{\pi^2 \sigma^3 t^2 (1 - \nu)}{q_0^2 a^2 S_n} \qquad (1.37)$$

where

t is leaflet thickness

ν is Poisson's ratio

q_0 is the pressure difference across the leaflet at time of occurrence of the closed MV vibration

S_n (the summation of a series) $= \displaystyle\sum_{n=1,3,5,\,\ldots} \left[\frac{2-n}{n(2+n)^3} \right]$

Based on Equations 1.36 and 1.37, the nondimensional constitutive parameter (m) of the MV, given by

$$m = \frac{dE}{d\sigma} = \frac{3\pi^6 f_2^4 \rho^2 t^2 a^2 (1 - \nu)}{q_0^2 S_n (K_{11}/2)^4} \qquad (1.38)$$

can be employed diagnostically to track the deterioration due to calcification of the MV, in terms of the change (Δm) in the value of the parameter m, according to the relationship:

$$\Delta m = (\partial m / \partial f) \Delta f + (\partial m / \partial q_0) \Delta q_0 \qquad (1.39)$$

$$\text{so that: } \frac{\Delta m}{m} = 4 \left(\frac{\Delta f_2}{f_2} \right) - 2 \left(\frac{\Delta q_0}{q_0} \right) \qquad (1.40)$$

$$\text{or, } \frac{m'(= m + \Delta m)}{m} = 1 + 4 \left(\frac{\Delta f_2}{f_2} \right) - 2 \left(\frac{\Delta q_0}{q_0} \right) \qquad (1.41)$$

While tracking a patient over a time period, the primary change will be in the FHS frequency $f_2(=\Delta f_2)$, due to progressive calcification of the MV. Hence, from Equation 1.40, we can adopt

$$\frac{\Delta m}{m} = \left(\frac{4\Delta f_2}{f_2}\right) \tag{1.42}$$

to represent the change (Δm) in the parameter (m), by merely monitoring the change in FHS frequency (Δf_2) with respect to the earlier value (f_2).

Section 2.5 provides a detail analysis of determination of in vivo properties of heart valves and their disease status.

1.3.8 Noninvasive Determination of Osteoporosis Index for Osteoporosis Detection

Osteoporosis refers to a group of metabolic bone diseases that are characterized by decreased density of normally mineralized bone. Literally, it is a condition of porous bones, which is characterized by decrease in mechanical strength and stiffness of the bone. Thus, the bone is subjected to fracture. The basic problem is that bone resorption outpaces bone formation.

The noninvasive measurements of bone density techniques now available are single-photon absorptiometry, dual-photon absorptiometry, dual x-ray absorptiometry, qualitative computed tomography, and ultrasound. However, a low-cost method for determination of osteoporosis index (OI) can be formulated in terms of the flexural stiffness (EI) of the ulna bone (where E is elastic modulus and I is moment of inertia of the bone cross-section) and bone density (ρ). The combined term of EI and the density (ρ) of ulna can be determined, because it can be shown to be proportional to the natural frequency (f) of ulna vibrations, which in turn can be obtained from its resonance excitation frequency.

In order to determine the resonance frequency of the ulna beam, it can be simply supported at its extremities and a vibrating probe can be pressed against the skin at the center of the forearm as carried out in [8]. Then, if the ulna bone (of average cross-sectional area A) has a weight W (or mass m) per unit length (such that $W = mg = \rho A g$, where ρ is the density), and length l, its primary-mode frequency f is given (in terms of its angular frequency p) by [9]

$$f = \frac{p}{2\pi} = \frac{1}{2\pi}\frac{\pi^2}{l^2}\left(\frac{EI}{\rho A}\right)^{1/2}$$
$$= 1.57\left(\frac{EI}{\rho A l^4}\right)^{1/2} = 1.57\left[\frac{EI}{(ulna\ mass)l^3}\right]^{1/2} \tag{1.43}$$

By altering the frequency of the vibrating probe, we set the ulna into resonance. The resonant frequency will be equal to the natural frequency. For f (resonance or primary-mode frequency) $= 400$ Hz, $A = 50 \times 10^{-4}$ m^2, $I = 3 \times 10^{-8}$ m^4, length (l) $= 0.17$ m, $\rho = 1.8 \times 10^3$ kg/m^3 [10] , we get $E = 20 \times 10^9$ N/m^2 from Equation 1.43.

It is seen that f is a measure of ulna stiffness (*EI*), mass and length. We can hence define and evaluate an osteoporosis index (OI) in terms of its resonance frequency (f), given by Equation 1.43. Thus, by modeling the ulna bone as a simply supported vibrating beam, and determining its resonance (or natural transverse-vibrational frequency), we can measure the bone mineralization content in terms of this OI.

1.4 Conclusion

Biomedical engineering needs to be a professional field, and hence biomedical engineering graduates need to be employed in hospitals. A field that biomedical engineers can also contribute to is hospital cost-effective management. In this concluding section, we will throw some light on this intriguing field, which can make biomedical engineers more versatile and enhance their employment in hospitals.

We have seen how we can formulate and evaluate nondimensional physiological numbers, to serve as physiological-system disorder indices, and hence represent health status. These nondimensional health-status indicators would also be indispensable for quantifying and evaluating performance indicators of healthcare units and hospital departments (as illustrated in the next section), and also lead to a more knowledgeable means of billing codes for hospital remuneration from state health departments and/or insurance companies.

A hospital has clinical service departments, medical supply and hospital service departments, and financial-management and administrative departments. Each of these five sets of departments has to function in a cost-effective fashion. Let us, for example, consider the intensive care unit (ICU) department. The human resource to an ICU department consists of physicians and nurses. Using activity-based costing, we can determine the human-resource strength, based on an assumed reasonable probability-of-occurrence of (for instance) two patients simultaneously (instead of just one patient) having life-threatening episodes.

1.4.1 Performance Index

We can then formulate the ICU performance indicator in terms of the amounts by which the physiological health index (PHI) values of patients were (1) enhanced in the ICU for those patients discharged into the ward from the ICU, and (2) diminished in the ICU in the case of patients who died in the ICU. Let us say that patients are admitted to the CCU if their PHI value falls below 50%. Thus, if the PHI of a patient improves from 30 to 50, the physiological health improvement index (PHII) for that patient is given by

$$\text{PHII} = 100 \left(\frac{50 - 30}{30} \right) = 67 \text{ (or 67\%)} \tag{1.44}$$

Thus, the PHII value is higher if a more seriously ill patient is discharged from the ICU, and lower if a not-so-seriously ill patient is discharged, i.e., if

$$\text{PHII} = 100\left(\frac{50-40}{40}\right) = 25 \text{ (or 25\%)} \tag{1.45}$$

We can then formulate the ICU performance index (PFI) for an ICU as follows.

ICU Performance Index (PFI)

$$= \frac{\Sigma\,\text{PHII of patients}}{\text{Number of those patients treated during a time period}} \tag{1.46}$$

Hence, the higher the value of ICU performance index, the better the performance of the ICU. If now a patient dies, as a result of the PHII becoming negative, i.e., slipping from (say) 30 to 10, then

$$\text{PHII} = 100\left(\frac{10-30}{30}\right) = -67 \tag{1.47}$$

As a result, Σ PHII (in Equation 1.46) will decrease, and the overall value of ICU performance index (namely, PFI, as calculated by means of Equation 1.46) will fall.

1.4.2   Cost-Effective Index

Now consider that (1) we have one physician and five nurses for a 10-bed CCU, based on the probability-of-occurrence of two patients having life-threatening events being, say, 0.2 (or 20%), and that (2) for this human resource/staffing, the ICU performance index value is (say) 40. If we increase the staffing, the ICU performance index value could go up to 50 or so, at the expense of more salary cost. So now we can come up with another indicator, namely, cost-effectiveness index (CEI), given by

$$\text{CEI} = \frac{\text{Performance index}}{\text{Total salary index (in salary units)}}$$

$$= \frac{\text{Performance index}}{\text{Resource index (in terms of salary units)}} \tag{1.48}$$

where, say a salary of $1000 = 0.1$ unit, $10,000 = 1$ unit, $20,000 = 2$ units, and so on.

Thus, if an ICU has one physician with a monthly salary of $20,000 (i.e., 2 salary units) and five nurses with a total monthly salary of $25,000 (i.e., total of 2.5 salary units), then from Equation 1.48

$$\text{CEI(ICU)} = \frac{\text{Performance index (of 40)}}{\text{Salary units index or resource index, } R_i [=(2+2.5)]} = \frac{40}{4.5} = 11.1$$

$$(1.49)$$

Now let us assume that we raise the PFI(ICU) to (say) 50 by augmenting the nursing staff, so as to have six nurses ($R_i = 3$ units) and 1.5 full-time equivalent physicians on duty ($R_i = 3$ units). Then,

$$\text{CEI(ICU)} = \frac{\text{PFI}}{R_i} = \frac{50}{(3+3)} = 8.3 \qquad (1.50)$$

Thus, while the PFI of ICU has gone up from 40 to 50, the CEI of ICU has gone down from 11.1 to 8.3.

1.4.3   Strategy of Operation

Our strategy would be to operate this "performance-resource" system in such a way that we can determine the resource index R_i for which we can obtain acceptable values of PFI and CEI.

Now let us formulate how a hospital budget can be optimally distributed. Let a hospital have "n" number of departments and a prescribed budget (or budget index, BGI). We would want to distribute the budget among the departments, such that none of the "n" departments has a PFI below the acceptable value of PFI_a and a CEI below the acceptable value of CEI_a.

So the operational problem is to be formulated as follows.

How to distribute or divide the given budget (or budget index value) into R_i ($i = 1, \ldots, n$), such that $\text{PFI}_i \geq \text{PFI}_a$ and $\text{CEI}_i \geq \text{CEI}_a$, for all i. This then is the prime task of a hospital administrator.

References

1. Ghista, D.N., Physiological systems numbers in medical diagnosis and hospital cost-effective operations. *Journal of Mechanics in Medicine and Biology*, Vol. 4, No. 4, 401–408, 2004.
2. Lim, G.H., Ghista, D.N., Koo, T.Y., Tan, J.C.C., Eng, P.C.T., and Loo, C.M., Cardiac fitness mathematical model of heart-rate response to VO_2 during and after stress-testing. *International Journal of Computer Application in Technology* (*Biomedical Engineering & Computing*), Vol. 21, No. 1–2, 38–45, 2003.
3. Ghista, D.N., Pasam, R., Vasudev, S.B., Bandi, P., and Kumar, R.V., Indicator for lung status in a mechanically ventilated COPD patient using lung-ventilation modeling and assessment, in *Human Respiration: Anatomy & Physiology, Mathematical Modeling, Numerical Simulation and Applications*, edited by V. Kulish, WIT Press, Southampton, U.K., 2006, pp. 169–185.

4. Ghista, D.N., Loh, K.M., and Damodaran, M., Lung ventilation modeling and assessment, in *Human Respiration: Anatomy & Physiology, Mathematical Modeling, Numerical Simulation and Applications*, edited by V. Kulish, WIT Press, Southampton, U.K., 2006, pp. 95–115.

5. Dittakavi, S.S. and Ghista, D.N., Glucose tolerance test modeling & patient simulation for diagnosis. *Journal of Mechanics in Medicine and Biology*, Vol. 1, No. 2, 193–223, Oct. 2001.

6. Zhong, L., Ghista, D.N., Ng, Y.K., and Chua, T.S.-J., Determination of aortic pressure-time profile, along with aortic stiffness and peripheral resistance. *Journal of Mechanics in Medicine and Biology*, Vol. 4, No. 4, 499–509, 2004.

7. Ghista, D.N. and Rao, A., Structural mechanics of the mitral valve: Stresses in the membrane, indirect determination of the instantaneous modulus of the membrane. *Journal of Biomechanics*, Vol. 5, No. 3, 295–307, 1972.

8. Steele, C.R. and Godon, A.F., Preliminary clinical results using 'SOBSA' for noninvasive determination of ulna bending stiffness, in *Advances in Bioengineering*, American Society of Mechanical Engineers Symposium, NY, 1978.

9. Timoshenko, S., Young, D.H., and Weaver, W., *Vibration Problems in Engineering*, John Wiley & Sons, New York, 1974, pp. 420–423.

10. Peterson, K., Noninvasive determination of bone stiffness, PhD thesis, Stanford University, 1977.

Section I

Cardiological Engineering Mechanics

2

Left Ventricular Mechanics

Dhanjoo N. Ghista and Liang Zhong

CONTENTS

2.1   Introduction and Scope

Biomechanics has been associated with major physiological advances and medicine. However, considerable insight into physiology and medicine can also be gained from innovative applications of even relatively basic engineering analyses. In this chapter, we are developing the concept of cardiology to demonstrate how even fundamental engineering disciplines can bring to bear enhanced logic to cardiology, to:

- Determine the pressure-drop across a stenotic aortic valve (AV)
- Demonstrate how AV disorders could lead to myocardial infarct
- Depict likely sites for myocardial ischemias and infarcts
- Explain how myocardial infarct impairs stroke volume and cardiac output
- Obtain quantifiable measures of left-ventricular (LV) stiffness and contractility, so as to provide a measure of impaired LV pumping capacity

2.2   Pressure-Drop across a Stenotic Aortic Valve

We start our journey in the heart, by analyzing the pressure-drop across a stenotic AV. The inlet and outlet to and from the left ventricle is regulated by heart valves. If the AV gets diseased and becomes stenotic, it will result in a big pressure-drop across the valve, which can be evaluated in terms of the LV outflow rate and the dimensions of the outflow tract, using Bernoulli theorem equation, as carried out in Figure 2.1.

$$P_1 - P_2 = \frac{\rho V_2^2}{2} - \frac{\rho V_1^2}{2} + \frac{\rho k_c V_2^2}{2} = \frac{\rho Q^2}{2} \left[\frac{1}{A_2^2} - \frac{1}{A_1^2} + \frac{k_c}{A_2^2} \right] \qquad (2.1)$$

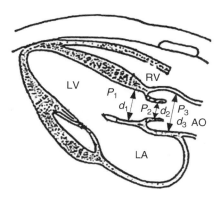

FIGURE 2.1
LV longitudinal cross-section showing the aortic outflow tract.

Pressure drop between sections 2 and 3 (refer Figure 2.1),

$$P_2 - P_3 = \frac{\rho Q^2}{2}\left[\frac{1}{A_3^2} - \frac{1}{A_2^2} + \frac{1}{A_2^2}\left(1 - \frac{A_2}{A_3}\right)^2\right] \tag{2.2}$$

Hence, $\quad \dfrac{P_1 - P_3}{\left(\frac{\rho Q^2}{2A_1^2}\right)} = \left(\dfrac{A_1}{A_3}\right)^2 - 1 + k_c\left(\dfrac{A_1}{A_2}\right)^2 + \left(\dfrac{A_1}{A_2}\right)^2\left(1 - \dfrac{A_2}{A_3}\right)^2 \tag{2.3}$

For a stenotic AV, let us take $Q = 1.6 \times 10^{-4}$ m^{3}s^{-1} (corresponding to a cardiac output of 4 L/min, HR $= 80$ min^{-1}, ejection period $= 0.31$ s), outflow-tract diameter $d_1 = 1.6$ cm, $d_2 = 0.8$ cm, $d_3 = 2$ cm, coefficient $k_c = 0.33$, blood density $\rho = 1000$ kg/m^3, we obtain from Equation 2.3, pressure-drop across AV $(P_1 - P_3) = 5071$ Pa or 38 mmHg.

2.3 Why Valvular Disorders Can Lead to Myocardial Infarcts?

The high pressure-drop across a stenotic AV will lead to increase in LV chamber pressure and LV hypertension (as illustrated in Figure 2.2). This is because the LV will have to contract and pump more vigorously, in order to overcome this pressure-drop $(P_1 - P_3)$ across the stenotic AV, and appropriately perfuse the systemic circulation. In other words, the LV chamber pressure (P_1) will increase (as schematized in Figure 2.2).

Now, by using the simplified Laplace law for wall stress in a pressurized thin-spherical shell, namely:

$$\text{wall stress }(\sigma) = \frac{\text{LV pressure }(P_1) \times \text{chamber radius }(R)}{2x \text{ wall thickness }(h)} \tag{2.4}$$

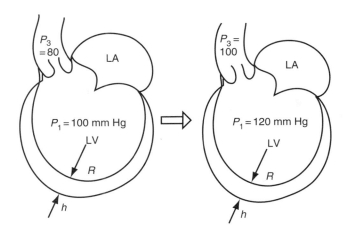

FIGURE 2.2
Why chronic AV stenosis can lead to MI? If $P_1 \uparrow$, then $\sigma \uparrow$, $\rightarrow$ LV-Work $\uparrow$, and LV O_2 demand $\uparrow$.

it can be readily seen that this leads to augmented wall stress (σ) and associated increased oxygen (O_2) demand, which can cause myocardial infarcts (due to O_2, supply–demand mismatch), as depicted in Figure 2.2. Based on Equation 2.4, the LV could compensate for this augmented wall stress, by increasing its wall thickness (h), to thereby contain the wall stress (σ) and hence the O_2 demand, and thereby prevent formation of myocardial ischemia and infarct. If now there is O_2 supply–demand mismatch, this phenomenon will cause myocardial ischemia, and eventually myocardial infarct (MI).

Where to look out for the presence of myocardial ischemic (or infarcted) segment?

In order to answer this question, we need to keep in mind that the circumferential (tensile) wall stress (σ_θ) in a pressurized thick-walled sphere, simulating the LV (Figure 2.3), is maximum at the inner (endocardial) wall [1], as expressed by Equations 2.5 and 2.6 and plotted in Figure 2.3. This is, in fact, where myocardial ischemias and infarcts mostly occur, because of the higher resistance to flow and myocardial perfusion near the inner wall. This appreciation can enable us to justify either coronary bypass grafting or even myocardial canalization for reperfusion from the intra-LV blood pool itself.

The expressions for σ_r and σ_θ are given by [1]

$$\frac{\sigma_r}{P} = \frac{\left[\left(\frac{r_i}{r_e}\right)^3 - \left(\frac{r_i}{r}\right)^3\right]}{\left[1 - \left(\frac{r_i}{r_e}\right)^3\right]} \tag{2.5}$$

$$\frac{\sigma_\theta}{P} = \frac{\left[\left(\frac{r_i}{r_e}\right)^3 + \frac{1}{2}\left(\frac{r_i}{r}\right)^3\right]}{\left[1 - \left(\frac{r_i}{r_e}\right)^3\right]} \tag{2.6}$$

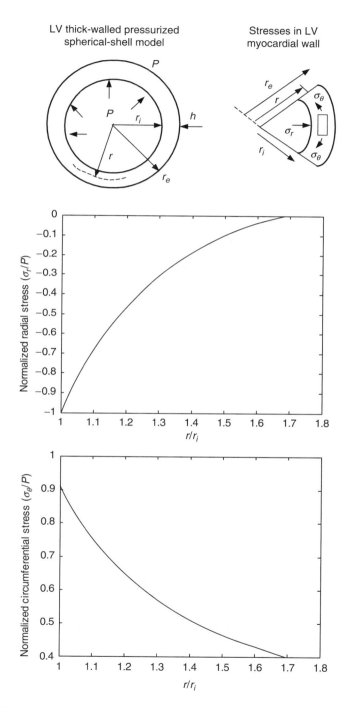

FIGURE 2.3

Normalized wall-stress variations for the LV spherical thick-walled model ($r_i = 1.5$ cm, $r_e = 2.5$ cm at an instant during the ejection phase).

2.4   How Myocardial Infarct Impairs Stroke Volume and Cardiac Output?

Figure 2.4 illustrates the wall motion of a contracting LV with an infarcted myocardial wall segment. We can apply the Bernoulli theorem between points 1 (on the lateral wall) and 4 (at the entrance to aorta), points 2 and 4, and points 3 and 4, as follows:

$$P_1 + (\rho V_1^2/2) = P_a + (\rho V_a^2/2) \quad \text{or} \quad P_a = P_1 + (\rho/2)(V_1^2 - V_a^2) \tag{2.7}$$

$$P_2 + (\rho V_2^2/2) = P_a + (\rho V_a^2/2) \quad \text{or} \quad P_a = P_2 + (\rho/2)(V_2^2 - V_a^2) \tag{2.8}$$

$$P_3 + (\rho V_3^2/2) = P_a + (\rho V_a^2/2) \quad \text{or} \quad P_a = P_3 + (\rho/2)(V_3^2 - V_a^2) \tag{2.9}$$

Therefrom we obtain from Equations 2.7 through 2.9:

$$P_a = \frac{1}{n}\sum_{i=1}^{n} P_i + \frac{\rho}{2}\left[\frac{1}{n}\sum_{i=1}^{n}(V_i^2 - V_a^2)\right] \tag{2.10}$$

for n points along the endocardial wall.

Let us say that a set of points $j\ (=1,\ldots,m)$ lie on the infarcted wall segment, and the remaining points $k\ (=1,\ldots,t)$ lie on the contracting endocardial wall. Then,

$$P_a = \frac{1}{n}\left[\sum_{j=1}^{m} P_j + \sum_{k=1}^{t} P_k\right] + \frac{\rho}{2}\left[\frac{1}{n}\left\{\sum_{j=1}^{m} V_j + \sum_{k=1}^{t} V_k\right\} - V_a^2\right] \tag{2.11}$$

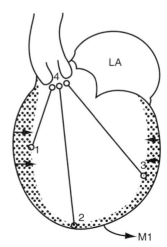

FIGURE 2.4
Contracting infarcted LV. In this figure, the regions associated with points 1 and 3 have normal wall contraction, whereas the wall region associated with point 2 is infarcted.

Now for these *j* points on the infarcted endocardial wall, V_j will be zero, and (as per Equation 2.11), P_a will be diminished. In other words, the output pressure will be diminished. With less LV pressure generated to overcome the aortic pressure (or LV afterload), the stroke volume (SV) and hence the cardiac output (CO) will also be impaired.

Thus, it can be seen that if an LV has a myocardial infarct, its myocardial wall will be stiffer and noncontractile in the infarcted region (Figure 2.4). Hence, the LV pressure generated will be impaired as illustrated by Equation 2.11. Thereby, the LV stroke volume and hence its cardiac output will be diminished.

2.5 Noninvasive Diagnosis of Diseased Heart (Mitral and Aortic) Valves, Based on Their Dynamics Modeling, Echo-, and Phonocardiography

2.5.1 Background

Diseases of heart valves (e.g., aortic and mitral) can result in valve leaflet thickening (qualitatively detected by auscultation and echocardiography), valvular stenosis, and high pressure-drop and valvular regurgitation (qualitatively detected by Doppler echocardiography). These detection methods are indirect and empirical, entailing measurement of influences of deterioration of valvular tissue properties on heart-sound frequency, pressure-gradient across the valve, and intracardiac flow, as opposed to direct measurement of in vivo tissue properties. As per these methods, if timely surgical-corrective or replacement intervention is not carried out, it can result in myocardial ischemia and even infarct.

Our approach to this issue is based on our findings that valvular disease affects the elastic constitutive properties of valvular tissues [2–6]. Hence, by developing methods for determination of the in vivo properties of healthy and diseased valves (in terms of leaflet modulus vs. stress), and by categorizing healthy and diseased tissues in terms of their parametric ranges, we can not only sensitively and quantitatively differentiate between healthy and diseased valves but also detect the severity of their diseased states.

2.5.2 Mitral Valve Biomechanical Model to Detect Diseased Valve

First heart sounds (FHS) are associated with the closure and ensuing vibrations of mitral valves (MV) and can be related to the resonant vibrational frequencies of MV [3]. Now the vibrational frequencies of MV are governed by and can be expressed in terms of valvular tissue elasticity [4]. Since a degenerating MV tissue has altered elasticity [2], it will have altered vibrational frequency.

Thus, by combining heart-sound power-spectral and valvular mechanics analyses with 2-d echocardiographic analysis (to determine MV dimensions), we can determine the in vivo elastic constitutive property of MV, as a relationship between the elastic modulus (E) and stress (σ) properties of the membrane. For this purpose, stress and vibrational model analyses of MV leaflets have been carried out based on semicircular models of the leaflet geometry obtainable from two-dimensional (2-D) echocardiography [2,4]. The MV leaflet-membrane static and vibrational analyses will yield expressions for the membrane leaflet stress (σ) and modulus (E) as functions of its primary vibrational frequency and its geometrical size parameter. From these expressions, we will formulate a new stress-based property (σ^*) and a modulus-based property (E^*), which can be determined in terms of the vibrational (and FHS) frequency (f) and valve dimension parameter (a). The parameters of E^* versus σ^* constitutive property can be employed to differentially diagnose normal and diseased MV(s).

2.5.2.1 *Analysis*

1. The MV forms a component and one segment of the boundary of the left ventricle (with the left atrium). Thus the LV primary-mode vibrational frequency (f_{lv}) will be lower than the MV primary-mode vibrational frequency (f_{mv}). The MV cusps can be modeled as semicircular membranes held along the valve ring as well as along the edges by the chordae tendineae [2,4]. The equation of equilibrium of an element of the MV leaflet membrane is represented (in polar coordinates) by

$$T(\nabla^2 W_s) = T\left(\frac{\partial^2 W_s}{\partial r^2} + \frac{\partial W_s}{r\partial r} + \frac{1}{r^2}\frac{\partial^2 W_s}{\partial \theta^2}\right) = -q = -\sum_{n=1,3,5}^{\infty}\frac{4q_0}{n\pi}\sin n\theta$$

(2.12)

where
q is the leaflet loading ($=$ differential pressure across the leaflet)
W_s is the membrane deflection
T is the membrane tension

For the boundary conditions of a semicircular leaflet membrane of radius a (held along its edges):

$$W_s(r = a, \theta) = 0; \quad W_s(r, \theta = 0) = 0; \quad W_s(r, \theta = \pi) = 0 \qquad (2.13)$$

Hence, we obtain the leaflet deflection as

$$W_s = \sum_{n=1,3,5}^{\infty} = \frac{4q_0}{n\pi T(4 - n^2)}\left[\left(\frac{r}{a}\right)^n - \left(\frac{r}{a}\right)^2\right]\sin n\theta \qquad (2.14)$$

It is noted that the above expression for W_s contains the tension term T. We can determine the tension T in the membrane from the condition that the change in membrane surface area due to its stretching by tension T in it (given by $dA_T = T(1 - v)\pi a^2/Et$) equals the change in its surface area (of thickness t) due to its being deflected by W_s (Equation 2.14) under the loading q_0 (given by $dA_w = q_0^2 a^4 S_n/T^2$), to obtain the following expression for the tension T and stress σ ($= T/t$) in the mitral valve leaflet membrane [2]:

$$\sigma = \frac{T}{t} = \left[\frac{E q_0^2 a^2 S_n}{\pi^2 t^2 (1 - v)}\right]^{1/3} \tag{2.15}$$

where E and v are the elastic modulus and Poisson's ratio of the valve leaflet material, and

$$S_n = \sum_{n=1,3,5,7}^{\infty} \frac{(2 - n)}{n(2 + n)^3} = 0.0234 \tag{2.16}$$

We have thus obtained an expression for the stress (σ) in the membrane, from the static analysis of the membrane under loading q_0 and tension T in the membrane.

2. In order to determine the constitutive property of the heart-valve membrane, we also need to determine the expression for its elastic modulus (E). For this purpose, we will now carry out a vibration analysis of the valve membrane associated with the FHS. Now at the instant of occurrence of the FHS, the closing MV membrane is vibrating about its deflected shape $W_s(r, \theta)$, due to the differential pressure q on it, by the amount $W_d(r, \theta)$, which is obtained from the solution of the following MV vibrational equation:

$$T\left(\frac{\partial^2 W_d}{\partial r^2} + \frac{1}{r}\frac{\partial W_d}{\partial r} + \frac{1}{r^2}\frac{\partial^2 W_d}{\partial \theta^2}\right) = \rho\frac{\partial^2 W_d}{\partial t^2} \tag{2.17}$$

By making the solution W_d satisfy the following boundary conditions:

$$W_d(r = a, \theta, t) = 0, \quad W_d(r, \theta = 0, t) = 0, \quad W_d(r, \theta = \pi, t) = 0$$

we obtain the primary-mode vibrating frequency of the semicircular MV membrane (as the frequency of the corresponding circular membrane vibrating about its diameter as a nodal line) as [2]:

$$f_{mv} = \frac{3.832}{2\pi a}\sqrt{\frac{\sigma}{d}}, \quad \text{or} \quad \sigma = \frac{\pi^2 f_{mv}^2 a^2 d}{(1.916)^2} \tag{2.18a}$$

where d is the density of the valve leaflet membrane. Then by combining the static and dynamic analyses results of Equations 2.15 and 2.18, we obtain the expression for the mitral valve leaflet modulus as

$$E = \frac{\pi^8 f_{mv}^6 d^3 t^2 a^4 (1 - v)}{(1.916)^6 q_0^2 S_n} \tag{2.18b}$$

wherein f_{mv} corresponds to the third peak of the FHS spectrum (since lower frequency peaks correspond to LV vibrations). As a matter of interest, for the data: $f_{mv} = 100$ Hz, $q_0 = 2$ mm Hg, $d = 1.02$ gm/cm^3, $a = 1$ cm, $t = 0.5$ mm, $v = 0.5$, and evaluating S_n (Equation 2.16), we get $\sigma = 2.75 \times 10^3$ N/m^2 and $E = 1.6 \times 10^5$ N/m^2.

3. Now, changes in MV pathology will affect its density (d) and thickness (t), and its modulus (E) vs. stress (σ) property which we want to determine by combining FHS power-spectrum analysis (to determine f_{mv}) and 2-d echocardiographic analysis (to determine the size parameter a).

We now designate a new stress-based property (σ^*) of mv (from Equation 2.18a), as

$$\sigma^* = \frac{\sigma}{d} = \frac{\pi^2 f_{mv}^2 a^2}{(1.916)^2} \qquad (2.19a)$$

as well as a new modulus-based property (E^*) of MV (from Equation 2.18b), as

$$E^* = \frac{Eq_0^2}{d^3 t^2} = \frac{\pi^8 f_{mv}^6 a^4 (1 - v)}{(1.916)^2 S_n} \qquad (2.19b)$$

We can now employ the E^* vs. σ^* relationship as a constitutive property of MV, to characterize and track its degeneration for timely intervention purpose.

This technology and methodology can provide the basis for timely surgical and/or replacement intervention for a diseased MV. In order to apply this analysis, we can determine the valvular leaflet size parameter from 2-D echocardiograms. The valvular leaflet vibrational frequency can be obtained from the frequency spectra of the FHS phonocardiographic signal associated with MV movement.

We can study a number of patients and determine the in vivo (E^*, σ^*) values of their valves, at regular intervals during their degeneration process. We can also simultaneously and regularly monitor cardiac symptoms and chamber sizes and correlate them with the valvular constitutive E^*–σ^* property. By means of these correlations, we can determine the critical (E^*–σ^*) boundary at which intervention will have to be made to replace the degenerated natural valve by means of a prosthetic flexible-leaflet MV [2].

2.5.3 Aortic Valve Biomechanical Model to Determine Normal and Diseased Aortic Valve Properties

Second heart sounds (SHSs) are associated with the closure and ensuing vibration of AVs. The heart-sound spectral frequencies can be related to the resonant vibrational frequencies of heart valves [5]. Now the vibrational frequencies of heart valves are governed by and can be expressed in terms of their tissue elasticity [6]. Since a degenerating tissue has altered elasticity

[3], it will have altered its vibrational (primary) frequency, and hence altered the heart-sound power-spectral frequency profile.

Thus, by combining 2-D echocardiography (to obtain valvular geometry), heart-sound power-spectral analysis (to obtain the valvular vibrational frequency), and valvular mechanics (stress-deformation and vibrational) analyses, we can determine the in vivo valvular properties of AVs [5,6].

The AV (as shown in Figure 2.5A) has three membrane sectors (each of angle $2\pi/3$). Each of these three membrane sectors deforms (by an amount W_s) under the differential pressure (P) across the valve. At the same time, toward the closure of the AV (and associated with the SHS), each membrane sector vibrates; it has been shown [6] that the second spectral peak (f) frequency of the SHS is best able to differentiate between normal and pathological valve.

Just as in the case of the MV analysis, we can also carry out both static-deformation and vibrational analyses of the AV membrane sectors [5–7]. From these analyses, we can obtain expressions for the elastic modulus and the stress in the valve leaflet, in terms of the SHS's second spectral peak frequency (f), the radius (a) of the valve ring, density (D) of the leaflet material, the pressure difference (P) across the valve leaflet, and W_s (the deflection of the valve leaflet under the differential pressure P) at the time of occurrence of the SHS. We will then express the valve leaflet constitutive property in terms of a modified modulus property (E^*) and stress (σ), both of which can be determined noninvasively in terms of f and a.

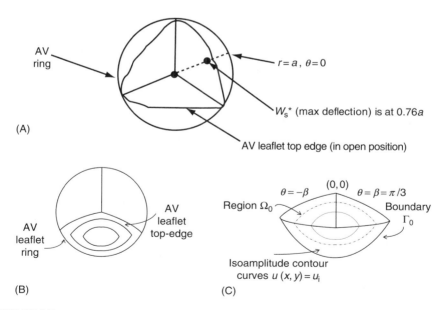

FIGURE 2.5
Aortic Valve geometry and deformation profile: (A) Schematic of the aortic valve geometry. (B) Aortic leaflet membrane analyzed as a circular 120° sector. (C) Schematic of the isoamplitude shape–function curves $u_i(x, y) = u_i$ (shown dotted in one of the leaflets), depicting the profile of the deflected surface of the AV leaflet when it vibrates.

2.5.3.1 Analysis (Figure 2.5)

1. An aortic valve leaflet membrane (shown in Figure 2.5A) is analysed as a circular 120° sector–shaped membrane of $\theta = \pm\beta = \pm\pi/3$ and radius $r\,(=a)$, as depicted in Figure 2.5B. Its boundary is represented by a simply connected plane curve Γ_0 enclosing a region Ω_0, as depicted in Figure 2.5C. When the AV vibrates at the time of its closure (associates with the SHS), its membrane deflection at a point in the region Ω_0 at any time t is denoted by $W(x, y, t)$.

2. Let us start with the deflection (W_s) of this 120° sector leaflet membrane (of $\theta = \pm\pi/3$ and radius $r = a$) at the time of occurrence of the SHS. When the membrane vibrates in one of its modes, the profile of its deflection surface can be described by a family of isoamplitude curves, which form a set of level curves $u(x, y) = u_i$ (constant) when projected on the (x, y) plane, as shown in Figure 2.5C. This family of isoamplitude curves is denoted by Γ_u (for $0 \leq u \leq u^*$, where u^* is the maximum value of u), wherein Γ_0 denotes the leaflet membrane boundary, and the region bounded by Γ_u is denoted by Ω_u. We hence need to scale $u(x, y)$ to the static deflection $W(x, y)$ of the pressure-loaded valve at the time of occurrence of SHS, by making u^* (the maximum value of $u(x, y)$) correspond to the maximum static deflection (W_s^*) of the valve membrane (at the location of u^*).

3. In the context of membrane analogy, when torsion is applied to a cylindrical beam with the same cross-section as the valve leaflet membrane (120° sector) boundary, the lines of constant shearing stress coincide with these isodeflection contours of the membrane. Hence, the function $u(x, y)$ satisfies the same Poisson equation as the Prandtl stress function (of torsion of a cylindrical beam of the same cross-section as the 120° valve leaflet section), given by [5,6]:

$$\nabla^2 u(x, y) = -2 \qquad (2.20a)$$

The static deflection W_s, under the pressure loading P (equal to aortic pressure minus LV pressure) and leaflet tension T, is given by [5,6]:

$$\nabla^2 W_s = P/T \qquad (2.20b)$$

However, we need to scale $u(x, y)$, such that its maximum value u^* corresponds to the maximum static deflection W_s^*, as follows [5,6]:

$$u^* = \max[u(x, y)] = \frac{-2TW_s^*}{P} \qquad (2.21)$$

so that (1) the W_s shape contours correspond to the $u(x, y)$ contours and (2) $u(x, y) = -2W_s(x, y)T/P$.

In order to specify u^* in equation (3), we need to first obtain the u (x, y) or $u(r, \theta)$ for a 120° $(2\pi/3)$ sector-shaped membrane. Herein, the boundary shape of an aortic leaflet is approximated as a circular sector of angle $2\pi/3$. Then, from the corresponding torsion problem, it can be shown [6] that for a sector of a circle, with boundary given by ($r = 0, r = a, \theta = \pm\beta$), we have the deformed deflection–amplitude shape function $u(r, \theta)$ as [5,6]

$$u = \frac{1}{2}r^2\frac{\cos 2\theta}{\cos 2\beta} + a^2\sum_{n=0}^{\infty}\left[A_{2n+1}\left(\frac{r}{a}\right)^{(2n+1)\frac{\pi}{2\beta}}\cos\left[(2n+1)\frac{\pi\theta}{2\beta}\right]\right] - \frac{1}{2}r^2$$

$$A_{2n+1} = (-1)^{n+1}\left[\frac{1}{(2n+1)\pi - 4\beta} - \frac{2}{(2n+1)\pi} + \frac{1}{(2n+1)\pi + 4\beta}\right]$$

$$(2.22a)$$

wherein r is the radial coordinate and a is the valve ring radius. The function u is a plane harmonic function which satisfies the condition $u = 0$ at the boundary given by ($r = 0, r = a, \theta = \pm\beta$). By symmetry, the maximum value of u will be found along the line $\theta = 0$. The computation of u along this line satisfies the profile of the membrane's deflection shape.

By choosing the aortic valve leaflet sector angle to be $\beta = \pi/3$ and by carrying out an iterative computer calculation [5,6], we obtain the maximum $u(r, \theta)$ value of $u^* = 0.11181r^2$ at $r = 0.76a$. In other words, for the aortic leaflet model, the maximum value of $u(r, \theta)$ is $u^* = 0.06458a^2$ at $r = 0.76a$, i.e.,

$$u^* = u_{\max} = 0.11181r^2(\text{at }\theta = 0 \text{ and } r = 0.76a) = 0.06458a^2 \quad (2.22b)$$

Likewise, W_s^* also occurs at $\theta = 0$ and $r = 0.76a$

4. When the aortic valve closes, its leaflets membranes are set into vibration. Hence, we next carry out a vibration analysis of the aortic valve leaflet. Consider a portion of the membrane boundary represented by a closed contour $u(x, y) = $ constant at an instant t. For the vibrating membrane, the differential equation of motion at any instant t can be written as follows [5,6]

$$T\left(\frac{\partial^2 W}{\partial x^2} + \frac{\partial^2 W}{\partial y^2}\right) = \rho\frac{\partial^2 W}{\partial t^2}$$

or, $\quad T\int_{\Gamma_u}\left(\frac{\partial^2 W}{\partial r^2} + \frac{1}{r^2}\frac{\partial^2 W}{\partial\theta^2}\right)ds - \rho\iint_{\Omega_u}\frac{\partial^2 W}{\partial t^2}d\Omega = 0 \quad (2.23a)$

where T denotes the membrane tension per unit length, W denotes the deflection of the membrane, ds denotes an element of the closed curve $u = $ constant, Γ_u corresponds to the earlier mentioned family of isoamplitude curves, $\int_{\Gamma_u}$ denotes integration along a closed contour $u(x, y) = $ constant, $\iint_{\Omega_u}$ denotes double integration over the interior of the closed region bounded by the contour $u(x, y) = $ constant, Ω_u denotes the region bounded by Γ_u, and ρ is the membrane mass per unit area. The solution to the above equation can be expressed as

$$W(u, t) = \sum_{i=1}^{\infty} w_i(u)f_i(t) = \sum w_i(u)e^{-i\omega t} \qquad (2.23b)$$

where the eigenfunctions $w_i(u)$, corresponding to this free vibration problem, are given in terms of the zeroth-order Bessel functions as

$$w_i(u) = AJ_0(\sqrt{2}\,kg) + BY_0(\sqrt{2}\,kg) \qquad (2.24)$$

where
$g = (u^* - u)^{1/2}$
$k^2 = (\rho/T)\omega^2$
A and B are arbitrary constants

In order to avoid infinite displacement of u at any point of the membrane, we set $B = 0$. Then considering the boundary condition of the membrane having zero displacement around the boundary $u(r, \theta) = 0$ or $u(x, y) = 0$, we obtain (from Equation 2.24)

$$J_0\sqrt{2u^*}k = 0 \qquad (2.25a)$$

$$\text{for which} \quad \sqrt{2u^*}k = B_i \qquad (2.25b)$$

where B_i is the ith zero of the zeroth order Bessel function J_0. Then the symmetric mode eigenvalues $\bar{\omega}$ (associated with frequency ω) are given by

$$\sqrt{2u^*}k = \bar{\omega} = 2.4048, 5.5201, 8.6537, \ldots \qquad (2.26)$$

From Equation 2.26, we can express the tension (T) and hence stress ($\sigma = T/h$) in the leaflet membrane in terms of its vibrational frequency ($f = \omega/2\pi$) and u^* ($= 0.06458a^2$). It has been determined by us [6] that when the data of the valve's SHS spectral peak frequency (f) and radius (a) are plotted on the (f–a) coordinate plane, employment of the second peak frequency of the SHS spectrum (corresponding to the second mode of vibration, for which $\bar{\omega} = 5.5201$) is best able to effectively separate normal and pathological AVs. Hence, we will adopt $\bar{\omega} = 5.5201$.

For the second mode frequency, we take $\bar{\omega} = 5.5201$, and obtain

$$k_2^2 = \frac{\rho}{T}\omega_2^2 = \frac{5.5201^2}{2u^*} \qquad (2.27a)$$

or

$$\omega_2 = (T/\rho)^{1/2} \frac{5.5201}{\sqrt{2u^*}} \tag{2.27b}$$

or

$$\sigma = \frac{T}{h} = \frac{8\pi^2 f^2 D u^*}{5.5201^2} \tag{2.28}$$

where $f = \omega_2/2\pi$, the density $D = \rho/h = 1\ \mathrm{g/cm^3}$, and $u^* = 0.06458a^2$.

5. We now obtain an expression for the tension (T) in the AV membrane under its pressure-loaded state. For this purpose, we equate (1) the change dA_w in the membrane surface area due to its taking up the deflected shape (W_s) under the pressure loading P to (2) the change dA_T in its surface area due its being stretched by the tension T in it.

The expression for dA_w is obtained as [6]

$$dA_w = \frac{P^2}{8T^2} A u^* \tag{2.29}$$

where A is the membrane area before deformation.
The expression for dA_T is obtained as [6]

$$dA_T = \frac{2T(1-\nu)A}{Eh} \tag{2.30}$$

Upon equating dA_w to dA_T, we obtain the following expression for the tension T in the leaflet:

$$T = \left[\frac{EhP^2 u^*}{16(1-v)} \right]^{1/3} \tag{2.31}$$

Keeping in mind that the static deflection satisfies the classical Poisson equation, and that

$$P = -2T \frac{W_s^*}{u^*} \quad \text{and} \quad u^* = \frac{-2T}{P} W_s^* = 0.06458a^2 \tag{2.32}$$

we obtain from Equations 2.31 and 2.32, the expression for the Young's modulus of the AV leaflet material, as

$$T^3 = \frac{Ehu^*}{16(1-v)} P^2 = \frac{Ehu^*}{16(1-v)} \left(2T \frac{W_s^*}{u^*} \right)^2$$

$$= \frac{Ehu^*}{16(1-v)} \left(2 \frac{W_s^*}{u^*} \right)^2 T^2 \tag{2.33a}$$

or

$$E = \frac{4u^*(1-v)T}{(W_s^*)^2 h} = \frac{4u^*(1-v)\sigma}{(W_s^*)^2} \tag{2.33b}$$

where $u^* = 0.06458a^2$ and W_s^* is the maximum deflection of the AV leaflet.

6. Now from the secondary vibrational mode of the AV leaflet, the stress in the aortic valve leaflet membrane can be written from Equation 2.28, as

$$\sigma = \frac{T}{h} = \frac{8\pi^2 f^2 D u^*}{(5.5201)^2} \tag{2.34}$$

where $u^* = 0.06458a^2$ and $D = 1$ g/cm^3 or 10^3 kg/m^3. Hence, from Equation 2.34

$$\sigma(\text{in N/m}^2) = 167.35 f^2 a^2 \tag{2.35}$$

where the vibrational frequency f ($=$ the SHS spectrum's second peak frequency) is in hertz (Hz) and a is in meters.

7. Now, we note that the valve leaflet modulus (as with most soft tissues) is a function of the stress σ within it; we will assume this to be a quasi-linear function in the (E, σ) operating range of the AV. At the instance of occurrence of the SHS, the pressure difference across the AV is small, and hence the corresponding leaflet stress $\sigma(=\sigma_1)$ and modulus $E(=E_1)$ are also small; accordingly, the leaflet deformation $W_s^*(=W_1^*)$ will also be small. However, at the start of the filling phase, there is a big pressure difference across the leaflet; the corresponding leaflet stress $\sigma(=\sigma_2)$ and modulus $E(=E_2)$ will also be bigger than σ_1 and E_1, and the corresponding leaflet deformation W_2^* will be measurable. However, based on our assumption (that E varies quasi-linearly with σ), we can state that

$$E_1/\sigma_1 = E_2/\sigma_2 \tag{2.36}$$

Hence, we can put down, from Equations 2.33b and 2.35,

$$\frac{167.35 f^2 a^2}{E_1} = \frac{(W_2^*)^2}{4u^*(1-\nu)} \tag{2.37}$$

so that

$$E_1 = \frac{669.4 f^2 a^2 u^*(1-\nu)}{(W_2^*)^2} \tag{2.38}$$

wherein $u^* = 0.06458a^2$

At the same time, we can employ Equation 2.35, as

$$\sigma_1 = 167.35 f^2 a^2 \tag{2.39}$$

wherein
σ_1 is in N/m^2, f is in Hz, and a is in m.

8. We can determine these pairs of values (σ_1, E_1) for different cycles, plot E versus σ, formulate the (E, σ) relationship as $E = p\sigma + E_0$, and evaluate the constitutive parameters p and E_0 for a patient. For patients with pathological (or calcified) AVs, we can hence determine the distributions of these parameters p and E_0, correlate them with valve leaflet pathology and pressure-drop across the valve, and arrive at their critical values for the replacement of the valve by a prosthetic valve.

Thus, we can distinguish valve leaflet pathology by means of the parameters p and E_0. For their determination, we evaluate E from Equation 2.38 and σ from Equation 2.39 at the instant of occurrence of SHS for each cardiac cycle, over a number of cycles. For evaluation of σ, we need to monitor the valve-ring radius (a) and the frequency (f) of the second spectral peak of SHS. For evaluation of E, we need to also monitor the maximum static deflection $W_2^*(W_s^*)$ of the valve leaflet, which happens to be on its symmetrical axis ($\theta = 0$) at $r = 0.76a$ at the start of the filling phase with respect to W_1^* at the instant of SHS occurrence (i.e. taking W_1^* to be the undeformed value; this can be done by ultrasound).

For $a = 0.8$ cm (or 0.008 m), $f = 100$ Hz, $D = 1$ g/cm^3 (or 10^3 kg/m^3), $v = 0.5$, $W_2^* = 3$ mm (or 0.003 m), we get $\sigma = 107$ N/m^2 (from Equation 2.39) and $E = 98.37$ N/m^2 (from Equation 2.38). At other LV cycles, we will obtain different values of σ and E, and hence of p and E_0. The values of p and E_0 can then be employed diagnostically to characterize and track pathological changes in the AV leaflet (from normal AV), and determine their critical values at which timely intervention can be carried out so as to avoid a big pressure-drop across the valve leaflet, leading to myocardial ischemia and infarction.

9. An alternative and perhaps more convenient approach (which would obviate the measurement of aortic leaflet deformation) would be to employ a modified leaflet elastic modulus parameter

$E^*(\text{in N}) = E(W_s^*)^2$, so that from Equation 2.33b

$$E^* = E(W_s^*)^2 = 4u^*(1 - v)\sigma = 0.13\sigma a^2, \quad \text{for } u^* = 0.06458a^2 \text{ and } v = 0.5$$

Then substituting for $\sigma(= 167.35f^2a^2)$ from Equation 2.35, we obtain

$$E^* = 21.6f^2a^4$$

We can then compute $E^*(= 21.6f^2a^4)$ and $\sigma(= 167.35f^2a^2)$ for different cardiac cycles, plot E^* vs. σ, and determine the regression curve based expression for E^* in terms of σ. The parameters of this expression would then represent the constitutive parameters for leaflet pathology.

2.6 Diseased LV Myocardial Segment Detection

For detection of the location and extent of ischemic and infarcted LV myocardial segments, we are invoking the concept that myocardial disease affects the elastic constitutive or modulus property of the myocardial tissue, and hence its ultrasonic echo-intensity distribution. The ultrasonic echotexture of the myocardium will be considerably heterogeneous in the case of an infarcted myocardium. Thus, by quantifying the ultrasonic echo-intensity distributions of LV myocardial segments, we can distinguish between normal, infected, and ischemic tissues.

Two-dimensional B-scan echocardiographic images of the heart can be scanned for tissue characterization and quantitative texture analysis of myocardial regions [8,9]. Each myocardial tissue component generates a grey scale pattern or texture related to the tissue density and fibrous content. In diseased states (such as myocardial ischemia and infarcts), changes in tissue density have been recognized. It has been found that hyper- reflectile echoes (HREs) correlate well with diseased myocardial tissue [8,9].

Figure 2.6a depicts the echo-intensity profile of an infant with visible scars in regions 1 and 2. The digitized echo-intensity profiles of these two regions are depicted in Figure 2.6b. In Figure 2.6b, the irreversibly damaged infarcted region is depicted in dark shade, the peri-infarcted ischemic border is depicted slightly lighter, while the normal tissue is depicted in light shade [9]. The aim of our drug therapy would be to convert the ischemic (slightly dark) region into normal (light) region.

In the neonatal infant patients who came to postmortem, their highly reflectile ultrasonic echozones were pathologically examined, and found to be necrotic and calcified [8–10]. We have thus demonstrated the capability of distinguishing scarred myocardial segments from healthy segments. Further, since the modulus property (and hence the echo-intensity) reflects myocardial disease severity, this concept can be further developed to provide the basis for differential detection of infarcted myocardial segments as well as the bordering ischemic myocardial segments.

Table 2.1 displays the echo-intensity values of myocardial segments of nine normal infants. The upper bound of the echo-intensity (of the pericardium region A) was set to 100% in each normal infant and the intensities from the rest of the image were referenced or normalized to this level. Thus, in the case of male patient W, for instance, the normaliszed echo-intensities of regions B, C, D, and E are 49.5, 31.3, 39.3, and 35.1, respectively. We note that the upper bound of echo-intensity value of healthy tissue (expressed as a percentage of pericardial echo-intensity) is 54.2 (in the case of patient TY).

Table 2.2 displays the echo-intensities of myocardial segments (normalized with respect to the pericardial echo-intensity, so as to make the results independent of instrumentation characteristics) of six infant patients with diseased hearts. It is noted (from the last column) that the normalized echo-intensities of the highly reflectile elements (HREs) from these six infant

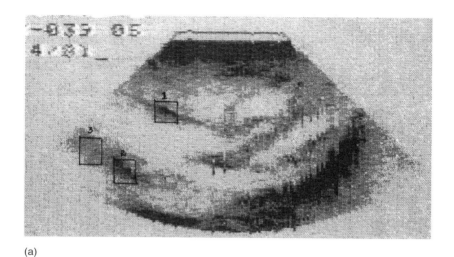

(a)

yfx	99	100	101	102	103	104	105	106	107	108	109	110	111	112	113	114
98	79	78	88	90	99	96	102	108	91	77	92	86	135	122	73	55
99	114	115	101	114	126	128	114	116	119	126	82	68	84	103	78	57
100	151	137	125	128	136	135	133	134	149	137	91	75	74	73	82	83
101	175	177	171	151	144	143	154	147	138	142	139	139	126	64	76	71
102	202	196	174	125	192	193	183	164	131	131	125	132	92	89	81	116
103	139	143	183	193	206	217	233	248	209	146	116	102	111	113	117	116
104	147	136	143	178	203	251	250	255	229	201	75	71	92	82	88	95
105	108	110	132	151	210	223	227	249	255	255	230	210	104	87	81	112
106	84	104	88	121	147	184	227	239	255	255	252	247	220	125	76	70
107	83	110	108	122	135	175	194	183	206	228	211	255	255	184	141	131
108	68	92	122	131	145	147	149	151	217	181	189	222	241	178	190	167
109	56	76	81	122	132	137	145	143	154	150	156	156	195	190	206	190
110	76	63	96	96	82	83	103	120	142	128	133	141	153	181	192	194
111	59	57	63	66	70	103	106	118	96	94	86	110	129	150	95	66
112	58	60	59	57	58	61	71	77	106	89	91	92	110	147	97	85
113	74	71	78	60	56	58	57	62	71	70	79	83	78	92	67	76
114	57	57	65	63	57	56	63	56	51	56	58	80	85	78	67	55
115	51	60	63	63	58	57	56	57	54	59	57	58	59	76	68	81

(b)

FIGURE 2.6 (See color insert following page 266.)
(a) Long axis 2-D ultrasonic view of a pediatric patient's heart, showing highly echoreflectile regions 1 and 2 and a healthy region 3. (b) Echocardiographic texture analysis, showing echo-intensity levels from myocardial region 1. (Adopted from Figure 2 of Kamath, M.V., Way, R.C., Ghista, D.N., Srinivasan, T.M., Wu, C., Smeenk, S., Marning, C., and Cannon, J., *Eng. in Med.*, 15, 137, 1986.)

TABLE 2.1

Echo-Intensity Values for Various Anatomic Regions of Normal Pediatric Hearts (Based on Long Axis Vies). (Adopted from Kamath, M.V., Way, R.C., Ghista, D.N., Srinivasan, T.M., Wu, C., Smeenk, S., Marning, C., and Cannon, J., *Eng. in Med.*, 15, 137, 1986.)

Patient (Sex)	Region A	Region B	Region C	Region D	Region E
W (M)	M: 227.75	M: 112.83	M: 71.21	M: 89.55	M: 80.06
	SD: 13.54	SD: 25.27	SD: 19.41	SD: 17.91	SD: 18.74
	N: 65	N: 84	N: 75	N: 31	N: 65
	P: 100	P: 49.5	P: 31.3	P: 39.3	P: 35.1
G (F)	M: 218.40	M: 98.30	M: 66.24	M: 103.10	M: 86.21
	SD: 14.74	SD: 15.79	SD: 20.20	SD: 21.13	SD: 32.62
	N: 67	N: 76	N: 41	N: 42	N: 98
	P: 100	P: 45	P: 30.3	P: 47.2	P: 39.5
S (M)	M: 212.01	M: 97.20	M: 42.09	M: 82.43	M: 92.43
	SD: 14.27	SD: 12.93	SD: 20.07	SD: 22.47	SD: 20.93
	N: 66	N: 84	N: 69	N: 21	N: 60
	P: 100	P: 45.8	P: 25.8	P: 38.9	P: 43.6
R (M)	M: 226.81	M: 92.23	M: 58.49	M: 89.50	M: 78.73
	SD: 12.46	SD: 16.42	SD: 19.10	SD: 14.69	SD: 17.89
	N: 78	N: 96	N: 55	N: 30	N: 56
	P:100	P: 40.7	P: 25.8	P: 39.8	P: 34.7
TY (M)	M: 195.85	M: 75.78	M: 57.93	M: 106.21	M: 93.78
	SD: 14.22	SD: 19.00	SD: 26.10	SD: 16.96	SD: 15.24
	N: 67	N: 74	N: 44	N: 24	N: 64
	P: 100	P: 38.7	P: 29.6	P: **54.2**	P: 47.9
O (F)	M: 204.11	M: 83.93	M: 57.79	M: 103.81	M: 84.77
	SD: 12.66	SD: 16.94	SD: 18.24	SD: 24.89	SD: 18.28
	N: 44	N: 93	N: 43	N: 43	N: 56
	P: 100	P: 42.8	P: 29.5	P: 53	P: 43.3
SG (M)	M: 209.53	M: 101.98	M: 63.921	M: 68.61	M: 94.18
	SD: 14.19	SD: 14.60	SD: 17.97	SD: 17.56	SD: 20.80
	N: 38	N: 65	N: 36	N: 18	N: 84
	P: 100	P: 48.7	P: 30.5	P: 32.7	P: 44.9
WB (F)	M: 237.22	M: 93.78	M: 75.27	M: 110.65	M: 106.85
	SD: 12.94	SD: 20.51	SD: 20.39	SD: 24.80	SD: 32.87
	N: 82	N: 86	N: 77	N: 17	N: 47
	P: 100	P: 39.5	P: 31.7	P: 46.6	P: 50.0
L (F)	M: 227.11	M: 89.09	M: 63.76	M: 108.95	M: 117.27
	SD: 12.56	SD: 20.08	SD: 25.99	SD: 27.24	SD: 16.46
	N: 72	N: 77	N: 76	N: 37	N: 67
	P: 100	P: 39.2	P: 28.1	P: 48.0	P: 51.6

Note: The numbers in the four rows represent mean (M), standard deviation (SD), number of pixels (N), and percentage of pericardial intensity (P). A = posterior pericardium, B = anterior myocardium, C = posterior myocardium, D = anterior mitral leaflet, and E = septum.

patients were distinctly higher than the echo-intensity range of normal myocardial tissue segments.

The normalized lower bound of the highly reflectile infarcted myocardial segments is 63.1% (Table 2.2), whereas the normalized upper bound of normal myocardial tissue is 54.2% (Table 2.1). Thus, the normalized echo

TABLE 2.2

Intensity Values for Various Anatomic Regions of diseased Pediatric Hearts (Based on Long Axis Vies). (Adopted from Kamath, M.V., Way, R.C., Ghista, D.N., Srinivasan, T.M., Wu, C., Smeenk, S., Marning, C., and Cannon, J., *Eng. in Med.*, 15, 137, 1986.)

Patient (Sex)	Region A	Region B	Region C	Region D	HRE and Its Location
B (M)	M: 167.44	M: 54.76	M: 51.02	M: 82.20	M: 105.74
	SD: 25	SD: 28.2	SD: 17.71	SD: 24.68	SD: 30.88
	N: 65	*N*: 84	*N*: 75	*N*: 31	*N*: 65
	P: 100	*P*: 32.7	*P*: 30.5	*P*: 49.1	*P*: **63.1**
					Septum
P (F)	M: 148.76	M: 61.73	M: 79.81	M: 61.7	M: 108.18
	SD: 26.78	SD: 23.02	SD: 22.05	SD: 24.2	SD: 13.03
	N: 50	*N*: 75	*N*: 47	*N*: 49	*N*: 40
	P: 100	*P*: 41.5	*P*: 53.8	*P*: 41.50	*P*: 72.6
					Septum
Br (M)	M: 141.65	M: 68.3	M: 69.3	M: 33.93	M: 89.412
	SD: 29.56	SD: 26.8	SD: 24.8	SD: 24.4	SD: 28.0
	N: 40	*N*: 40	*N*: 49	*N*: 44	*N*: 79
	P: 100	*P*: 41.5	*P*: 53.8	*P*: 41.5	*P*: 73.1
					Septum
F (F)	M: 157.34	M: 50.1	M: 60.8	M: 53.8	M: 112.1
	SD: 30.0	SD: 29.5	SD: 18.8	SD: 22.7	SD: 10.3
	N: 35	*N*: 45	*N*: 49	*N*: 44	*N*: 31
	P: 100	*P*: 31.8	*P*: 38.6	*P*: 34.2	*P*: 71.2
					Right ventricle
HI (M)	M: 168.1	M: 54.7	M: 58.2	M: 62.4	M: 96.4
	SD: 21.35	SD: 21.8	SD: 16.9	SD: 20.0	SD: 14.7
	N: 47	*N*: 36	*N*: 33	*N*: 37	*N*: 49
	P: 100	*P*: 32.5	*P*: 34.6	*P*: 37.1	*P*: 47.3
					Left ventricle
G (M)	M: 117.7	M: 46.9	M: 45.5	M: 42.7	M: 85.3
	SD: 20.6	SD: 19.0	SD: 20.6	SD: 19.1	SD: 22.6
	N: 45	*N*: 44	*N*: 40	*N*: 49	*N*: 37
	P: 100	*P*: 39.8	*P*: 38.7	*P*: 36.2	*P*: 72.5
					Right ventricle

Note: The numbers in the four rows represent mean (M), standard deviation (SD), number of pixels (*N*), percentage of pericardial intensity (*P*). A = posterior pericardium, B = anterior myocardium, C = posterior myocardium, and D = septum.

intensity range of ischemic myocardial segments is 54.2%–63%. If a myocardial segment has normal echo-intensity in this range, and if after pharmacological drug therapy its echo-intensity becomes less than 54.2, then we can say that this drug is effective in normalizing the ischemic myocardial segment on the border of the infarcted myocardial region.

2.6.1 Comments

We have noted that diseased (qualified) heart valves can cause high LV wall stress leading to myocardial ischemia. Hence, we need a more definitive index for diseased heart valve (based on the biomechanical property of the valve leaflets) as well as a more reliable guideline for intervention before the onset of myocardial ischemia.

For the purpose of determining the valvular leaflet biomechanical property, to characterize valvular degeneration, we have shown that (1) we can make use of noninvasive determination of valve leaflet vibrational frequency (from spectral phonocardiography) and valve leaflet deformation (by processing 2-D echocardiograms), (2) express the valve leaflet modulus (E) and stress (σ) in terms of valve leaflet vibrational frequency and deformation, and (3) then develop an analytic expression for E as a function of σ.

The parameters of this $E-\sigma$ expression can then be employed to characterize valve leaflet degeneration, which in turn is manifested as valvular stenosis and/or incompetency, and thereafter lead to myocardial ischemia. If we were to determine the pressure-drop across the heart valve or LV backflow, and use it as a measure of candidacy for intervention, then it is quite possible that at that stage the patient's LV wall stress could have become high enough to cause myocardial ischemia or infarct. Thus, we could employ LV wall stress as a criterion for intervention, except that it requires knowledge of intra-LV pressure, which cannot be determined noninvasively.

Hence, we suggest (based on our research) that we employ echocardiographic texture analysis (i.e., LV myocardial echo-intensity profile determination) as a criterion for intervention. In other words, we can keep postponing intervention until myocardial ischemia becomes discernable, on the 2-D echo-intensity profile of the LV, as a normalized echo-intensity segment in the 54.2%–64% range.

We then want to propose that (1) we determine the valve leaflet's biomechanical (E vs. σ) property, as described herein, (2) correlate the biomechanical E versus σ property (displayed graphically and/or analytically) with the intervalvular pressure-drop and/or backflow (due to valvular regurgitation), and (3) also correlate this biomechanical (E vs. σ) property with myocardial 2-D echo-intensity profile (as demonstrated in this section).

This will reveal to us the following:

- Sensitivity of valve leaflet (E vs. σ) biomechanical property to depict valvular degeneration and dysfunction
- New early warning signal for timely intervention, before the onset of LV myocardial ischemia, based on its detection of the myocardial 2-D echo-intensity profile
- Effectiveness of a particular drug therapy in converting an ischemic peri-infarcted myocardial segment (in the normalized echo-intensity range of 54.2–64.2) into a normal myocardial segment (having normalized echo-intensity <54.2)
- Comprehensive profile of the LV heart-valve property, associated with LV heart-valve dysfunction (characterized by stenosis or incompetency), and LV myocardial property distribution, represented by its echo-intensity distribution

2.7 Analysis of Blood Flow in the LV (Using Monitored LV Wall-Motion Data to Determine Intra-LV Flow Velocity and Pressure-Gradient Distributions)

2.7.1 Finite-Element Analysis of Blood Flow in the LV

The data required for the finite-element analysis (FEA) consist of

- LV 2-D long-axis frames during LV diastolic and systolic phases
- LV pressure versus time, associated with these LV frames
- Computation of LV instantaneous wall velocities as well as instantaneous velocity of blood entering the LV during the filling phase and leaving the LV during the ejection phase

From this FEA, we determine the instantaneous distributions of intra-LV blood-flow velocity and differential pressure during filling and ejection phases, to intrinsically characterize LV resistance-to-filling (RTF) and LV contractility (CONT), respectively [11].

The FEA employed for computing intra-LV flow velocity and pressure distributions entails solution of the potential equation:

$$\nabla^2 \Phi = 0 \tag{2.40}$$

where ∇^2 is the Laplacian operator, Φ is the velocity potential, and $\nabla\Phi$ is the velocity vector. For FEA, the governing differential equation

$$\frac{\partial^2 \Phi}{\partial x^2} + \frac{\partial^2 \Phi}{\partial y^2} = 0 \tag{2.41}$$

for a 2-D planar flow domain, is transformed to a finite-element equation form, by making use of the Galerkin-weighted residual procedure.

The resulting stiffness-matrix system of equations

$$[K]\{\Phi\} = \{F\} \tag{2.42}$$

can be solved for Φ at those point(s) in the flow domain, by specifying $\partial\Phi/\partial n(V_n)$ along the endocardial boundary, and Φ at those point(s) on the boundary where V_n is not specified. By specifying Φ to be constant along the open boundaries, the flow can be constrained to be normal to that boundary; this constraint also allows the solution to obtain a flow balance. The value of the constant Φ is arbitrary, and $\Phi=0$ is specified along the open boundary.

The matrix system $[K]$, in Equation 2.42, is symmetrical and banded. Equation 2.42 is solved for Φ using a Gaussian elimination method, which transforms the matrix system $[K]$. From the computed values of Φ at each internal point, we determine the velocity components at each internal point of the LV chamber and hence obtain instantaneous maps of intra-LV blood-flow velocity patterns.

2.7.2 Analysis for Intra-LV Pressure Distribution

Once the intra-LV flow velocities are determined, the intracardiac pressure distribution at any point inside the LV chamber can be obtained from the Bernoulli equation for unsteady potential flow as

$$P + (1/2)\rho V^2 + \rho(\partial\Phi/\partial t) = C(t) \tag{2.43}$$

where
    P is the pressure
    $(1/2)\rho V^2$ is the dynamic pressure term
    ρ is the density of blood
    V is the velocity of blood
    $\rho(\partial\Phi/\partial t)$ is the effect due to acceleration
    $C(t)$ represents the total pressure as sensed by a pressure probe facing
      the oncoming fluid.

$C(t)$ is a constant and the gravitational or hydrostatic effects are neglected. The partial derivative, $(\partial\Phi/\partial t)$, is computed from the value of Φ at the same point at successive instants, using the finite difference scheme.

Since we want the procedure not to utilize the catheter-pressure data, we can obtain the pressure distribution relative to a reference point in the chamber, say at the center of the aortic or mitral orifice. Hence the differential pressure field at a point s, in terms of the pressure P_0 at the inlet (during diastole) or outlet (during the ejection phase) of the ventricle, is given by

$$P_s - P_0 = (1/2)\rho(V_0^2 - V_s^2) + \rho(\partial\Phi/\partial t|_0 - \partial\Phi/\partial t|_s) \tag{2.44}$$

where V_0 and V_s are, respectively, the velocity of blood flow at the center of the orifice (i.e., at the aortic or mitral orifice during systolic or diastolic phase) and at a point s inside the LV chamber. The differential pressure $(P_s - P_0)$ can be expressed and displayed in nondimensional form, as

$$C_p = (P_s - P_0)/\frac{1}{2}\rho V_0^2 \tag{2.45}$$

where C_p is the nondimensional pressure coefficient.

This instantaneous graphical display of the relative pressure distribution in the LV chamber can provide an indication of the RTF as well as of the effectiveness of the LV contraction in setting up the appropriate pressure distribution in the chamber, so as to promote adequate emptying.

By comparing intra-LV pressure-gradients before and after administration of nitroglycerin (a myocardial perfusing agent, and hence a quasi-simulator of coronary bypass surgery), we can infer how the myocardium is going to respond and how these LV functional indices will improve after coronary bypass surgery.

2.7.3   Intra-LV Flow during Diastolic Filling

How well and how easily the LV fills is depicted by the instantaneous intra-LV flow distribution and the interframe variations in flow distribution, which are governed by the segmental stiffness of the LV, and are manifestations of RTF. In general, the flow is highest during the first half of the diastole (20–30 cm s^{-1}) in all patients, and the relative flow during all phases of diastole is at a maximum in the inflow segment of the LV, just below the MV.

The results suggest that the early filling phase could possibly be due to the actively relaxing LV wall setting up a pressure-gradient, conducive to filling, instead of the LV wall motion responding passively to blood flow. Subsequently, during late-filling phases, the increasing stiffness of the LV wall (due to increasing LV volume) provides increased resistance to LV filling in the form of reduced flow.

The ideal situation is for the wall contraction to be so graded that adequate flow is generated in the apical region and a near-uniform flow is maintained throughout the LV chamber. The factors contributing to adequate intra-LV flow and cardiac output, with a smooth washout, are strong LV wall contraction and uniformly accelerating wall motion. If following administration of nitroglycerin, the LV wall can contract more uniformly and thereby set up a more favorable intra-LV velocity field, instead of a pattern of compensatory regional hypercontractility (and associated high wall tension and oxygen demand) to make up for a region of hypocontractility, then such a patient would be a good candidate for coronary bypass surgery.

The results of the analysis are displayed in Figure 2.7; for a typical patient with a myocardial infarct

- Figure 2.7a1 and a2 depict superimposed LV outlines during diastole and systole, before nitroglycerin administration (a1) and after nitroglycerin administration (a2)
- Figure 2.7b1 and c1 depict intra-LV blood-flow velocity distributions during diastole and systole, before nitroglycerin administration
- Figure 2.7b2 and c2 depict intra-LV blood-flow velocity distributions during diastole and systole, after nitroglycerin administration
- Figure 2.7d1 and e1 depict intra-LV blood-flow pressure distributions during diastole and systole, before nitroglycerin administration
- Figure 2.7d2 and e2 depict intra-LV blood-flow pressure distribution during diastole and systole, after nitroglycerin administration

From a computational viewpoint, the intra-LV flow is determined from the LV wall-motion boundary condition to the potential-flow equation (Equation 2.30), and the intra-LV pressure-gradient can in turn be computed from the flow by employing Equation 2.33. However, we could interpret the phenomenon as if the LV wall stiffness were providing the resistance to wall motion for filling during diastole, and the contracting LV were facilitating emptying of the LV during systole, thereby setting up the requisite intra-LV pressure-gradients and velocity distributions.

For this patient, Figure 2.7d1 and e1 demonstrate poor LV RTF and LV CONT in terms of adverse intra-LV blood pressure-gradients during filling and ejection phases, respectively. However, following administration of nitroglycerin, these filling and ejection phases' pressure-gradients (and hence LV RTF and LV CONT) are improved (Figure 2.7d2 and e2),

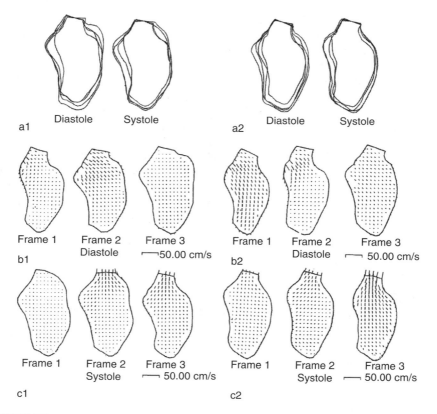

FIGURE 2.7
Results for a typical patient with a myocardial infarct: (a) Superimposed sequential diastolic and systolic endocardial frames (whose aortic valve centers and the long axis are matched) (1) before and (2) after administration of nitroglycerin. (b) Instantaneous intra-LV distributions of velocity during diastole (1) before and (2) after administration of nitroglycerin. (c) Instantaneous intra-LV distributions of velocity during ejection phase (1) before and (2) after administration of nitroglycerin.

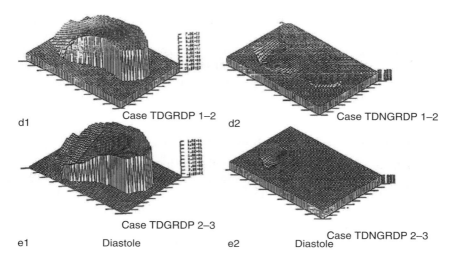

d1   Case TDGRDP 1–2   d2   Case TDNGRDP 1–2

Case TDGRDP 2–3

e1   Diastole   e2   Case TDNGRDP 2–3   Diastole

FIGURE 2.7 (continued)
(d) Instantaneous intra-LV distributions of pressure differential during diastole (1) before and (2) after administration of nitroglycerin. (e) Instantaneous intra-LV distributions of pressure differential during ejection phase (1) before and (2) after administration of nitroglycerin. (Adopted from Figure 5 of Subbaraj, K., Ghista, D.N., and Fallen, E.L., *J. Biomed. Eng.*, 9, 206, 1987.)

thereby providing the basis for advocating coronary bypass surgery for this patient.

2.8 Explaining Left-Ventricular Pressure Dynamics in Terms of LV Passive and Active Elastances (as Measures of LV Pressure Dynamics Response to LV Volume Change and LV Contractility)

2.8.1 Scope

There has been a lot of characterization of the heart as a pump by means of models based on elastance and compliance. In this section, we are presenting a somewhat new concept of time-varying passive and active elastances. The biomechanical basis of LV cyclical elastances is presented. We have defined elastance in terms of the relationship between ventricular pressure and volume, as $dP = E dV + V dE$, where E includes passive (E_p) and active (E_a) elastance. By incorporating this concept in LV models to simulate diastolic (filling) and systolic phases, we have obtained the time-varying expression for E_a and the LV-volume-dependent expression for E_p.

These two elastances of E_a and E_p can be deemed to represent intrinsic LV properties. The active elastance (E_a) can be used to characterize the LV contractile state, while passive elastance (E_p) can represent a measure of LV pressure response to LV filling and emptying. Further, we have demonstrated how the LV pressure dynamics (and LV pressure response to LV volume) can be explained in terms of E_a and E_p [12,13].

2.8.2 Concepts of Passive and Active Elastances

At the start of the diastolic filling phase, the LV incremental pressure dP_{LV} is the response (1) to LV E_a continuing to decrease due to the sarcomere continuing to relax well into the filling phase and (2) to the rapid inflow of blood and the corresponding increase in LV volume, along with increase in LV E_p. The corresponding governing differential equation, relating LV pressure and volume, can be put down as [12]

$$M(d\dot{V}) + d(EV) = M(d\dot{V}) + V dE + E dV = dP_{LV} \qquad (2.46)$$

where
   $\dot{V}$ represents the time derivative of V (dV/dt, in which t is measured from the start of filling phase)
   V represents the volume of the LV (mL) during the filling phase
   P_{LV} represents the pressure of the LV in mmHg (hereafter symbolized by P)
   M represents the inertia term $= $ [LV wall density $(\rho)/$(LV surface area/ wall thickness)] $= \rho h/4\pi R^2$ for a spherical LV model (in mmHg/(mL/s^2))
   E represents LV elastance (mmHg/mL)

Likewise during ejection, the LV pressure variation (dP_{LV}) is caused by both E_a variation as well as E_p decrease (due to LV volume decrease). The instantaneous time-varying ventricular elastance (E) is the sum of (1) the volume-dependent passive elastance (E_p) and (2) the active elastance (E_a) due to the activation of the LV sarcomere. Hence,

$$E = E_a + E_p \qquad (2.47)$$

We will now provide the expressions for E_p and E_a, and then their formulations.

2.8.2.1 *Expression for Passive Elastance (E_p) of the LV*

The passive (inactivated) myocardium exhibits properties of an elastic material, developing an increasing stress as strain increases, as occurs during ventricular filling. The passive stress–strain relation of a myocardial muscle strip is nonlinear and follows an exponential relationship [14–16].

Likewise, the relation between LV passive pressure and volume is adopted to be exponential, as

$$P = P_0 e^{z_p V} \tag{2.48}$$

$$\text{so that,} \quad E_p = (dP/dV) = E_{p0} e^{z_p V} \tag{2.49}$$

where E_{p0} is the passive elastance coefficient ($= P_0 z_p$), z_p is the passive elastance exponent parameter, and V is the LV volume; its evaluation for a clinical case is provided in a subsequent section. During the latter part of the diastolic phase, we use Equation 2.49 to fit the LV pressure–volume relation to determine the corresponding parameters, P_0 and z_p (or E_{p0} and z_p), and hence obtain the passive elastance E_p.

2.8.2.2 *Expression for Active Elastance (E_a) of the LV*

During isovolumic contraction, $dV = 0$. Hence $d\dot{V} = 0$, and E_p is constant and equal to E_{ped} (the value of E_p at end-diastole). As a result, the governing Equation 2.46 becomes $VdE = dP_{LV}$, which can be discretized as

$$
\begin{aligned}
V_i(E_i - E_{i-1}) &= V_i[(E_{a,i} + E_{p,i}) - (E_{a,i-1} + E_{p,i-1})] \\
&= V_i(E_{a,i} + E_{ped} - E_{a,i-1} - E_{ped}) = dP_{LV,i} = P_i - P_{i-1}
\end{aligned}
$$

$$\text{Hence,} \quad E_{a,i} = \frac{(P_i - P_{i-1})}{V_i} + E_{a,i-1} \tag{2.50}$$

where i is a time instant during the isovolumic contraction and relaxation, V_i and $P_{LV,i}$ are the monitored LV volume and pressure at this instant, and E_{ped} is the passive elastance at the end-diastolic phase.

During the ejection phase, the governing equation (Equation 2.46) can be discretized, in a similar way, as

$$E_{a,i} = \frac{(P_i - P_{i-1}) - Md\dot{V}_i - V_i(E_{p,i} - E_{p,i-1}) - E_{p,i}(V_i - V_{i-1}) + V_i E_{a,i-1}}{2V_i - V_{i-1}} \tag{2.51}$$

Also, during isovolumic relaxation, because $dV = 0$, $d\dot{V} = 0$, and E_p is constant and equal to its end-systolic value of E_{pes}, the governing Equation 2.46 again becomes $VdE = dP_{LV}$, which can be represented as

$$
\begin{aligned}
V_i[(E_{a,i} + E_{p,i}) - (E_{a,i-1} + E_{p,i-1})] &= V_i(E_{a,i} + E_{pes} - E_{a,i-1} - E_{pes}) = dP_{LV,i} \\
&= P_i - P_{i-1}
\end{aligned}
$$

$$\text{Therefore,} \quad E_{a,i} = \frac{P_i - P_{i-1}}{V_i} + E_{a,i-1} \tag{2.52}$$

where E_{pes} is the passive elastance at the end of systole.

During the diastolic phase, the formula for computing active elastance is the same as Equation 2.51. Hence, from Equations 2.50 through 2.52, we can calculate the values of active elastance from LV pressure–volume data during the cardiac cycle. After calculating the values of active elastance (E_a), we adopt the following expression for E_a [12]:

$$E_a = E_{a0}\left[1 - e^{-\left(\frac{t}{\tau_C}\right)^{Z_c}}\right]\left[e^{-\left(\frac{(t-d)u(t-d)}{\tau_R}\right)^{Z_R}}\right] \tag{2.53}$$

where (1) t is measured from the start of isovolumic contraction, (2) the parameter E_{a0} is the active elastance coefficient, (3) the time coefficient (τ_C) describes the rate of elastance rise during the contraction phase, while (τ_R) describes the rate of elastance fall during the relaxation phase, (4) the exponents "Z_c" and "Z_R" are introduced to smoothen the curvatures of the E_a curve during isovolumic contraction and relaxation phases, (5) the parameter d is a time constant whose (to be determined) value is during the ejection phase, and (6) $u(t-d)$ is the unit step function, so that $u(t-d)=0$ for $t < d$.

The rationale for the formulation of Equation 2.53 is based on E_a incorporating (1) parameters (Z_c and τ_C) reflecting the generation of LV pressure during isovolumic contraction, (2) parameters (Z_R and τ_R) reflecting the decrease of LV pressure during isovolumic relaxation and early filling, and (3) all of these parameters (Z_c, τ_C, Z_R, and τ_R) representing the LV pressure–volume relationship during filling and ejection phases. We can determine the values of these parameters by fitting Equation 2.53 to the computed values of E_a (from Equations 2.50 through 2.52), and employing the parameter-identification procedure to evaluate these above-mentioned parameters.

2.8.3 Clinical Application

Data Measurements. The subjects in this study (satisfying appropriate ethics procedures) were studied in a resting recumbent state, after premedication with 100–500 mg of sodium pentobarbital by retrograde aortic catheterization. LV chamber pressure was measured by a pigtail catheter and Statham P23Eb pressure transducer; the pressure was recorded during ventriculography. Angiography was performed by injecting 30–36 mL of 75% sodium diatrizoate into the LV at 10 to 12 mL/s. It has been found, by using biplane angiocardiograms, that orthogonal chamber diameters are nearly identical [17]. These findings are used to justify the use of single-plane cine techniques, which allow for beat-to-beat analysis of the chamber dimensions.

For our study, monoplane cineangiocardiograms were recorded in an RAO 30° projection from a 9 in. image intensifier using 35 mm film at

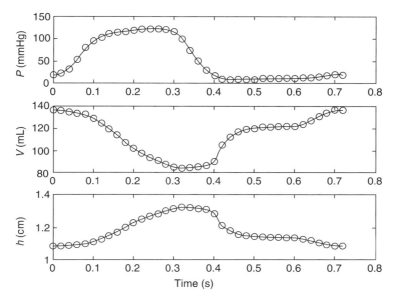

FIGURE 2.8

An example of a patient (HEL) measured LV pressure, volume, and wall thickness during a cardiac cycle; $t = 0$–0.08 s is the isovolumic contraction phase, $t = 0.08$–0.32 s is the ejection phase, $t = 0.32$–0.40 s is the isovolumic relaxation phase, and $t = 0.40$–0.72 s is the filling phase. Note that even after 0.4 s, the LV pressure still continues to decrease from 17 (at 0.4 s, at start of filling) to 8 mmHg at 0.44 s. (Adopted from Figure 2 of Zhong, L., Ghista, D.N., Ng, E.Y.K., and Lim, S.T., *Biomed. Eng. Online*, 4, 10, 2005.)

50 frames/s, using INTEGRIS Allura 9 system at the National Heart Centre (NHC), Singapore. Therefrom, automated analysis was carried out to calculate LV volume and myocardial wall thickness. The LV data of a typical patient, employed for this analysis, consist of measured volume and myocardial thickness of the chamber as well as the corresponding pressure (Figure 2.8). All measurements are corrected for geometric distortion due to the respective recording systems. Figure 2.9 displays pressure versus volume for this patient.

This figure shows that although LV volume must remain constant during the isovolumic phases, it does not actually do so. The changes in volume cause changes in pressure due to bulk-modulus effect. It is noted that the LV volume increases slightly during isovolumic relaxation, instead of being constant. Similarly, LV volume increases slightly during isovolumic contraction. This slight increase in LV volume during isovolumic contraction is what causes LV pressure to increase.

Case Study. Here, we show one case study. The LV cineangiographic data (depicted in Figure 2.8), consists of measured LV volume and corresponding pressure. When LV pressure and volume are plotted in Figure 2.9, it is noted that during the early filling phase, LV pressure decreases even though

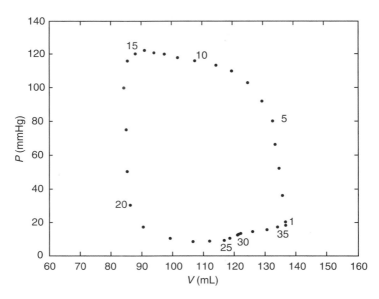

FIGURE 2.9

Relationship between LV volume and pressure for one sample data. Points (21–36) constitute
the filling phase, (1–5) constitute the isovolumic contraction phase, (5–17) constitute the ejection
phase, and (17–21) constitute the isovolumic relaxation phase. Note that after point 21 (the start
of LV filling), the LV pressure decreases; this characterizes LV suction effect. (Adopted from
Figure 3 of Zhong, L., Ghista, D.N., Ng, E.Y.K., and Lim, S.T., *Biomed. Eng. Online*, 4, 10, 2005.)

LV volume increases. This phenomenon is defined as the "LV suction
effect," which can be explained by using our concepts of active and passive
elastances.

From the data in Figure 2.9, we will now compute E_p and E_a, by employing the analysis in Section 2.8.2.

Evaluation of E_p. By fitting Equation 2.48 to the pressure and volume data,
as shown in Figure 2.10, we obtain the values of the parameters P_0 and z_p, as

$$z_p = 0.040 \text{ mL}^{-1}, \quad P_0 = 0.080 \text{ mmHg} \tag{2.54}$$

and the E_p function (corresponding to its expression given by Equation 2.49)
as follows:

$$E_p = 3.20 \times 10^{-3} e^{0.040V} \tag{2.55}$$

We now propose to adopt E_p as a measure of LV RTF. During ejection and
filling phases, E_p can be calculated at any time using Equation 2.55.

Evaluation of E_a. Using Equations 2.50 through 2.52, we can calculate the
active elastance E_a during isovolumic contraction, ejection, isovolumic
relaxation, and diastolic filling phases, respectively. The values of E_a during

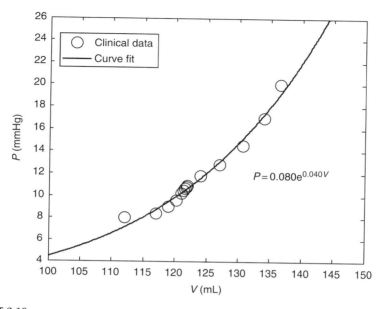

FIGURE 2.10
Here we have used Equation 2.48 to fit the pressure–volume data during filling phase. The volume 100 mL corresponds to the start of the filling phase, and the volume 150 mL corresponds to the end of the filling phase. (Adopted from Figure 4 of Zhong, L., Ghista, D.N., Ng, E.Y.K., Lim, S.T., Tan, R.S., and Chua, L.P., *Proc. Inst. Mech. Eng. Part H, J. Eng. Med.*, 220, 647, 2006.)

a cardiac cycle are shown in Figure 2.11. Then the parameters in Equation 2.53 can be determined by fitting the computed values of E_a, which are listed in Figure 2.11 caption as well as in Table 2.3.

Upon substituting these computed values of the parameters (E_{a0}, τ_c, Z_c, τ_R, Z_R) into Equation 2.53, we obtain the $E_a(t)$ function as follows:

$$E_a = 2.20\left[1 - e^{-\left(\frac{t}{0.17}\right)^{1.96}}\right]\left[e^{-\left(\frac{(t-0.3)u(t-0.3)}{0.12}\right)^{0.96}}\right] \tag{2.56}$$

2.8.4 Results of Case Studies

Depicting the Computed Values of E_p and E_a (in Figure 2.12). The variations of model-derived nonlinear passive and active elastances for the subject HEL are shown in Figure 2.12. For this particular subject (HEL), the maximum active elastance is 2.10 mmHg/mL. Herein, it is noted that the LV pressure decreases immediately after opening of the MV at frame 21 and then starts increasing. This is because of the effect of E_a.

The period of LV pressure depression from frames 21 to 23 during early filling enables passive filling of the LV by suction. This suction period of 0.04 s (=12.5% filling period) from time frames 21 to 23 also corresponds to the period during which active elastance persists (due to the continued

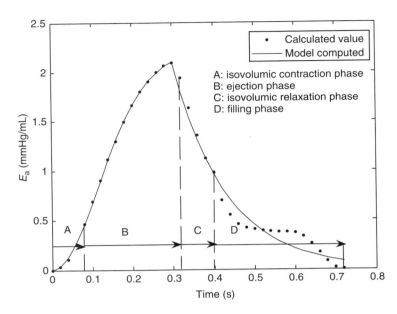

FIGURE 2.11

Calculated values of active elastance E_a during cardiac cycle. Using Equation 2.53 to fit the calculated values, we have: $E_{a0} = 2.20\,\text{mmHg/mL}$, $\tau_C = 0.17\,\text{s}$, $Z_c = 1.96$, $d = 0.3\,\text{s}$, $\tau_R = 0.12\,\text{s}$, and $Z_R = 0.96$. Note that E_a reaches its maximum value at frame. Also note the drastic decrease in E_a after frame 21, which contributes to LV suction effect. (Adopted from Figure 5 of Zhong, L., Ghista, D.N., Ng, E.Y.K., Lim, S.T., Tan, R.S., and Chua, L.P., *Proc. Inst. Mech. Eng. Part H, J. Eng. Med.*, 220, 647, 2006.)

activation of the contractile element of the myocardial sarcomere into the filling phase) but keeps decreasing.

2.8.4.1 Pressure Dynamics during Filling Phase

The pressure variation during filling is a combination of pressure changes due to the action of both active elastance (E_a) and passive elastance (E_p)

TABLE 2.3

Computed Values of Parameters in E_a Expression (Equation 2.53), for the Subject Whose LV Data Are Given in Figure 2.9

Parameters	Values	Unit
E_{a0}	2.20	mmHg/mL
τ_C	0.17	S
Z_c	1.96	Nondimensional
d	0.3	S
τ_R	0.12	S
Z_R	0.96	Nondimensional

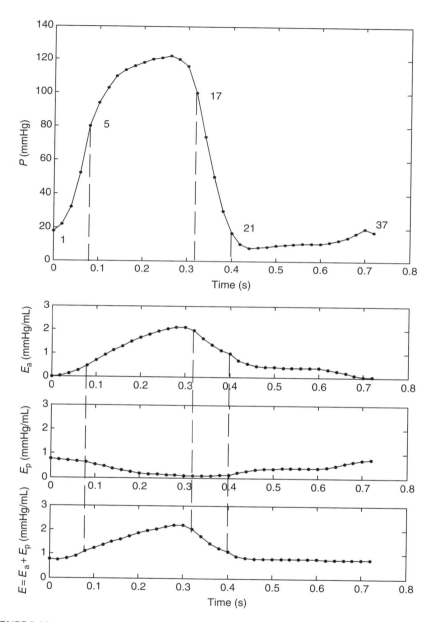

FIGURE 2.12

Pressure, active elastance E_a, passive elastance E_p, and total $E = (E_a + E_p)$ for the sample subject data shown in Figure 2.2. In this figure, frames 1–5 represent the isovolumic contraction phase, frames 5–17 represent the ejection phase, frames 17–21 represent the isovolumic relaxation phase, and frames 21–37 represent the diastolic filling phase. Note the drastic decrease in E_a after frame 21, which offsets the increase in E_p (due to LV volume increase) and contributes to the LV suction effect. (Adopted from Figure 6 of Zhong, L., Ghista, D.N., Ng, E.Y.K., Lim, S.T., Tan, R.S., and Chua, L.P., *Proc. Inst. Mech. Eng. Part H, J. Eng. Med.*, 220, 647, 2006.)

response to blood filling caused by LA contraction. In Equation 2.46, if we neglect the term $Md\dot{V}$ (as being small compared to the other terms in Equation 2.51), the pressure dynamics is expressed as

$$P_i - P_{i-1} = (E_{p,i} + E_{a,i})(V_i - V_{i-1}) + V_i(E_{a,i} - E_{a,i-1} + E_{p,i} - E_{p,i-1}) \quad (2.57)$$

By employing the monitored LV volume values and the computed values of E_p and E_a, we can compute the values of LV pressure. In other words, if we obtain the LV volume values, and if somehow the E_p and E_a functions (as given by Equations 2.55 and 2.56) were known as intrinsic properties of the LV, then we could compute the LV pressure variation from Equation 2.57. Let us take the computed values of E_a and E_p, and V_i and V_{i-1} during early filling, and compute $(P_{22} - P_{21})$ as follows:

$$P_{22} - P_{21} = (E_{p,22} + E_{a,22})(V_{22} - V_{21}) + V_{22}(E_{a,22} - E_{a,21} + E_{p,22} - E_{p,21})$$
$$= -6.7 \text{ mmHg} \qquad\qquad (2.58)$$

We can see that $(P_{22} - P_{21})$ is negative, thereby demonstrating the suction effect. Now, we take the computed values of elastances and LV volumes during late filling and compute $(P_{34} - P_{33})$:

$$P_{34} - P_{33} = (E_{p,34} + E_{a,34})(V_{34} - V_{33}) + V_{34}(E_{a,34} - E_{a,33} + E_{p,34} - E_{p,33})$$
$$= 1.7 \text{ mmHg} \qquad\qquad (2.59)$$

We note that $(P_{34} - P_{33})$ is positive.

In Figure 2.13, these pressure differences are plotted from the beginning of the isovolumic contraction phase. It can be seen that the computed pressure difference closely approximates the monitored LV pressure difference.

2.8.4.2 *Pressure Dynamics during Ejection Phase*

We can likewise determine the pressure variation during ejection phase, as

$$P_i - P_{i-1} = (E_{p,i} + E_{a,i})(V_i - V_{i-1}) + V_i(E_{a,i} - E_{a,i-1} + E_{p,i} - E_{p,i-1}) \quad (2.60)$$

Let us take the computed values of $(E_p$ and E_a, V_i and $V_{i-1})$ during early and late ejection, and compute $(P_7 - P_6)$ and $(P_{16} - P_{15})$, as follows:

$$P_7 - P_6 = (E_{p,7} + E_{a,7})(V_7 - V_6) + V_7(E_{a,7} - E_{a,6} + E_{p,7} - E_{p,6}) = 9 \text{ mmHg}$$
$$(2.61)$$

$$P_{16} - P_{15} = (E_{p,16} + E_{a,16})(V_{16} - V_{15}) + V_{16}(E_{a,16} - E_{a,15} + E_{p,16} - E_{p,15})$$
$$= -4 \text{ mmHg} \qquad\qquad (2.62)$$

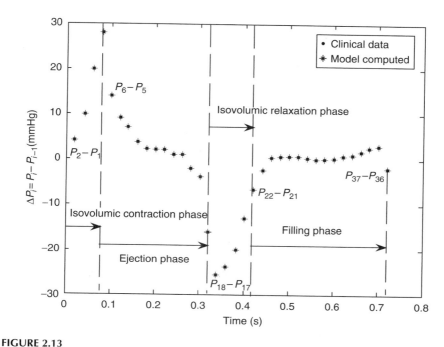

FIGURE 2.13

Pressure dynamics during ejection and filling phases. Note the pressure decrease (i.e., negative ΔP_i) during early filling (from frames 21 to 23), representing LV suction phenomenon even though the LV volume is increasing from frames 21 to 23 (in Figure 2.9). Also, LV pressure increase keeps decreasing during the first third of ejection phase, remains constant in the middle third phase of ejection, and becomes negative in late ejection phase. (Adopted from Figure 7 of Zhong, L., Ghista, D.N., Ng, E.Y.K., Lim, S.T., Tan, R.S., and Chua, L.P., *Proc. Inst. Mech. Eng. Part H, J. Eng. Med.*, 220, 647, 2006.)

We note that $(P_7 - P_6)$ is positive, while $(P_{16} - P_{15})$ is negative, indicating that the LV pressure has already started decreasing because of the E_p effect.

In Figure 2.13, these computed pressure differences $(P_i - P_{i-1})$ are plotted. This graph illustrates how (1) E_a increase (due to force development in the myocardial sarcomere) and constant E_p during isovolumic contraction contribute to LV pressure increase, (2) E_a increase during ejection (due to increase in sarcomeric force development) and E_p decrease (due to blood volume decrease) contribute to LV pressure dynamics during the ejection phase, and (3) E_a decrease and E_p increase (due to blood volume increase) contribute to the pressure dynamics during the filling phase.

2.8.5 Active Elastance as a New Contractility Index

The basis of E_a is that the LV chamber wall is comprised of helically wound myocardial fibers. When these fibers contract at the start of isovolumic contraction, the LV chamber is deformed and the LV pressure increases.

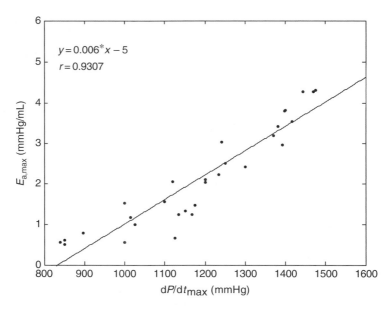

FIGURE 2.14

Relating our contractility index $E_{a,max}$ to the traditional contractility index dP/dt_{max}, with r being the correlation coefficient. (Adopted from Figure 8 of Zhong, L., Ghista, D.N., Ng, E.Y.K., Lim, S.T., Tan, R.S., and Chua, L.P., *Proc. Inst. Mech. Eng. Part H, J. Eng. Med.*, 220, 647, 2006.)

A myocardial fiber comprises of the sarcomere, containing actin and myosin filaments. It is the development of interfilament bonds that causes (1) the development of contractile force and fiber shortening and (2) LV pressure generation, as depicted by Figure 2.8.

Thus, the operation of this myocardial fiber sarcomeres, in conjunction with the helical orientation of the myocardial fibers, gives rise to the concept of LV active elastance (E_a). This is why E_a is deemed to be an intrinsic property of the LV; this is also why we now propose E_a to represent a contractility index. However, because $E_a(t)$ is a cyclic time-varying function, we have decided to adopt the maximum value of E_a ($E_{a,max}$) during cardiac cycle to represent a new contractility index. In Figure 2.14, we have depicted the computed traditional contractility index $(dP/dt)_{max}$ as well as $E_{a,max}$ for a number of subjects studied by us. The good correlation between $E_{a,max}$ and $(dP/dt)_{max}$ agrees well for the employment of $E_{a,max}$ as a contractility index.

2.8.6   Discussion

LV Suction Phenomena. Diastolic suction is defined as that property of the left ventricle that tends to cause it to refill itself during early diastole independent of any contractile effort from the left atrium (LA). Physiologists have been intrigued by the observation that the relaxing ventricle seems somehow to suck blood into its chamber. With pioneering physical intuition, Katz [18] proposed that the early rapid filling of the heart is due to mechanical suction of blood by the ventricle. The concept of the heart as a suction pump has over the years been suggested by many researchers [18–23], and is no longer questioned; however, the underlying mechanisms have not been clarified.

Many researchers have believed that LV suction was caused by elastic recoil during relaxation, or by a sudden stretching resulting from the filling of the coronary arteries, or by a marked asynchronous cessation of contraction [19]. Earlier in 1981, Sabbah and Stein [24] had indicated that early rapid filling of the LV is due to forces within the ventricular wall that act to restore the ventricle to its diastolic dimensions. This means that the suction phenomenon is resulting from the contributions of elastic recoil and restoring loads due to the compression of sarcomeres.

Later in 1986, Robinson and colleagues [25] proposed that the connective tissue matrix of the heart as a whole (including its connection to the vessel) stores contractile strain during systole. This strain is recovered as elastic recoil in diastole, and performs external work on the left-atrial (LA) blood, by sucking it into the LV. This physiological phenomenon of the heart muscle shortening, storing elastic strain in its structures, and then relengthening to cause suction has also been expressed by Prewitt et al. [26] and Kovács [27].

However, these explanations are quite ambiguous. It is in fact the rapid decrease in LV wall E_a that causes a decrease in LV pressure below the value of LA pressure, as shown by our Equation 2.58 to thereby create a suction effect on the LA blood. In other words, the LV pressure continues to decrease during early filling, because of continuing sarcomere relaxation (and decrease in LV elastance) and rapid filling of blood, resulting in volume acceleration. The rapid decrease in elastance during isovolumic relaxation extends into the filling phase, and can explain the decrease in LV pressure (in Figure 2.13) even after LV filling has commenced. In Equation 2.47, it is seen that P_i can be less than P_{i-1} (or that $P_i - P_{i-1} < 0$) only if $(E_{a,i} - E_{a,i-1})$ is negative, i.e., active elastance is decreasing. The pressure dynamics during filling (calculated using Equation 2.57), as depicted in Figure 2.13, confirms the decrease of pressure during early filling.

Hence, it is our novel concept of "decreasing E_a during the early phase of filling" that enables us to explain the phenomenon of decreasing LV pressure during the early stage of filling. In other words, it is suggested that the sarcomere actin–myosin activity continues into the filling phase. The decreasing E_a during the filling phase reflects decreasing sarcomeric activity during filling. Likewise, the increase in E_a during isovolumic contraction is responsible for increase in LV pressure at constant volume, as demonstrated by means of Equation 2.50. This concept is in fact indirectly supported by several works [21,22,28–39] relating diastolic suction to negative LV pressure.

$E_{a,max}$ *as a Contractility Index.* Earlier, we have seen how Figure 2.14 shows the correlation between $E_{a,max}$ and the invasive measures of LV contractility.

It is noted that $E_{a,max}$ has a high degree of correlation with $(dP/dt)_{max}$. It is interesting to compare our correlation-coefficient value (0.9307) with the value of 0.89 obtained by Mehmel et al. [38] by their computing elastance as an extrinsic property $= [P/(V - V_d)]_{es}$. The difference between the two indices is that dP/dt_{max} is an extrinsic index based on the LV pressure response, while our $E_{a,max}$ is an intrinsic index which in fact governs the LV pressure response.

Pressure Variation Phenomena. Both the active and passive elastances (given by Equations 2.61 and 2.62 and depicted in Figures 2.10 and 2.11) can explain LV pressure variation during ejection, as indicated by Equations 2.51 and 2.52. In other words, the increase of E_a during ejection (due to increase in sarcomeric contractile force) and decrease of E_p during ejection (due to LV volume decrease) together causes the pressure variation during ejection, as shown in Figure 2.8. Likewise, the combined action of E_a and E_p contributes to the change of LV pressure during LV filling phase, as demonstrated by Equations 2.58 and 2.59. What is implied here is that the intrinsic property of LV (represented by E_a and E_p) contributes to the manner in which the pressure varies during ejection and filling and in fact during the cardiac cycle.

2.9 Conclusion

Let us recapitulate the organization and contents of this chapter. We started with pointing out how a stenotic AV could augment LV wall stress and cause an oxygen supply–demand mismatch. We then developed the basis of how echocardiography and phonocardiography can be combined to determine the stiffness properties of MV and AV as well as the passive stiffness property of the LV. Then, we showed how echotexture determination can help to detect ischemic and infarcted segments, and how pharmacological treatment could help us to noninvasively determine the restoration of myocardial ischemic segments to normality.

Perhaps the best way to illustrate the effect of LV myocardial infarct is to determine the intra-LV flow velocity and pressure distributions from LV wall-motion data obtained echocardiographically. We have demonstrated this for a typical patient, and shown how the adverse blood pressure-gradient from the apex to the base of the LV can be improved by administration of nitroglycerin (a vasodilator), so as to justify the candidacy of that potential's impaired LV to be improved by coronary bypass surgery.

Finally, we have formulated a new concept of dual passive and active elastances operating throughout the cardiac cycle. These passive and active elastances values are evaluated separately and individually. Our definitions of E_p and E_a enable us to explain the phenomena of (1) LV suction during early filling, (2) LV pressure rise during isovolumic contraction, (3) LV pressure variation during the ejection phase, and (4) LV pressure-drop during the relaxation phase. From the viewpoint of intrinsic indices of LV assessment, E_p can represent LV myocardial stiffness property and resistance to LV filling. On the other hand, E_a has been shown to correspond to LV contractility, as depicted by Figure 2.14. Herein, we have shown a high degree of correlation between $E_{a,max}$ and $(dP/dt)_{max}$.

References

1. Eringen C, *Mechanics of Continua*, Wiley, 1967.
2. Ghista DN and Rao AP, Mitral valve mechanics—Stress/strain characteristics of excised leaflets, analysis of its functional mechanics and medical applications, *Medical and Biological Engineering and Computing*, 11(6):691–702, 1973.
3. Ghista DN et al., Biomechanical analysis of the cardiovascular system, in *Mathematical Methods in Medicine*, edited by Ingram and Bloch, Wiley, pp. 247–293, 1985.
4. Mazumdar J, Hearn T, and Ghista DN, Determination of in vivo constitutive property and normal- pathogenic states of mitral valve leaflets, in *Applied Physiological Mechanics*, edited by Dhanjoo N. Ghista, Hardwood Academic Publishers, pp. 275–306, 1979.
5. Hearn TC, Gopal DN, Ghista DN, and Mazumdar J, Assessment of atrio-ventricular valve leaflet property-based pathology from heart-valve vibrational & heart-sound spectral analyses, *Automedica*, 5:317–328, 1984.
6. Nandagopal D, Ghista DN, Heam TC, and Wu C, Paediatric aortic valve assessment from phono-echo cardiograms and aortic valve vibration analysis, *Automedica*, 6(4):249–267, 1986.
7. Ghista DN, Way RC, Subbaraj K, Kamath MV, Kitabatake A, and Hearn T, Noninvasive biomedical engineering diagnostic technology, in *Handbook of Biomedical Engineering*, edited by Kline J, published by Academic Press, pp. 685–711, 1998.
8. Kamath MV, Way RC, Ghista DN et al., Detection of Cardiac Tissue Lesions by Echocardiac Texture Analysis, *Proceedings of the XIV International Conference in Medical and Biological Engineering*, 1985.
9. Kamath MV, Way RC, Ghista DN, Srinivasan TM, Wu C, Smeenk S, Marning C, and Cannon J, Detection of myocardial scars in neonatal infants from computerized echocardiographic texture analysis, *Engineering in Medicine*, 15(3):137–141, 1986.
10. DeSa DJ and Donnelly WH, Myocardial necrosis in the newborn, *Perspectives in Pediatric Pathology*, 4(4):295–311, 1984.
11. Subbaraj K, Ghista DN, and Fallen EL, Intrinsic indices of the left ventricle as a blood pump in normal and infarcted left ventricles, *Journal of Biomedical Engineering*, 9(3):206–215, 1987.
12. Zhong L, Ghista DN, Ng EYK, and Lim ST, Passive and active ventricular elastances of the left ventricle, *Biomedical Engineering Online*, 4:10, 2005.
13. Zhong L, Ghista DN, Ng EYK, Lim ST, Tan RS, and Chua LP, Explaining left ventricular pressure dynamics in terms of LV passive and active elastances, *Proceedings of the Institution of Mechanical Engineers. Part H, Journal of Engineering in Medicine*, 220:647–655, 2006.
14. Mirsky I, Assessment of passive elastic stiffness of cardiac muscle: Mathematical concepts, physiologic and clinical considerations, directions of future research, *Progress in Cardiovascular Diseases*, XVIII(4):277–308, 1976.

15. Mirsky I and Pasipoularides A, Clinical assessment of diastolic function, *Progress in Cardiovascular Diseases*, 32:291–318, 1990.
16. Gilbert JC and Glantz SA, Determinants of left ventricular filling and of the diastolic pressure–volume relation, *Circulation Research*, 64(5):827–852, 1989.
17. Sandler H and Dodge HT, The use of single plane angiocardiograms for the calculation of left ventricle volume in man, *American Heart Journal*, 75:325–334, 1968.
18. Katz L, The role played by the ventricular relaxation process in filling the ventricle, *American Journal of Physiology Heart Circulatory Physiology*, 95:542–553, 1930.
19. Brutsaert DL, Rademakers FE, and Sys SU, Triple control of relaxation: Implication in cardiac disease, *Circulation*, 69:190–196, 1984.
20. Covell JW, Nikolic S, LeWinter MM, Ingels NB, Yellin et al., Restoring forces, in *Systolic and Diastolic Function of the Heart*, edited by Ingels NB Jr, Daughters GT, Baan J, Covell JW, Reneman RS, and Yin FCP, IOS press and Ohmsha, 1995.
21. Ingels NB Jr, Daughters GTH, Nikolic SD, DeAnda A, Moon MR et al., Left atrial pressure-clamp servomechanism demonstrated LV suction in canine hearts with normal mitral valves, *American Journal Physiology Heart Circulatory Physiology*, 267:H354–H362, 1994.
22. Ingels NB Jr, Daughters GTH, Nikolic SD, DeAnda A, Moon MR et al., Left ventricular diastolic suction with zero left atrial pressure in open-chest dogs, *American Journal of Physiology Heart Circulatory Physiology*, 270:H1217–H1224, 1996.
23. Bleasdale RA, Turner MS, and Mumford CE, Left ventricular suction: An explanation for the restrictive filling pattern observed in patients with heart failure, *European Heart Journal*, 24:613–623, 2003.
24. Sabbah HN and Stein PD, Negative diastolic pressure in the intact canine right ventricle. Evidence of diastolic suction, *Circulation Research*, 49:108–113, 1981.
25. Robinson FR, Factor SM, and Sonnenblick EH, The heart as a suction pump, *Scientific American*, 254(6):84–91, 1986.
26. Prewitt D, Wickline SA, and Kovács SJ, Characterization of abnormal left ventricular filling in hypertrophic hearts, using a novel kinematic model, *Journal of American College in Cardiology*, 21(A):2–14, 1993.
27. Kovács SJ, Diastolic function from transmitral flow: The kinematic paradigm, in *Systolic and Diastolic Function of the Heart*, edited by Ingels NB Jr, Daughters, GT, Baan, J, Covell, JW, Reneman RS, and Yin, FCP, IOS press and Ohmsha, pp. 125–135, 1995.
28. Bell SP, Fabian J, Higashiyama A, Chen Z et al., Restoring forces assessed with left atrial pressure clamps, *American Journal of Physiology Heart Circulatory Physiology*, 270:H1015–H1020, 1996.
29. Bell SP, Fabian J, and LeWinter MM, Effect of dobutamine on left ventricular restoring forces, *American Journal of Physiology Heart Circulatory Physiology*, 275: H190–H194, 1998.
30. Bloom WL, Diastolic filling of the beating excised heart, *American Journal of Physiology Heart Circulatory Physiology*, 187:143–144, 1955.
31. Bloom WL and Kovács SJ, Negative ventricular diastolic pressure in beating heart studied in vitro and in vivo, *Proceedings of the Society for Experimental Biology and Medicine*, 93:451–454, 1956.
32. Brecher GA, Experimental evidence of ventricular diastolic suction, *Circulation Research*, 4:513–518, 1956.

33. Brecher GA, Critical review of recent work on ventricular diastolic suction, *Circulation Research*, 6:554–566, 1958.
34. Brecher GA and Kissen AT, Relation of negative intraventricular pressure to ventricular volume, *Circulation Research*, 5:157–162, 1957.
35. Fowler NO, Bloom WL, and Ferris EB, Systolic–diastolic pressure relationships in the isolated beating heart, *Circulation Research*, 5:485–488, 1957.
36. Fowler NO, Shabetai R, and Braunstein JR, Transmural ventricular pressure in experimental cardiac tamponade, *Circulation Research*, 7:733–739, 1959.
37. Hori M, Yellin EL, and Sonnenblick EH, Left ventricular diastolic suction as a mechanism of ventricular filling, *Japanese Circulation Journal*, 146:124–129, 1982.
38. Mehmel HC, Stochins B, Ruffmann K, Olshausen K, Schuler G, and Kubler W, The linearity of the end-systolic pressure–volume relationship in man and its sensitivity for assessment of left ventricular function, *Circulation*, 63(6):1216–1222, 1981.
39. Shoucri RM, Active and passive stresses in the myocardium, *American Journal of Physiology Heart Circulatory Physiology*, 279:H2519–H2528, 2000.

3

Left Ventricular Contractility Indices

Dhanjoo N. Ghista and Liang Zhong

CONTENTS

3.1 Scope

Contractility is the key mechanism of left ventricular pumping role. Hence, indices of contractility are important for differentiating poorly contracting left ventricles (LVs) from normally contracting LVs. In this chapter, we provide the theory and application of contractility indices based on (1) the left ventricular shape factor, in terms of the LV wall stress normalized with respect to the LV internal pressure, and (2) the spirally wound myocardial fiber's sarcomere characteristics of contractile element force versus shortening velocity. These contractility indices values are compared to the values of the traditional contractility index of $(dP/dt)_{max}$, and good correlations are observed between our new indices and the traditional index of $(dP/dt)_{max}$.

3.2 Left Ventricular Shape Factor Based Contractility Index

Over the past decades, while several indices for estimating the left ventricular contractile state have been proposed, very few studies have been dedicated to the influence of the LV shape factor on its contractility. It has been observed that the shape of the LV is of clinical relevance for prognosis of heart patients [1–5]. In this regard, some investigators have associated a more spherically shaped and less-ellipsoidal shaped LV with the failing heart [6]. Invasive animal experiments have indicated that the shape of the LV is somewhat like a prolate ellipsoid [7]. From cineventriculography, we can obtain the two-dimensional shape of the LV, and therefrom the ellipsoidal shape of LV. This information has been applied, herein, to develop a left-ventricular ellipsoidal geometry model and its wall stress. We can then define an LV shape-based index to represent the capacity of the LV to generate necessary and sufficient intramyocardial stress (σ) to provide necessary and sufficient pressure and kinetic energy to the ejected blood. Further, we can normalize this wall stress with respect to the LV pressure ($\sigma^* = \sigma/P$), and gauge LV contractile capability in terms of the maximum value of generated normalized intramyocardial stress, or $(d\sigma^*/dt)_{max}$. Thus it can help provide more insight into the LV shape-based contractile stress for its ejection function [8].

Our concept of an LV shape-based contractility index is that it is a measure of the capacity of the LV myocardial sarcomere to contract and generate the wall stress that will adequately raise intra-LV pressure to eject the blood. Now since the LV wall stress depends on its shape, the LV contractile capacity also depends on the LV shape. This is the rationale behind the LV shape-based index. On the basis of clinical observations, a healthy LV shape factor is more akin to the optimal-ellipsoidal shape factor, but transforms into a more spherical shape in a poorly contracting LV as well as in LV failure. Hence, our LV shape-based index, expressed as $(d\sigma^*/dt)_{max}$, is meant to quantitatively express this clinical observation.

3.2.1 LV Model Geometry Development

Herein, the LV is treated as a prolate spheroid, truncated 50% of the distance from equator to base, as suggested by Streeter and Hanna [9] (Figure 3.1). The LV shape can be defined by the major and minor radii of its two surfaces: the endocardium of the LV and the septum, and a surface defined by the epicardium of the free wall. The overall longitudinal distance from the base to apex ($= 3LA/2$) is thus 1.5 times the major radius of the ellipse. Left ventricular cavity and wall volumes are calculated from the epicardial anterior–posterior (AP) and base-apex lengths according to equations:

$$V_M = \frac{9\pi[(LA + h)(SA + h)^2 - LA \times SA^2]}{8} \qquad (3.1)$$

$$V = \frac{9\pi(SA)^2 LA}{8} \qquad (3.2)$$

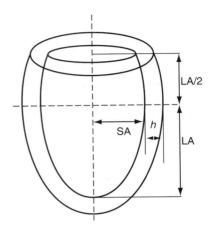

FIGURE 3.1
Left ventricle (LV) model geometry, showing the
major and minor radii of the inner surface of the LV
(LA & SA) and the wall-thickness (h).

wherein

V is LV volume
V_M is myocardial volume
h is wall-thickness
LA and SA are endocardial major and minor radii

Simplifying Equation 3.1 by neglecting the $9\pi(\text{LA} \times h^2 + 2\text{SA} \times h^2 + h^3)/8$
term, we obtain

$$V_M = \frac{9\pi\,(2\text{LA} \times \text{SA}^2 + \text{SA}^2)h}{8} \tag{3.3}$$

Now, LV volume (V), wall-thickness (h), and myocardial volume (V_M) are
measured by cineventriculography. Hence, by using Equations 3.2 and 3.3,
we can calculate the major and minor radii LA and SA. Firstly, from
Equation 3.2, we have

$$\text{LA} = \frac{8V}{(9\pi\text{SA}^2)} \tag{3.4}$$

Then, by substituting Equation 3.4 into Equation 3.3, we get

$$V_M = \frac{9\pi}{8}\left(\frac{16V}{9\pi\text{SA}} + \text{SA}^2\right)h$$

$$\text{SA}^2 + \frac{16V}{9\pi\text{SA}} - \frac{8V_M}{9\pi h} = 0$$

which gives us an equation in SA to obtain the value of SA.

3.2.2 Determination of LV Model Wall Stress

The generated wall stress (GWS) in the LV is a measure of the effectiveness
of the sarcomere contractile machinery of the LV myocardium. The GWS is

adjusted to be necessary and sufficient for carrying out its ejection function. Hence, we deem that the normalized wall stress σ/P ($=\sigma^*$) developed for blood ejection can provide a more intrinsic measure of its contractile capacity than, for instance, $(dP/dt)_{max}$, because the developed LV pressure is in fact a consequence of the generated LV wall stress.

For an ellipsoidal shell, the circumferential wall stress σ_θ (referred to as σ) at the waist of the LV ellipsoidal model is given by Mirsky [10] as

$$\sigma = P\frac{SA}{h}\left[1 - \frac{SA(SA/LA)^2}{(2SA + h)}\right] = P\frac{SA}{h}\left[1 - \frac{(SA/h)(SA/LA)^2}{2(SA/h) + 1}\right] \quad (3.5)$$

From Equations 3.2 and 3.3, we have by putting $S = SA/LA$

$$\frac{SA}{h} = \frac{V}{V_M}\left(2 + \frac{SA}{LA}\right) = \frac{V}{V_M}(2 + S) = \frac{2 + S}{V^*} \quad (3.6)$$

where
$S = SA/LA$ constitutes the LV shape factor
$V^* = V_M/V$ represents the volume ratio

Combining Equations 3.5 and 3.6, we can express the normalized stress σ^* as

$$\sigma^* = \frac{\sigma}{P} = \frac{2 + S}{V^*}\left[1 - \frac{S^2(2 + S)}{2(2 + S) + V^*}\right] = f(V^*, S) \quad (3.7)$$

Equation 3.7 is a function of S for a given V^*. We can now compute the time-variation of σ^* during ejection, in terms of V^* and S. Figure 3.2 indicates the cyclic variations of h, S ($= SA/LA$), and σ^* versus time during the ejection phase, for three of our patients.

3.2.3 Normalized Wall Stress based Shape Factor Index

A well-known definition of contractility is $(dP/dt)_{max}$. However, we can more intrinsically characterize contractility in terms of the max rate of generation of the LV normalized stress σ^* ($=\sigma/P$) using Equation 3.7, as

$$SFI1 = \left|\frac{d\sigma^*}{dt}\right|_{max}$$

$$= \left|\frac{[\dot{V}(2+S)/V] + \dot{S}}{V^*} - \frac{S\left(\begin{array}{c}S\dot{V}[16 + 8V^* + (24 + 8V^*)S + (12 + 2V^*)S^2 + 2S^3]\\ +\dot{S}[32 + 8V^* + (56 + 12V^*)S + (32 + 4V^*)S^2 + 6S^3]\end{array}\right)}{V^*(4 + 2S + V^*)^2}\right|_{max}$$

$$= F(S, \dot{S}, V, \dot{V}, V^*) \quad (3.8)$$

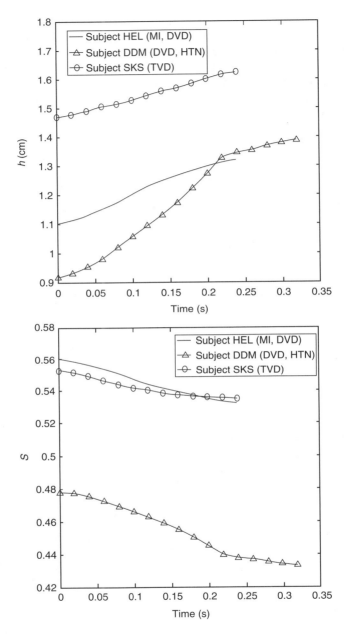

FIGURE 3.2
Variation of h, S, and σ^* versus time during the ejection phase for subject HEL with myocardial infarct (MI) and double vessel disease (DVD), subject DDM with DVD and hypertension (HTN), and subject SKS with triple vessel disease (TVD), during the ejection phase. Herein, $t = 0$ represents the start-of-ejection. Subject SKS has the minimum generated σ^*, while subject DDM has the maximum σ^* during the ejection phase. (Adopted from Zhong, L., Ghista, D.N., Eddie, Y.K.Ng., Lim, S.T., Chua, T., and Lee, C.N., *J. Biomech.*, 39, 2397, 2006.)

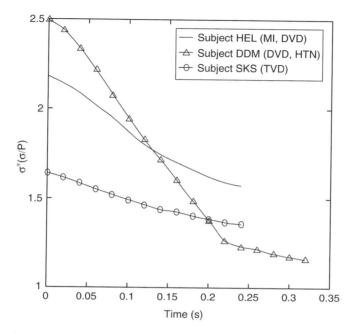

FIGURE 3.2 (continued)

Equation 3.8 indicates that corresponding to a patient's $V(t)$ and $\dot{V}(t)$ variations, the shape factor index $SFI1$ value for that patient is a function of the shape factor (S) of an LV. Now cardiologists have been observing that an infarcted LV becomes less ellipsoidal as compared to a normally contracting LV shape [6]. This resultant distorted shape of an impaired LV does not allow it to contract and deform in an optimal twisting mode [11], so as to perform its pumping function and deliver the requisite cardiac output efficiently. In accordance with this clinical observation, our shape factor index $SFI1$ (Equation 3.8) incorporates the LV shape factor ($S = SA/LA$), and the influence of the distorted shape of an infarcted LV to its impaired pumping function.

3.2.4 Optimal Left Ventricle Shape Factor and Corresponding Shape Factor Index 2

Let us designate the optimal shape factor S ($=SA/LA$) to be that value for which the generated myocardial wall stress σ^* for a given LV volume (at the start-of-ejection $V = V_{se} = V_{ed}$) is maximum for a specific value of V^*. The concept of optimizing the shape factor is based on the rationale that LV pressure $P = \sigma/\sigma^*$. During systole, the interaction of the actin–myosin filaments causes contraction of the myocardial fibers and generation of myocardial wall stress (σ). The resultant LV pressure generation is given by σ/σ^*, where σ^* is purely dependent on LV geometry and is a function of the shape factor (S) and volume ratio (V^*). For a particular V^*, as S increases (i.e., as the LV becomes more spherical and less ellipsoidal), σ^* decreases, hence the LV pressure increases, as seen in Figure 3.3. For an adequate

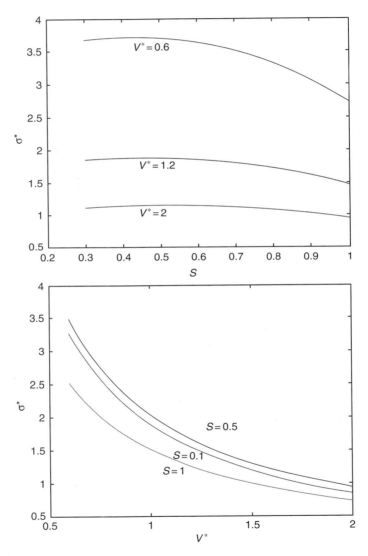

FIGURE 3.3
Variation of σ^* with (a) S for different values of $V^*(= V_M/V)$ and (b) V^* for different values of S. (Adopted from Zhong, L., Ghista, D.N., Eddie, Y.K.Ng., Lim, S.T., Chua, T., and Lee, C.N., *J. Biomech.*, 39, 2397, 2006.)

amount of LV wall stress (σ) generated, we want the LV pressure to be maintained low, so that its oxygen demand is minimal. Hence, we want that (for a specific V^*) σ^* be as high as possible and correspondingly S be as low as possible, i.e., the LV must be more ellipsoidal in shape.

From Equation 3.7, we can maximize σ^* with respect to S, as

$$\frac{d\sigma^*}{dS} = \frac{1}{V^*} - \frac{1}{V^*}\frac{[(8S+12S^2+4S^3)(V^*+4+2S)-(8S^2+8S^3+2S^4)]}{(V^*+4+2S)^2} = 0 \quad (3.9)$$

Simplifying Equation 3.9, we have

$$6S^4 + (4V^*+32)S^3 + (12V^*+52)S^2 + 4(V^*+4)S - (V^*+4)^2 = 0 \quad (3.10)$$

from which we obtain the optimal shape factor S as a function of V^*, as shown in Figure 3.4. It appears that S is linearly proportional to the V^*, as given by

$$S^{\mathrm{op}} = 0.053V^* + 0.39 \quad (3.11)$$

This line can be called the optimal-S line.

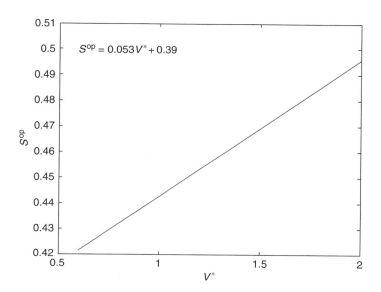

FIGURE 3.4
Optimal shape factor S versus V^* at the start-of-ejection, represented by: $S^{\mathrm{op}} = 0.053V^* + 0.39$. (Adopted from Zhong, L., Ghista, D.N., Eddie, Y.K.Ng., Lim, S.T., Chua, T., and Lee, C.N., *J. Biomech.*, 39, 2397, 2006.)

If we substitute Equation 3.11 into $d^2\sigma^*/dS^2$, we get $d^2\sigma^*/ds^2$ to be negative. In other words, this optimal S function (of V^*) maximizes σ^*, in accordance with our rationale. The significance of Equation 3.11 is that one can adjudge the cardiac health state of a patient merely in terms of how close the shape factor S ($=SA/LA$) corresponding to a patient's V^* value (at the start-of-ejection) is to the optimal value obtained from Figure 3.4. We do not even need to compute σ^* or $d\sigma^*/dt$ in order to evaluate how efficiently a particular LV is pumping.

Hence, another way to define LV contractility would be in a nondimensional form at the start-of-ejection (se), as follows:

$$SFI2 = \frac{(S_{se} - S_{se}^{op})}{S_{se}^{op}} \tag{3.12}$$

where

S_{se} is the measured shape factor value

S_{se}^{op} is the corresponding optimal value at the start-of-ejection

So, as $SFI2$ value increases, the LV contractility becomes poorer. Then, from Equation 3.12, the $SFI2$ value for the patient's data shown in Figure 3.2 is obtained to be 0.21 for subject HEL, 0.057 for subject DDM, and 0.11 for subject SKS, as shown in Table 3.1. Note that for both our new indices of the 3 subjects, the patient DDM has the higher EF; correspondingly $SFI1$ is maximum for DDM, and $SFI2$ is minimum for DDM.

Below the optimal line (Figure 3.4), the shape of the LV becomes physiologically unnatural (i.e., too narrow), in order to support a reasonable value of V_{se}, the volume at the start-of-ejection. We can postulate that if the shape factor S is located in the A zone of Figure 3.5, it can be a tolerable shape to provide a reasonable LV contractility; then, the B zone of Figure 3.5 can represent a poorly contracting LV, while the C zone can represent a failing heart. The three cases are depicted on the (S–V^*) plane, in Figure 3.5. This is further validated by the good correlation of $SFI1$ and $SFI2$ with the traditional contractility indices of EF and $(dP/dt)_{max}$, as discussed in the next section.

TABLE 3.1

Clinical History, Calculated S_{se}, $SFI1$, and $SFI2$
from Subjects (HEL, DDM, and SKS)

Subject	Disease	S_{se}	$SFI1$	$SFI2$	EF
HEL	MI, DVD	0.56	3.84	0.21	0.36
DDM	DVD, HTN	0.48	6.90	0.057	0.66
SKS	TVD	0.55	1.72	0.11	0.24

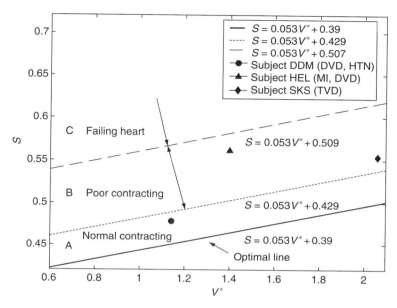

FIGURE 3.5

We can postulate LVs to be normal contracting, poorly contracting, and failing heart, as illustrated in the above figure. Subject DDM, shown on the $S–V^*$ plane, can hence be depicted to have normally contracting heart. On the other hand, subjects HEL and SKS have poorly contracting hearts. The corresponding *SFI2* values of these 3 subjects are shown in Table 3.1, based on the location of the calculated (S, V^*) plots on this plot. (Adopted from Zhong, L., Ghista, D.N., Eddie, Y.K.Ng., Lim, S.T., Chua, T., and Lee, C.N., *J. Biomech.*, 39, 2397, 2006.)

3.2.5 Optimally Shaped LV(s) Compared to Abnormally Shaped LV(s) for Different Age Groups

Now let us see how an optimally shaped LV looks like, for different values of normal LV volume V_{se} (volume of the start-of-ejection) variation with age. In other words, for various age groups, we have taken normal V_{se} values for Asian and American populations, based on the data shown in Table 3.2. Figure 3.6 illustrates the LV(s) for the same value of A in each age group, for 3 different values of V_{se} at the start-of-ejection. For each such V_{se} value, we compute the value of SA corresponding to $S = S^{op}$, determine S, and then plot the LV shape.

The anatomically abnormal LV(s) (to the left of the optimal LVs) have less than normal end-diastolic volume (EDV) as well as less than optimal contractility. Hence, they would not be able to meet the SV demand of the circulatory and organ systems. The physiologically abnormal LV(s) (to the right of the optimal LVs) have bigger "SA" values and bigger "S" values, for the same value of "LA" as the optimal LVs. Hence, as shown in Equation 3.7, these enlarged LVs will have lower values of σ^* and hence higher LV pressure for the same amount of myocardial-wall σ generated. They will hence be prone to becoming hypertensive.

TABLE 3.2

Normal Values of Left Ventricle (LV) Volumes and Mass for Adults and Children

End-Diastolic Volume (mL/m²)	Volume (mL/m²)	End-Systolic Volume (mL/m²)	Ejection Fraction (SV/EDV)	Thickness (mm)	Left Ventricle Mass (gm/m²)
Adults					
70 ± 20	45 ± 13	24	0.67 ± 0.08	10.9 ± 2.0	92 ± 11
Children and infants less than 2 years of age					
42 ± 10	28.6	13.4	0.68 ± 0.05		96 ± 11
More than 2 years of age					
73 ± 11	44 ± 5	27 ± 7	0.63 ± 0.05		86 ± 11

Source: From Dodge, H.T. and Sandler, H., in *Cardiac Mechanics*, I. Mirsky, D.N. Ghista, and H. Sandler, New York, Wiley, 1973, 171–201.

3.2.6 Clinical Applications

3.2.6.1 *Measurements*

All subjects included in this study were in resting recumbent state, after premedication. The LV chamber pressure was measured by a pigtail catheter and Statham P23Eb pressure transducer; the pressure was recorded immediately before or during the angiocardiography in all cases. Single plane cineangiocardiograms were recorded in a posterior–anterior projection from an image intensifier at 50 frames/s using INTEGRIS Allura 9 with Dynamic Flat Detector (Philips Inc.). For a sample subject (HEL), the LV ellipsoidal model's pressure, volume, wall thickness (as derived from the cineventriculography films) are presented in Figure 3.7, along with the calculated ellipsoid major and minor axis (LA and SA from Equations 3.1 and 3.3), and calculated absolute value of $d\sigma^*/dt$ (from Equation 3.8).

3.2.6.2 *Subjects*

Ten subjects with $EF = 0.63 \pm 0.05$ and $(dP/dt)_{max} = 1406 \pm 51$ mmHg/s were selected to comprise group 1. They did not use nicotine, caffeine or alcohol. The age profiles were similar and their anthropometric data, blood pressure, heart rate, and ejection fraction (EF) were within the expected range.

Ten other patients (with coronary and/or valvular disease) with $EF = 0.49 \pm 0.13$ and $(dP/dt)_{max} = 1183 \pm 62$ mmHg/s were classified into group 2, having mean-age of 57.4 years. Finally, we have group 3 of hospitalized patients (of having $EF = 0.38 \pm 0.12$ and $(dP/dt)_{max} = 948 \pm 78$ mmHg/s) with poor (clinically assessed) contractility. These subjects are listed in Table 3.3.

3.2.6.3 *Results*

For each subject, the chamber pressure and dimensions are monitored at 20 ms intervals during the cardiac cycle. A typical set of pressure and

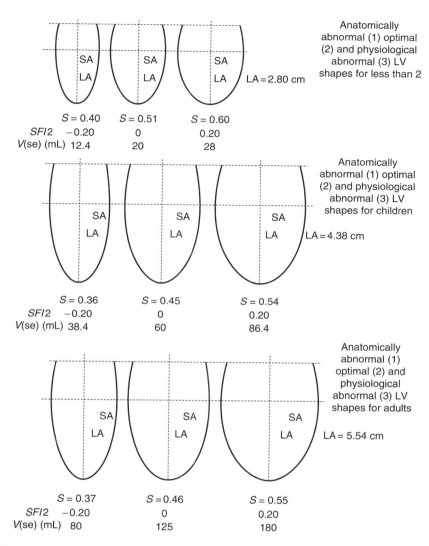

FIGURE 3.6

Schematics of (1) anatomically abnormal, (2) optimal, and (3) physiological abnormal LV shapes, for children less than 2 years of age, more than 2 years of age, and adults, based on data in Table 3.2. (Adopted from Zhong, L., Ghista, D.N., Eddie, Y.K.Ng., Lim, S.T., Chua, T., and Lee, C.N., *J. Biomech.*, 39, 2397, 2006.)

chamber variations for the subject HEL is shown in Figure 3.7. For this subject, the time-derivative of normalized stress (σ^*) and the shape factor (S) are calculated for each 20 ms during the cardiac cycle (Figure 3.7e). Figure 3.7g also depicts the cyclic variation of absolute value of ($d\sigma^*/dt$) during the ejection phase. During ejection, the maximum value of *SFI*1 is found to be 3.84 s^{-1}.

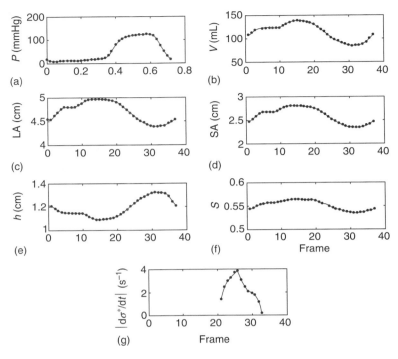

FIGURE 3.7
Pressure (*P*), volume (*V*), and dimensions (*A*, *B*, *h*, and *s*) during a cardiac cycle (using LV ellipsoidal model), along with the absolute value of (dσ*/d*t*) calculated using Equation 3.8 during the ejection phase for subject HEL. (Adopted from Zhong, L., Ghista, D.N., Eddie, Y.K.Ng., Lim, S.T., Chua, T., and Lee, C.N., *J. Biomech.*, 39, 2397, 2006.)

Considering that $S > S^{op}$ is associated with poor contractile heart, Table 3.3 summarizes the patients' history, which includes patient age, heart rate (HR), EF, myocardial volume of LV (V_M), start-of-ejection volume V(se), and end-ejection volume V(ee).

TABLE 3.3

Clinically Monitored Data and Computed Parameters for Three Groups: Group 1 (Normal Contractility), Group 2 (Inadequate Contractility), and Group 3 (Poor Contractility)

	Group 1	Group 2	Group 3
Age (years)	58.70 ± 6.65	57.40 ± 5.85	58.20 ± 9.11
$(dP/dt)_{max}$	1406.00 ± 51.00	1183.00 ± 62.00^a	948.00 ± 78.00^a
HR (beats/min)	72.69 ± 9.20	67.70 ± 10.04	74.02 ± 10.09
V_M (ml)	146.00 ± 43.00	189.00 ± 78.00	216.00 ± 80.00^a
V(se) (ml)	119.26 ± 31.75	148.70 ± 68.32	177.41 ± 90.00
V(ee) (ml)	43.64 ± 9.87	79.45 ± 53.75^a	116.73 ± 54.01^a
EF	0.63 ± 0.05	0.49 ± 0.13^a	0.38 ± 0.12^a

[a] $p < .05$ compared with normal contractility group.

Figure 3.8 depicts the mean and standard deviation of V_M, EF, and V(se), and V(ee) for all the patients analyzed by us. There exists a substantial difference ($p < .05$) between the average values of EF, V_M, V(ee) in normal

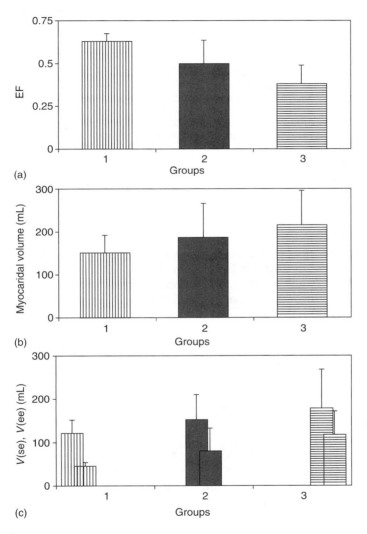

(a)

(b)

(c)

FIGURE 3.8

Comparison of the EF, MV (wall volume), and V(se), V(ee) in groups 1 (normal contractility), 2 (inadequate contractility), and 3 (poor contractility). In figure (c), the first bar corresponds to start-of-ejection (se) and the second bar to end ejection (ee). Figure (d) provides the mean values of V, S, S^{OP}, and $SFI2$. The S^{OP} was calculated using Equation 3.11. In Figure (a), the mean parameter values are given for the 3 groups and values that were statistically different ($P < .05$) from group 1 are indicated (*). (Adopted from Zhong, L., Ghista, D.N., Eddie, Y.K.Ng., Lim, S.T., Chua, T., and Lee, C.N., *J. Biomech.*, 39, 2397, 2006.)

(continued)

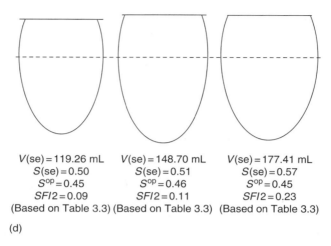

$V(se) = 119.26$ mL $V(se) = 148.70$ mL $V(se) = 177.41$ mL
$S(se) = 0.50$ $S(se) = 0.51$ $S(se) = 0.57$
$S^{op} = 0.45$ $S^{op} = 0.46$ $S^{op} = 0.45$
$SFI2 = 0.09$ $SFI2 = 0.11$ $SFI2 = 0.23$
(Based on Table 3.3) (Based on Table 3.3) (Based on Table 3.3)

(d)

FIGURE 3.8 (continued)

patients compared to patients with inadequate and poor contractility. In Figure 3.8d, we take the average values of $V(se)$ and $S(se)$ for each group, and then show how the corresponding LV shape looks like for these three groups. The LV in group 3 (with poor contractility) has a bigger S value compared to group 1. Thus, based on Figure 3.5, group 1 has normal contractility, group 2 has poor contractility, while group 3 represents a failing heart. Thus, it can be concluded that a more-spherical shape is associated with poor systolic function and decreased contractility of the LV.

Ranges for shape factor indices: Figure 3.9 illustrates the values of $S(se)$, $S(ee)$, $(dP/dt)_{max}$, $SFI1$, and $SFI2$ for the three groups: group 1 (normal subjects), group 2 with inadequate contractility, and hospitalized group 3 with poor contractility. The values of $SFI1$ and $SFI2$ in group 1 are considered to be normal contractility. Group 3 patients with poor contractility have comparatively lower values of $SFI1$ ($p < .05$) and bigger $SFI2$ as compared to those of normal group ($p < .05$).

The average value of $SFI1$ decreases and of $SFI2$ increases in group 2 and group 3, in relation to $SFI1$ and $SFI2$ for normal group 1. The average values of $SFI1$ and $SFI2$ for normal group are 8.75 ± 2.30 s^{-1} and 0.09 ± 0.07 (Table 3.4). In the group of patients with poor contractility (group 3) the values of the indices are significantly different compared to group 1 ($p < .05$). The index $SFI2$ is biggest in group 3, suggesting that this group is having a more spherical or abnormal LV shape. Therefore it can again be concluded that a less ellipsoidal and more-spherical shape is associated with poor systolic function and decreased contractility of the LV, which is also in agreement with the $SFI1$ values in Figure 3.9d and the values of EF in Figure 3.8a. This supports our premise that an infarcted LV is less ellipsoidal compared to a normally contracting LV.

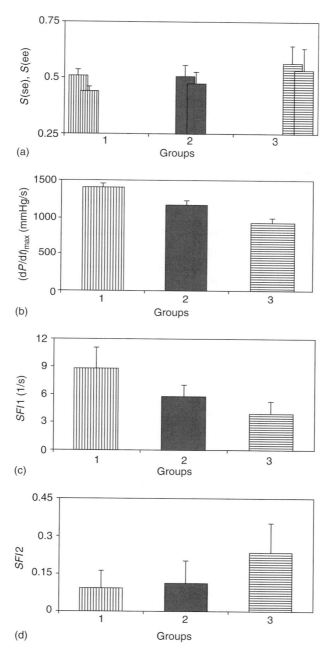

FIGURE 3.9

Comparison of S at end systole, $(dP/dt)_{max}$, $SFI1$, $SFI2$, for group 1 (normal), group 2 (mild heart failure), and group 3 (severe heart failure). (Adopted from Zhong, L., Ghista, D.N., Eddie, Y.K.Ng., Lim, S.T., Chua, T., and Lee, C.N., *J. Biomech.*, 39, 2397, 2006.)

TABLE 3.4

Mean Values with Standard Deviations of S(se), S(ee), $SFI1$, and $SFI2$ for Group 1 (Normal Contractility), Group 2 (Inadequate Contractility), and Group 3 (Poor Contractility)

	Group 1	Group 2	Group 3
S(se)	0.50 ± 0.03	0.51 ± 0.05	0.57 ± 0.08^a
S(ee)	0.44 ± 0.02	0.47 ± 0.05	0.53 ± 0.10^a
$SFI1$ (s^{-1})	8.75 ± 2.30	5.78 ± 1.30^a	3.90 ± 1.30^a
$SFI2$	0.09 ± 0.07	0.11 ± 0.09	0.23 ± 0.12^a

[a] $p < .05$ compared with normal contractility group.

3.2.7 Comparison with Traditional Invasive LV $(dP/dt)_{max}$

For the 3 groups of patients, the comparisons between the indices ($SFI1$ and $SFI2$) are summarized in Table 3.4. Further comparisons between the $SFI1$ and $(dP/dt)_{max}$, $SFI2$ and $(dP/dt)_{max}$ are displayed in Figure 3.10a and b. Figure 3.10a shows a fair correlation of $r = .713$ between $SFI1$ and $(dP/dt)_{max}$ as $SFI1 = 0.0096\,(dP/dt)_{max} - 5.1$, $r = .73$, $p < .01$, while Figure 3.10b suggests a fair correlation:

$$SFI2 = 0.00033\,(dP/dt)_{max} + 0.54, \quad r = .60, \quad p < .01.$$

3.2.8 Discussion and Conclusion

On the basis of Figure 3.10, we can conclude that our new $SFI1$ (evaluated for different groups of patients, i.e., normal, mild, and severe heart failures) compares favorably with that of conventional index $(dP/dt)_{max}$, in distinguishing patients with poor contractility from normal patients. Further, $SFI1$ can be determined noninvasively and is also demonstrated to be potentially more sensitive to changes in the LV shape. Thus the new index $SFI1$ can be an excellent substitute to $(dP/dt)_{max}$ for contractility measure. Concerning the second index $SFI2$, although its correlation with $(dP/dt)_{max}$ is not so good as that of $SFI1$, it is more convenient to compute it and hence use it clinically to diagnose the heart disease.

The shape of LV has intrigued physiologists as well as clinicians in attempting to gain a better understanding of its mode of operation, and trying to obtain diagnostic information on its performance [2,5,13–15]. In this study, we developed new $SFIs$, $SFI1$ and $SFI2$, based on the LV wall stress and hence on the LV shape. We have further shown that the $SFI1$ and $SFI2$ compare well with $(dP/dt)_{max}$. This confirms that the function and contractility of the LV are closely related to its shape changes. Hence, an evaluation of its shape permits early prediction of both physiological and pathophysiological changes in LV functionality.

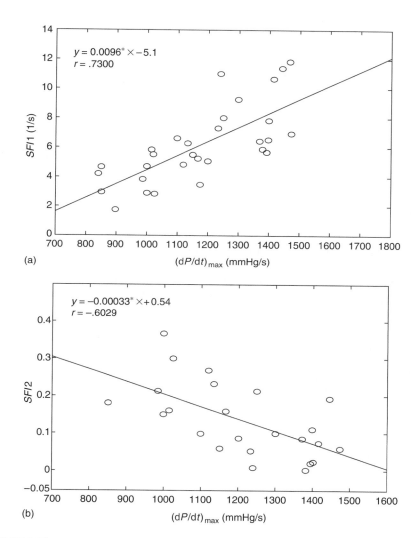

FIGURE 3.10

Relating new developed contractility indices *SFI*1, *SFI*2 to traditional contractility index $(dP/dt)_{max}$. r, correlation coefficient.

3.3 Left Ventricular Sarcomere Contractile Characteristics and Associated Power Index

3.3.1 Left Ventricle Cylindrical Model (Incorporating the Myocardial Fibers within Its Wall)

We represent the left ventricle (LV) as a thick-walled cylindrical shell. Transverse isotropy is assumed with respect to the axis of the cylinder [16–18].

FIGURE 3.11
(a) Schematic of LV myocardial structure, (b) LV cylindrical model, depicting a typical myocardial fiber arranged as a helix within the LV model wall; L, R_i, and R_o are the length, inner, and outer radii of the LV cylindrical model. (Adopted from Ghista, D.N., Zhong, L., Eddie, Y.K.Ng., Lim, S.T., Tan, R.S., and Chua, T., *Mol. Cell. Biomech.*, 2, 217, 2005.)

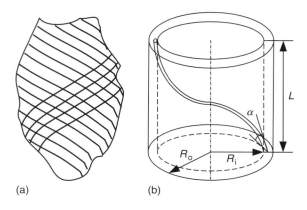

(a)   (b)

In Figure 3.11, we depict this LV model cylinder wall to be composed of N myocardial fiber units, oriented as helixes of pitch angle α. Half of these (i.e., $N/2$) fibers are wrapped in a clockwise fashion, and the other $N/2$ fibers in counter-clockwise fashion. The biomechanical model ultra-structure of each fiber is the MSU, as depicted later in Figure 3.11.

For our LV cylindrical model [19], we assume that each myocardial model fiber is helically wrapped within the LV cylindrical model wall (as illustrated in Figure 3.11), and composed of two in-series MSUs, as illustrated in Figure 3.12. In actuality, there will be many MSUs along any one myocardial fiber from bottom to top. However, herein, for convenience of analysis, we adopt each myocardial fiber to be composed of two MSUs in series.

Now although there are a number of myocardial fibers across the LV wall thickness, it is assumed that, within the wall of our LV model, one set ($N/2$ number) of fibers are oriented in a clockwise fashion, while another equal number ($N/2$) of fibers are oriented counterclockwise. Hence, across the LV wall thickness, we have two fibers; in other words, the LV wall thickness equals to the thickness of two fibers thickness.

The geometric parameters of the LV cylindrical model are defined in Figure 3.11. The volumes of myocardial wall (MV) and of the LV are given as

$$V_M = \pi(R_o^2 - R_i^2)L = \pi(2R_i + h)hL \qquad (3.13)$$

$$V = \pi R_i^2 L \qquad (3.14)$$

where
  R_i and R_o are inside and outside radii of the cylindrical model
  L and h are the length and wall-thickness of the model

Herein, the LV volume (V), wall thickness (h), and myocardial volume (V_M) are obtained by cineventriculography. Using Equations 3.13 and 3.14, we can calculate the instantaneous radii $R_i(t)$ and length $L(t)$ (at any time instant t) in terms of the measured V_M, V, and h, as

$$R_i = \frac{2Vh/V_M + \sqrt{(2Vh/V_M)^2 + 4Vh^2/V_M}}{2}, \quad L = \frac{V}{\pi R_i^2} \quad (3.15)$$

Then

$$R_o = R_i + h \quad \text{and} \quad R_m = \frac{(R_o + R_i)}{2}$$

3.3.2 Myocardial Structural Unit (MSU) Model

In Figure 3.12a, the sarcolemma of the MSU is shown to consist of overlapping myosin and actin filaments. The myosin filament is symmetrical about

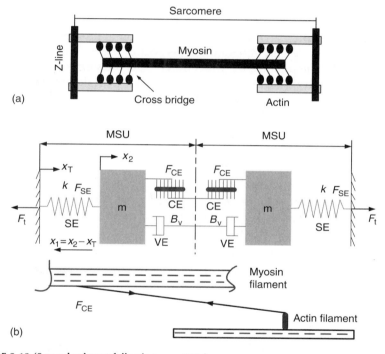

FIGURE 3.12 (See color insert following page 266.)
(a) The actin and myosin filaments constituting the contractile components of the myocardial fibril; (b) Myocardial fibril model composed of two symmetrical myocardial structural units (MSUs), which are mirror images of each other. Each MSU is composed of (i) an effective mass (m) that is accelerated; (ii) connective-tissue series element having parameter k (elastic modulus of the series element) and the force F_{SE}; (iii) the parallel viscous element of the sarcolemma having viscous damping parameter B_v and force F_{VE}; (iv) the contractile element (CE), which generates contractile force F_{CE} between the myosin (thick) and actin (thin) filaments. When the contractile element shortens (by amount x_2), the series element lengthens (i.e., x_1 increases). During ejection, the MSU x_T decreases, and during filling the MSU x_T increases. (Adopted from Ghista, D.N., Zhong, L., Eddie, Y.K.Ng., Lim, S.T., Tan, R.S., and Chua, T., *Mol. Cell. Biomech.*, 2, 217, 2005.)

its midpoint, and contains two sets of regular arrays of myosin heads. Muscle contraction is driven by the motor protein II, which binds transiently to an actin filament, generates a unitary filament displacement or "working stroke," then detaches and repeats the cycle [20]. Sarcomere shortening is generated by the relative sliding of the two filaments, driven by the working stroke of the myosin head. In Figure 3.12b, we define the myocardial fibril model to be composed of two myocardial structural units (MSUs) in series. On the basis of MSU three-element model [21] and Huxley cross bridge theory [22,23], the sarcomere actin–myosin filaments can be represented by the contractile element (CE), the connective tissue can be represented by the series-elastic element (SE), while the sarcolemma can be represented by a parallel viscous element (VE), as illustrated in Figure 3.12b.

Hence, the biomechanical model of the myocardial structural unit (MSU) consists of the MSU mass, a series-elastic element (SE), a parallel-viscous element (VE), and a contractile element (CE) [24]. The sarcomere represents the fundamental functional structure of contraction of the MSU. It makes the muscle fiber contract, and generates stress within the wall.

In Figure 3.12b, m denotes the MSU mass; B_v is the viscosity parameter; k is the connective tissue elasticity parameter; x_T is the displacement of the MSU relative to the center line; x_2 is the displacement of the MSU mass due to contraction and resulting shortening of its CE; x_1 is the displacement of SE $= (x_2 - x_T)$; F_{CE} denotes the force generated by the CE; F_{VE} denotes the force in the VE; and F_t denotes the resulting total MSU force, which is related to the chamber pressure of LV.

3.3.3 Determination of Fiber Density, Length, and Force; Fiber Angle α and MSU Force (F_t); Torque Produced on the LV due to Fiber Activation

It is known that the LV twists during systole and unwinds thereafter. This twist is due to the contraction of the myocardial fibers. We also acknowledge that the fiber angle will vary across the wall thickness and also with time during a cycle. Nevertheless, for convenience of theoretical analysis, we have adopted that in our LV model there are two adjacent sets of fibers within the wall thickness, one set oriented clockwise and another set oriented counter clockwise. Each myocardial fiber is assumed to be oriented helically within the LV myocardial wall, at a pitch angle α (as illustrated in Figure 3.11), with $N/2$ fibers are oriented helically clockwise at pitch angle α, and the other $N/2$ fibers are oriented anticlockwise at the same pitch angle, analogous to that adopted by Pietrabissa et al. [25]. We now determine this fiber angle for our LV model.

3.3.3.1 Fiber Density, Length, and Force

During filling, the fibers will extend as the LV cylindrical model fills with blood. During systole, the fibers will contract and shorten, deform and twist the LV cylindrical model. Thus, the LV will twist and unwind during a cardiac cycle.

In this cylindrical model, there are N number of myocardial fibers within the LV wall (as shown in figure). Hence:

$$\frac{N}{2} = \frac{A_{\text{cylinder}}}{2A_{\text{msu}}} \tag{3.16}$$

wherein

A_{cylinder} (the cross-section area of cylindrical model myocardium) $= \pi(R_o^2 - R_i^2)$

A_{msu} (the cross-section area of MSU) approximately equals 7.85×10^{-5} cm^2 [26]

While A_{cylinder} varies during a cycle, the number of fibers remains constant. Hence, we can determine the value of N at the start of isovolumic contraction, from Equation 3.16.

The activation of these fibers develops an active force (F_{CE}) in the sarcomere unit of MSU, which in turn generates wall stress and thereby raises the intraventricular pressure. When the pressure exceeds the pressure in the aorta, then the aortic valve opens, the LV shortens (and its wall thickens) to pump an appropriate stroke volume.

The instantaneous length (l_t) of each myocardial fiber (or myofiber) is given by

$$l_t = \frac{L_t}{\sin \alpha_t} \tag{3.17}$$

where l_t is the instantaneous length of the LV model.

For instantaneous LV pressure $P(t)$, the force in a myofiber is given by (with reference to Figure 3.11)

$$F_t = \frac{\pi R_i^2 P_t}{(N/2) \sin \alpha_t} \tag{3.18}$$

where R_i denotes the instantaneous value of the inner radius of the model obtained from Equations 3.13 and 3.14.

Because we have two MSUs in series along each myofiber, the axial displacement x_T of an MSU (shown in Figure 3.12) can be related to the change of length (Δl_t) of an MSU, and hence to the change in length (ΔL_t) of the LV cylinder model as

$$x_T = \frac{\Delta l_t}{2} = \frac{\Delta L_t}{(2 \sin \alpha_t)} \tag{3.19}$$

where

$\Delta L = L_{t+1} - L_t$, L_{t+1} and L_t refer to successive time instants

L_t is given by Equation 3.15

3.3.3.2 Determining the Fiber Pitch Angle α

We refer to the paper of Pietrabissa et al. [25], wherein it is shown that the fiber angle for a cylindrical model can be shown to be independent of the LV instantaneous dimensions, and hence can be assumed to be constant throughout the cycle. We now determine this fiber angle α.

At any instant, it is assumed that the depolarization wave is traveling along one set of myocardial fibers, i.e., either along $N/2$ clockwise or $N/2$ anticlockwise oriented fibers. The contraction of one set $(N/2)$ of these fibers hence results in a clockwise or anticlockwise twist of the LV. The distance (d) along a circumference between two adjacent fibers arranged in the same direction (i.e., clockwise or counter-clockwise) is given by (with reference to Figure 3.13):

$$d = \frac{2\pi R_i}{(N/2)} = \frac{4\pi R_i}{N} \tag{3.20}$$

The axial pitch (u) between the fibers arranged in the same direction intersected by a cylinder generator is given by

$$u = \frac{L}{(n/2)} = \frac{2L}{n} \tag{3.21}$$

wherein $n/2$ is the number of fibers arranged in the same direction intersected by a cylinder generator.

From Equations 3.20 and 3.21

$$\frac{u}{d} = \tan \alpha = \frac{LN}{2\pi R_i n} \tag{3.22}$$

In Figure 3.14a, the equilibrium of axial forces in one set of fibers arranged in the same direction, acting on the bottom or top circular plane surface of the LV model cylinder, requires that the sum of the vertical components of the fiber forces equilibrates the force due to LV pressure acting on the LV top (or bottom) surface. Hence, as indicated before (by Equation 3.18):

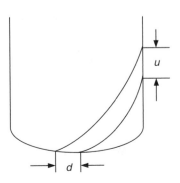

FIGURE 3.13
Calculation of distances u and d. (Adopted from Ghista, D.N., Zhong, L., Eddie, Y.K.Ng., Lim, S.T., Tan, R.S., and Chua, T., *Mol. Cell. Biomech.*, 2, 217, 2005.)

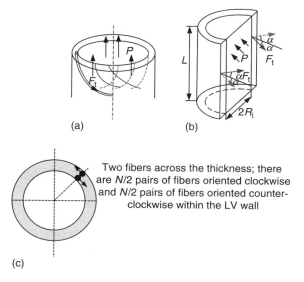

(a)

(b)

(c)

Two fibers across the thickness; there are *N*/2 pairs of fibers oriented clockwise and *N*/2 pairs of fibers oriented counterclockwise within the LV wall

FIGURE 3.14 (See color insert following page 266.)
(a) Equilibrium of fiber force and LV pressure on the top circular plane of the LV cylindrical model. (b) Equilibrium of fiber force and LV pressure in the circumferential direction. (c) Location of two sets of fibers across the LV wall thickness. (Adopted from Ghista, D.N., Zhong, L., Eddie, Y.K.Ng., Lim, S.T., Tan, R.S., and Chua, T., *Mol. Cell. Biomech.*, 2, 217, 2005.)

$$\pi R_i^2 P = (N/2)(F_t \sin \alpha) \tag{3.23}$$

where

P is the LV cavity pressure
F_t is the force within each of the N fibers

In Figure 3.14b, the force equilibrium of the cylinder in the circumferential direction, under the action of fiber forces arranged in the same direction, requires that:

$$(2LR_i)P = nF \cos \alpha \tag{3.24}$$

Upon substituting Equations 3.23 and 3.24 into Equation 3.22, we obtain the equivalent fiber angle (α) for the LV as follows:

$$\tan \alpha = \frac{1}{\sqrt{2}} \tag{3.25}$$

which yields $\alpha = 35.26°$.

3.3.3.3 Torque Imparted to the LV by Fiber Contraction

At this point, it is noteworthy that (on the basis of Figure 3.14a), while the vertical components of the fiber forces cause pressure on the top and bottom surfaces of the LV chamber (as per Equation 3.23), their horizontal components produce a torque (T) in the LV, given by

$$T_t = (N/2)F_t \cos \alpha_t = \frac{(N/2)\pi R_i^2 P_t \cos \alpha_t}{N/2 \sin \alpha_t} = \pi R_i^2 P_t ctg\alpha_t \tag{3.26}$$

This torque (T_t) will result in a twist of the LV by angle θ_t given by

$$\theta_t = \frac{T_t L_t}{JG} = \frac{\pi R_i^2 P_t L_t (ctg\alpha_t)}{JG} \tag{3.27}$$

where
  L_t is the instantaneous length of the LV cylindrical model
  α_t ($=\alpha$) is given by Equation 3.25
  J (the polar moment of inertia) $= \pi(R_o^4 - R_i^4)/2$
  G (the shear modulus of the LV myocardium) $\cong 100$ Gpa [27]

This means that for, say, a 60 mmHg pressure rise during isovolumic contraction, an LV model (having $R_i = 2$ cm, $R_o = 3$ cm, and $L = 14$ cm) will twist by an amount of 10°; then, it will twist more up to 20° to the instant when the LV pressure becomes maximum. After that, the LV will rewind. These calculated twist angles correspond to the monitored values [28], thereby lending some credibility to our model.

Equation 3.27 relates the twist angle (θ) to the fiber angle (α). It indicates that if we can measure the twist angle θ (of the apex of the LV with respect to its base) by MRI-tagging, then we can also determine the value of the fiber angle α corresponding to the monitored LV pressure. So, we do not need to adopt α to stay constant during a cardiac cycle. Hence, although in this chapter, we have taken α to be constant during a cardiac cycle, we can subsequently compute the instantaneous value of α from Equation 3.27. However, at this stage, we are in a position to only obtain data on LV pressure and volume and not simultaneously on the twist angle. It has been shown that this twist angle θ varies by about 10° during systole [28], which corresponds to the value obtained from Equation 3.27.

It can be conceptually noted that for certain instantaneous dimensions R_i and L_i, the in vivo value of θ during a cardiac cycle influences the value of P_t generated. However, for the sake of demonstrating how we can relate the sarcomere contractile force and shortening velocity to LV pressure and volume data (and compute these sarcomere parameters), we adopt the angle α to remain constant throughout the cycle (even though we concede that this is not true in practice).

3.3.4 Dynamics of a Myocardial Structural Unit

3.3.4.1 *Governing Equation of MSU Dynamics and Its Solution*

From Figure 3.15, the governing differential equation for an MSU dynamics, due to the generated contractile force (F_{CE}), can be expressed as

$$m\ddot{x}_2 + B_v \dot{x}_2 - F_{CE} + kx_1 = 0 \tag{3.28}$$

or

$$m\ddot{x}_1 + B_v \dot{x}_1 + kx_1 = F_{CE} - B_v \dot{x}_T - m\ddot{x}_T \tag{3.29}$$

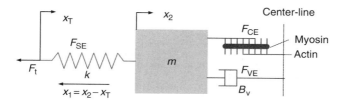

FIGURE 3.15 (See color insert following page 266.)
Dynamic model of MSU having effective mass m; k is the elastic modulus of series element; B_v is the viscous-damping parameter of parallel viscous element; F_t denotes the total generated force caused by the contractile stress F_{CE}; F_{SE} is the force in the series element $[= k(x_1 + x_{1ed})]$, where x_{1ed} is the deformation of the SE at end diastole; F_{VE} is the force in the viscous element $(= B_v \dot{x}_2)$; x_1 then represents the added deformation of the SE during systole (over and above its deformation during the filling phase) due to the development of F_{CE}. (Adopted from Ghista, D.N., Zhong, L., Eddie, Y.K.Ng., Lim, S.T., Tan, R.S., and Chua, T., *Mol. Cell. Biomech.*, 2, 217, 2005.)

where
   F_{CE} is the applied force exerted by the contractile-element of MSU
   m is the muscle mass per unit cross-section area $= \pi(R_o^2 - R_i^2)L\rho/2N$
   ρ is muscle density
   B_v is the viscous damping parameter of the parallel viscous element (VE)
   k is the elastic stiffness (or modulus) of the series-elastic element (SE)
   x_T is the shortening displacement of the myocardial-fiber unit relative to its center-line
   x_1 is the stretch of the SE element $= x_2 - x_T$
   x_2 is the displacement of muscle mass m (relative to center-line) due to CE contraction $= x_T + x_1$ (positive sign represents shortening)

$$F_{VE} = B_v \dot{x}_2 \quad \text{and} \quad F_{SE} = k(x_1 + x_{1ed}) \tag{3.30}$$

wherein
   x_{1ed} is x_1 at end diastole $(= F_{ted}/k)$
   F_{ted} is the fiber force at end diastole, obtainable from Equation 3.24 corresponding to R_i and P at end diastole

Because the terms $m\ddot{x}_1$ and $m\ddot{x}_T$ can be neglected due to their small values compared to other terms; for instance, $m\ddot{x}_1$ and $m\ddot{x}_T$ are of the order of 10^{0-1}, while the other terms are of the order of 10^{3-4} [24]. Equation 3.29 can thus be rewritten as

$$B_v \dot{x}_1 + kx_1 = F_{CE} - B_v \dot{x}_T \tag{3.31}$$

Now, let us consider myocardial contraction during the systolic phase. The systolic contraction can be considered to comprise of two temporal phases,

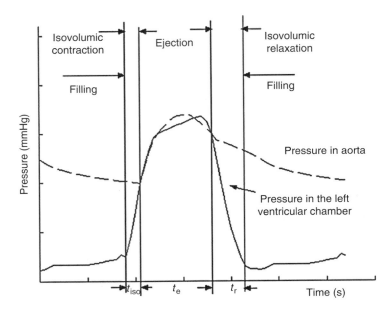

FIGURE 3.16
Schematic of LV pressure and aortic pressure variation during a cardiac cycle. (Adopted from Ghista, D.N., Zhong, L., Eddie, Y.K.Ng., Lim, S.T., Tan, R.S., and Chua, T., *Mol. Cell. Biomech.*, 2, 217, 2005.)

as depicted in Figure 3.16. Phase I, denoted by t_{iso} (and measured in seconds), corresponds to isovolumic contraction; it comprises the interval from the closing of the mitral valve until the opening of the aortic valve. Phase II, denoted by t_e, corresponds to the ejection phase.

As shown in Figure 3.12, each myocardial fiber from bottom to top edge of the LV myocardial model is composed of two MSUs. The governing differential equation for this model is given by Equations 3.28 and 3.29. Now let us discuss the terms on the right-hand side of Equation 3.31. As the MSU (and LV) depolarizes, excitation–contraction coupling leads to sarcomere contraction and the development of ventricular wall stress along with a rapid increase in intraventricular pressure, as shown in Figure 3.15. During this phase of systolic contraction, we express the generated MSU-CE force (F_{CE}) function (analogous to the LV pressure wave shape) as

$$F_{CE} = F_{CE0}\sin(\omega_{ce}t)e^{-z_{ce}t} \qquad (3.32)$$

where
$\omega_{ce} = \pi/t_s$; t_s is the contraction duration, to be determined
F_{CE0} and z_{ce} are the additional parameters, to be determined
$t = 0$ corresponds to the start of isovolumic contraction phase

It should be noted that this expression for F_{CE} is similar to that for the active elastance of our earlier paper [29].

Let us now discuss the x_T term on the right hand side of Equation 3.31. During the filling phase, the MSU will stretch passively due to LV enlargement. Concerning the x_T term (depicted in Figures 3.12 and 3.15), during the filling phase, it will be negative and its absolute value will increase due to passive stretching of the myocardial fibers caused by LV volume increase. During this phase, with reference to Figure 3.15, the SE element will stretch while $x_2 = 0$, and hence $x_1 = -x_T$. At the end of filling phase, we denote x_1 by x_{1ed}. Further increase in x_1 now occurs during isovolumic contraction due to the development of F_{CE} and the generation of CE shortening (x_2). However, in this phase $x_T = 0$, and hence x_1^{iso} is only due to x_2 caused by F_{CE}.

During the ejection phase, x_T is positive and is caused by LV ejection and volume decrease. At the same time, x_2 is being generated by CE contraction, resulting in F_{CE} development. Somewhere during the ejection phase, x_2 will reach its maximum value and thereafter decrease. Now during the isovolumic relaxation phase, x_2 keeps decreasing, while x_T does not change from its end-ejection value.

Then when the filling phase starts, x_T again becomes negative and $|x_T|$ starts increasing as the LV volume increases. Meanwhile x_2 keeps decreasing and reaches a zero value, a short while after the start of filling phase at $t = t_o$. This time period t_o is designated as the LV suction phase caused by deceasing F_{CE}, before the left atrium starts to contract and pump blood with the LV. Herein, we will also demonstrate this suction effect in terms of the time period t_o.

3.3.4.2 Phase I: Solving Equation 3.31 for Isovolumic Contraction Phase (during $0 < t < t^{iso}$)

Since both the mitral and aortic valves are closed, the volume of blood in the ventricle is constant. Yet the pressure inside LV is increasing due to the sarcomere contraction, i.e., due to F_{CE} generation. Hence, putting $x_T = \dot{x}_T = \ddot{x}_T = 0$, and employing F_{CE} from Equation 3.32, we can rewrite Equation 3.31 as

$$B_v \dot{x}_1 + k x_1 = F_{CE0} \sin(\omega_{ce} t) e^{-z_{ce} t} \qquad (3.33)$$

The solution of Equation 3.33 is given by $x_1 (= x_1^{iso})$, as follows:

$$x_1(t) = x_1^{iso}(t) = C_1 e^{-k/B_v t} + [a \sin(\omega_{ce} t) + b \cos(\omega_{ce} t)] e^{-z_{ce} t} \qquad (3.34)$$

where

$$a = \frac{F_{CE0}(k - z_{ce} B_v)}{(k - z_{ce} B_v)^2 + (B_v \omega_{ce})^2}, \quad b = -\frac{F_{CE0} B_v \omega_{ce}}{(k - z_{ce} B_v)^2 + (B_v \omega_{ce})^2}$$

For this phase of contraction, the initial condition that we will impose is

$$x_1^{iso}(0) = C_1 + b = 0 \qquad (3.35)$$

from which

$$C_1 = -b \tag{3.36}$$

Hence, $x_1 (= x_1^{iso})$, during the isovolumic contraction phase, is given by

$$x_1(t) = x_1^{iso}(t) = (-b)e^{-k/B_v t} + [a \sin(\omega_{ce} t) + b \cos(\omega_{ce} t)]e^{-z_{ce} t} \tag{3.37}$$

3.3.4.3 *Phase II: Expression for x_T and Solving Equation 3.19 for the Ejection Phase to Determine Parameters x_{T0} and z_e*

For mathematical convenience, we make a shift in the time variable and redefine it as $t_a = t - t_{iso}$, such that

$$0 \le t_a \le t_e \tag{3.38}$$

where t_e is the ejection phase duration.

In this phase, x_T is no longer zero, and hence we need to relate it to the LV dimensional change, as per Equation 3.19, as

$$x_T = \frac{\Delta l_t}{2} = \frac{(L_{(t+1)} - L_t)}{2 \sin \alpha} \tag{3.39}$$

wherein

L_t and L_{t+1} refer to successive time instants t_i and t_{i+1}
L_i (or L) is given by Equation 3.15 in terms of V, MV, and h

We now adopt for $\dot{x}_T$ a function to correspond to that of the LV flow rate $\dot{V}$ [30], as follows:

$$\dot{x}_T = x_{T0} \sin(\omega_e t_a)e^{-z_e t_a} \tag{3.40}$$

where

$\omega_e = \pi/t_e$, t_e is the duration of ejection as shown in Figure 3.19
x_{T0} and z_e are the (to-be-determined) parameters
$t_a = 0$ corresponds to the start-of-ejection phase

By integrating Equation 3.40, and employing the initial condition of $x_T (t_a = 0) = 0$, we get

$$x_T = -\frac{x_{T0}}{z_e^2 + \omega_e^2} [z_e \sin(\omega_e t_a) + \omega_e \cos(\omega_e t_a)]e^{-z_e t_a} + \frac{x_{T0}\omega_e}{z_e^2 + \omega_e^2} = \frac{\Delta l_t}{2}$$

$$= \frac{(L_{i+1} - L_i)}{2 \sin \alpha} \tag{3.41}$$

Now, based on Equation 3.41, x_T can be evaluated in terms of Δl_t and hence in terms of monitored LV wall thickness $h(t)$, LV volume $V(t)$, and myocardial volume MV. Hence, the parameters x_{T0} and z_e (of $\dot{x}_T$ in Equation 3.40) can be obtained by matching the x_T expression of Equation 3.41 with the clinically obtained MSU length-change $\Delta l_t/2$, as indicated by Equation 3.41. This then enables us to also determine the expression $\dot{x}_T$ (Equation 3.40) in terms of its now evaluated parameters x_{T0} and z_e.

Then, by substituting Equations 3.32 and 3.40 into the governing Equation 3.31, we have

$$B_v \dot{x}_1 + k x_1 = F_{CE0} \sin[\omega_{ce}(t_a + t_{iso})]e^{-z_{ce}(t_a + t_{iso})} - B_v x_{T0} \sin(\omega_e t_a)e^{-z_e t_a} \quad (3.42)$$

where t_a is the time variable. The solution of Equation 3.42 is given by $x_1(=x_1^e)$, as follows:

$$x_1(t_a) = x_1^e(t_a) = C_2 e^{-k/B_v t_a} + \{a \sin[\omega_{ce}(t_a + t_{iso})] + b \cos[\omega_{ce}(t + t_{iso})]\}e^{-z_{ce}(t_a + t_{iso})}$$
$$+ [c \sin(\omega_e t_a) + d \cos(\omega_e t_a)]e^{-z_e t_a} \quad (3.43)$$

where

$$a = \frac{F_{CE0}(k - z_{ce}B_v)}{(k - z_{ce}B_v)^2 + (B_v \omega_{ce})^2}, \quad b = -\frac{F_{CE0}B_v \omega_{ce}}{(k - z_{ce}B_v)^2 + (B_v \omega_{ce})^2}$$
$$c = -\frac{B_v x_{T0}(k - B_v z_e)}{(k - B_v z_e)^2 + (B_v \omega_e)^2}, \quad d = \frac{B_v^2 x_{T0}\omega_e}{(k - B_v z_e)^2 + (B_v \omega_e)^2}$$

In Equation 3.43, the unknown parameters are k, B_v, F_{CE0}, ω_{ce}, and z_{ce}.

Now $x_1(t)$ between phases I and II is continuous, i.e., $x_1^e(t_a = 0) = x_1^{iso}(t = t_{iso})$. This determines the initial condition for phase II. Hence, from Equations 3.37 and 3.43, we get

$$x_1^e(0) = C_2 + [a \sin(\omega_{ce}t_{iso}) + b \cos(\omega_{ce}t_{iso})]e^{-z_{ce}t_{iso}} + d$$
$$= x_1^{iso}(t = t_{iso})$$
$$= -be^{-k/B_v t_{iso}} + [a \sin(\omega_{ce}t_{iso}) + b \cos(\omega_{ce}t_{iso})]e^{-z_{ce}t_{iso}} \quad (3.44)$$

Solving Equation 3.44, we get

$$C_2 = -be^{-k/B_v t_{iso}} - d \quad (3.45)$$

Hence, the total SE deformation $x_1(=x_1^e)$ during the ejection phase (on top of x_1 at t^{iso} the end of isovolumic contraction, given by Equation 3.37) can be written as

$$x_1(t_a) = x_1^e(t_a) = (-be^{-k/B_v t_{iso}} - d)e^{-k/B_v t_a}$$
$$+ \{a \sin[\omega_{ce}(t_a + t_{iso})] + b \cos[\omega_{ce}(t_a + t_{iso})]\}e^{-z_{ce}(t_a + t_{iso})}$$
$$+ [c \sin(\omega_e t_a) + d \cos(\omega_e t_a)]e^{-z_e t_a} \quad (3.46)$$

3.3.4.4   Evaluating the Model Parameters (k, B_v, F_{CE0}, ω_{ce}, z_{ce})

Having determined the parameters x_{T0} and z_e from Equation 3.41, by matching x_T with Δl, we will now determine the remaining parameters k, B_v, F_{CE0}, ω_{ce}, and z_{ce} (in Equation 3.33). On the basis of Figures 3.12 and 3.14 and Equations 3.30 and 3.31, we put down

$$F_{SE} = F_t = k(\text{total SE deformation}) = k(x_{led} + x_1^e) = F_{ted} + kx_1^e \qquad (3.47)$$

where (1) x_1^e during the ejection phase is given by Equation 3.46, (2) x_{led} (x_1 at end diastole) is given by Equation 3.30, and (3) F_t and F_{ted} are obtained in terms of LV pressure, model geometry, fiber angle (α) and N from Equation 3.24. Hence,

$$k \cdot x_1^e = F_t - F_{ted} = \frac{2\pi R_i^2 (P - P_{ed})}{(N \sin \alpha)} \qquad (3.48)$$

wherein the x_1^e expression is given by Equation 3.46. We now match the expression for kx_1^e (of Equation 3.46) with the evaluated value of the right-hand side term of Equation 3.48 in terms of clinical-derived data (of LV pressure, as well as R_i and N from Equations 3.15 and 3.16) of the subject. By carrying out parameter-identification, we can determine the corresponding parameters k as well as B_v, F_{CE0}, ω_{ce}, and z_{ce} (in Equation 3.46). Once we know the values of these parameters, we can determine the values of the x_1^e during the ejection phase.

3.3.5   Sarcomere Force (F_{CE}), Shortening Velocity ($\dot{x}_2$), and Power

3.3.5.1   Determining Sarcomere Contractile F_{CE} and x_2, and Their Physiological Implications

Having evaluated the parameters (k, B_v, F_{CE0}, ω_{ce}, z_{ce}) in the earlier section, we can determine the CE contractile force F_{CE} from Equation 3.32, as well as x_1 during isovolumic contraction (from Equation 3.37) and during ejection (from Equation 3.46). The shortening displacement of CE, x_2 ($= x_1 + x_T$), can also be computed by employing (1) x_1 from Equation 3.46 in terms of its evaluated parameters k, B_v, F_{CE0}, ω_{ce}, and z_{ce}, and (2) x_T from Equation 3.41 in terms of its evaluated parameters x_{T0} and z_e. Now, for a total variation of x_2 during a cardiac cycle, we adopt the x_2 expression as

$$x_2 = x_{20} \sin(\omega_{ce}t)e^{-z_2 t} = x_1 + x_T \qquad (3.49)$$

where
  x_1 can be computed from Equation 3.46
  x_T from Equation 3.41
  $t = 0$ corresponds to the start of isovolumic contraction phase

Hence, in Equation 3.49, we can now evaluate the additional parameters x_{20} and z_2 by parameter-identification.

Myocardial fiber shortening x_2 is an important cardiac performance parameter. On the basis of Equation 3.49, it starts at the end of LV filling and initiates LV contraction. Somewhere during the ejection phase, it reaches its maximum value. It then starts decreasing and continues to do so into the filling phase, causing the phenomenon of LV suction. It would be interesting to determine the instant (t_o) within the filling-phase, when x_2 becomes zero. From a cardiac physiological viewpoint, during this time-interval (from the start of filling up to the instant t_o), the left ventricular pressure value will be below its value at the start of filling phase. This time period from the start of filling phase up to t_o is denoted as the LV suction phase [31].

Hitherto, it has been difficult to provide an explanation for this suction phase. However, it can be explained in terms of the continuing activation of the contractile element into the filling phase from Equations 3.32, 3.48, and 3.49, as follows:

$$\frac{2\pi R_i^2(P_t - P_{ed})}{N \sin \alpha} = F_t - F_{ted} = F_{SE} - F_{ted} = F_{CE} + F_{VE} - F_{ted}$$

$$= F_{CE0} \sin(\omega_{ce}t)e^{-z_{ce}t} + B_v[x_{20} \sin(\omega_{ce}t)]e^{-z_2t} - F_{ted} \quad (3.50)$$

In Equation 3.50, we have determined the parameters of F_{CE} and F_{VE} from the monitored instantaneous LV pressure (P_t), and the LV geometry (defined by R_o and R_i). As per Equation 3.50, it is F_{CE} (due to sarcomere contraction) that intrinsically governs the generation of this pressure P_t. Hence, it is the continuing activation of CE into the filling phase that causes LV suction and a temporal dip in LV pressure before the left atrium contracts and pumps blood into the LV. Later on, we will demonstrate the suction effect in terms of the time instant t_0 in the filling phase.

3.3.5.2 Power Generated by the Sarcomere Contractile-Element

Now, because we have incorporated 2 MSU(s) in each myocardial fiber (as illustrated in Figures 3.12), we now define the LV myocardial sarcomere instantaneous power (MSP) in terms of the MSU-CE force causing shortening by amount x_2 and shortening rate of $\dot{x}_2$, as

$$\text{MSP} = 2 \times \left(\frac{N}{2}\right)(F_{CE} \times \dot{x}_2) = N(F_{CE} \times \dot{x}_2) \quad (3.51)$$

where
both F_{CE} and $\dot{x}_2$ are functions of time
F_{CE} is the contractile force generated by each contractile element
$\dot{x}_2$ is the shortening velocity of the CE element

In this equation, N is computed from Equation 3.16; F_{CE} is computed from Equation 3.32, with its parameters (F_{CE0}, ω_{ce}, and z_{ce}) obtained from

Equation 3.48 by parameter-identification scheme; $x_2 = x_1 + x_T$, with x_1 and x_T computed from Equations 3.46 and 3.41, respectively. The total myocardial sarcomere power (TMSP) is then obtained as

$$\text{TMSP} = N \int F_{CE}\, d\dot{x}_2 \tag{3.52}$$

i.e., by the area under the F_{CE} versus $\dot{x}_2$ curve.

3.3.5.3 Defining a Contractile Power Index

Herein, in quantifying the contractile performance of the LV, we define contractility (corresponding to the traditional contractility index of $(dP/dt)_{max}$) as the ability of the LV myocardium to produce a contractile force with a high shortening-velocity capability, so as to exert maximum contractile power. In order to compare power among patients of differing LV size and mass, we normalize MSP with respect to myocardial volume, as

$$\text{MSPI} = \text{MSP}_{max}/V_M \tag{3.53}$$

where V_M is LV wall volume (mL), MSPI is in Watt/liters.

3.3.6 Clinical Application and Results

The analysis is now applied to the clinically obtained data of the subject's left ventricular (instant-of-instant) dimensions (obtained by cineangiocardiograph) and chamber pressure (obtained by cardiac catheterization). In so doing, for each subject's left ventricular data (of pressure, volume, wall thickness, and myocardial volume), we evaluate the model parameters F_{CE} and x_2, the contractile power input and the new contractility index MSPI (given by Equation 3.53).

Table 3.5 lists the measured hemodynamic variables for three subjects (subject HEL, DDM, and TPS). Subject HEL serves as a representative of a

TABLE 3.5

Clinical History, Measured Hemodynamic Data from Subjects (HEL, DDM, and TPS)

Subject	HEL	DDM	TPS
Disease	MI, DVD	DVD, HTN	LAD and ischemia
LVP (mmHg)	122/18	170/24	147/22
AOP (mmHg)	125/75	169/99	140/71
EDV/ESV (ml)	132.5/84.3	121.7/41.3	112/35.5
EF	0.36	0.66	0.68

Note: LVP, left ventricle chamber pressure; AOP, aortic pressure; EDV, end-diastolic volume; ESV, end-systolic volume; EF, ejection fraction; MI, myocardial infarct; DVD, double vessel disease; HTN, hypertension; LAD, left artery disease.

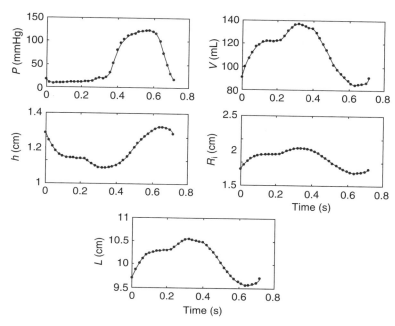

FIGURE 3.17

LV Pressure variation and LV (thick-wall cylinder) model dimensions variations, during a cardiac cycle for subject HEL. $V_M = 185$ mL. (Adopted from Ghista, D.N., Zhong, L., Eddie, Y.K.Ng., Lim, S.T., Tan, R.S., and Chua, T., *Mol. Cell. Biomech.*, 2, 217, 2005.)

patient with myocardial infarct. Subject DDM is an example of a patient with double vessel disease (DVD) and hypertension, treated with PTCA. Subject TPS corresponds to a patient with native LAD, ischemia in anterior territory, and mitral regurgitation (MR). These three subjects have also been studied by our earlier analysis of passive and active elastance computation [29]. Figure 3.17 depicts one-sample cineangiocardiographically derived LV dimensions and the derived cylindrical model dimensions during a cardiac cycle.

3.3.6.1 Evaluation of the Model Parameters

From the clinical data shown in Figure 3.17, we calculate the LV model x_T, using Equation 3.39. This "x_T versus time" function during ejection is shown in Figure 3.18, as illustrated by the round points. We then use the expression of x_T given by Equation 3.41 to fit this clinical-derived data of "x_T versus t", and determine the parameters x_{T0} and z_e, as shown in Figure 3.18. The model-computed x_T matches the x_T ($= (L_{i+1} - L_i)/2 \sin \alpha$) clinical data very well, with R-square $= 0.9944$ and RMS $= 0.02$ cm. The solid line is the model-computed displacement x_T (Equation 3.41), while the round points constitute the clinical-derived x_T.

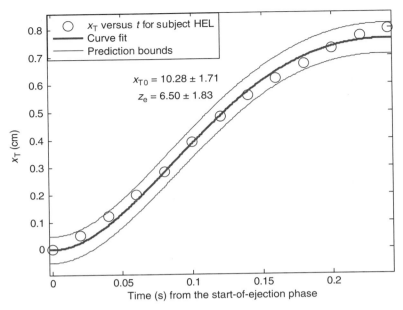

FIGURE 3.18 (See color insert following page 266.)
Computed $x_T(t)$ during the ejection phase ($t = 0$ corresponds to start-of-ejection): From the data shown in Figure 3.17, we calculate the model x_T during the ejection phase by using Equation 3.39, as shown by the round points in the figure. This data is now fitted with Equation 3.41. The resulting values of the parameters (x_{T0} and z_e) are shown in the figure and also listed in Table 3.4. Here $t = 0$ corresponds to the start-of-ejection. (Adopted from Ghista, D.N., Zhong, L., Eddie, Y.K.Ng., Lim, S.T., Tan, R.S., and Chua, T., *Mol. Cell. Biomech.*, 2, 217, 2005.)

Now, we use the LV pressure and R_i data in Figure 3.17, along with calculating N (by Equation 3.16), to obtain the right-hand side of the Equation 3.48, and to hence evaluate the term kx_1^e. Since the expression for x_1^e is given by Equation 3.46, we can now employ the parameter-identification scheme to make the kx_1^e expression (Equation 3.46) fit the values of $kx_1^e (= 2\pi R_i^2 (P - P_{ed})/N \sin \alpha)$, and compute the other parameters k, B_v, F_{CE0}, ω_{ce}, and z_{ce}, in Equation 3.50 (as listed in Table 3.6). In Figure 3.19, we have shown how the kx_1^e expression (Equation 3.46) matches the computed values of kx_1^e, to evaluate the parameters k, B_v, σ_{CE0}, ω_{ce}, and z_{ce}.

From the data shown in Figure 3.17, we calculate the LV model myocardial force F_t and F_{ted} using Equation 3.22. Then we compute $kx_1^e (= (F_t - F_{ted})/N) = 2\pi R_i^2 (P - P_{ed})/N \sin \alpha$, as shown by the round points in this figure, with $N = 2.24 \times 10^5$ from Equation 3.16. This clinical-derived data of $(F_t - F_{ted})/N$ is now fitted with kx_1^e expression (based on Equation 3.46), to obtain the values of kx_1^e parameters (F_{CE0}, z_{ce}, k, B_v, ω_{ce}) listed in the figure as well as in Table 3.4. Herein, $t = 0$ corresponds to the start-of-ejection.

TABLE 3.6

Computed Values of MSU Terms x_T and x_1, and of Their Parameters Related to the Subject HEL Shown in Figure 3.20, during Ejection Phase ($N = 2.24 \times 10^5$)

Variable	Parameters	Values	How Obtained	RMS
x_T	x_{T0} (cm)	10.28 ± 1.71	x_T fit using Equation 3.29	0.02 cm
	z_e (s^{-1})	6.50 ± 1.83		R-square
				0.99
x_1^e	F_{CE0} (Pa cm^2)	5.66 ± 1.61	kx_1^e fit using Equation 3.36	RMS
				0.028 Pa cm^2
	z_{ce} (s^{-1})	3.95 ± 1.06		R-square
	B_v (Pa cm s)	0.12 ± 0.069		
	k (Pa cm)	3.95 ± 1.28		0.97
	ω_{ce} (s^{-1})	7.14 ± 1.95		
x_2	x_{20}	0.55 ± 0.01	x_2 fit using Equation 3.37	0.01
	z_2	-3.03 ± 0.12		0.99

3.3.6.2 Determination of CE Force F_{CE} and Shortening x_2 Characteristics, with Determination of the LV Suction Effect

Shown in Figure 3.20 are the computed values of MSU dynamics terms for subject HEL. Figure 3.20a provides the measured data of LV pressure in one cardiac cycle. By means of the values of the parameters (k, B_v, σ_{CE0}, ω_{ce}, and z_{ce}) in Table 3.4, we have determined and plotted x_1 versus time, x_2 versus

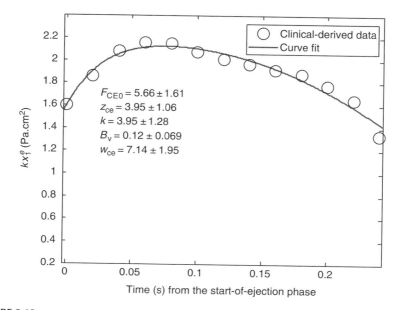

FIGURE 3.19
Computed kx_1^e and its parameters during the ejection phases ($t = 0$ corresponds to start-of-ejection). (Adopted from Ghista, D.N., Zhong, L., Eddie, Y.K.Ng., Lim, S.T., Tan, R.S., and Chua, T., *Mol. Cell. Biomech.*, 2, 217, 2005.)

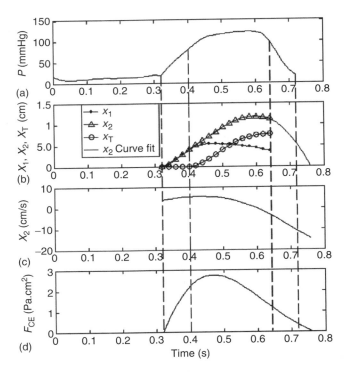

FIGURE 3.20 (See color insert following page 266.)
Computed results of MSU model-dynamics terms x_1, x_2, x_T, $\dot{x}_2$, and F_{CE}, for subject HEL. Diastolic phase: 0–0.32 s; isovolumic contraction phase: 0.32–0.4 s; ejection phase: 0.4–0.64 s; isovolumic relaxation phase: 0.64–0.72 s. Here $t = 0$ corresponds to the start-of-filling. Note that F_{CE} and x_2 extend into the filling phase; $t_0 = 0.04$ s. (Adopted from Ghista, D.N., Zhong, L., Eddie, Y.K.Ng., Lim, S.T., Tan, R.S., and Chua, T., *Mol. Cell. Biomech.*, 2, 217, 2005.)

time, and x_T versus time, in Figure 3.20b. The computed CE shortening-velocity ($\dot{x}_2$) and force (F_{CE}) are shown in Figures 3.20c and 3.23d, respectively. Notice that the CE force variation during systole is similar to that of LV active elastance in our earlier paper [29].

Now we adopted the expression x_2 given by Equation 3.49 in order to project the time-duration (t_0) of sarcomere shortening continuing into the filling phase. We can now see, from Figure 3.20b, that this duration is 0.04 s. This validation and quantification of the LV suction effect is an important added finding of our model analysis.

3.3.6.3 Computing TMSP and MSPI (Equations 3.52 and 3.53)

Next, we also plot the "force versus shortening" and the "force versus shortening velocity" for the CE after the initiation of isovolumic contraction phase, as shown in Figures 3.21 and 3.22. As seen in Figure 3.21 for patient HEL, the CE shortening (x_2) reaches its maximum value late in the ejection

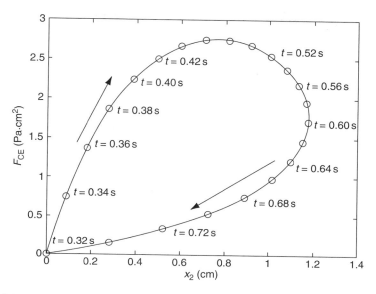

FIGURE 3.21

CE Force (F_{CE}) versus displacement (x_2) relationship for subject HEL. The arrow direction indicates progression of time, starting from the diastolic-filling phase. Here $t = 0.32$ s corresponds to the end-of-filling, the time $t = 0.32$ s corresponds to the start of isovolumic-contraction phase. The CE shortening (x_2) becomes zero at $t = 0.76$ s, about 0.04 s into the filling phase. (Adopted from Ghista, D.N., Zhong, L., Eddie, Y.K.Ng., Lim, S.T., Tan, R.S., and Chua, T., *Mol. Cell. Biomech.*, 2, 217, 2005.)

phase. The area encircled by force-displacement curve and x-axis represents the CE energy input.

In Figure 3.22 for patient HEL, the CE shortening velocity increases, along with increasing CE force. They both reach their maximum values at about one-third ejection, and thereafter decrease. The area encircled by the curve, multiplied by the number of fibers (N) gives us the value of the index TMSP. From this figure, we can again note that the contractile element stays active for 0.04 ($= 0.76 - 0.72$) s into the filling phase. This causes LV suction of blood, even prior to the initiation of left atrial contraction. From Figure 3.22, we calculate TMSP to be 5.40 W. The maximum value of instantaneous power, MSP_{max} (Equation 3.51), is computed to be 3.32 W. Using this value, we now calculate the contractility index MSPI (Equation 3.53) to be 17.94 W/L.

3.3.7 Discussion: Comparison of CE Performance Characteristics for Three Patients, and Correlation of MSPI with $(dP/dt)_{max}$

3.3.7.1 *Computation of CE Performance Characteristics for Other Subjects*

This analysis is now carried out for two other subjects (DDM and TPS) listed in Table 3.5, and the results are provided in Table 3.5. For these subjects, the

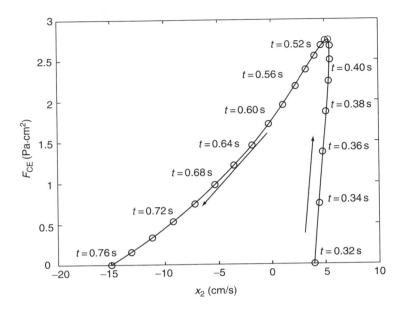

FIGURE 3.22
CE force (F_{CE}) versus shortening-velocity ($\dot{x}_2$) relationship for subject HEL. The arrow direction indicates progression of time, starts from diastolic filling phase. Here $t = 0.32$ s corresponds to the end-of-filling, the time $t = 0.32$ s corresponds to the start of isovolumic-contraction phase. The next-cyclic filling phase starts at $t = 0.72$ s, while F_{CE} becomes zero at $t = 0.76$ s. In other words, the LV suction effect lasts for about 0.04 s into the filling phase. If we observe the LV pressure variation in Figure 3.21a, we can note that the LV pressure in fact decreases after initiation of filling phase and recovers to the level of the start of filling phase after about 0.04 s. To determine the LV power-input, we determine the area under this curve and multiply it by N. This gives LV TMSP = 5.40 W. (Adopted from Ghista, D.N., Zhong, L., Eddie, Y.K.Ng., Lim, S.T., Tan, R.S., and Chua, T., *Mol. Cell. Biomech.*, 2, 217, 2005.)

TMSPs are 5.18 W and 5.48 W. Figures 3.23 and 3.24 depict the computed "CE force versus shortening" and "CE force versus shortening-velocity" characteristics for subjects HEL (with MI, DVD), DDM (DVD, HTN), and TPS (LAD, MR, ischemia). Figure 3.23 shows the CE force–shortening for these three subjects with different heart diseases. The area encircled by the curve and the x-axis indicates the amount of energy generated by the CE.

In Figure 3.24 it is seen that the CE force–shortening-velocity curve follows the same trend for all the subjects. The CE force and shortening velocity both reach their maximal values at about one-third ejection. However, the loop made by HEL has the least area encircled within it, and correspondingly has the least contractile power input of the three subjects (as seen in Table 3.7).

3.3.7.2 Computation of MSPI, in Comparison with $(dP/dt)_{max}$

Finally, we compute the traditional indices of contractility (EF and $(dP/dt)_{max}$), and compare them with our proposed contractility index

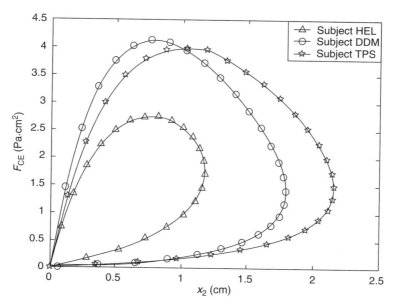

FIGURE 3.23

CE Force (F_{CE})–displacement (x_2) for subjects HEL, DDM, and TPS. Of the three subjects, TPS has the biggest area encircled by the F_{CE} versus x_2 curve; hence, this subject's CE is generating higher energy relative to the other two subjects. (Adopted from Ghista, D.N., Zhong, L., Eddie, Y.K.Ng., Lim, S.T., Tan, R.S., and Chua, T., *Mol. Cell. Biomech.*, 2, 217, 2005.)

MSPI. Figures 3.25 and 3.26 show the correlations between MSPI and EF, and between MSPI and $(dP/dt)_{max}$. The respective correlations are as follows: $MSPI = 55EF - 2.4$, $(r = .8905)$; $MSPI = 0.04\ (dP/dt)_{max} - 22$, $(r = .9054)$. These good correlations hence add credence to our newly formulated contractility index.

3.3.8 Highlights

We have analyzed the LV systolic performance by means of an LV mechatronic cylindrical model of myocardial fiber located within the LV model wall. The myocardial fibers are helically oriented within the LV model wall. Each myocardial fiber sarcomere unit is composed of three elements: series element (analogue to connective tissue), viscous element (analogue to sarcolemma), and contractile element (analogue to sarcomere). The sarcomere contraction is associated with the relative sliding of the actin–myosin filaments.

The contractile force F_{CE} and shortening x_2 of the LV myocardial-sarcomere unit are related to the LV pressure and volume data, and evaluated in terms of the model's parameters (k, B_v). After that, we determine the in vivo characteristics of the LV sarcomere (CE), in terms of "F_{CE} versus x_2" and "F_{CE} versus $\dot{x}_2$", as well as the power generated by the sarcomere (CE). Both

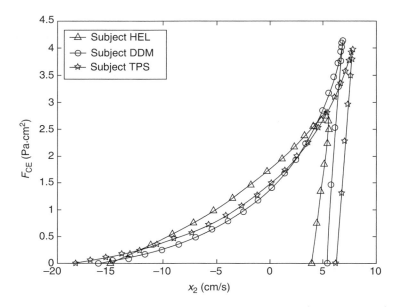

FIGURE 3.24 (See color insert following page 266.)
CE force (F_{CE})–velocity ($\dot{x}_2$) relationships for subjects HEL, DDM, and TPS. Of the three subjects, the subject TPS has the biggest area encircled within the F_{CE} versus $\dot{x}_2$ curve, and hence has the bigger contractile power input. (Adopted from Ghista, D.N., Zhong, L., Eddie, Y.K.Ng., Lim, S.T., Tan, R.S., and Chua, T., *Mol. Cell. Biomech.*, 2, 217, 2005.)

F_{CE} versus $\dot{x}_2$ characteristics and the total myocardial sarcomere power input (TMSP) can be regarded as important LV functional indices.

Our evaluated in vivo CE force versus shortening and CE force versus shortening-velocity characteristics are seen to depict LV contractile

TABLE 3.7

Clinical History: $(dP/dt)_{max}$, Maximal Contractile Force, F_{CE}, Shortening Velocity $\dot{x}_2$ of CE, Area under F_{CE} versus x_2, Calculated Maximum Power (Power$_{max}$), TCPI, Myocardial Volume (MV), and Left Ventricular Contractility Index (MSPI) from Subjects (HEL, DDM, and TPS)

Subject	HEL	DDM	TPS
Disease	MI, DVD	DVD, HTN	LAD, ischemia
EF	0.36	0.66	0.68
$(dP/dt)_{max}$ (mmHg/s)	984	1475	1478
MV (mL)	185	138	140
Maximum F_{CE} (Pa cm^2)	2.74	4.12	3.98
Maximal shortening velocity $\dot{x}_2$ (cm/s)	5.55	6.84	7.82
TCPI (W)	5.40	5.97	6.33
Power$_{max}$ (W)	3.32	5.18	5.48
MSPI (W/L)	17.94	37.53	39.14

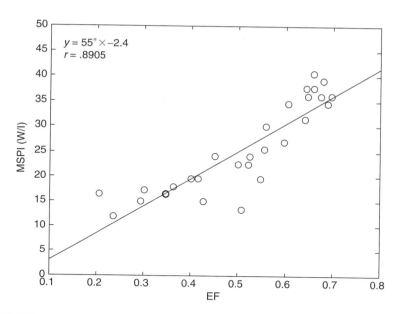

FIGURE 3.25
Correlation of the myocardial sarcomeric power index (MSPI) to EF, the correlation coefficient $r = .8905$.

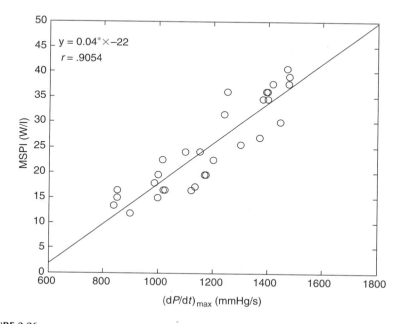

FIGURE 3.26
Correlation of the myocardial sarcomeric power index MSPI to the traditional contractility index $(dP/dt)_{max}$, with the correlation coefficient $r = .9054$.

function features. Less area encircled within the force-shortening velocity curve is associated with less contractility; this indicates that an LV with impaired contractility is not able to generate as much power required to provide adequate EF and stroke volume as a properly-contracting LV.

Subject HEL has myocardial infarct, and hence has a weaker contracting myocardium. This is manifested by a lower CE maximal force and shortening velocity, in comparison with subjects DDM and TPS (shown in Table 3.5). Correspondingly, its values of maximum power generated by CE and the contractility index (MSPI) are lower than for the other two subjects. Also, the area of CE force-displacement curve for subject HEL is significantly less compared with the other two subjects. These results quantify how myocardial infarct impairs the left ventricular performance in terms of our model's contractile power generated and contractility indices. Subject TPS (with myocardial ischemia) has the maximal area encircled within its F_{CE} and $\dot{x}_2$ curve. This could reflect an adaptive mechanism attempting to restore the LV performance, which is in agreement with its ejection fraction value (EF = 0.68).

Table 3.7 summarizes all of these results. Figure 3.26 enables us to compare our MSPI with the traditional $(dP/dt)_{max}$. On the basis of case studies for 30 subjects, our new power index MSPI correlates well with the traditional contraction index $(dP/dt)_{max}$, and hence may merit clinical employment.

References

1. Tischler MD, Niggel J, Norowski DT, and LeWinter MM, Relationship between left ventricular shape and exercise capacity in patients with left ventricular dysfunction, *Journal of the Amercian College of Cardiology*, 22:751–757, 1993.

2. Devereus RB, Left ventricular geometry, pathophysiology and prognosis, *Journal of the American College of Cardiology*, 25:885–887, 1995.

3. Krumholz HM, Larson M, and Levy D, Prognosis of left ventricular geometric patterns in the Framingham Heart Study, *Journal of the American College of Cardiology*, 25:879–884, 1995.

4. Juznic SCJE, Juznic G, and Knap B, Ventricular shape: Spherical or cylindrical? In: *Analysis and Assessment of Cardiovascular Function*, Drzewiecki G and Li JK-J (Eds.), New York: Springer-Verlag, pp. 156–171, 1998.

5. Knap B, Juznic G, Bren AF, Drzewiecki G, and Noordergraaf A, Elongation as a new shape index for the left ventricle, *International Journal of Cardiovascular Imaging*, 18:421–430, 2002.

6. Lee TH, Hamilton MA, Stevenson LW, et al., Impact of left ventricular cavity size on survival in advanced heart failure, *American Journal of Cardiology*, 72:601–612, 1993.

7. De Anda A Jr, Moon MR, Nikolic S, et al., A method to assess endocardial regional longitudinal curvature of the left ventricle, *American Journal of Physiology*, 268:2553–25560, 1995.

8. Zhong L, Ghista DN, Ng EYK, Lim ST, Chua T, and Lee CN, LV shape-based contractility indices, *Journal of Biomechanics*, 39:2397–2409, 2006.

9. Streeter DD Jr and Hanna WT, Engineering mechanics for successive states in canine left ventricular myocardium: I. Cavity and wall geometry, *Circulation Research*, 33:639–655, 1973.

10. Mirsky I, Basic terminology and formulae for left ventricular wall stress. In: *Cardiac Mechanics*, Mirsky I, Ghista DN, and Sandler H, New York: Wiley, pp. 3–10, 1973.

11. Yeo SY, Mechanics of intra-left ventricular pressure development during systole, Master Dissertation, Nanyang Technological University, 2004.

12. Dodge HT and Sandler H, Clinical applications of angiocardiography. In: *Cardiac Mechanics*, Mirsky I, Ghista DN, and Sandler H (Eds.), New York: Wiley, pp. 171–201, 1973.

13. Ghista DN and Sandler H, An analytic plastic-visco-elastic model of the shape and forces of the left ventricle, *Journal of Biomechanics*, 2:35–47, 1969.

14. Rankin JS, McHale PA, Arentzen CE, Ling D, Greenfield JC Jr, and Anderson RW, The three-dimensional dynamic geometry of the left ventricle in conscious dog, *Circulation Research*, 39:304–313, 1976.

15. Gaudron P, Eilles C, Kugler I, and Trtl G, Progressive left ventricular dysfunction and remodeling after myocardial infarction, *Circulation*, 87:755–763, 1993.

16. Shoucri RM, The pressure-volume relation and mechanics of left ventricular contraction, *Japanese Heart Journal*, 31:713–729, 1990.

17. Shoucri RM, Studying the mechanics of left ventricle contraction, *IEEE Transactions on Biomedical Engineering*, 17:95–101, 1998.

18. Shoucri RM, Active and passive stresses in the myocardium, *American Journal of Physiology Heart Circulatory Physiology*, 279:H2519–H2528, 2000.

19. Ghista DN, Zhong L, Ng EYK, Lim ST, Tan RS, and Chua T, Systolic modeling of the left ventricle as a mechatronic system: Determination of myocardial fiber's sarcomere contractile characteristics and new performance indices, *Molecular and Cellular Biomechanics*, 2(4):217–233, 2005.

20. Reconditi M, Linari M, Lucii L, Stewart L, and Sun YB, The myosin motor in muscle generates a smaller and slower working stroke at higher load, *Nature* 428:78–81, 2004.

21. Hill AV, The heat of shortening and dynamic constants of muscle, *Proceedings of the Royal Society of London Series B*, 126–136, 1938.

22. Huxley AF, Niedergerke R, Structural changes in muscle during contraction, *Nature*, 173:971–973, 1954.

23. Huxley AF, Muscular contraction: A review lecture, *Journal of Physiology*, 243:1–43, 1974.

24. Zhong L, Ghista DN, and Ng EYK, Contractility of the left ventricle in terms of its sarcomere power generation, *Proceedings of the 1st International Bioengineering Conference (IBEC)*, Singapore, pp. 267–270, 2004.

25. Pietrabissa R, Montevecchi FM, and Funero R, Ventricle mechanics based on sarcomere and fiber models. In: *Computational Methods in Bioengineering*, Spilker RL and Simon BR (Eds.), New York: ASME, BED 9, pp. 399–410, 1988.

26. Paladino JL and Noordergraaf A, Muscle contraction mechanism from ultra-structural dynamics. In: *Analysis and Assessment of Cardiovascular Function*, Drzewiecki GM and Li JKJ (Eds.), New York: Springer, pp. 33–57, 1998.

27. Ionescu I, Guilkey J, Berzins M, Kirby RM, and Weiss J, Computational simulation of penetrating trauma in biological soft tissues using the material point method, *Studies in Health Technology and Informatics*, 111:213–218, 2005.

28. Xia L, Huo MM, Wei Q, Liu F, and Crozier S, Analysis of cardiac ventricular wall motion based on a three-dimensional electromechanical biventricular model, *Physics in Medicine and Biology*, 50:1901–1917, 2005.
29. Zhong L, Ghista DN, Ng EYK, and Lim ST, Passive and active ventricular elastance of the left ventricle, *Biomedical Engineering Online*, 4:10, 2005.
30. Zhong L, Ghista DN, Ng EYK, Lim ST, and Chua T, Determination of aortic pressure-time profile, along with aortic stiffness and peripheral resistance, *Journal of Mechanics in Medicine and Biology*, 4(4):499–509, 2004.
31. Zhong L, Ghista DN, Ng EYK, Lim ST, Tan RS, and Chua LP, Explaining left ventricular pressure dynamics in terms of LV passive and active elastance, *Journal of Engineering in Medicine*, 220(5):647–655, 2006.

4

Vascular Biomechanics

Dhanjoo N. Ghista and Liang Zhong

CONTENTS

4.1 Scope

In this chapter, we analyze the various phenomena that occur in the cardio-vascular tree [1], namely

1. how the aortic pressure waveform can be obtained from the left-ventricular (LV) outflow rate into the aorta and the auscultatory diastolic and systolic blood pressures [2], as well as

2. other vascular phenomena associated with pulsatile blood flow in the arterial bed to

   (a) determine the aortic constitutive property,

   (b) develop the concepts of arterial impedance and peripheral resistance [3],

   (c) study wave reflection,

   (d) analyze what happens at arterial branching, and

   (e) study how wave reflection influences the composite pressure wave and the hydraulic load on the heart [4].

4.2 Determination of Aortic Pressure–Time Profile along with Aortic Stiffness and Peripheral Resistance

4.2.1 Introduction

In both Ayurvedic medicine and traditional Chinese medicine, the pressure-pulse shape is felt to provide diagnostic information concerning diseases and disorders. Conventionally, accurate measurement of blood pressure waveform requires insertion of a catheter into the artery. Some of the noninvasive methods that are currently utilized for blood pressure are the auscultatory measurement and the oscillometric measurement methods [5]. In this regard, a precise evaluation of the aortic pressure–time profile and correlation of its shape parameters with diseases (using traditional Chinese and Ayurvedic medical knowledge-base system) would constitute a signifi-cant contribution to medicine.

During the LV ejection phase, as the blood is pumped into the aorta, the aortic pressure rises and the aorta distends [6]. Thus, not all of the blood pumped into the aorta is distributed into the peripheral circulation imme-diately, and a portion of it is stored in the distended central aorta. The equation governing the modulation of aortic pressure can be formulated by considering that the rate of change of aortic pressure is governed by the product of (1) the volume elasticity (or distensibility of the aorta) (dP/dV) and (2) the difference between and the rate-of-inflow $I(t)$ into the aorta created by the ventricular pump (stroke volume) and the rate-of-outflow $Q(t)$ from the aorta into the systemic circulation (peripheral blood flow).

After closure of the aortic valve, no more blood enters the aorta, but the distended vessel now recoils according to its volume elasticity, and the blood is propelled into the peripheral circulation. Thus, the rate of fall of aortic pressure, in the elastic aortic chamber during this diastolic phase, is a function of the volume elasticity of the aorta and the flow resistance. We will now analyze how we can determine the arterial pressure pulse profile [2].

4.2.2 Analysis of Aortic Diastolic and Systolic Pressure

For the blood control volume, shown in Figure 4.1, we have

$$\frac{\mathrm{d}V}{\mathrm{d}t} = I(t) - Q(t) = I(t) - \frac{P(t)}{R} \tag{4.1}$$

where $I(t)$ and $Q(t)$ are inflow and outflow rates of the aorta, respectively, and R is the resistance to flow in the aorta.

We can also put down

$$\frac{\mathrm{d}P}{\mathrm{d}t} = \frac{\mathrm{d}P}{\mathrm{d}V} \cdot \frac{\mathrm{d}V}{\mathrm{d}t} = m\frac{\mathrm{d}V}{\mathrm{d}t} \tag{4.2}$$

where m is volume elasticity of the aorta.

By combining Equations 4.1 and 4.2, we obtain

$$\frac{\mathrm{d}P}{\mathrm{d}t} + \frac{m}{R}P = mI(t) \tag{4.3}$$

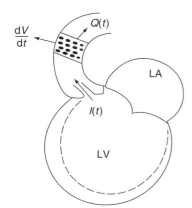

FIGURE 4.1

Control volume analysis to derive Equations 4.1 and 4.4 for aortic pressure response to left-ventricular flow rate into the aorta. (Adopted from Figure 1 of Zhong, L., Ghista, D.N., Ng, E.Y.K., Lim, S.T., and Chua, T., *J. Mech. Med. Biol.*, 4, 499, 2004.)

By putting $\lambda = m/R$, Equation 4.3 becomes

$$\frac{dP}{dt} + \lambda P = mI(t) \tag{4.4}$$

where λ is a parameter representing the aortic volume elasticity and the flow resistance during the diastolic phase.

The left ventricle pumps blood into the aorta only during the systolic phase. The aortic inflow rate $I(t)$ is schematized in Figure 4.2. Based on the $I(t)$ profile, the inflow rate into the aorta can be approximated by the following function:

$$
\begin{aligned}
I(t) &= ae^{-bt}t(t_s - t) \quad \text{for} \quad 0 < t < t_s \text{ (systole)} \\
&= 0 \quad \text{for} \quad t_s < t < T \text{ (diastole)}
\end{aligned}
\tag{4.5}
$$

where a and b are constants related to the rate-of-inflow and t_s is the duration of the cardiac ejection phase. Then, by carrying out integration of Equation 4.5 with respect to time, we can formulate the LV volume ejected into the aorta (or blood volume input into the aorta) during the systolic phase, as

$$V = \frac{a}{b}e^{-bt}\left[t^2 + t\left(\frac{2}{b} - t_s\right) + \frac{2/b - t_s}{b}\right] + \text{constant} \tag{4.6}$$

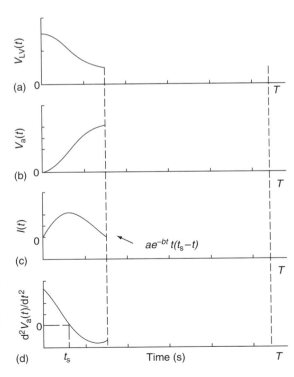

FIGURE 4.2
Schematics for (a) $V_{LV}(t)$ the LV volume, (b) $V_a(t)$ volume input into the aorta, (c) aortic inflow rate, and (d) volume acceleration d^2V_a/dt^2 during the LV ejection phase. The period for the systolic phase is $t = 0 - t_s$, and $(t_s - T)$ is the period for the diastolic phase. (Adopted from Figure 2 of Zhong, L., Ghista, D.N., Ng, E.Y.K., Lim, S.T., and Chua, T., *J. Mech. Med. Biol.*, 4, 499, 2004.)

In Equation 4.6, by imposing the initial condition, such that at the start of the ejection phase the aortic volume $V(t=0)=0$, we obtain the following expression for the volumes ejected into the aorta:

$$V = \frac{a}{b}e^{-bt}\left[t^2 + t\left(\frac{2}{b} - t_s\right) + \frac{(2/b - t_s)}{b}\right] + \frac{a(t_s - 2/b)}{b^2} \tag{4.7}$$

In Figure 4.2d, the volume acceleration of blood ejected by the LV into the aorta is shown. The maximum value of this volume acceleration d^2V_a/dt^2, can be employed as an LV contractility index.

4.2.3 Diastolic Pressure $P_d(t)$ Analysis

During diastole, the aortic valve is closed with zero inflow into the aorta. Hence, in Equation 4.4, $I(t)=0$, and we get

$$\frac{dP}{dt} + \lambda P = 0 \tag{4.8}$$

The solution of Equation 4.8 is

$$P = P_d(t) = C_2 e^{-\lambda t} \tag{4.9}$$

From Figure 4.3, we have

$$P_d(t = t_s) = P_2 \tag{4.10}$$

$$P_d(t = T) = P_3 \tag{4.11}$$

where P_2 is the aortic pressure at the end of cardiac ejection or the start of aortic diastole. Equation 4.9 describes the variation of aortic pressure $P_d(t)$ during the ventricular diastolic phase.

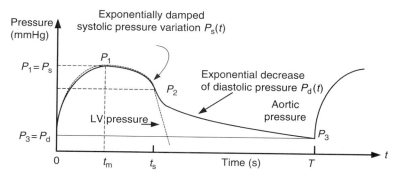

FIGURE 4.3
Schematic variation of aortic pressure during a cardiac cycle. Herein, $(0 - t_s)$ is the aortic systolic phase and $(t_s - T)$ is the aortic diastolic phase. P_1 is the aortic systolic pressure and P_3 is the aortic diastolic pressure. (Adopted from Figure 3 of Zhong, L., Ghista, D.N., Ng, E.Y.K., Lim, S.T., and Chua, T., *J. Mech. Med. Biol.*, 4, 499, 2004.)

Combining Equations 4.9 through 4.11, we obtain the expression for the parameter λ in terms of the end-diastolic pressure P_3 and end-systolic pressure P_2, in the form of

$$\lambda = \frac{\ln (P_2/P_3)}{(T - t_s)} = \frac{m}{R} \tag{4.12}$$

From Equations 4.9 and 4.11, we have

$$C_2 = P_3 e^{\lambda T} \tag{4.13}$$

Hence, Equation 4.9, for diastolic pressure becomes

$$P_d(t) = P_3 e^{\lambda(T-t)} \tag{4.14}$$

where the λ expression is given by Equation 4.12.

4.2.4   Systolic Pressure $P_s(t)$ Analysis

During systole, when the ventricle pumps blood into the aorta, the inflow rate is represented by the function $I(t) = ae^{-bt}t(t_s - t)$, given by Equation 4.5. Hence, from Equations 4.4 and 4.5, we obtain

$$\frac{dP}{dt} + \lambda P = mae^{-bt}t(t_s - t) \tag{4.15}$$

Upon solving Equation 4.15, we obtain an expression for the (exponentially damped) variation of the aortic pressure $P_s(t)$ during systole,

$$P = P_s(t) = C_1 e^{-\lambda t} + \frac{ma}{b - \lambda} e^{-bt} \left[t^2 + \left(\frac{2}{b - \lambda} - t_s \right) t + \frac{\left(\frac{2}{b - \lambda} - t_s \right)}{b - \lambda} \right] \tag{4.16}$$

where m and λ are the model parameters, which can be determined by making Equation 4.16 match the arterial tonometry data. However, herein, we will employ systolic and diastolic arterial pressure obtained from cuff auscultatory method to evaluate the model parameter's m and λ.

From Figure 4.3, we have

$$P_s(t = 0) = P_3, \text{ the aortic diastolic pressure} \tag{4.17}$$

Therefore, we have from Equation 4.16

$$C_1 = P_3 - \frac{ma(2/(b - \lambda) - t_s)}{(b - \lambda)^2} \tag{4.18}$$

Upon substituting the expression for C_1 into Equation 4.16, we get the total expression for $P_s(t)$.

4.2.5 Determination of P_2

Now, we need to evaluate the coefficients a and b in Equation 4.7, and then evaluate the model parameters m and λ in Equation 4.4. For this purpose, we note that we can determine the pressures P_1 and P_3 by the cuff auscultation procedure. However, in order to determine λ, we also need to know the value of P_1, the maximum value of $P_s(t)$, which is the auscultatory systolic pressure, as shown in Figure 4.3. Hence by differentiating Equation 4.16 and equating it to zero, we can obtain the time t_m, when $P_s(t)$ is maximum, and equal to P_1. Now,

$$\frac{dP_s(t)}{dt} = -C_1\lambda e^{-\lambda t} + \frac{ma}{b-\lambda}e^{-bt}\left[-bt^2 + \frac{-2\lambda + bt_s(b-\lambda)}{b-\lambda}t + \frac{\lambda t_s(b-\lambda) - 2\lambda}{(b-\lambda)^2}\right]$$

(4.19)

By putting $dP_s(t)/dt = 0$ at $t = t_m$, we obtain

$$-C_1\lambda e^{-\lambda t_m} + \frac{ma}{b-\lambda}e^{-bt_m}\left[-bt_m^2 + \frac{2\lambda + bt_s(b-\lambda)}{b-\lambda}t_m + \frac{\lambda t_s(b-\lambda) - 2\lambda}{(b-\lambda)^2}\right] = 0$$

(4.20)

Then, based on Equation 4.16, we have

$$P_1 = C_1e^{-\lambda t_m} + \frac{ma}{b-\lambda}e^{-bt_m}\left[t_m^2 + \left(\frac{2}{b-\lambda} - t_s\right)t_m + \frac{\left(\frac{2}{b-\lambda} - t_s\right)}{b-\lambda}\right]$$

(4.21)

where C_1 is given by Equation 4.20.

In addition, for compatibility between $P_d(t)$ and $P_s(t)$ expressions, the diastolic pressure must be equal to the systolic pressure at time t_s, so that

$$P_2 \overset{\text{Equation 14}}{=} P_d(t = t_s) \overset{\text{Equations 16 and 18}}{=} P_s(t = t_s)$$

(4.22)

This equation involves the aortic diastolic pressure (P_3) and the systolic pressure (P_1), which can be obtained from the noninvasive cuff sphygmomanometry method, with sufficient accuracy.

4.2.6 Determination of Coefficients a and b in Equation 4.7

In order to evaluate the coefficients a and b, we need to know the aortic volume $V_a(t)$ during LV ejection. For this purpose, the LV geometry and hence the LV volume data can be obtained from cineangiography measurements. In other words, from the dynamic geometry of the left ventricle,

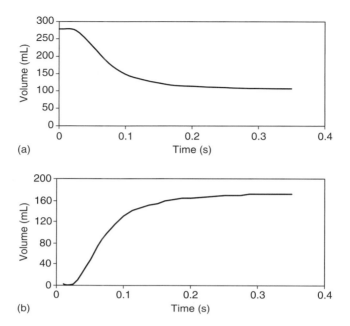

FIGURE 4.4

(a) Cineangiography-derived data of LV volume versus time during ejection. (b) Volume input into the aorta from the LV as derived from Figure 4.4a during the systolic phase. (Adopted from Zhong, L., Ghista, D.N., Ng, E.Y.K., Lim, S.T., and Chua, T., *J. Mech. Med. Biol.*, 4, 499, 2004.)

we can get the volume–time curve of the aorta during the systolic phase. In Figure 4.4b, we present the data on cineangiographically derived aortic volume during systole, derived from LV volume versus time of Figure 4.4a. Then, by using Equation 4.7 to match the derived aortic volume variation from the measurable LV volume (as shown in Figure 4.5), we can determine the coefficients a and b, as given in Table 4.1. However, if we can monitor LV volume geometry from LV echocardiography, then the coefficients a and b can be obtained noninvasively and the procedure for the determination of the aortic pressure profile can be noninvasive.

4.2.7 Determination of the Model Parameters m and λ

We note that we can monitor P_1 (Equation 4.21) and P_3 (Equation 4.18) by cuff sphygmomanometry. In order to determine m and λ, we also need to evaluate t_m in Equation 4.20. We thus have three unknowns: m, λ, and t_m. The corresponding three equations are Equations 4.20 through 4.22, which involve the three to-be-determined parameters (m, λ, and t_m) based on our knowledge of the monitored values of P_3 and P_1. For the subject, whose LV volume $V_{LV}(t)$ is displayed in Figure 4.4a, the corresponding monitored auscultatory pressures are $P_1 = 120$ mmHg and $P_3 = 83.57$ mmHg. Hence by solving Equations 4.20 through 4.22, we evaluate the

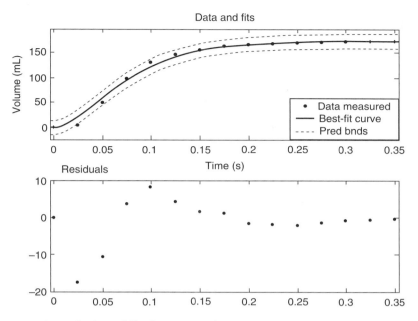

FIGURE 4.5 (See color insert following page 266.)
Plot of computed aortic volume versus time during the systolic phase of the aorta. The round points represent the measured data, while the solid line represents the filled computed volume curve for values of *a* and *b* given in Table 4.1. The prediction bounds define the width of the interval with a level of 95%. The values of these parameters and the RMS 1 (root-mean-square error) are given in Table 4.1.

parameters: $t_m = 0.175$ s, $m = 0.2796$ mmHg/mL, $\lambda = 0.4680$ s^{-1}, and $R = 0.5974$ mmHg·s/mL; these values are also tabulated in Table 4.1. Then, by substitution of these values of m and λ, into $P_d(t)$ and $P_s(t)$ expression (given by Equations 4.14 and 4.16), we obtain the complete aortic pressure–time profile, as shown in Figure 4.6.

TABLE 4.1

Parameters Evaluated in the Case Study

Parameters	Values	Unit
m	0.2796	mmHg/mL
λ	0.468	s^{-1}
t_m	0.175	S
a	2.827×10^5	mL/s^3
b	20.37	s^{-1}
RMS 1	0.012	mL
RMS 2	1.78	mmHg

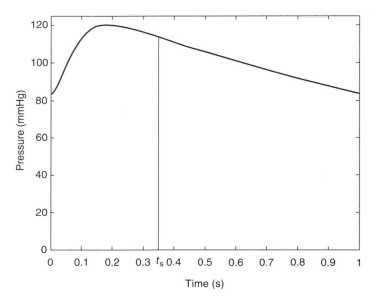

FIGURE 4.6

Plot of the computed aortic pressure during one cardiac cycle, $t_s = 0.35$ s. The time period $(0 - t_s)$ represents the systolic phase, and $(t_s - 1)$ represents the diastolic phase. (Adopted from Zhong, L., Ghista, D.N., Ng, E.Y.K., Lim, S.T., and Chua, T., *J. Mech. Med. Biol.*, 4, 499, 2004.)

4.2.8   Validation of the Computed Aortic Pressure Profile

For proper validation, we need to know the aortic beat-to-beat pressure profile obtained by tonometry. However, for this subject, we only know P_1 and P_3 from cuff sphygmomanometry. So let us compare the aortic pressure $P_s(t)$ with the cineangiographically monitored $P_{LV}(t)$ during the ejection phase. The result is shown in Figure 4.7, where the LV pressure is greater than the aortic pressure during the early systolic phase due to the pressure-drop across the aortic valve. However, the aortic pressure is greater than LV pressure in the latter phase of ejection due to the backflow during the late systolic phase. However, if we are interested in determining aortic stiffness parameter m and the peripheral resistance R, then we match our expression for aortic pressure given by Equation 4.16 with the actual monitored beat-to-beat aortic pressure by catheterization, and evaluate the parameters m, λ, and R.

4.2.9   Application

If we can obtain the LV volume from echocardiography as well as the diastolic pressure (P_3) and systolic pressure (P_1) by cuff sphygmomanometry, we can get the aortic pressure profile as well as the derived aortic parameters (λ, R, and m). Hence, we can also determine the LV contractility (the maximum value of the LV volume ejected into the aorta) as $[d^2 V_a / dt^2]_{max}$ in terms of noninvasively monitored LV flow rate from its volume data.

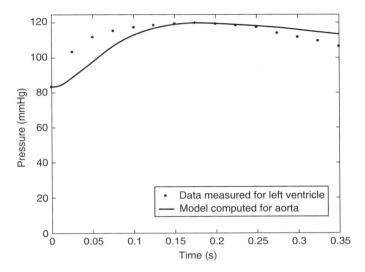

FIGURE 4.7
Plot of model-computed pressure versus time during the systolic phase of the aorta. Round points represent the measured LV data, while the solid line represents the model-computed aortic pressure curve. The RMS 2 value for this match is given in Table 4.1. (Adopted from Zhong, L., Ghista, D.N., Ng, E.Y.K., Lim, S.T., and Chua, T., *J. Mech. Med. Biol.*, 4, 499, 2004.)

Alternatively, if we know the beat-to-beat aortic pressure by arterial tonometry, which also requires information about cuff sphygmomanometry-derived $P_d(t)$ and $P_s(t)$, we can match these data with the expressions (given by Equations 4.14, 4.16, and 4.8) of $P_d(t)$ and $P_s(t)$, and hence get the values of parameters (λ, R, m, and a, b). For reduction of the peripheral (vascular) resistance R (or λ parameter), to in turn reduce blood pressure, we can administer administration drugs to reduce stress-induced peripheral vasoconstriction.

Our determination of the aortic characteristics is based on aortic parameters (λ, R, and m), as well as its pressure–time profile during the systolic and diastolic filling phases. The ability to determine the aortic pressure noninvasively may be deemed to be significant. The volume ejected into the aorta and the inflow rate into the aorta (Equations 4.7 and 4.5), associated with the parameters a and b, also have physiological implications. For an aorta with different pathological conditions, these parameters would of course change.

With a large enough database for different patients ($N > 30$), we can determine the distribution of aortic stiffness parameter (m) and peripheral (vascular) resistance R (or parameter λ), from which we can categorize normal patients and patients with arteriosclerosis (with high value of m) and vasoconstriction (with high value of R). This analysis can hence enable the clinician to decide on appropriate drug administration, to alter the aortic pressure profile, and to in fact treat hypertension.

4.3 Aortic Stiffness and Arterial Impedance

4.3.1 Measure of Aortic Stiffness (or Arteriosclerosis)

Now that we have stepped out from the heart into the aorta, we can recognize that the blood flow in the aorta is pulsatile. This pulsatile flow phenomenon can be employed [3] to measure and obtain the formula for pulse-wave velocity (PWV) and arterial wall stress.

The pulse-wave velocity (PWV or V_p), $V_p = (Eh/2a\rho)^{1/2}$, is obtained from the governing pulse-wave propagation differential equation

$$\partial^2 p/\partial x^2 = (1/V_p^2)\partial^2 p/\partial t^2 \tag{4.23}$$

and the arterial wall stress

$$\sigma = pa/h \tag{4.24}$$

where
  E = elastic modulus of the arterial wall
  a = aortic inner radius
  h = arterial wall thickness

We can then better characterize arteriosclerosis (or hardened artery disease) in terms of the aortic constitutive property of its wall elastic modulus (E) versus wall stress (σ), by determining (for each cardiac cycle) E and σ, as follows:

$$E = \frac{2(\text{arterial radius, } a)\,(\text{PWV or } V_p)^2\,(\text{blood density, } \rho)}{\text{arterial wall thickness, } h},$$
$$\sigma = \frac{(\text{auscultatory diastolic pressure, } P_d)\,(\text{arterial radius, } a)}{\text{arterial wall thickness, } h} \tag{4.25}$$

to obtain the relationship:

$$E = m\sigma + E_0 \tag{4.26}$$

Figure 4.8 illustrates schematically the measurement of the pulse-wave velocity V_p.

Now in Equations 4.25 and 4.26, we can ultrasonically monitor the arterial dimensions a and h, the auscultatory diastolic pressure P_d, as well as the PWV (V_p) from the time taken by a pulse to traverse between two aortic cross-sections. Thus, from Equations 4.25 and 4.26, we can evaluate the E versus σ relationship, as shown in Table 4.2.

The in vivo constitutive relationship "E versus σ'," provides a clinical measure of arterial stiffness and arteriosclerotic disease. For instance, for the following monitored data in Table 4.2, the aortic constitutive relationship is given by the following expression: E (N/m^2) $= 4.2\sigma + 0.5 \times 10^5$ (N/m^2).

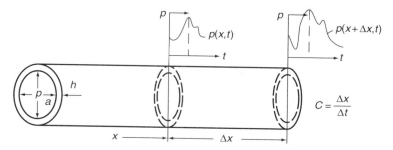

FIGURE 4.8

Schematic of measurement of pulse-wave velocity V_p.

4.3.2 Determination of Arterial Impedance (Another Parameter for Arteriosclerosis) from Pulse-Wave Velocity and Arterial Cross-Sectional Area

In this section, we will study the concept of arterial impedance and how it influences the pulse-wave reflection and the composite arterial pulse pressure [4,5].

The arterial pulse waveform is derived from the complex interaction of the LV stroke volume, the physical properties of the arterial tree, and the characteristics of the fluid in the system [7]. The principal components of blood pressure (p), flow rate (q) and velocity (u) comprise both a steady component (mean arterial pressure and flow rate) and a pulsatile component (pulse pressure and flow rate) [8], as shown in Figure 4.9 for pressure waveform.

$$p = <p> + \Delta p$$
$$q = <q> + \Delta q \tag{4.27}$$

The pulsatile component of pressure is determined by the pattern of LV ejection, the stroke volume, and the compliance characteristics of the arterial circulation [9]. Arterial compliance is defined as the change in area or volume of an artery or arterial bed for a given change in pressure [10]. The pulse pressure for a given ventricular ejection and heart rate will depend on arterial compliance, as well as the timing and magnitude of peripheral pulse-wave reflection.

TABLE 4.2

Clinical Measure of Arterial Stiffness from Monitoring the Arterial Dimensions (a and h), the Auscultatory Pressure (P), and Pulse-Wave Velocity (V_p)

P (mmHg)	V_p (m/s)	a (mm)	h (mm)	E (N/m^2) = $2c^2 a\rho/h$	S (N/m^2) = Pah
80	5.3	4.1	1.10	2.13×10^5	3.88×10^4
85	5.4	4.5	1.00	2.60×10^5	4.97×10^4
90	5.42	4.8	0.94	3.01×10^5	5.97×10^4
95	5.5	5.0	0.90	3.38×10^5	6.86×10^4

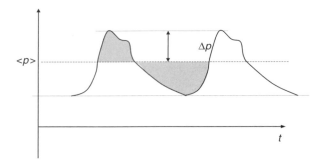

FIGURE 4.9 (See color insert following page 266.)
Schematic of a typical arterial pressure waveform.

Impedance, a term borrowed from electrical engineering theory, describes the opposition to flow presented by a system. The impedance load of the arterial tree can be quantified by analyzing pulse pressure–flow relationships produced through the effects of disease on the structural and functional components of the arterial system [11,12]. Input impedance relates simultaneously recorded arterial pressure and flow waveforms.

4.3.3 Peripheral Resistance (TPR) and Impedance (z_0)

$$\text{We define } <q> = \frac{<p>}{R}, \quad \text{where } R = \frac{8\mu L}{\pi a^4} = \text{TPR} \tag{4.28}$$

With atherosclerosis, the arterial radius (a) decreases and resistance (R) increases markedly. Hence, for a given $<q>$, $<p>$ is increased a lot.

$$\text{We define } \Delta q = \frac{\Delta p}{z_0}, \quad \text{where } z_0(\text{impedance}) = \frac{\rho c}{A} = \frac{\rho}{A}\sqrt{\frac{Eh}{2a\rho}} \tag{4.29}$$

For a given Δq, Δp is high if z_0 (impedance) is high. In other words, the impedance (z_0) is a direct measure of arterial hardening or stiffness or arteriosclerosis. We will now derive Equation 4.29.

Based on Figure 4.10, for force equilibrium:

$$-A\frac{\partial \Delta p}{\partial x}dx - \rho A dx \frac{\partial \bar{u}}{\partial t} = 0 \tag{4.30}$$

$$A\rho\frac{\partial \bar{u}}{\partial t} = -A\frac{\partial \Delta p}{\partial x} \tag{4.31}$$

Now, $\Delta q(x,t) = A\bar{u}(x,t)$

$$\rho\frac{\partial \Delta q}{\partial t} = -A\frac{\partial \Delta p}{\partial x} \tag{4.32}$$

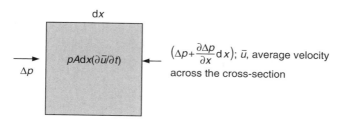

FIGURE 4.10
Equilibrium of a fluid element due to differential pressure pulse across it.

Now, let

$$\Delta p = f_1(x - ct) \quad \left(\text{since } \frac{\partial^2 \Delta p}{\partial x^2} = \frac{1}{(Eh/2a\rho)} \frac{\partial^2 \Delta p}{\partial t^2} = \frac{1}{c^2} \frac{\partial^2 \Delta p}{\partial t^2}\right)$$

$$\Delta q = \phi_1(x - ct) \quad \left(\text{since } \frac{\partial^2 \Delta q}{\partial x^2} = \frac{1}{c^2} \frac{\partial^2 \Delta q}{\partial t^2}\right) \tag{4.33}$$

For instance, we can have $\Delta p = \Delta p_1 \sin \frac{2\pi}{\lambda}(x - ct)$, and $\Delta q_1 \sin \frac{2\pi}{\lambda}(x - ct)$.
Substituting Equation 4.33 in Equation 4.32, we get

$$-\rho c \phi_1'(x - ct) = -A f_1'(x - ct) \tag{4.34}$$

Integrating, we get

$$\rho c \phi_1 = A f_1 \tag{4.35}$$

Hence, for right-propagating waves $f(x - ct)$ and $\phi(x - ct)$; the impedance is given by

$$z_{01} = \frac{\Delta \vec{p}_1}{\Delta \vec{q}_1} = \frac{\rho c}{A} = \frac{\rho}{A}\left(\frac{Eh}{2a\rho}\right)^{1/2} = \frac{4\rho}{\pi}\left(\frac{Eh}{2a^5\rho}\right)^{1/2} \tag{4.36}$$

Now, for a left-propagating wave $\Delta p_2 = f_2(x + ct)$ and $\Delta q_2 = \phi_2(x + ct)$; we have

$$-\rho c \phi_2 = +A f_2$$

Hence, for a left-traveling wave,

$$z_{02}\left(= \frac{\rho c}{A}\right) = -\frac{f_2}{\phi_2} = -\frac{\Delta \vec{p}_2}{\Delta \vec{q}_2} \tag{4.37}$$

4.3.4 Implication

If the arterial stiffness E is high (as in arteriosclerosis), then (as per Equations 4.29 and 4.36) both z_0 and Δp will be high. Based on Equations 4.28 and 4.36 (1) if a person smokes or has atherosclerosis, $<p>$ will be elevated and

(2) if a person has hardened artery (arteriosclerosis), Δp will be elevated. Note that R (in Equation 4.28) is increased by high μ and low a. On the other hand, z_0 is increased by high E as well as low a and high h.

Reiterating, for a right-propagating wave, in Equation 4.23, let

$$\Delta p = f_1 = \Delta p_1 \sin \frac{2\pi}{\lambda}(x - ct),$$

$$\Delta q = \phi_1 = \Delta q_1 \sin \frac{2\pi}{\lambda}(x - ct)$$

Then from Equation 4.35

$$\rho c \Delta q_1 \sin \frac{2\pi}{\lambda}(x - ct) = A \Delta p_1 \sin \frac{2\pi}{\lambda}(x - ct) \tag{4.38}$$

and hence,

$$z_{01} = \frac{\rho c}{A} = \frac{\Delta p_1}{\Delta q_1} \tag{4.39}$$

For a left-propagating wave, let

$$\Delta p = f_2 = \Delta p_2 \sin \frac{2\pi}{\lambda}(x + ct)$$

$$\Delta q = \phi_2 = \Delta q_2 \sin \frac{2\pi}{\lambda}(x + ct)$$

Since from Equation 4.32,

$$\rho \frac{\partial \Delta q}{\partial t} = -A \frac{\partial \Delta p}{\partial x}$$

we have corresponding to Equation 4.35:

$$\frac{2\pi \rho c}{\lambda} \Delta q_2 \sin \frac{2\pi}{\lambda}(x + ct) = -\frac{A 2\pi}{\lambda} \Delta q_2 \sin \frac{2\pi}{\lambda}(x + ct) \tag{4.40}$$

$$\text{Therefore, } \rho c \Delta q_2 = -A \Delta p_2 \tag{4.41}$$

$$\text{and } z_{02} = \frac{\rho c}{A} = -\frac{\Delta p_2}{\Delta q_2} \tag{4.42}$$

Hence, the arterial pressure characteristics in normal, atherosclerosis, and arteriosclerosis states will be as shown in Figure 4.11.

Summarizing, we have $<p> = R <q>$, where R is the arteriolar bed resistance; and $\Delta p = Z_0 \Delta q$, where Z_0 is the arterial impedance. Thus, in a hypertensive person, we can reduce $<p>$ by reducing R, and reduce Δp by reducing Z_0 (or increasing compliance) by appropriate medication.

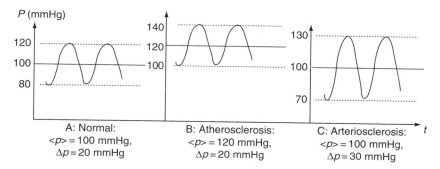

FIGURE 4.11
Schematics of arterial pressure profile for normal, atherosclerotic, and arteriosclerotic subjects.

4.4 Wave Reflection (due to Step Change in Arterial Impedance)

Let us now study the phenomenon of wave reflection of this aortic pressure wave. The arteriolar system may be assumed to act as a total absorber of energy generated by the heart, so that the pressure (p) will be dissipated into it. However, in the arterial segment, there are variations in vessel wall elasticity and cross-sectional area, which can give rise to pulse-wave reflection. In order to develop the analysis to elucidate these effects, let us start with a simple idealised model (Figure 4.12) of a step variation in the characteristic arterial impedance (Z_0) due to a step change in the cross-sectional area (A) and/or the arterial elasticity (E).

When the incident pressure wave (Δp_1) reaches one such junction of impedance change characterized by proximal and distal vessels on either side of the junction, part of the wave ($\Delta p'$) is transmitted and part (Δp_2) reflected. Let $\Delta p\ (= \Delta p_1 + \Delta p_2)$ denote the composite wave in the proximal

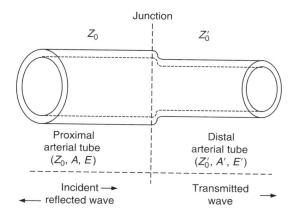

FIGURE 4.12
Model for determination of the reflection coefficient due to a step change in arterial impedance from Z_0 to Z_0'. When the pulse wave arrives at the junction, part of it is transmitted and part of it is reflected.

tube. Then, since on either side of the junction the pressure fluctuations must be identical, we have

$$\Delta p(x, t) = \Delta p_1(x - ct) + \Delta p_2(x + ct) = \Delta p' \tag{4.43}$$

Also, for continuity of flow, we must have

$$\Delta q_1 + \Delta q_2 = \Delta q' \tag{4.44}$$

Now, based on Equations 4.36 and 4.37, we have

$$\Delta q_1 = \frac{\Delta p_1}{Z_0}, \quad \Delta q_2 = -\frac{\Delta p_2}{Z_0}, \quad \Delta q' = \frac{\Delta p'}{Z_0'} \tag{4.45}$$

Upon substituting Equation 4.45 in Equation 4.44, and invoking Equation 4.43, we obtain

$$\frac{Z_0'}{Z_0}(\Delta p_1) - \frac{Z_0'}{Z_0}(\Delta p_2) = \Delta p' = \Delta p_1 + \Delta p_2 \tag{4.46}$$

thereby yielding the following expression for the reflection coefficient (R_f):

$$R_f = \frac{\Delta p_2}{\Delta p_1} = \frac{\text{reflected wave pressure}}{\text{incident wave pressure}} = \frac{Z_0' - Z_0}{Z_0' + Z_0} = \frac{1 - \lambda_z}{1 + \lambda_z} \tag{4.47}$$

where

$$\lambda_z = Z_0/Z_0'$$

Now the following possible outcomes can take place: if $\lambda_z < 1$, then there will be reflection with no phase change, so that, for instance, a compression wave will be reflected as a compression wave. If $\lambda_z = 1$, $R_f = 0$ or $\Delta p_2 = 0$; this constitutes a matched vessel junction and there will be no reflection at the junction ($x = 0$). If $\lambda_z > 1$, there will be reflection with 180° phase change.

It can be noted that as λ_z increases R_f decreases. Since $\lambda_z = Z_0(\text{proximal})/Z_0'$ (distal), if due to distal vasodilation Z_0' decreases (due to increase in vessel diameter), then λ_z would increase and R_f would decrease, i.e., there would be less reflection.

4.5   What Happens at an Arterial Bifurcation?

Let us now determine what happens at an arterial bifurcation [4]. Based on Figure 4.13, we have

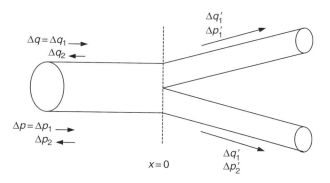

FIGURE 4.13
Schematics of transmitted and reflected pressure and flow pulses at a bifurcation.

$$\Delta p = \Delta p' = \Delta p'_1 = \Delta p'_2 \tag{4.48}$$

$$\Delta q' = \Delta q'_1 + \Delta q'_2 = \frac{\Delta p'}{z'_{01}} + \frac{\Delta p'}{z'_{02}} = \Delta p' \left(\frac{1}{z'_{01}} + \frac{1}{z'_{02}} \right)$$

$$= \frac{2\Delta p'}{z'_0} \text{ (if } z'_{01} = z'_{02}) \tag{4.49}$$

Also,

$$\Delta p = \Delta p_1 + \Delta p_2 = \Delta p' = \Delta p'_1 = \Delta p'_2 \tag{4.50}$$

and

$$\Delta q = \Delta q_1 + \Delta q_2 = \Delta q' = \Delta q'_1 + \Delta q'_2 \tag{4.51}$$

Hence, from Equation 4.51,

$$\frac{\Delta p_1}{z_0} - \frac{\Delta p_2}{z_0} = \frac{\Delta p'}{z'_0} + \frac{\Delta p'}{z'_0} = \frac{2\Delta p'}{z'_0} \tag{4.52}$$

Then, from Equations 4.52 and 4.50,

$$\frac{z'_0}{z_0} \Delta p_1 - \frac{z'_0}{z_0} \Delta p_2 = 2\Delta p' = 2\Delta p_1 + 2\Delta p_2$$

or

$$\Delta p_1 \left(\frac{z'_0}{z_0} - 2 \right) = \Delta p_2 \left(\frac{z'_0}{z_0} + 2 \right) \tag{4.53}$$

If λ_z be defined as

$$z_0/z_0' \tag{4.54}$$

we have

$$\frac{\Delta p_2}{\Delta p_1} = \frac{z_0' - 2z_0}{z_0' + 2z_0} = \frac{1 - 2\lambda_z}{1 + 2\lambda_z} \tag{4.55}$$

Let R_f (reflection coefficient) be defined as

$$R_f = \frac{\Delta p_2}{\Delta p_1} = \frac{1 - 2\lambda_z}{1 + 2\lambda_z} \tag{4.56}$$

So, for no reflection (as ideally expected in nature), $\Delta p_2 = 0$, $R_f = 0$

$$\text{i.e., } \lambda_z = \frac{z_0}{z_0'} = 0.5 \tag{4.57}$$

$$\text{i.e., } \left(\frac{E}{E'} \times \frac{h}{h'}\right)^{0.5} \times \left(\frac{a'}{a}\right)^{2.5} = 0.5 \tag{4.58}$$

Suppose

$$a' = 0.75a \tag{4.59}$$

This could be a reasonable assumption as the arterial radius keeps decreasing with arterial branchings. Then, if we can assume conservation of mass between the parent vessel and its bifurcations, we will have

$$2\pi a h = 2(2\pi a' h') \tag{4.60}$$

Hence, from Equations 4.58 through 4.60,

$$\frac{h}{h'} = 2\frac{a'}{a} = 2 \times (0.75) = 1.5 \tag{4.61}$$

$$\text{and } \frac{E}{E'} = \frac{0.25}{(a'/a)^5 \times (h/h')} \tag{4.62}$$

Now since based on Equations 4.59 and 4.61 $(a'/a)^5 = 0.237$ and $h/h' = 1.5$, respectively, then

$$E/E' = 0.703 \qquad (4.63)$$

It would be interesting to experimentally validate this relationship, based on the appropriate experimentally determined value of a'/a, as the basis of intrinsic optimality condition at arterial bifurcations.

Hence, we have

(i) For $R_f = 0$, $\Delta p_2 = 0$ or no reflection, and $\lambda_z = 0.5$

i.e., $\lambda_z = 0.5$ means $\Delta p_2 = 0$, and $R_f = 0$ (i.e., no reflection) $\qquad (4.64)$

(ii) For $\lambda_z < 0.5$, $\Delta p_2 = R_f \Delta p_1$, and $0 < R_f < 1$ $\qquad (4.65)$

and there is reflection with no phase change, i.e., an incident expansion wave at a site will be superimposed by a reflected expansion wave (of less magnitude) at that site. Let us represent Equation 4.65 by

$$A_2 \sin(x + ct) = R_f A_1 \sin(x - ct); \quad 0 < R_f < 1$$

and adopt $R_f = 0.3$, so that

$$A_2 \sin(x + ct) = 0.3 A_1 \sin(x - ct) \qquad (4.66)$$

This means that the reflected pulse-wave amplitude will add to the amplitude of the incident wave.

(iii) For $\lambda_z > 0.5$, $\Delta p_2 = -R_f \Delta q_2$, and $R_f < 0$

Let $R_f = -0.2$; then, $A_2 \sin(x + ct) = -\frac{1}{5} A_1 \sin(x - ct) = \frac{1}{5} A_1 \sin[(x - ct) - \pi]$, i.e., $A_2 = A_1/5$.

This means that the reflected pulse wave will be 180° out of phase with the incident way and contribute to decreasing the amplitude of the combined incident and reflected wave. An interesting implication of this analysis is that some persons, in whom the reflected wave is in phase with the incident wave, could be intrinsically hypertensive.

4.6 Artery Dividing into n Branches

In nature and in the body, there is a lot of inbuilt and intrinsic optimization, as to the design of cardiovasculature, pulmonary bronchioles, heart valves, etc. To investigate these phenomena could indeed be fascinating. We will briefly illustrate this concept with respect to arterial branching.

Let an artery of impedance Z_{op} divide into n identical branches, each one having impedance Z_{od}. Then at the branch site, we have the following governing equations:

Δp_1(incident pressure pulse) $+ \Delta p_2$ (reflected pressure pulse)
$$= \Delta p_d \text{ (the transmitted pressure pulse in each}$$
$$\text{of the branched vessels)} \quad (4.67)$$

Δq_1(incident flow-rate pulse) $+ \Delta q_2$ (reflected flow-rate pulse)
$$= n\Delta q_d \text{ (sum of the flow-rate pulse in the } n$$
$$\text{branch vessels)} \quad (4.68)$$

Now since

$$\Delta q_1 = \Delta p_1/Z_{op}$$
$$\text{and } \Delta q_2 = -\Delta p_2/Z_{op} \quad (4.69)$$

where the arterial impedance $z_0 = \rho(\text{PWV})/A = \rho c/A$ (the cross-sectional area) and the pulse-wave velocity

$$\text{PWV(or } c) = (Eh/2a\rho)^{1/2} = (Eh/2\rho)^{1/2}(\pi/A)^{1/4} \quad (4.70)$$

We then obtain from Equations 4.68 and 4.69,

$$\frac{\Delta p_1}{Z_{op}} - \frac{\Delta p_2}{Z_{op}} = n\frac{\Delta p_d}{Z_{od}} \quad (4.71)$$

and

$$\Delta p_1\left(\frac{z_{od}}{nZ_{op}}\right) - \Delta p_2\left(\frac{z_{od}}{nZ_{op}}\right) = \Delta p_d \quad (4.72)$$

Then, from Equations 4.67 and 4.72, we obtain

$$\Delta p_1\left(1 - \frac{Z_{od}}{nZ_{op}}\right) + \Delta p_2\left(\frac{z_{od}}{nZ_{op}} + 1\right) = 0 \quad (4.73)$$

Therefore,

$$\frac{\Delta P_2}{\Delta P_1} = \frac{Z_{od} - nZ_{op}}{Z_{od} + nZ_{op}} \quad (4.74)$$

Now for no reflection at the branched junction, we need to put the pulse-reflection coefficient $R_f = (\Delta P_2/\Delta P_1)$ equal to zero, giving us (from Equation 4.74)

$$R_f = (1 - n\lambda_z)/(1 + n\lambda_z) = 0;$$

$$\text{wherein } \lambda_z = z_{op}/z_{od}, \quad (4.75)$$

$$\text{i.e., } \lambda_z = z_{op}/z_{od} = 1/n \quad (4.76)$$

Then, based on Equations 4.76, 4.29, and 4.70, we obtain the relationship

$$(PWV_p/PWV_d)(A_d/A_p) = 1/n \quad (4.77)$$

$$\text{or } \left(\frac{E_p}{E_d} * \frac{h_p}{h_d}\right)^{1/2} \left(\frac{A_d}{A_p}\right)^{5/4} = 1/n \quad (4.78)$$

It would be so interesting to check if this relationship holds good at branched arterial sites, as an index of optimal design of arterial branched junctions, for minimizing pulse-wave reflections at the branched junctions.

4.7 Wave Reflection Influence on the Composite (Arterial) Pressure Wave

Now that we have seen how wave reflection occurs, let us study the influence of the wave reflection on the composite pressure and flow-rate waves. Then we can understand how wave reflection affects the amplitude and time course of the composite (or integrated) pressure wave which in turn constitutes the ventricular afterload [4].

Consider a composite wave $\Delta p(x,t)$ made up of an incident wave, $\Delta p_1 = A\cos(kx - wt + \phi_1)$ and a reflected wave, $\Delta p_2 = B\cos(kx - wt + \phi_2)$ where w = wave frequency and ϕ is the phase angle. We can rewrite the expressions for the two pressure waves as follows:

$$\Delta p_1 = A\cos[wt - (kx + \phi_1)]$$
$$\Delta p_2 = B\cos[wt + (kx + \phi_2)] \quad (4.79)$$

We can analyze for the following alternative situations:

(a) Incident and reflected waves (Δp_1 and Δp_2) will be in phase at points x_n ($= x_0, x_2, \dots$), if, based on Equation 4.79:

$$kx_n + \phi_2 = -(kx_n + \phi_1) + n\pi; \quad n = 0, 2, 4, \dots \quad (4.80)$$

Then, the amplitude of the composite wave will be $A + B$, and the incident and reflected waves will be in phase. These points of amplification of pressure fluctuation are referred to as antinodal points of pressure.

(b) Incident and reflected waves (Δp_1 and Δp_2) will be 180° out of phase at x_n ($=x_1, x_3, x_5$, etc.), if, based on Equation 4.79

$$kx_n + \phi_2 = -(kx_n + \phi_1) + n\pi; \quad n = 1, 3, 5, \ldots \qquad (4.81)$$

The composite wave will then have an amplitude of $A - B$, and will be in phase with the incident wave if $A > B$. These sites of minimal pressure fluctuation are referred to as nodal points.

Equations 4.80 and 4.81 can be restated as

$$2kx_n + \phi_1 + \phi_2 = n\pi; \quad n = 0, 1, 2, \ldots; \text{ reflected wave in phase with}$$
$$\text{incident wave}$$

$$2kx_{n+1} + \phi_1 + \phi_2 = (n+1)\pi; \quad n = 0, 1, 2, \ldots; \text{ reflected wave 180° out of}$$
$$\text{phase with incident wave}$$

Upon subtracting these two equations, we obtain

$$2k(x_{n+1} - x_n) = 2k\Delta x = \pi; \quad \Delta x = \pi/2k$$

and since k, the wave number, is equal to $2\pi/\lambda$, where λ is the wavelength, we have the distance between nodal and antinodal points:

$$\Delta x = \lambda/4 \qquad (4.82)$$

This situation is pictorially depicted in Figure 4.14. It can be seen that the ideal location of the heart with respect to the vascular system is at a nodal point and not at an antinodal point (of maximal pressure fluctuation), where it would have to do a lot more work.

Let us continue on, and develop the expression for the composite pressure wave Δp. This is given from Equation 4.79 by

$$\begin{aligned}
\Delta p &= A\cos[wt - (kx + \phi_1)] + B\cos[wt + (kx + \phi_2)] \\
&= A\cos(kx + \phi_1)\cos\ wt + A\sin(kx + \phi_1)\sin\ wt \\
&\quad + B\cos(kx + \phi_2)\cos\ wt - B\sin(kx + \phi_2)\sin\ wt \\
&= a_1(x)\cos\ wt + a_2(x)\sin\ wt \qquad (4.83)
\end{aligned}$$

where

$$a_1(x) = A\cos(kx + \phi_1) + B\cos(kx + \phi_2) \text{ and}$$
$$a_2(x) = A\sin(kx + \phi_1) - B\sin(kx + \phi_2)$$

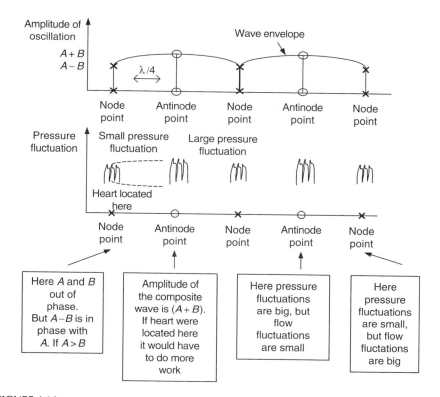

FIGURE 4.14
Variation of pressure amplitude oscillations along the arterial segment showing the location of nodal and antinodal points.

Expression 4.83 for the composite pressure wave can be more concisely expressed as

$$\Delta p = a(x)[\cos wt \cos \phi - \sin wt \sin \phi] = a(x)\cos[wt + \phi(x)]$$

where

$$
\begin{aligned}
a(x) &= [a_1^2 + a_2^2]^{1/2} = [A^2 + B^2 + 2AB\cos(2kx + \phi_1 + \phi_2)]^{1/2} \\
&= A + B, \text{ when } 2kx + \phi_1 + \phi_2 = n\pi;
\end{aligned}
$$

for n even, corresponding to an antinodal point

$$= A - B, \text{ when } 2kx + \phi_1 + \phi_2 = n\pi;$$

for n odd, corresponding to a nodal point (4.84a)

and $\phi(x) = \tan^{-1}\left(-\dfrac{a_2(x)}{a_1(x)}\right) = \tan^{-1}\left[-\dfrac{A\sin(kx + \phi_1) - B\sin(kx + \phi_2)}{A\cos(kx + \phi_1) - B\cos(kx + \phi_2)}\right]$

(4.84b)

Thus, if we study the composite pressure wave, the following observations can be made:

1. At any site n in the vascular tree, the oscillation $\Delta p(=a(x) \cos[wt + \phi(x)])$, is sinusoidal with a frequency ω and a constant amplitude $a(x)$ at x.

2. At different sites x, the amplitude of the oscillation Δp will be different.

3. Phase of the oscillation $\Delta p(=a(x) \cos[\omega t + \phi(x)])$ will vary along x, as given by

$$\tan \phi(x) = -\frac{A \sin(kx + \phi_1) - B \sin(kx + \phi_2)}{A \cos(kx + \phi_1) + B \cos(kx + \phi_2)} \tag{4.85}$$

Now, the relation of ϕ with the phase angle ϕ_1 of the incident wave is given by

$$\tan [\phi(x) - \phi_1] = \frac{AB \sin(2kx + \phi_1 + \phi_2)}{A^2 - AB \cos(2kx + \phi_1 + \phi_2)} \tag{4.86}$$

and in general ϕ will be different from ϕ_1.

Now let us see what happens to the composite flow wave or oscillation. According to Equations 4.36 and 4.37 we have for the composite flow pulse-wave oscillation

$$\Delta q = \frac{1}{Z_0}(\Delta p_1 - \Delta p_2) \tag{4.87}$$

Substituting for Δp_1 and Δp_2 from Equation 4.79, we obtain

$$\Delta q = \frac{1}{Z_0}[A \cos[\omega t - (kx + \phi_1)] - B \cos[\omega t + (kx + \phi_2)]]$$
$$= b(x) \cos[\omega t + \psi(x)] \tag{4.88}$$

where

$$b(x) = \sqrt{\frac{A^2 + B^2 - 2AB \cos(2kx + \phi_1 + \phi_2)}{Z_0}}$$

$$= \frac{A - B}{Z_0} \text{ for } 2kx + \phi_1 + \phi_2 = n\pi; \text{ for } n \text{ even, corresponding}$$

to an antinodal point

$$= \frac{A + B}{Z_0} \text{ for } 2kx + \phi_1 + \phi_2 = n\pi; \text{ for } n \text{ odd, corresponding}$$

to a nodal point $\tag{4.89a}$

TABLE 4.3

Pressure Pulse, Flow Pulse, and Impedance at Nodal and Antinodal Points

n	Pressure Fluctuation Amplitude	Flow Rate Fluctuation Amplitude	Z, Impedance = Pressure Fluctuation/Flow Fluctuation
Even (antinodal point)	$A + B$ (max, if A and B positive)	$(A - B)/Z_0$ (min, if A and B positive)	$\frac{(A+B)}{(A-B)} \cdot Z_0$
Odd (nodal point)	$A - B$ (min, if A and B positive)	$(A + B)/Z_0$ (max, if A and B positive)	$\frac{(A-B)}{(A+B)} \cdot Z_0$

Note: A and B are the amplitudes of the incident and reflected pressure waves.

$$\text{and } \psi(x) = \tan^{-1}\left[-\frac{A\sin(kx + \phi_1) - B\sin(kx + \phi_2)}{A\cos(kx + \phi_1) + B\cos(kx + \phi_2)} \right] \tag{4.89b}$$

Thus, from Equations 4.84 and 4.88, we develop Table 4.3.

Thus, at nodal points, we have low pressure fluctuations and high flow fluctuations. On the other hand, at antinodal points, we have high pressure fluctuations and low flow fluctuations. At the heart, we have large fluctuations of flow and ideally low fluctuations of pressure, while at the other end of the arterial tree we have low fluctuations of flow but high fluctuations of pressure.

Now in the case of the human arterial system, the wavelength for the first harmonic of the pulse wave is of the order of $\lambda = 400$ cm, so that $\lambda/4 = 100$ cm. Thus, if the heart be deemed to be naturally located at the nodal point, the first subsequent antinodal point will be 100 cm away from it. Thus, from the heart toward the terminus of the arterial tree, the pressure pulse will increase. In other words, the terminal or distal pressure pulse amplitude will exceed the proximal pressure pulse amplitude, which would be minimum at the entrance to the aorta. Correspondingly, the distal flow pulse amplitude will decrease, as schematically shown in Figure 4.15.

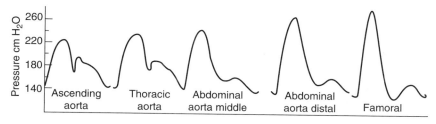

FIGURE 4.15

Variation of the pressure pulse along with aorta showing that the amplitude of pressure oscillation increases distally from the heart. It is to be noted that the corresponding flow velocity oscillation amplitude will, on the other hand, decrease distally from the heart.

4.8   Preventing Myocardial Infarction (by Reducing the Hydraulic Load on the LV)

We have recognized that the hydraulic load on the LV would be due to (1) the aortic steady pressure $<p>$ and (2) the composite pulse pressure of amplitude $(A - B)$ at a nodal point, due to arterial branchings. The hydraulic load on the heart can be manipulated by reducing arteriolar resistance, increasing arterial compliance, and by reducing or delaying reflected pressure waves arriving back at the heart during systole. Most vasodilator agents are considered to exert their beneficial effect by arteriolar dilatation. It is also likely that some vasodilators such as nitroglycerin may reduce wave reflections by a direct effect on the compliance of peripheral arteries. Agents, which increase arterial compliance and yet have little effect on arteriolar resistance, are likely to be of value in reducing the left ventricular pulsatile load, yet maintain mean vascular perfusion pressures unchanged.

References

1. Ghista DN, Biomechanical analysis of the cardiovascular system, in *Mathematical Methods in Medicine*, edited by Ingram D and Bloch RF, John Wiley & Sons Ltd., pp. 247–291, 1986.
2. Zhong L, Ghista DN, Ng EYK, Lim ST, and Chua T, Determination of aortic pressure–time profile, along with aortic stiffness and peripheral resistance, *Journal of Mechanics in Medicine and Biology*, 4(4):499–509, 2004.
3. Ghista DN, Determination of the in-vivo elasticity of a blood vessel & detection of arterial disease, *Automedica*, 1(3):151–164, 1974.
4. Fitchett D and Ghista DN, Mechanism & pharmological manipulation of the hydraulic load on the left ventricle, in *Clinical Cardiac Assessment: Interventions & Assist Technology*, edited by Ghista DN, in Advances in Cardiovascular Physics, Karger, 7:39–71, 1990.
5. Brinton TJ, Cotter B, Kailasam MT, Brown DL, Chio S-S, O'Connor DT, and DeMaria AN, Development and validation of a noninvasive method to determine arterial pressure and vascular compliance, *American Journal of Cardiology*, 80:323–330, 1997.
6. Sato T, Nishinaga M, Kawamoto A, Ozawa T, and Takatsuji H, Accuracy of a continuous blood pressure monitor based on arterial tonometry, *Hypertension*, 21(6):866–874, 1993.
7. McVeigh GE, Bank AJ, and Cohn JN, Arterial compliance, in *Cardiovascular Medicine*, edited by Willerson JT and Cohn JN, Churchill Livingstone, Philadelphia, pp. 1479–1496, 2000.
8. Smulyan H and Safar ME, Systolic blood pressure revisited, *Journal of American College in Cardiology*, 29:1407–1413, 1997.
9. McVeigh GE, Finklestein SM, and Cohn JN, Pulse contour and impedance parameters derived from arterial waveform analysis, in *Functional Abnormalities*

of the Aorta, edited by Boudoulas H, Toutouzas P, and Wooley CF, Futura, Armonk, pp. 183–193, 1996.

10. Simon A and Levenson J, Use of arterial compliance for evaluation of hypertension, *American Journal of Hypertension,* 4:97–105, 1991.
11. Finkelstein SM and Collins VR, Vascular hemodynamic impedance measurement, *Progress in Cardiovascular Diseases,* 24:401–418, 1982.
12. Nichols WW, Pepine CJ, Geiser EA, and Conti CR, Vascular load defined by the aortic input impedance spectrum, *Federation Proceedings,* 39:196–201, 1980.

Section II

Pulmonary Engineering Mechanics

5

Lung Ventilation Modeling for Lung Disease Diagnosis

Dhanjoo N. Ghista and Meena Sankaranarayanan

CONTENTS

5.1 Scope

This chapter deals with modeling of lung ventilation, first, in the form of a first-order differential equation of the lung volume (V) response to lung driving pressure (P_N), in terms of the lung compliance (C) and resistance to airflow rate (R). The solution of this equation is derived in terms of compliance (C) and the resistance to flow rate (R), which can then be employed to diagnose lung disease states. The parameters R and C are also combined to formulate a nondimensional index, whose ranges of values would differ with lung disease states. Thus, this index makes it more convenient to diagnose lung diseases.

The determination of the model parameters R and C from lung volume and driving pressure requires intubation of the patient. However, the solution of the governing equation also contains terms involving a combination of pressure and compliance as well as of τ ($= RC$). Hence, when this solution is made to match the monitored lung volume response, we can evaluate these terms without requiring to know the driving pressure independently; this avoids intubation of the patient. We then formulate another corresponding nondimensional index involving these terms, and demonstrate that this index in fact involves the pressure terms, as well as R and C independently. This provides validation of this index, based on the lung volume response to driving pressure in terms of R and C.

We next formulate a second-order differential equation for lung volume response to lung driving pressure. We demonstrate how the new parameters of this governing equation can be determined. These parameters also involve R and C, and hence can also be employed to diagnose lung disease states.

Now, it is possible that one lobe of the lung be normal and the other diseased. For this purpose, we develop a two-lobe lung model in terms of the response of their volumes to lung driving pressure. This two-lobe model is formatted using the first-order differential equation model. The model involves the compliance and flow resistances of the two lobes. We then demonstrate how this two-lobe model can be employed to separately evaluate the parameters of the two lobes, and hence assure the normality or diseased states of the two lobes separately.

This chapter is developed along similar lines to our previous Chapter 4 in Ref. [1], and the figures employed are adopted from this chapter in the afore-mentioned book.*

5.2 Ventilatory Function Represented by a First-Order Differential Equation Model

Lung mechanics involves inhalation and exhalation pressure and volume changes. Three pressures are involved in the ventilatory function, namely

---

* With permission from the publisher WIT Press, Southampton, U.K.

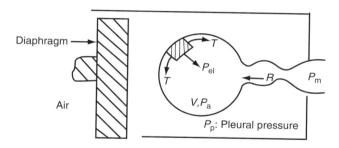

FIGURE 5.1

Lung ventilation lumped parameter model. In the figure, P_a is the alveolar pressure; P_m is the pressure at the mouth; P_p is the pleural pressure; $P_{el} = P_a - P_p = 2h\sigma/r = 2T/r$, the lung elastic recoil pressure; r is the radius of the alveolar chamber and h is its wall thickness; T is the wall tension in the alveolar chamber; V is the lung volume; R is the resistance to airflow; and C is the lung compliance. Adopted from Ghista, D.N., Loh, K.M., and Damodaran, M., in *Human Respiration; Anatomy and Physiology, Mathematical Modeling, Numerical Simulation and Applications*, V. Kulish (Ed), WIT Press, Southampton, U.K., 2006.)

atmospheric pressure or pressure at the mouth (P_m), alveolar pressure (P_a), and pleural pressure (P_p). The pressure gradient between the atmospheric and alveolar pressures causes respiration to occur. During inspiration, $P_a < P_m$, and air enters the lungs. During expiration, $P_a > P_m$, and air is expelled out of the lungs passively. This pressure differential between P_m and P_p provides the driving pressures (P_L) for gas flow, in terms of the elastic recoil pressure of the lumped alveolar chamber and the pressure differential between P_m and P_a (expressed as $R\dot{V}$). Thus, the assessment of respiratory mechanics involves the measurements of flows, volumes, pressure-gradients, and their dynamic interrelationships. The lung ventilatory model (LVM) then enables computation of lung compliance (C) and airway resistance-to-airflow (R), which are the parameters of the governing equation. Lung ventilatory dysfunction due to various diseases is characterized by the altered values of R and C, or in terms of an index involving a combination of R and C. Hence, the LVM can be employed to detect and characterize lung disease states.

The lung ventilation model is based on the equilibrium differential equation, expressing lung volume response to driving pressure across the lung. This dynamic relationship includes lung compliance (C) and the resistance-to-flow (R) offered by the airways during inspiration and expiration. In this model, the pressure generated by the respiratory muscles on the chest wall, namely the mouth pressure minus the pleural pressure, represents the driving force for the operation of the respiratory pump (for lung filling and expiration), as depicted in Figure 5.1.

5.2.1 Simulating Clinical Data

The clinical data consists of lung volume and driving pressure (mouth pressure – pleural pressure) in intubated subjects. The lung volume can be measured by integrating the airflow velocity–time curve, where the airflow velocity can be measured by means of a ventilator pneumatograph. Inhalation and exhalation pressures are measured by means of a pressure transducer connected to the ventilatory tubing. The pleural pressure is measured by placing a balloon catheter transducer through the nose into the esophagus, assuming that the esophageal tube pressure equals the pressure in the pleural space surrounding it.

5.2.2 Derivation of the Governing Differential Equation for Lung Volume (*V*) Response to Driving Pressure

For developing the lung ventilatory model, we employ a typical sample clinical data on both lung pressure (mouth pressure – pleural pressure) and volume, as depicted in Figure 5.2.

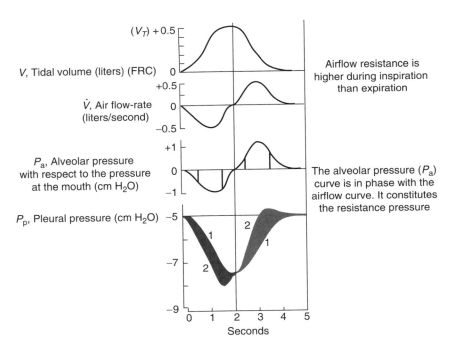

FIGURE 5.2 (See color insert following page 266.)
Lung ventilatory model and lung-volume and pleural-pressure data. In the bottom figure, graph 1 represents $(P_p - P_a) = -P_{el}$ (the pressure required to overcome lung elastance plus lung elastic recoil pressure at the end of expiration $= V/C + P_{el0}$); graph 2 represents P_p, obtained by adding $(P_a - P_m)$ to graph 1. The driving pressure $P_N(t)$ in Equation 5.1 equals P_p minus P_{el0} at the end of expiration. (Adopted from Ghista, D.N., Loh, K.M., Damodaran, M., in *Human Respiration; Anatomy and Physiology, Mathematical Modeling, Numerical Simulation and Applications*, V. Kulish (Ed), WIT Press, Southampton, U.K., 2006.)

Based on Figure 5.1, we get

$(P_a - P_p) = P_{el}$ (elastic recoil pressure)

$P_{el} = (2h\sigma)/r = 2\,T/r = V/C + P_{el0}$ (at the end of expiration)

$(P_m - P_a) = R(dV/dt)$

$P_L = P_m - P_p = (P_m - P_a) + (P_a - P_p)$

From the above equations, we get

$$R(dV/dt) + V/C = P_L - P_{el0} = P_N \qquad (5.1)$$

where P_L is the total (positive) pressure across the lung and $(P_L - P_{el0}) = P_N$ (the driving pressure). Equation 5.1 is the governing differential equation of lung volume (V) response to driving pressure P_N.

5.3 Lung Ventilation Performance Using the Linear First-Order Differential Equation Model

We first analyze lung ventilation function by means of Equation 5.1. A sample clinical pressure–volume data is displayed in Figure 5.2.

Graph 2, at the bottom of Figure 5.2, represents P_p (with respect to the pressure at the mouth) $= P_p - P_m$. According to our analysis (Equation 5.1),

$$-P_L = (P_p - P_m) = (P_p - P_a) + (P_a - P_m) = -\left(\frac{V}{C} + P_{el0}\right) - R\frac{dv}{dt}$$

$$\therefore P_L - P_{el0} = P_N(t) = R\dot{V} + \frac{V}{C}$$

which is in fact Equation 5.1.

In the model governing Equation 5.1, for lung volume (V) response to driving pressure (P_N):

$$R\overset{\circ}{V} + \frac{V}{C} = P_L(t) - P_{el0} = P_N(t) \qquad (5.2)$$

(1) The values of the net driving pressure are obtained from the P_N data, provided in Figure 5.3a, where $P_N(t)$ values are with respect to P_{el0}; (2) the parameters of this governing-differential equation are lung compliance (C) and airflow resistance (R) (in the equation, both R and C are instantaneous values); (3) lung $V = V(t) - V_0$ (the lung volume at the end expiration); the lung volume date is provided in Figure 5.3b; (4) P_{el0} is the lung elastic recoil pressure at the end of expiration, and (5) $P_{el0} = P_{el} - V/C$.

The frequency of the lung ventilatory cycle is ω, and T is the period of one lung inspiration–expiration cycle. For our sample data $\omega = 1.26$ rad/s. At the end of expiration when $\omega\,t = \omega\,T$, $P_L = P_{el0}$.

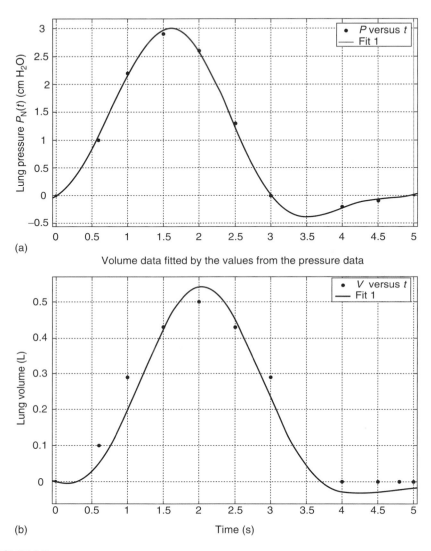

FIGURE 5.3
(a) The pressure curve $P_N(t)$ represented by Equation 5.5 matched against the pressure data (represented by dots). (b) The volume curve $V(t)$ represented by Equation 5.8, for $C_a = 0.2132\,1$ $(\text{cm H}_2\text{O})^{-1}$ and $R_a = 2.275 \text{ cm H}_2\text{O s L}^{-1}$ matched against the volume data represented by dots. In this figure, the lung volume and lung pressure are indicated with respect to the end-expiratory volume and pressure, respectively. (Adopted from Ghista, D.N., Loh, K.M., and Damodaran, M., in *Human Respiration; Anatomy and Physiology, Mathematical Modeling, Numerical Simulation and Applications.* V. Kulish (Ed), WIT Press, Southampton, U.K., 2006.)

Let the driving pressure $P_N(t)$ be represented by means of the following expression:

$$P_N(t) = \sum_{i=1}^{3} P_i \sin(\omega_i t + c_i)$$

Then, the governing Equation 5.2 (for lung volume response to driving pressure) becomes

$$R\overset{\circ}{V} + \frac{V}{C} = P_N(t) = \sum_{i=1}^{3} P_i \sin(\omega_i t + c_i) \tag{5.3}$$

where the right-hand side represents the net driving pressure, $P_N = (P_m - P_p) - P_{el0}$. This P_N is in fact the driving pressure $(P_m - P_p)$ normalized with respect to its value at the end of expiration. Equation 5.3 can be rewritten as follows:

$$\overset{\circ}{V} + \frac{V}{RC} = \frac{1}{R} \sum_{i=1}^{3} P_i \sin(\omega_i t + c_i) \tag{5.4}$$

where the $P_N(t)$ clinical data (displayed in Figure 5.3) is assumed to be represented by

$$P_N(t) = \sum_{i=1}^{3} P_i \sin(\omega_i t + c_i) \tag{5.5a}$$

where

$$
\begin{aligned}
&P_1 = 1.581 \text{ cm H}_2\text{O} \quad P_2 = -5.534 \text{ cm H}_2\text{O} \quad P_3 = 0.5523 \text{ cm H}_2\text{O} \\
&\omega_1 = 1.214 \text{ rad/s} \quad\quad \omega_2 = 0.001414 \text{ rad/s} \quad \omega_3 = 2.401 \text{ rad/s} \\
&c_1 = -0.3132 \text{ rad} \quad\quad c_2 = 3.297 \text{ rad} \quad\quad\quad c_3 = -2.381 \text{ rad}
\end{aligned}
\tag{5.5b}
$$

The clinical driving pressure $P_N(t)$ data, as represented by Equation 5.5, is depicted in Figure 5.3a. The corresponding lung volume $V(t)$ response is also represented in Figure 5.3b. If, in Equation 5.4, we designate R_a and C_a as the average values (R and C) for the ventilatory cycle, then the solution of Equation 5.4 (as derived in Appendix A) is given by

$$V(t) = \sum_{i=1}^{3} \frac{P_i C_a [\sin(\omega_i t + c_i) - \omega_i R_a C_a \cos(\omega_i t + c_i)]}{(1 + \omega_i^2 (R_a C_a)^2)} + He^{-t/R_a C_a} \tag{5.6}$$

where $\tau_a = R_a C_a$. An essential condition is that the flow rate (dV/dt) be zero at the beginning and end of expiration. In other words, we want that $dV/dt = 0$ at $t = 0$, in order to determine the constant H.

From Equation 5.6, we have

$$\frac{dV}{dt} = \sum_{i=1}^{3} \frac{P_i C_a \left[\omega_i(\cos \omega_i t + c_i) + \omega_i^2 \tau_a \sin(\omega_i t + c_i)\right]}{\left(1 + \omega_i^2 \tau_a^2\right)} - \frac{H}{\tau_a} e^{-t/\tau_a} = 0 \text{ at } t = 0$$

$$\therefore \sum_{i=1}^{3} \frac{P_i C_a \left[\omega_i \cos c_i + \omega_i^2 \tau_a \sin c_i\right]}{\left(1 + \omega_i^2 \tau_a^2\right)} - \frac{H}{\tau_a} = 0$$

$$\therefore H = \sum_{i=1}^{3} \frac{P_i C_a \tau_a \left[\omega_i \cos c_i + \omega_i^2 \tau_a \sin c_i\right]}{\left(1 + \omega_i^2 \tau_a^2\right)} \tag{5.7a}$$

When this expression of H is substituted in Equation (5.6), the resulting expression of $V(t = 0) = \sum_{i=1}^{3} P_i C_a \sin c_i$. Its value becomes very small for the data values in Equation (5.5b) and the order of magnitude of C_a. Hence, we can assume that $V(t = 0) \simeq 0$. Also, based on the values of (P_i, c_i, ω_i) and of the general order of magnitude of τ_a and C_a, the value of H during a respiratory cycle is small, so that we can afford to neglect the exponential term in Equation 5.6. Hence, the lung volume response to driving pressure can be represented by Equation 5.6 without the exponential term.

Now, by matching the above $V(t)$ in Equation 5.6, without the exponential term, with the given $V(t)$ data (in liters) in Figure 5.3b, and carrying out parameter identification, we can determine the in vivo values of R_a and C_a to be

$$C_a = 0.23132 \text{ L(cm H}_2\text{O)}^{-1}, \quad R_a = 2.275 \text{ (cm H}_2\text{O)s L}^{-1}, \quad \text{and} \tag{5.7b}$$
$$\tau_a = 0.485 \text{ s}$$

The computed $V(t)$ curve, represented by Equation 5.6, without the exponential term, for the above values of C_a and R_a, is shown in Figure 5.3b.

We can, however, also analytically evaluate R_a and C_a by satisfying some conditions. For this purpose, we first note that V is maximum (= tidal volume, TV) at $t = t_v = 2.02$ s. At $t = t_v$, the exponential term e^{-t/τ_a} in Equation 5.6 becomes of the order of e^{-10}, and hence negligible. Then, by putting $\overset{\circ}{V}(t = 2.02) = 0$ in Equation 5.6 (without the exponential term), we obtain

$$\overset{\circ}{V}\Big|_{t=2.02} = \sum_{i=1}^{3} \frac{P_i C_a \left[\omega_i \cos(\omega_i \times 2.02 + c_i) + \omega_i^2 \tau_a \sin(\omega_i \times 2.02 + c_i)\right]}{\left(1 + \omega_i^2 \tau_a^2\right)} = 0 \tag{5.8}$$

in which the values of P_i, ω_i, and c_i are given by Equation 5.5b. Then by solving Equation 5.8, we get $\tau_a = 0.522$ s, which is of the same order of magnitude as τ_a evaluated earlier and given by Equation 5.7b.

Then, we also note that at $t_v = 2.02$ s (at which $dV/dt = 0$), $V = 0.55$ L. Hence, upon substituting the parametric values from Equation 5.5b into

Equation 5.6, and neglecting the exponential term, we get the following algebraic equation:

$$V(t)|_{t=2.02} = \sum_{i=1}^{3} \frac{P_i C_a[\sin(\omega_i t + c_i) - \omega_i \tau_a^2 \cos(\omega_i t + c_i)]}{(1 + \omega_i^2 \tau_a^2)} = 2.55\, C_a \qquad (5.9)$$

by employing the values of P_i, ω_i, and c_i from Equation 5.5b. Now since $V(t = 2.02\text{ s}) = 0.55$ L, we get

$$2.55\, C_{(at\ t=2.02\ s)} = 0.55 \text{ L}$$
$$C_{(at\ t=2.02\ s)} = 0.22 \text{ L(cm H}_2\text{O)}^{-1} \qquad (5.10)$$

which is also of the same order of magnitude as the average compliance C_a given by Equation 5.9.

Since lung disease will influence the values of R and C, these parameters can be employed to diagnose lung diseases. For instance, in the case of emphysema, the destruction of lung tissue between the alveoli produces a more compliant lung, and hence results in a larger value of C. In asthma, there is increased airway resistance (R) due to contraction of the smooth muscle around the airways. In fibrosis of the lung, the membranes between the alveoli thicken and hence lung compliance (C) decreases. Thus, by determining the normal and diseased ranges of the parameters R and C, we can employ this simple lung-ventilation model for differential diagnosis.

5.4 Ventilatory Index

Although R_a and C_a have by themselves diagnostic values, let us formulate a nondimensional index to serve as a ventilatory performance index VTI_1 (to characterize ventilatory function) as

$$VTI_1 = [(R_a C_a)(\text{Ventilatory rate in s}^{-1})60]^2 = \tau_a^2 (BR)^2 60^2 \qquad (5.11)$$

where BR is the breathing rate.

Now, let us obtain its order of magnitude by adopting representative values of R_a and C_a in normal and disease states. Let us take the earlier computed values of $R_a = 2.275$ (cm H$_2$O) s L^{-1}, $C_a = 0.2132$ L (cm H$_2$O)$^{-1}$, and BR $= 12$ m^{-1} or 0.2 s^{-1}, computed by simulating the data of Figure 5.3 by means of Equation 5.6, as given by Equation 5.7b.

Then, in a supposed normal situation, the value of VTI_1 is of the order of 33.88. In the case of obstructive lung disease (with increased R_a), let us take

$R_a = 5$ (cm H$_2$O) s L^{-1}, $C_a = 0.12$ L (cm H$_2$O)$^{-1}$, and BR $= 0.3$ s^{-1}; then we get VTI$_1 = 116.6$. For the case of emphysema (with enhanced C_a), let us take $R_a = 2.0$ (cm H$_2$O) s L^{-1}, $C_a = 0.5$ L (cm H$_2$O)$^{-1}$, and BR $= 0.2$ s^{-1}; then we obtain VTI$_1 = 144$. In the case of lung fibrosis (with decreased C_a), we take $R_a = 2.0$ (cm H$_2$O) s L^{-1}, $C_a = 0.08$ L (cm H$_2$O)$^{-1}$, and BR $= 0.2$ s^{-1}; then we obtain VTI$_1 = 3.7$.

We can hence summarize that VTI$_1$ would be in the range of 2–5 in the case of fibrotic lung disease, 5–50 in normal persons, 50–150 in the case of obstructive lung disease, and 150–200 for the case of emphysema. This would of course need more validation by analyzing a big patient population.

Now, all of this analysis requites pleural pressure data, for which the patient has to be intubated. If now we evaluate the patient in an outpatient clinic, in which we can only monitor lung volume and not the pleural pressure, then can we develop a noninvasively obtainable ventilatory index?

5.4.1 Noninvasively Determinable Ventilatory Index

In order to formulate a noninvasively determinable ventilatory index based on the governing Equation 5.1, we need to recognize that in this case $P_N(t)$ (and hence P_i, ω, and c_i) will be unknown, and we hence need to redesignate the model parameters and indicate their identification procedure. For this purpose, we fit Equation 5.6 (without the exponential term) to the $V(t)$ data in Figure 5.3a, and obtain

$$P_1 C = 0.3223 \quad P_2 C = 0.3143 \quad P_3 C = -0.02269 \tag{5.12}$$

$$\omega_1 = -1.178 \quad \omega_2 = 0.5067 \quad \omega_3 = 1.855 \tag{5.13}$$

$$c_1 = 90{,}223 \quad c_2 = 0.2242 \quad c_3 = -3.961 \tag{5.14}$$

$$\tau_a = 0.5535 \tag{5.15}$$

We can now also formulate another noninvasively determinable nondimensional ventilatory index (VTI$_2$) in terms of these parameters as follows:

$$\mathrm{VTI}_2 = \frac{(BR)\tau[TV]^2}{|P_1 C||P_2 C||P_3 C|} = \frac{(BR)R[TV]^2}{|P_1 P_2 P_3 C^2|} \tag{5.16}$$

It is seen that VTI$_2$ can in fact be expressed in terms of P_1, P_2, P_3, and the lung parameters R, C. Then, after evaluating VTI$_2$ for a number of patients, its distribution can enable us to categorize and differentially diagnose patients with various lung disorders and diseases.

5.5 Work of Breathing

Just like the nondimensional ventilatory indices, the work of breathing (WOB) is also an important diagnostic index. The premise for determining WOB is that the respiratory muscles expand the chest wall during inspiration, thereby lowering the pleural pressure (i.e., making it more negative) below the atmospheric pressure to create a pressure differential from the mouth to the alveoli during inspiration. Then, during expiration, the lung recoils passively.

Hence, the work done (during a respiratory life cycle) is given by the area of the loop generated by plotting lung volume (V) versus net driving pressure (P_p). This plot is shown in Figure 5.4. Its area can be obtained graphically, as well as analytically as shown below:

$$\text{WOB} = \int_{0}^{t=T} P_N \frac{dV}{dt}\, dt \qquad (5.17a)$$

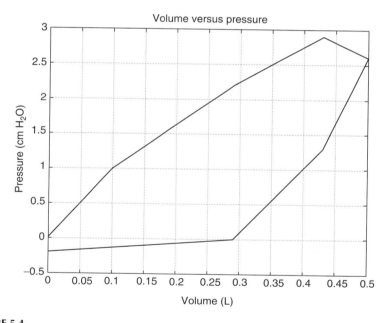

FIGURE 5.4

Plot of lung pressure versus volume. The area under the curve provides the work done. (Adopted from Ghista, D.N., Loh, K.M., and Damodaran, M., in *Human Respiration; Anatomy and Physiology, Mathematical Modeling, Numerical Simulation and Applications*. V. Kulish (Ed), WIT Press, Southampton, U.K., 2006.)

Upon substituting the expressions for P_N (from Equation 5.5a) and V (from Equation 5.6, without the exponential term), we get:

$$WOB = \int_0^{t=T} \sum_{i=1}^{3} P_i \sin(\omega_i t + c_i) \sum_{j=1}^{3} \frac{P_j C_a \left[\omega_j \cos(\omega_j t + c_j) + \omega_j^2 R_a C_a \sin(\omega_j t + c_j) \right]}{1 + \omega_j^2 R_a^2 C_a^2} dt$$

$$= \sum_{i=1}^{3} \sum_{j=1}^{3} \int_0^{t=T} \frac{P_i \sin(\omega_i t + c_i) \left\{ P_j C_a \left[\omega_j \cos(\omega_j t + c_j) + \omega_j^2 R_a C_a \sin(\omega_j t + c_j) \right] \right\}}{1 + \omega_j^2 R_a^2 C_a^2} dt$$

$$= \sum_{i=1}^{3} \sum_{j=1}^{3} \int_0^{t=T} \frac{P_i P_j C_a \omega_j \sin(\omega_i t + c_i) \cos(\omega_j t + c_j) + P_i P_j C_a \sin(\omega_i t + c_i) \omega_j^2 R_a C_a \sin(\omega_j t + c_j)}{1 + \omega_j^2 R_a^2 C_a^2} dt$$

$$= \sum_{i=1}^{3} \sum_{j=1}^{3} \left[\begin{array}{l} \int_0^{t=T} \frac{P_i P_j C_a \omega_j}{1 + \omega_j^2 R_a^2 C_a^2} \sin(\omega_i t + c_i) \cos(\omega_j t + c_j) dt \\ + \int_0^{t=T} \frac{P_i P_j \omega_j^2 R_a C_a^2}{1 + \omega_j^2 R_a^2 C_a^2} \sin(\omega_i t + c_i) \sin(\omega_j t + c_j) dt \end{array} \right]$$

$$= \sum_{i=1}^{3} \sum_{j=1}^{3} \left[\begin{array}{l} \frac{P_i P_j C_a \omega_j}{1 + \omega_j^2 R_a^2 C_a^2} \int_0^{t=T} \frac{\sin(\omega_i t + c_i + \omega_j t + c_j) + \sin(\omega_i t + c_i - \omega_j t - c_j)}{2} dt \\ + \frac{P_i P_j \omega_j^2 R_a C_a^2}{1 + \omega_j^2 R_a^2 C_a^2} \int_0^{t=T} \frac{\cos(\omega_i t + c_i - \omega_j t - c_j) - \cos(\omega_i t + c_i + \omega_j t + c_j)}{2} dt \end{array} \right] \quad (5.17b)$$

Therefore,

$$WOB = \sum_{i=1}^{3} \sum_{j=1}^{3} \left\{ \begin{array}{l} \frac{P_i P_j C_a \omega_j}{1 + \omega_j^2 R_a^2 C_a^2} \left[\begin{array}{l} \frac{\cos(c_i + c_j) - \cos(\omega_i T + c_i + \omega_j T + c_j)}{2(\omega_i + \omega_j)} \\ + \frac{\cos(c_i - c_j) - \cos(\omega_i T + c_i - \omega_j T - c_j)}{2(\omega_i - \omega_j)} \end{array} \right] \\ + \frac{P_i P_j \omega_j^2 R_a C_a^2}{1 + \omega_j^2 R_a^2 C_a^2} \left[\begin{array}{l} \frac{\sin(\omega_i T + c_i - \omega_j T - c_j) - \sin(c_i - c_j)}{2(\omega_i - \omega_j)} \\ - \frac{\sin(\omega_i T + c_i + \omega_j T + c_j) - \sin(c_i + c_j)}{2(\omega_i + \omega_j)} \end{array} \right] \end{array} \right\} \quad (5.17c)$$

The above expression for WOB can be evaluated, with the computed values of P_i, c_i, ω_i, and τ (as indicated by Equation 5.5b). So let us substitute into this equation the following values associated with Equation 5.5b:

$$\begin{array}{lll} P_1 = 1.581 \text{ cm H}_2\text{O} & P_2 = -5.534 \text{ cm H}_2\text{O} & P_3 = 0.5523 \text{ cm H}_2\text{O} \\ \omega_1 = 1.214 \text{ rad/s} & \omega_2 = 0.001414 \text{ rad/s} & \omega_3 = 2.401 \text{ rad/s} \\ c_1 = -0.3132 \text{ rad} & c_2 = 3.297 \text{ rad} & c_3 = -2.381 \text{ rad} \\ \tau_a = 0.485 \text{ s} \end{array} \quad (5.18)$$

We compute the value of WOB to be 0.9446 L (cm H$_2$O) in 5 s, or 0.19 L (cm H$_2$O)s^{-1} or 0.14 L mmHg s^{-1} or 0.02 W. This value can be verified by calculating the value of the area of the pressure–volume loop in Figure 5.4 which is equal to 0.8 L (cm H_2O).

5.6 Second-Order Model for Single-Compartment Lung Model

Let us now consider the dynamic (instead of static) equilibrium of the spherical alveolar chamber of the lung model in Figure 5.1, obtained as

$$m_s \ddot{u} + (P_p - P_a) + P_{el} = 0 \tag{5.19}$$

where
P_a and P_p are the alveolar and pleural pressures
u is the alveolar-wall displacement
m_s is lung mass (M) per unit surface area $= M/4\pi r^2$

and

$$P_{el} = \frac{2\sigma h}{r} = \frac{V}{C} + P_{el0} \tag{5.20}$$

where
C (lung compliance) is in L (cm H_2O)$^{-1}$
m_s (wall mass per unit surface area or surface density) $= \rho h$, where ρ is the density (mass per unit volume)
σ is the wall stress
h and r are the wall thickness and radius of the alveolar-lung chamber
P_{el0} is the elastic recoil pressure at the end of expiration

Now, the corresponding oscillating alveolar volume, $V = \frac{4}{3}\pi(r+u)^3$, from which we get

$$\ddot{V} \approx 4\pi r^2 \ddot{u} \tag{5.21}$$

Now, by putting

$$P_p - P_a = (P_m - P_a) + (P_p - P_m) \text{ and } P_L = P_m - P_p$$

so that,

$$P_p - P_a = (P_m - P_a) - P_L = R\dot{V} - P_L \tag{5.22}$$

$$m_s \ddot{u} = \left(\frac{M}{4\pi r^2}\right)\left(\frac{\ddot{V}}{4\pi r^2}\right) = \frac{M\ddot{V}}{16\pi^2 r^4} = M^* \ddot{V}; \quad M^* = \frac{M}{16\pi^2 r^4} = \frac{m_s}{4\pi^2 r^2} \tag{5.23}$$

we obtain from Equations 5.19 through 5.23:

$$M^* \overset{\circ\circ}{V} + (P_\mathrm{m} - P_\mathrm{a}) + \frac{V}{C} = P_\mathrm{L} - P_{\mathrm{el},0}; \quad M^* = \frac{M}{16\pi^2 r^4} \left(= \frac{m_s}{4\pi^2 r^2} \right). \qquad (5.24)$$

Now, putting $P_\mathrm{m} - P_\mathrm{a} = R\overset{\circ}{V}$, we obtain the governing equation, for lung volume (V) response to driving pressure (P_N), as

$$M^* \overset{\circ\circ}{V} + R\overset{\circ}{V} + \frac{V}{C} = P_\mathrm{L} - P_{\mathrm{el}0} = \sum_{i=1}^{3} P_i \sin(\omega_i t + c_i) - P_{\mathrm{el}0} = P_\mathrm{N} \qquad (5.25)$$

where $M^* = m_s/4\pi\, r^2$ is in $\mathrm{kg\, m}^{-4}$.

The clinical data for P_N in Figure 5.3, is assumed to be represented by

$$P_\mathrm{N}(t) = \sum_{i=1}^{3} P_i \sin(\omega_i t + c_i) \qquad (5.26)$$

where

$P_1 = 1.581$ cm H_2O	$P_2 = -5.534$ cm H_2O	$P_3 = 0.5523$ cm H_2O
$\omega_1 = 1.214$ rad/s	$\omega_2 = 0.001414$ rad/s	$\omega_3 = 2.401$ rad/s
$c_1 = -0.3132$ rad	$c_2 = 3.297$ rad	$c_3 = -2.381$ rad

Then we can rewrite Equation 5.25 for lung volume response to driving pressure, as

$$\overset{\circ\circ}{V} + \left(\frac{R}{M^*} \right) \overset{\circ}{V} + \frac{V}{CM^*} = \sum_{i=1}^{3} \frac{P_i}{M^*} \sin(\omega_i t + c_i) \qquad (5.27)$$

or as

$$\overset{\circ\circ}{V} + 2n\overset{\circ}{V} + p^2 V = \sum_{i=1}^{3} Q_i \sin(\omega_i t + c_i) \qquad (5.28)$$

where

the damping coefficient, $2n = R/M^*$

the natural frequency of the lung-ventilatory cycle, $p^2 = 1/CM^*$

and

$$Q_i = P_i/M^* \qquad (5.29)$$

So the governing Equation 5.28 of the lung ventilatory response to the inhalation pressure has three parameters M^*, n, and p^2 or M^*, R, and C, if the lung pressure is also monitored by intubating the patient. The solution of this equation (as derived in Appendix B) is given by

$$V(t) = \sum_{i=1}^{3} \left\{ \frac{Q_i(-2\omega_i \cos(\omega_i t + c_i)n + \sin(\omega_i t + c_i)p^2 - \sin(\omega_i t + c_i)\omega_i^2)}{4n^2\omega_i^2 + p^4 - 2p^2\omega_i^2 + \omega_i^4} \right\}$$

$$- \sum_{i=1}^{3} Q_i/2 \left\{ \left[-(n^2 - p^2)^{\frac{1}{2}}\omega_i^2 \sin c_i + p^2(n^2 - p^2)^{\frac{1}{2}} \sin c_i - 2\omega_i n^2 \cos c_i \right. \right.$$

$$+ p^2 n \sin c_i - 2\omega_i n(n^2 - p^2)^{\frac{1}{2}} \cos c_i - \omega_i^3 \cos c_i + \omega_i^2 n \sin c_i + \omega_i p^2 \cos c_i \right]$$

$$\times e^{\left(-n + \sqrt{n^2 - p^2}\right)t} \left/ \left[(n^2 - p^2)^{\frac{1}{2}}(4n^2\omega_i^2 + p^4 - 2p^2\omega_i^2 + \omega_i^4) \right] \right\}$$

$$+ \sum_{i=1}^{3} Q_i/2 \left\{ \left[-p^2(n^2 - p^2)^{\frac{1}{2}} \sin c_i + np^2 \sin c_i + \omega_i \cos c_i p^2 \right. \right.$$

$$+ \omega_i^2 n \sin c_i - 2\omega_i n^2 \cos c_i + 2\omega_i n(n^2 - p^2)^{\frac{1}{2}} \cos c_i$$

$$\left. + \omega_i^2(n^2 - p^2)^{\frac{1}{2}} \sin c_i - \omega_i^3 \cos c_i \right] e^{\left(-n - \sqrt{n^2 - p^2}\right)t}$$

$$\left/ \left[(n^2 - p^2)^{\frac{1}{2}}(4n^2\omega_i^2 + p^4 - 2p^2\omega_i^2 + \omega_i^4) \right] \right\} \qquad (5.30)$$

We will ignore the exponential terms and perform parameter identification by matching the above expression for $V(t)$ to the clinical data, shown in Figure 5.3. The matching is illustrated in Figure 5.5, where both the first- and second-order differential equation solutions for $V(t)$ are depicted. The computed values of the model parameters (M^*, n, p^2) are also shown in the table below the figure, along with the values of R and C computed from these dynamic model parameters. Further, the first- and second-order model values of R and C are compared in the table.

Figure 5.5 illustrates how the volume curve obtained by parametric identification compares with the actual lung volume. Figure 5.5 also illustrates how closely the first- and second-order differential equation models compare with each other in matching the actual clinical lung volume data. The values of model parameters M^*, n, and p^2 (and the values of the intrinsic parameters R and C obtained from them) are also given in Figure 5.5, which then enables us to compare the values of R and C obtained by the the two models.

5.7 Two-Compartmental First-Order Ventilatory Model

Now, it is possible that only one of the two lungs (or lung lobes) may be diseased. Thus, let us develop a procedure to distinguish between the normal lung and the pathological lung. We hence employ a two-compartment model (based on our first-order differential equation model of lung ventilatory function) to solve the problem of a two-lung model (schematized in Figure 5.6).

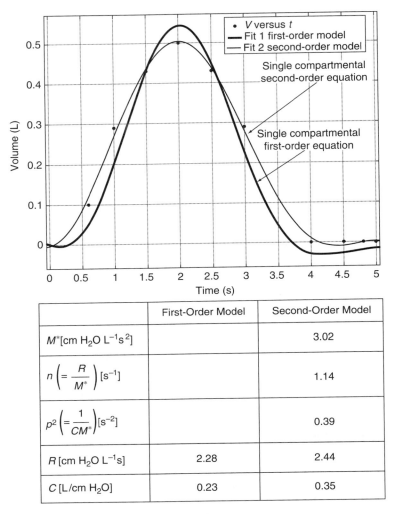

	First-Order Model	Second-Order Model
$M^*[\text{cm H}_2\text{O L}^{-1}\text{s}^2]$		3.02
$n\left(=\dfrac{R}{M^*}\right)[\text{s}^{-1}]$		1.14
$p^2\left(=\dfrac{1}{CM^*}\right)[\text{s}^{-2}]$		0.39
$R\,[\text{cm H}_2\text{O L}^{-1}\text{s}]$	2.28	2.44
$C\,[\text{L/cm H}_2\text{O}]$	0.23	0.35

FIGURE 5.5 (See color insert following page 266.)
Results of Second-order Single-compartment model (based on differential equation formulation), compared with the First-order model. (Adopted from Ghista, D.N., Loh, K.M., and Damodaran, M., in *Human Respiration; Anatomy and Physiology, Mathematical Modeling, Numerical Simulation and Applications,* V. Kulish (Ed), WIT Press, Southampton, U.K., 2006.)

For this purpose we make the subject breathe at a particular frequency ω_i, and monitor the total lung pressure $P^T(t)$ [or $P_N^T(t)$] and total lung volume $V^T(t)$, analogous to the lung pressure and volume data, depicted in Figure 5.3. Correspondingly, we have $P^L(t)$ and $V^L(t)$, and $P^R(t)$ and $V^R(t)$ for the left and right lungs, respectively. The governing equations will be as follows (refer Figure 5.3):

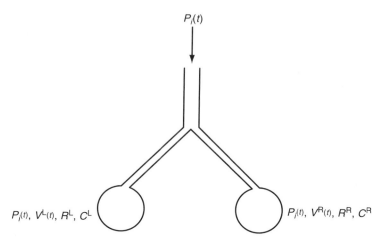

$P_i(t)$

$P_i(t)$, $V^L(t)$, R^L, C^L

$P_i(t)$, $V^R(t)$, R^R, C^R

FIGURE 5.6
Schematic of the two-compartment first-order lung-ventilation model. (Adopted from Ghista, D.N., Loh, K.M., and Damodaran, M., in *Human Respiration; Anatomy and Physiology, Mathematical Modeling, Numerical Simulation and Applications*. V. Kulish (Ed), WIT Press, Southampton, U.K., 2006.)

$$P^T = P^L = P^R \tag{5.31}$$

$$V^T = V^L + V^R \tag{5.32}$$

where

$$V^l(t) = f(R^L, C^L, P^T(t)) \tag{5.33}$$

$$V^R(t) = f(R^R, C^R, P^T(t)) \tag{5.34}$$

In these equations,

   (i) the variables $P^T(t)$, $V^T(t)$ are deemed to be known, i.e., monitored.
  (ii) the parameters R^L, C^L, and R^R, C^R are to be evaluated.

Using the first-order differential equation model, presented in Section 5.3, as given by Equation 5.6 (without the exponential term):

$$V(t) = \sum_{i=1}^{3} \frac{(P_i C_a)[\sin(\omega_i t + c_i) - \omega_i \tau_a \cos(\omega_i t + c_i)]}{(1 + \omega_i^2 \tau^2)} \tag{5.35}$$

we put down the expression for $V^T(t) = V^L(C_L, \tau_L) + V^R(C_R, \tau_R)$, match it with the total lung volume data (using a parameter-identification technique software), to obtain the values of (C_L, τ_L) and (C_R, τ_R), by means of which we can differentially diagnose left and right lung lobes' ventilatory capabilities and associated disorders (or diseases).

5.7.1 Two-Compartmental Lung Ventilatory Model Analysis

Using Equation 5.6 without the exponential term, we put down the expression for the total lung volume equal to the sum of left and right lung volumes, as follows:

$$V(t) = \sum_{i=1}^{3} \frac{P_i C^L \left[\sin(\omega_i t + c_i) - \omega_i \tau_L \cos(\omega_i^t + c_i)\right]}{(1 + \omega_i^2 \tau_L^2)}$$
$$+ \sum_{i=1}^{3} \frac{P_i C^R \left[\sin(\omega_i t + c_i) - \omega_i \tau_R \cos(\omega_i t + c_i)\right]}{(1 + \omega_i^2 \tau_R^2)} \qquad (5.36)$$

where, for the clinical data, we have

$P_1 = 1.581$ cm H_2O $P_2 = -5.534$ cm H_2O $P_3 = 0.5523$ cm H_2O
$\omega_1 = 1.214$ rad/s $\omega_2 = 0.001414$ rad/s $\omega_3 = 2.401$ rad/s
$c_1 = -0.3132$ rad $c_2 = 3.297$ rad $c_3 = -2.381$ rad

Now, in order to develop a measure of confidence in our analysis, we first generate the total lung volume data by assuming different values of C and R for left and right lung lobes. We then use Equation 5.36 along with the above data on pressure, to generate the total lung volume data. We adopt this generated lung volume data as the clinical-volume data.

We now make our volume solution expression (Equation 5.36) match this generated clinical volume data, by means of the parameter-identification procedure, to evaluate C and R for the left and right lung lobes and hence VTI_1 and VTI_2 (Equations 5.11 and 5.16) for these lobes. Based on the values of VTI_1 and VTI_2, we can differentially diagnose the left and right lung lobes.

5.7.2 Simulation of a Stiff Right Lung (with Compliance Problems)

We now simulate a normal left lung and stiff right lung, represented by

$$R^L = R^R = 1.14 \, (\text{cm } H_2O) L^{-1} \quad \text{and} \quad C^L = 0.11, \quad C^R = 0.05 \, L/(\text{cm } H_2O) \quad (5.37)$$

Substituting these parametric values into Equation 5.36, we generate the total lung volume data, as illustrated in Figure 5.7.

Now our clinical data for this two-compartment model comprises of the pressure data of Figure 5.3a and the generated total lung volume data of Figure 5.7. For this clinical data, we match the volume solution given by Equation 5.36 with the generated volume data, illustrated in Figure 5.7,

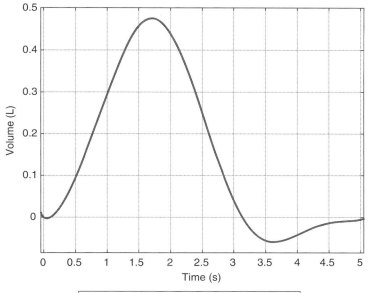

Two-Compartmental Model		
	First-Order Model	
	Left Lung	Right Lung
R [cm H_2O L^{-1}s]	1.137 (1.14)	1.137 (1.14)
C [L/cm H_2O]	0.1066 (0.11)	0.0533 (0.05)
VTL_1	2.115	0.5289
VTL_2	0.2198	1.0320

FIGURE 5.7

Results of the two-compartment model, based on the first-order differential equation model. In the table, the assumed values of R and C (in Equation 5.37) are indicated in brackets. It is noted that the calculated values of R and C (by parameter-identification technique) match the assumed values of R and C employed to generate the total lung volume data (shown in the figure). (Adopted from Ghista, D.N., Loh, K.M., and Damodaran, M., in *Human Respiration; Anatomy and Physiology, Mathematical Modeling, Numerical Simulation and Applications*. V. Kulish (Ed), WIT Press, Southampton, U.K., 2006.)

and carry our parameter identification. The computed values of R and C, for the left and right lungs, listed in the table of Figure 5.7, are in close agreement with the initially assumed parametric values of Equation 5.37. This lends credibility to our model and to our use of parameter-identification method.

Now for differential diagnosis, we also compute the lung-ventilatory indices (given by Equations 5.11 and 5.16), as shown in the table of Figure 5.7.

5.7.3   Simulation of a Right Lung with Flow-Rate Resistance Problems

Now, we simulate a lung with an obstructive right lung, as represented by

$$R^L = 1.14 \, (\text{cm H}_2\text{O}) \, L^{-1}\text{s} \quad \text{and} \quad R^R = 2.28 \, (\text{cm H}_2\text{O}) \, L^{-1}\text{s}, \quad \text{and}$$
$$C^L = C^R = 0.11 \, L/\text{cm H}_2\text{O} \tag{5.38}$$

We substitute these parameter values into Equation 5.36 and generate the total lung volume, as depicted in Figure 5.8. We then carry out a parameter-identification procedure and determine the values of R^L, R^R, and C^L, C^R, for the total lung volume (Equation 5.35) to match the generated lung volume data. The simulation results are shown in Figure 5.8, and the computed parametric values (along with the indices values) are depicted in the table below the figure.

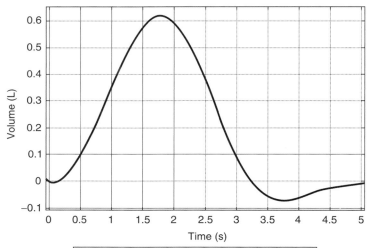

Two-Compartmental Model		
	First-Order Model	
	Left Lung	Right Lung
R [cm H$_2$O L^{-1}s]	1.138 (1.14)	2.276 (2.28)
C [L/cm H$_2$O]	0.1066 (0.11)	0.1066 (0.11)
VTL$_1$	2.1192	8.4766
VTL$_2$	0.3553	0.8341

FIGURE 5.8
Results of the two-compartment model, based on first-order differential equation model. In the table, the assumed values of R and C (Equation 5.38) are shown in brackets. It is noted that the calculated values of R and C (by parametric-identification method) closely match the assumed values of R and C, employed to generate the total volume data (as shown in this figure).

5.7.4 Determining Left and Right Lung Parameters and Indices without Requiring Intubation of the Patient

We could also employ as model parameters P_iC^L, P_iC^R, τ_L, and τ_R in Equation 5.36 representation of the total lung volume expression. This would not require us to intubate the patient. We could then employ the ventilatory index VTI_2 for each lung lobe, to diagnose its disease state.

Appendix A: Solution of Equation 5.4

Equation (5.4) is

$$\overset{\circ}{V} + \frac{V}{RC} = \frac{1}{R}\sum_{i=1}^{3} P_i \sin(\omega_i t + c_i) \tag{A.1}$$

Its auxiliary equation is given by

$$m + \frac{1}{RC} = 0; \quad \therefore m = -\frac{1}{RC} \tag{A.2}$$

Hence the complimentary solution (CF) is given by

$$V(t) = He^{-\frac{1}{RC}t} \tag{A.3}$$

where H is a constant.

The particular integral (PI) is given by

$$\text{PI} = \sum_{i=1}^{3}\left[\frac{1}{\left(D + \frac{1}{RC}\right)}\frac{1}{R}P_i \sin(\omega_i t + c_i)\right] \tag{A.4}$$

$$= \sum_{i=1}^{3}\left[\frac{\left(D - \frac{1}{RC}\right)}{\left(D^2 - \frac{1}{R^2C^2}\right)}\frac{1}{R}P_i \sin(\omega_i t + c_i)\right] \tag{A.5}$$

$$= \sum_{i=1}^{3}\left[\frac{\left(D - \frac{1}{RC}\right)}{\left(-\omega_i^2 - \frac{1}{R^2C^2}\right)}\frac{1}{R}P_i \sin(\omega_i t + c_i)\right] \tag{A.6}$$

$$= -\sum_{i=1}^{3}\left[\frac{RC^2}{\left(R^2C^2\,\omega_i^2 + 1\right)}\left(D - \frac{1}{RC}\right)P_i \sin(\omega_i t + c_i)\right] \tag{A.7}$$

$$= -\sum_{i=1}^{3}\left\{ \frac{RC^2 P_i}{\left(R^2C^2\,\omega_i^2 + 1\right)}\left[\omega_i \cos(\omega_i t + c_i) - \frac{1}{RC}\sin(\omega_i t + c_i)\right]\right\} \tag{A.8}$$

Therefore, $\text{PI} = -\sum_{i=1}^{3} \left[\frac{RC^2 P_i}{(R^2 C^2 \omega_i^2 + 1)} \omega_i \cos(\omega_i t + c_i) - \frac{CP_i}{(R^2 C^2 \omega_i^2 + 1)} \sin(\omega_i t + c_i) \right]$

$$(A.9)$$

Hence the General Solution of Equation (A.1) or (5.4) is:

$$V(t) = He^{-\frac{1}{RC}t} - \sum_{i=1}^{3} \left[\frac{RC^2 P_i}{(R^2 C^2 \omega_i^2 + 1)} \omega_i \cos(\omega_i t + c_i) - \frac{CP_i}{(R^2 C^2 \omega_i^2 + 1)} \sin(\omega_i t + c_i) \right]$$

$$(A.10)$$

$$V(t) = He^{-\frac{1}{RC}t} + \sum_{i=1}^{3} \frac{P_i C[\sin(\omega_i t + c_i) - RC\omega_i \cos(\omega_i t + c_i)]}{(R^2 C^2 \omega_i^2 + 1)}$$

$$(A.11)$$

Now we impose the initial condition of $dV/dt = 0$ at $t = 0$. From Equation (A.11), we get:

$$\frac{dV}{dt} = -\frac{He^{-\frac{1}{RC}t}}{RC} + \sum_{i=1}^{3} \frac{P_i C\omega_i [\cos(\omega_i t + c_i) + RC\omega_i \sin(\omega_i t + c_i)]}{(R^2 C^2 \omega_i^2 + 1)}$$

$$(A.12)$$

Putting $dV/dt = 0$ at $t = 0$, we get:

$$0 = -\frac{H}{RC} + \sum_{i=1}^{3} \frac{P_i C\omega_i (\cos c_i + RC\omega_i \sin c_i)}{(R^2 C^2 \omega_i^2 + 1)}$$

$$(A.13)$$

This yields

$$H = \sum_{i=1}^{3} \frac{P_i C\omega_i \tau (\cos c_i + \omega_i \tau \sin c_i)}{(\omega_i^2 \tau^2 + 1)}$$

$$(A.14)$$

where $\tau = RC$.

Upon substituting this expression for H into Equation (A.11), let us now determine $V(t = 0)$. From Equations (A.14) and (A.11), we get:

$$V(t = 0) = H + \sum_{i=1}^{3} \frac{P_i C(\sin c_i - RC\omega_i \cos c_i)}{(R^2 C^2 \omega_i^2 + 1)}$$

$$(A.15)$$

$$= \sum_{i=1}^{3} \frac{P_i C\omega_i \tau (\cos c_i + \omega_i \tau \sin c_i)}{(\omega_i^2 \tau^2 + 1)} + \sum_{i=1}^{3} \frac{P_i C(\sin c_i - \omega_i \tau \cos c_i)}{(\omega_i^2 \tau^2 + 1)}$$

$$(A.16)$$

where $\tau = RC$

$$\therefore V(t = 0) = \sum_{i=1}^{3} \frac{P_i C \left(\omega_i^2 \tau^2 \sin c_i + \sin c_i \right)}{\left(\omega_i^2 \tau^2 + 1 \right)}$$

$$= \sum_{i=1}^{3} \frac{P_i C \sin c_i \left(\omega_i^2 \tau^2 + 1 \right)}{\left(\omega_i^2 \tau^2 + 1 \right)} \tag{A.17}$$

$$= \sum_{i=1}^{3} P_i C \sin c_i \tag{A.18}$$

Appendix B: Solution of Equation 5.28

Equation 5.28 is:

$$\overset{\circ\circ}{V} + 2n\overset{\circ}{V} + p^2 V = \sum_{i=1}^{3} Q_i \sin(\omega_i t + c_i) \tag{B.1}$$

The auxiliary equation of Equation (B.1) is:

$$m^2 + 2nm + p^2 = 0 \tag{B.2}$$

for which

$$m = \frac{-2n \pm \sqrt{4n^2 - 4p^2}}{2}$$

or,

$$m = -n \pm \sqrt{n^2 - p^2} \tag{B.3}$$

Hence the complimentary solution (CF) is:

$$V(t) = k_1 e^{\left(-n - \sqrt{n^2 - p^2} \right)t} + k_2 e^{\left(-n + \sqrt{n^2 - p^2} \right)t} \tag{B.4}$$

The particular solution (PI) of Equation (B.1) is given by:

$$\text{PI} = \sum_{i=1}^{3} \frac{1}{(D^2 + 2nD + p^2)} [Q_i \sin(\omega_i t + c_i)] \tag{B.5}$$

Replacing D^2 by $-\omega_i^2$, we get:

$$PI = \sum_{i=1}^{3} \left\{ \frac{1}{(-\omega_i^2 + 2nD + p^2)} \left[Q_i \sin(\omega_i t + c_i) \right] \right\} \qquad (B.6)$$

$$= \sum_{i=1}^{3} \left\{ \frac{1}{(p^2 - \omega_i^2) + 2nD} \left[Q_i \sin(\omega_i t + c_i) \right] \right\} \qquad (B.7)$$

Upon multiplying by the conjugate, we get:

$$PI = \sum_{i=1}^{3} \left\{ \frac{\left[(p^2 - \omega_i^2) - 2nD \right] \left[Q_i \sin(\omega_i t + c_i) \right]}{\left[(p^2 - \omega_i^2) + 2nD \right] \left[(p^2 - \omega_i^2) - 2nD \right]} \right\} \qquad (B.8)$$

$$= \sum_{i=1}^{3} \left\{ \frac{\left[(p^2 - \omega_i^2) - 2nD \right] \left[Q_i \sin(\omega_i t + c_i) \right]}{(p^2 - \omega_i^2)^2 - 4n^2 D^2} \right\} \qquad (B.9)$$

Upon replacing D^2 by $-\omega_i^2$ in the denominator, we get:

$$PI = \sum_{i=1}^{3} \left\{ \frac{\left[(p^2 - \omega_i^2) - 2nD \right] \left[Q_i \sin(\omega_i t + c_i) \right]}{(p^2 - \omega_i^2)^2 + 4n^2 \omega_i^2} \right\} \qquad (B.10)$$

$$= \sum_{i=1}^{3} \left\{ \frac{Q_i \left[(p^2 - \omega_i^2) \sin(\omega_i t + c_i) - 2n\omega_i \cos(\omega_i t + c_i) \right]}{(p^2 - \omega_i^2)^2 + 4n^2 \omega_i^2} \right\} \qquad (B.11)$$

Hence the general solution (GS) of Equation (B.1) or (5.28) as given by

$$GS = CF + PI$$

is:

$$V(t) = k_1 e^{\left(-n - \sqrt{n^2 - p^2} \right)t} + k_2 e^{\left(-n + \sqrt{n^2 - p^2} \right)t}$$
$$+ \sum_{i=1}^{3} \left\{ \frac{Q_i \left[(p^2 - \omega_i^2) \sin(\omega_i t + c_i) - 2n\omega_i \cos(\omega_i t + c_i) \right]}{(p^2 - \omega_i^2)^2 + 4n^2 \omega_i^2} \right\} \qquad (B.12)$$

Now to find the constants k_1 and k_2, we apply the initial conditions:

$$V(t) = 0 \text{ at } t = 0; \quad \frac{dV}{dt} = 0 \text{ at } t = 0 \qquad (B.13)$$

Putting $V(t) = 0$ at $t = 0$ in Equation (B.12), gives:

$$0 = k_1 + k_2 + \sum_{i=1}^{3} \left\{ \frac{Q_i \left[(p^2 - \omega_i^2) \sin c_i - 2n\omega_i \cos c_i \right]}{(p^2 - \omega_i^2)^2 + 4n^2 \omega_i^2} \right\} \qquad (B.14)$$

Now, we want to impose the condition $dV/dt = 0$ at $t = 0$ in Equation (B.12). From Equation (B.12), we get:

$$\frac{dV}{dt} = k_1\left(-n - \sqrt{n^2 - p^2}\right)e^{\left(-n - \sqrt{n^2 - p^2}\right)t} + k_2\left(-n + \sqrt{n^2 - p^2}\right)e^{\left(-n + \sqrt{n^2 - p^2}\right)}$$
$$+ \sum_{i=1}^{3}\left\{\frac{Q_i\left[(p^2 - \omega_i^2)\omega_i\cos(\omega_i t + c_i) + 2n\omega_i^2\sin(\omega_i t + c_i)\right]}{(p^2 - \omega_i^2)^2 + 4n^2\omega_i^2}\right\} \tag{B.15}$$

Making $\frac{dV}{dt} = 0$ at $t = 0$, we get:

$$0 = k_1\left(-n - \sqrt{n^2 - p^2}\right) + k_2\left(-n + \sqrt{n^2 - p^2}\right) + \sum_{i=1}^{3}\left\{\frac{Q_i\left[(p^2 - \omega_i^2)\omega_i\cos c_i + 2n\omega_i^2\sin c_i\right]}{(p^2 - \omega_i^2)^2 - 4n^2\omega_i^2}\right\}$$
$$\tag{B.16}$$

$$\text{Let } C_i = \left[(p^2 - \omega_i^2)\omega_i\cos c_i + 2n\omega_i^2\sin c_i\right] \tag{B.17}$$

Hence, Equation (B.16), can be rewritten as:

$$-n(k_1 + k_2) - \sqrt{n^2 - p^2}(k_1 - k_2) + \sum_{i=1}^{3}\frac{Q_i C_i}{(p^2 - \omega_i^2)^2 + 4n^2\omega_i^2} = 0 \tag{B.18}$$

Substituting for $(k_1 + k_2)$ from Equation (B.14) in the above Equation (B.18), we get:

$$k_1 - k_2 = \frac{1}{\sqrt{n^2 - p^2}}\left\{-n(k_1 + k_2) + \sum_{i=1}^{3}\frac{Q_i C_i}{(p^2 - \omega_i^2)^2 + 4n^2\omega_i^2}\right\} \tag{B.19}$$

$$= \sum_{i=1}^{3}\left\{\begin{array}{l}\dfrac{nQ_i}{\sqrt{n^2 - p^2}}\dfrac{\left[(p^2 - \omega_i^2)\sin c_i - 2n\omega_i\cos c_i\right]}{(p^2 - \omega_i^2)^2 + 4n^2\omega_i^2} \\ + \dfrac{1}{\sqrt{n^2 - p^2}}\dfrac{Q_i C_i}{(p^2 - \omega_i^2)^2 + 4n^2\omega_i^2}\end{array}\right\} \tag{B.20}$$

Now solving for k_1 and k_2 from Equations (B.14) and (B.20), we get:

$$2k_1 = \sum_{i=1}^{3}\left\{\begin{array}{l}\dfrac{-Q_i\left[(p^2 - \omega_i^2)\sin c_i - 2n\omega_i\cos c_i\right]}{(p^2 - \omega_i^2)^2 + 4n^2\omega_i^2} \\ + \dfrac{nQ_i}{\sqrt{n^2 - p^2}}\dfrac{\left[(p^2 - \omega_i^2)\sin c_i - 2n\omega_i\cos c_i\right]}{(p^2 - \omega_i^2)^2 + 4n^2\omega_i^2} \\ + \dfrac{Q_i}{\sqrt{n^2 - p^2}}\dfrac{\left[(p^2 - \omega_i^2)\omega_i\cos c_i + 2n\omega_i^2\sin c_i\right]}{(p^2 - \omega_i^2)^2 + 4n^2\omega_i^2}\end{array}\right\} \tag{B.21}$$

Therefore,

$$k_1 = \sum_{i=1}^{3} \left\{ \frac{Q_i}{2\sqrt{n^2-p^2}\left[\left(p^2-\omega_i^2\right)^2+4n^2\omega_i^2\right]} \begin{bmatrix} -\sqrt{n^2-p^2}\left(p^2-\omega_i^2\right)\sin c_i \\ +2n\sqrt{n^2-p^2}\,\omega_i\cos c_i \\ +n\left(p^2-\omega_i^2\right)\sin c_i - 2n^2\omega_i\cos c_i \\ +\left(p^2-\omega_i^2\right)\omega_i\cos c_i + 2n\omega_i^2\sin c_i \end{bmatrix} \right\}$$

(B.22)

Then from Equations (B.14) and (B.22), we get the expression for k_2 as:

$$k_2 = -k_1 - \sum_{i=1}^{3} \left\{ \frac{Q_i\left[\left(p^2-\omega_i^2\right)\sin c_i - 2n\omega_i\cos c_i\right]}{\left(p^2-\omega_i^2\right)^2+4n^2\omega_i^2} \right\}$$

(B.23)

Simplifying Equation (B.22) for the constant k_1, we get:

$$k_1 = \sum_{i=1}^{3} \left\{ \frac{Q_i}{2\sqrt{n^2-p^2}\left[\left(p^2-\omega_i^2\right)^2+4n^2\omega_i^2\right]} \begin{pmatrix} -\sqrt{n^2-p^2}\,p^2\sin c_i + \sqrt{n^2-p^2}\,\omega_i^2\sin c_i \\ +2n\sqrt{n^2-p^2}\,\omega_i\cos c_i + np^2\sin c_i \\ +n\omega_i^2\sin c_i - 2n^2\omega_i\cos c_i \\ +p^2\omega_i\cos c_i - \omega_i^3\cos c_i \end{pmatrix} \right\}$$

(B.24)

Then k_2 can be obtained from Equation (B.14) or (B.23), as follows:

$$k_2 = \sum_{i=1}^{3} \left\{ \frac{-Q_i}{2\sqrt{n^2-p^2}\left[\left(p^2-\omega_i^2\right)^2+4n^2\omega_i^2\right]} \begin{pmatrix} -\sqrt{n^2-p^2}\,p^2\sin c_i + \sqrt{n^2-p^2}\,\omega_i^2\sin c_i \\ +2n\sqrt{n^2-p^2}\,\omega_i\cos c_i + np^2\sin c_i \\ +n\omega_i^2\sin c_i - 2n^2\omega_i\cos c_i \\ +p^2\omega_i\cos c_i - \omega_i^3\cos c_i \end{pmatrix} \right.$$
$$\left. - \frac{Q_i\left[\left(p^2-\omega_i^2\right)\sin c_i - 2n\omega_i\cos c_i\right]}{\left(p^2-\omega_i^2\right)^2+4n^2\omega_i^2} \right\}$$

(B.25)

$$= -\sum_{i=1}^{3} \left\{ \frac{Q_i}{2\sqrt{n^2-p^2}\left[\left(p^2-\omega_i^2\right)^2+4n^2\omega_i^2\right]} \begin{bmatrix} -\sqrt{n^2-p^2}\,p^2\sin c_i + \sqrt{n^2-p^2}\,\omega_i^2\sin c_i \\ +2n\sqrt{n^2-p^2}\,\omega_i\cos c_i + np^2\sin c_i + n\omega_i^2\sin c_i \\ -2n^2\omega_i\cos c_i + p^2\omega_i\cos c_i - \omega_i^3\cos c_i \\ +2\sqrt{n^2-p^2}\left(p^2-\omega_i^2\right)\sin c_i - 4n\sqrt{n^2-p^2}\,\omega_i\cos c_i \end{bmatrix} \right\}$$

(B.26)

Equations (B.26) and (B.24) can be rewritten as:

$$k_2 = -\sum_{i=1}^{3} \left\{ \frac{Q_i}{2\sqrt{n^2-p^2}\left[\left(p^2-\omega_i^2\right)^2+4n^2\omega_i^2\right]} \begin{pmatrix} \sqrt{n^2-p^2}\,p^2\sin c_i - \sqrt{n^2-p^2}\,\omega_i^2\sin c_i \\ -2n\sqrt{n^2-p^2}\,\omega_i\cos c_i + np^2\sin c_i \\ +n\omega_i^2\sin c_i - 2n^2\omega_i\cos c_i \\ +p^2\omega_i\cos c_i - \omega_i^3\cos c_i \end{pmatrix} \right\}$$

(B.27)

and

$$
k_1 = \sum_{i=1}^{3} \left\{ \frac{Q_i}{2\sqrt{n^2 - p^2}\left[\left(p^2 - \omega_i^2\right)^2 + 4n^2\omega_i^2\right]} \left(\begin{array}{l} -\sqrt{n^2 - p^2}\,p^2\sin c_i + \sqrt{n^2 - p^2}\,\omega_i^2\sin c_i \\ +2n\sqrt{n^2 - p^2}\,\omega_i\cos c_i + np^2\sin c_i \\ +n\omega_i^2\sin c_i - 2n^2\omega_i\cos c_i \\ +p^2\omega_i\cos c_i - \omega_i^3\cos c_i \end{array} \right) \right\}
$$

(B.28)

Upon substituting the expression for k_1 and k_2 from Equations (B.28) and (B.27) into Equation (B.12), we get the general solution of Equation (B.1) or (5.28) as:

$$
V(t) = \sum_{i=1}^{3} \left\{ \begin{array}{l} \dfrac{Q_i e^{\left(-n-\sqrt{n^2-p^2}\right)t}}{2\sqrt{n^2 - p^2}\left[\left(p^2 - \omega_i^2\right)^2 + 4n^2\omega_i^2\right]} \left(\begin{array}{l} -\sqrt{n^2 - p^2}\,p^2\sin c_i + \sqrt{n^2 - p^2}\,\omega_i^2\sin c_i \\ +2n\sqrt{n^2 - p^2}\,\omega_i\cos c_i + np^2\sin c_i \\ +n\omega_i^2\sin c_i - 2n^2\omega_i\cos c_i \\ +p^2\omega_i\cos c_i - \omega_i^3\cos c_i \end{array} \right) \\[2em] -\dfrac{Q_i e^{\left(-n+\sqrt{n^2-p^2}\right)t}}{2\sqrt{n^2 - p^2}\left[\left(p^2 - \omega_i^2\right)^2 + 4n^2\omega_i^2\right]} \left(\begin{array}{l} \sqrt{n^2 - p^2}\,p^2\sin c_i - \sqrt{n^2 - p^2}\,\omega_i^2\sin c_i \\ -2n\sqrt{n^2 - p^2}\,\omega_i\cos c_i + np^2\sin c_i \\ +n\omega_i^2\sin c_i - 2n^2\omega_i\cos c_i \\ +p^2\omega_i\cos c_i - \omega_i^3\cos c_i \end{array} \right) \\[2em] +\dfrac{Q_i\left[\left(p^2 - \omega_i^2\right)\sin(\omega_i t + c_i) - 2n\omega_i\cos(\omega_i t + c_i)\right]}{\left(p^2 - \omega_i^2\right)^2 + 4n^2\omega_i^2} \end{array} \right\}
$$

(B.29)

Reference

1. D.N. Ghista, K.M. Loh, and M. Damodharan, Lung ventilation modeling and assessment, in *Human Respiration: Anatomy and Physiology, Mathematical Modeling, Numerical Simulation and Application*, edited by V. Kulish, WIT Press, Southampton, U.K., 2006, Chapter 4, pp. 95–115.

6

Lung Gas-Transfer Performance Analysis

Dhanjoo N. Ghista and Kah Meng Loh

CONTENTS

6.1   Objectives

The primary function of the lung is to (1) oxygenate the blood and thereby provide oxygen to the cells for metabolization purposes, and (2) to remove the collected CO_2 from the pulmonary blood. Herein, we will analyze the compositions of the inspired and expired air per breath, and from there compute the O_2 consumption and CO_2 production rates. Next, we derive expressions for diffusion coefficients D_{O_2} and D_{CO_2} in terms of the evaluated cardiac-output, O_2 and CO_2 concentrations in arterial and venous blood, alveolar and blood O_2 and CO_2 partial-pressures. We then take up a typical case study, and demonstrate the computation of D_{O_2} and D_{CO_2}, to represent the lung performance capability to oxygenate the blood.

This chapter (along with the figures) is based on our earlier Chapter 3 on lung gas composition and transfer analysis in *Human Respiration* edited by V. Kulish and published by WIT Press [1].*

6.2   Respiratory System

The respiratory system is the system of the body that deals with breathing. When we breathe, the body takes in the oxygen that it needs and removes the carbon dioxide that it does not need.

First, the body breathes in the air which is sucked through the nose or mouth and down through the trachea (windpipe). The trachea is a pipe shaped by rings of cartilage. It divides into two tubes called bronchi (Figure 6.1), which carry air into each lung.

Inside the lung, the airway tubes divide into smaller and smaller tubes called bronchioles. At the end of each of these tubes are small air sacs called alveoli (Figure 6.2). Capillaries are wrapped around these alveoli. The walls are so thin and close to each other that the air easily seeps through. In this way, oxygen diffuses into the bloodstream, and carbon dioxide (in the bloodstream) diffuses into the alveoli and is then removed from the body when we breathe out (Figure 6.3).

6.3   Gas Transfer between Lung Alveoli
and Pulmonary Capillaries

The lung is designed for gas exchange. Its prime function is to allow oxygen to diffuse from the alveolar air into the venous blood and to remove carbon

---

* With the permission of the publisher WIT Press, Southampton, U.K.

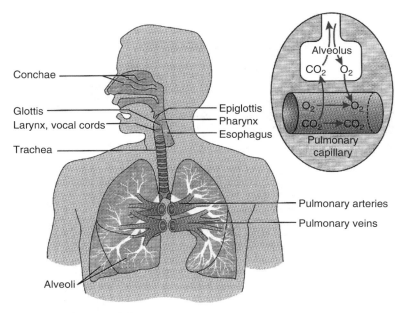

Conchae

Glottis
Larynx, vocal cords
Trachea

Epiglottis
Pharynx
Esophagus

Alveolus
CO_2 O_2

$O_2 \rightarrow O_2$
$CO_2 \rightarrow CO_2$
Pulmonary
capillary

Pulmonary arteries
Pulmonary veins

Alveoli

FIGURE 6.1 (See color insert following page 266.)
Respiratory passages. (Adopted from Guyton, A.C., *Text Book of Medical Physiology*, Saunders, Philadelphia, 1991. With permission from Elsevier.)

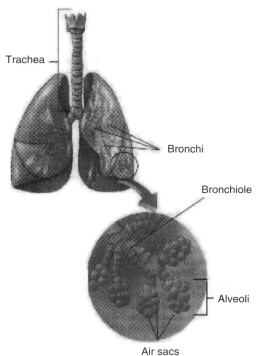

Trachea

Bronchi

Bronchiole

Alveoli

Air sacs

FIGURE 6.2 (See color insert following page 266.)
Trachea, bronchi, and alveoli.

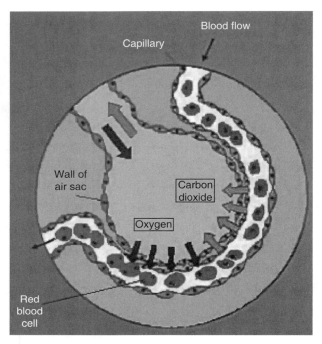

FIGURE 6.3 (See color insert following page 266.)
Exchange of carbon dioxide and oxygen.

dioxide from the capillary blood into the alveoli to be exhaled. It also metabolizes some compounds, filters toxic materials from the circulation, and acts as a reservoir for blood. But its key function is gas exchange.

Oxygen and carbon dioxide move between the alveolar air and capillary blood by simple diffusion. Fick's law of diffusion states that the amount of gas which moves across a sheet of tissue is proportional to the area of sheet but inversely proportional to its thickness. The blood–gas barrier is about 0.5 μm and has an area of between 50 and 100 m^2.

The prodigious surface area for diffusion from inside the limited alveolar cavity is obtained by wrapping the capillaries around an enormous number of alveoli. There are about 300 million alveoli in the human lung, each about 1/3 mm in diameter. If they were spherical, their total surface area would be 85 m^2, but their volume is only 4 L.

6.4   Lung Air Composition Analysis (O_2 Consumption and CO_2 Production Rates)

In this section, we will determine the O_2 consumption and CO_2 production rates, from inspired and expired air compositions [1]. The lung functional

TABLE 6.1

Inspired Air Composition and Partial Pressures

Respiratory Gases	Atmospheric Air		Expired Air	
	mmHg	mL/%	mmHg	mL/%
N_2	597	393.1	566	393.1
		78.55%		74.5%
O_2	159	104.2	120	80.6
		20.84%		15.7%
CO_2	0.3	0.2	27	19.1
		0.04%		3.6%
H_2O	3.7	2.5	47	32.6
		0.49%		6.2%
Total	760	500	760	525.3
		100%		100%

performance is characterized by (1) its ventilatory capacity, to bring air (and hence O_2) into the alveoli, and (2) its capacity to transfer O_2 and CO_2 into and from the pulmonary capillary bed. Hence, the O_2 and CO_2 diffusion coefficients as well as the O_2 consumption rate and the CO_2 production rate represent the lung performance indices. We will deal with this item in the next section.

We carry out a mass balance analysis, involving:

(1) Compositions of air breathed in and out
(2) Consumption or losses of O_2, CO_2, and H_2O

Table 6.1 provides a typical sample clinical data on partial pressures and volumes of N_2, O_2, CO_2, and H_2O of atmospheric air breathed in and expired air, of one breath cycle. The monitored breathing rate (BR) $= 12$ breaths/min, at each breath 500 mL of air is inspired, and we assume P_{H_2O} at 37°C $= 47$ mmHg.

We first compute the volume of expired air compositions, and their pressures based on percent volume. It can be noted (from Table 6.1) that the expired air volume slightly exceeds the inspired air volume for this particular breath cycle. The H_2O loss of 30.1 mL ($= 32.6 - 2.5$ mL) contributes to the major portion of this difference. Note that the water vapor in atmospheric air or inspired air is 0.49% and that in the expired air is 6.2%, while the volume of water vapor in the expired air $= (47 \text{ mmHg}/713 \text{ mmHg}) \times 492.8 = 32.6$ mL

6.4.1 Calculation of O_2 Consumption Rate and CO_2 Production Rate

We now determine the O_2 consumption rate and CO_2 production rates from the inspired and expired gases.

Assuming the patient breathes at 12 times per minute (and 500 mL of air at each breath), we have

$$O_2 \text{ consumption rate} = (\text{Inspired } O_2 - \text{Expired } O_2) \times 12$$
$$= (104.2 - 80.6) \times 12$$
$$= 283.2 \text{ mL/min}$$

$$CO_2 \text{ production rate} = (\text{Expired } CO_2 - \text{Inspired } CO_2) \times 12$$
$$= (19.1 - 0.2) \times 12$$
$$= 226.8 \text{ mL/min}$$

The amount of water vapor in the humidified expired air amounts to 6.2% of the expired air (compared to 0.49% of the dry inspired air), corresponding to partial pressure ratio of water vapor in the expired air ($=47/760$). The volume of the dry expired air $= (525.3 - 32.6) \text{ mL} = 492.7 \text{ mL}$.

Now, let us assume that out of 500 mL of inspired air, the dead space air volume (not taking part in gas-transfer process) is 150 mL, and the alveolar air volume is 350 mL. We can then compute the dead space air volume composition (refer Figure 6.4).

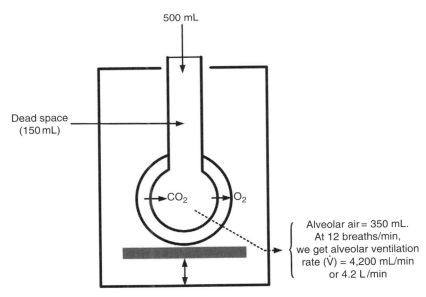

FIGURE 6.4
Dead space volumes. (Modified from Ghista, D.N., Loh, K.M., and Ng, D., in *Human Respiration: Anatomy and Physiology, Mathematical Modeling, Numerical Simulation and Applications*, V. Kulish (Ed), WIT Press, Southampton, U.K., 2005.)

6.4.2 Dead Space Air Composition

The clinical data of expired air composition is

$N_2 = 393.1$ mL, $O_2 = 83.36$ mL, $CO_2 = 16.87$ mL, $H_2O = 34.15$ mL
Total $= 527.49$ mL

Now, the dead space air (Figure 6.4) will be made up of (1) dry air portion from the inspired air (assumed to be of amount 141 mL) plus (2) the water vapor taken up by the dry air (estimated to be 9 mL). Since the expired air portion of 141 mL will not have undergone O_2 and CO_2 transfer, its composition is the same as that of inspired air, and given by

$N_2 = 111$ mL (78.55%), $O_2 = 29.40$ mL (20.84%), $CO_2 = 0.06$ mL (0.04%), $H_2O = 0.69$ mL (0.49%); total volume $= 141$ mL

When this inspired air (in the dead space) of 141 mL is fully humidified, it will take up a further X mL of H_2O vapor, in the ratio of the partial-pressures, as follows:

$$\frac{X}{141} = \frac{47}{713} = 0.0659$$

$$\therefore X = 0.0659 \times 141 = 9.29 \text{ mL of } H_2O \text{ vapor}$$

(which is close to our estimated value of 9 mL).

So, by adding 9.29 mL of H_2O vapor to 0.69 mL of water vapor in the inspired portion of dead space air volume of 141 mL, the total water vapor in the dead space air is 9.98 mL. The humidified dead space air composition will hence be (as tabulated in column 2 of Table 6.2):

$N_2 = 111.00$ ml $(= 73.78\%)$,
$O_2 = 29.40$ ml $(= 19.55\%)$,

TABLE 6.2

The Alveolar Air Composition

	Expired Air (mL)	Dead Space Air (mL)	Alveolar Air (mL)	Alveolar Air Partial Pressure (mmHg)
N_2	393.1	111.00	282.1	569.41
O_2	80.53	29.40	51.13	103.21
CO_2	19.12	0.06	19.06	38.47
H_2O	34.21	9.98	24.23	48.91
Total	526.96	150.44	376.52	760

$CO_2 = 0.06$ ml $(= 0.04\%)$,
$H_2O = 9.98$ ml $(= 6.63\%)$,
Total $= 150.44$ ml

6.4.3 Alveolar Air Composition and Partial Pressures

We can now compute the alveolar air volume composition, by subtracting the dead space air composition volumes from the expired air. These values are tabulated in column 3 of Table 6.2.

Finally, we compute the partial pressures of O_2 and CO_2 (as well as of N_2 and H_2O), as tabulated in column 4 of Table 6.2, so that we can then determine the diffusion coefficients of O_2 and CO_2 based on the monitoring of arterial and venous blood concentrations.

6.5 Lung Gas-Exchange Model and Parametric Analysis

In this section we will characterize O_2 and CO_2 exchange capacities between the lung alveoli and pulmonary capillary bed in terms of O_2 and CO_2 diffusion coefficients D_{O_2} and D_{CO_2} [1].

6.5.1 Expressions for (i) O_2 and (ii) CO_2 Diffusion Coefficients (D_{O_2} and D_{CO_2}), Alveolar and Blood Partial Pressures

The gas exchange between the alveolar air and pulmonary capillary blood is represented by the following O_2 and CO_2 conservation equations (Figure 6.5):

$$Q^{VE}C_{O_2}^{VE} = Q^{AE}C_{O_2}^{AE} + \overset{\circ}{V}_{O_2} \text{ (transfer rate of } O_2 \text{ from the alveolar}$$

$$\text{air to capillary blood)}$$

$$= Q^{AE}C_{O_2}^{AE} + (\Delta P_{av}^{O_2})D_{O_2}; \quad P_{O_2}^{cap} = P_{O_2}^{AE}; \quad D_{O_2} = \frac{\overset{\circ}{V}_{O_2}}{\Delta P_{av}^{O_2}} \quad (6.1)$$

in which $P_{O_2}^{cap} = P_{O_2}^{PRB}$ (O_2 concentration of the pre-oxygenated blood)

$$Q^{VE}C_{CO_2}^{VE} = Q^{AE}C_{CO_2}^{AE} - \overset{\circ}{V}_{CO_2} \text{ (from capillary blood to alveoli)}$$

$$= Q^{AE}C_{CO_2}^{AE} - (\Delta P_{av}^{CO_2})D_{CO_2}; \quad P_{CO_2}^{cap} = P_{CO_2}^{AE}; \quad D_{CO_2} = \frac{\overset{\circ}{V}_{CO_2}}{\Delta P_{av}^{CO_2}} \quad (6.2)$$

in which $P_{CO_2}^{cap} = P_{CO_2}^{PRB}$ (CO_2 concentration of the pre-oxygenated blood), and wherein

(1) $C_{O_2}^{VE} = C_{O_2}^{AB}, \ C_{O_2}^{AE} = C_{O_2}^{VB}; \ C_{CO_2}^{VE} = C_{CO_2}^{AB}, \ C_{CO_2}^{AE} = C_{CO_2}^{VB}.$

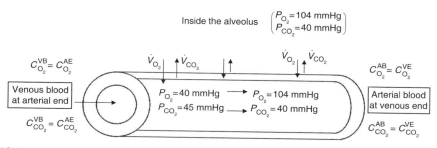

FIGURE 6.5
Schematic of blood–gas concentrations in the pulmonary capillary. Note that $P_{O_2}^{cap} = P_{O_2}^{AE}$; VB, venous blood; AB, arterial blood; AE, arterial end; VE, venous end. (Modified from Ghista, D.N., Loh, K.M., and Ng, D., in *Human Respiration: Anatomy and Physiology, Mathematical Modeling, Numerical Simulation and Applications*, V. Kulish (Ed), WIT Press, Southampton, U.K., 2005.)

We define diffusing capacity of gas across the alveolar-capillary membrane as

$$D = \frac{\text{Volume-rate of transfer of gas}}{\text{Average partial pressure difference}} = \frac{\overset{\circ}{V} \text{ of gas (mL/min)}}{\Delta P_{av} \text{ of the gas (mmHg)}}$$

For instance for O_2, $D_{O_2} = \dfrac{\overset{\circ}{V}_{O_2} \text{ from alveolar air to blood (mL/min)}}{\Delta P_{av}^{O_2} \text{ (mmHg)}}$

We will refer to D_{O_2} as diffusion coefficient of O
$_2$. V_{O_2} is the O_2 consumption rate, which can either be obtained from inspired and expired compositions (as shown earlier) or from
$$Q(C_{O_2}^{AB} - C_{O_2}^{VB}) = Q(C_{O_2}^{VE} - C_{O_2}^{AE}).$$

Typically, for example, $D_{O_2} = \dfrac{252 \text{ mL/min}}{12 \text{ mmHg}} = 21 \text{ mL/min mmHg}.$

(2) Q^{AB} and Q^{VB} are arterial and venous blood flow-rates, $Q^{AB} = Q^{VE}$ (at venous end), $Q^{VB} = Q^{AE}$ (at arterial end); also, $Q^{AB} = Q^{VB} = Q$.

(3) $P_{O_2}^{al}$ and $P_{O_2}^{cap}$ are the alveolar and capillary O_2 partial pressures.

(4) $P_{CO_2}^{al}$ and $P_{CO_2}^{cap}$ are the alveolar and capillary CO_2 partial pressures.

(5) D_{O_2} and D_{CO_2} are the O_2 and CO_2 diffusion coefficients, defined in the caption of Figure 6.5.

(6) $\Delta P_{av}^{O_2} = $ average of $(P_{O_2}^{al} - P_{O_2}^{cap})$ over the capillary length
$\Delta P_{av}^{CO_2} = $ average of $(P_{CO_2}^{cap} - P_{O_2}^{al})$ over the capillary length.

(7) $\overset{\circ}{V}_{O_2}$ is the O_2 transfer rate from alveolar air to capillary blood ($= O_2$ consumption rate), V_{CO_2} is the CO_2 transfer-rate from capillary blood (at arterial end) to alveolar air ($= CO_2$ production rate).

Now we can equate the arterial and venous blood flow rates, as

$$Q^{AB} = Q^{VB} = Q = \frac{SV}{EP} = \frac{CO}{60}$$

wherein SV, EP, and CO being the stroke-volume (in cc), ejection-period (in s), and cardiac-output (in cc or mL/s), respectively. Hence, the above equations can be rewritten, as indicated by the following Equations 6.3 and 6.4.

From Equation 6.1:

$$Q^{AB}C_{O_2}^{AB} = Q^{VB}C_{O_2}^{VB} + (\Delta P_{av}^{O_2})D_{O_2}; \quad P_{O_2}^{cap} = P_{O_2}^{AE} = P_{O_2}^{VB}; \quad C_{O_2}^{VE} = C_{O_2}^{AB}$$

$$QC_{O_2}^{AB} = QC_{O_2}^{VB} + (\Delta P_{av}^{O_2})D_{O_2}$$

$$D_{O_2} = \frac{Q(C_{O_2}^{AB} - C_{O_2}^{VB})}{(\Delta P_{av}^{O_2})} = \frac{\overset{\circ}{V}_{O_2}}{\Delta P_{av}^{O_2}}$$

$$(6.3)$$

wherein $\overset{\circ}{V}_{O_2}$ is the oxygen consumption rate.

From Equation 6.2:

$$Q^{AB}C_{CO_2}^{AB} = Q^{VB}C_{CO_2}^{VB} + (\Delta P_{av}^{CO_2})D_{CO_2}; \quad P_{CO_2}^{cap} = P_{CO_2}^{AE} = P_{CO_2}^{VB}; \quad C_{CO_2}^{VE} = C_{CO_2}^{AB}$$

$$QC_{CO_2}^{AB} = QC_{CO_2}^{VB} + (\Delta P_{av}^{O_2})D_{CO_2}$$

$$D_{CO_2} = \frac{Q(C_{O_2}^{VB} - C_{O_2}^{AB})}{(\Delta P_{av}^{O_2})} = \frac{\overset{\circ}{V}_{CO_2}}{\Delta P_{av}^{CO_2}}$$

$$(6.4)$$

wherein $\overset{\circ}{V}_{CO_2}$ is the carbon dioxide production rate.

In Equations 6.3 and 6.4:

(1) Q, $C_{O_2}^{AB}$ and $C_{O_2}^{VB}$, $C_{CO_2}^{AB}$ and $C_{CO_2}^{VB}$ can be monitored.

(2) D_{O_2} and D_{CO_2} represent the lung gas-exchange parameters.

(3) $\Delta P_{av}^{O_2}$ and $\Delta P_{av}^{CO_2}$ need to be determined in terms of ($P_{O_2}^{al}$ and $P_{O_2}^{VB}$) and ($P_{CO_2}^{al}$ and $P_{CO_2}^{VB}$), respectively, in order to be able to evaluate D_{O_2} and D_{CO_2}.

(4) $P_{CO_2}^{al}$ itself depends on $\overset{\circ}{V}_{O_2}$ and ventilation rate $\overset{\circ}{V}$, $P_{CO_2}^{al}$ itself depends on $\overset{\circ}{V}_{CO_2}$ and ventilation rate $\overset{\circ}{V}$.

(5) $P_{O_2}^{VB}$ depends on $C_{O_2}^{VB}$, and $P_{CO_2}^{VB}$ depends on $C_{CO_2}^{VB}$.

Hence, from Equations 6.3 and 6.4, if we want to evaluate the diffusion coefficients D_{O_2} and D_{CO_2}, we need to determine $\Delta P_{av}^{O_2}$ and $\Delta P_{av}^{CO_2}$, for which we need to also express $P_{O_2}^{al}$, $P_{O_2}^{cap}$ and $P_{CO_2}^{al}$, $P_{CO_2}^{cap}$ in terms of the monitorable quantities of $\overset{\circ}{V}_{O_2}$ and $C_{O_2}^{cap}$ and $\overset{\circ}{V}_{CO_2}$ and $C_{CO_2}^{cap}$. In this regard,

(1) Alveolar $P_{O_2}^{al}$ can be expressed in terms of $\overset{\circ}{V}$ (the ventilation rate) and $\overset{\circ}{V}_{O_2}$ (the O_2 consumption rate), based on Figure 6.6 [2], as

$$P_{O_2}^{al} \text{ (in mmHg)} = k_1 \left[1 - e^{-k_2 \left[(\overset{\circ}{V}/\overset{\circ}{V}_m)/\overset{\circ}{V}_{O_2} \right]} \right]$$

$$(6.5)$$

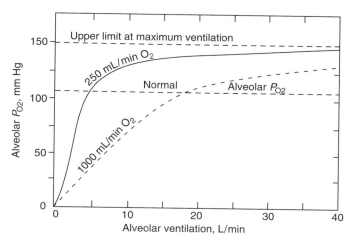

FIGURE 6.6

Effect on alveolar P_{O_2} of (i) Alveolar ventilation ($\overset{\circ}{V}$), and (ii) rate of Oxygen absorption from alveoli P_{O_2} or O_2 consumption rate, $\overset{\circ}{V}_{O_2}$. This relationship is expressed by Equation 6.5. (Adopted from Guyton, A.C., *Text Book of Medical Physiology*, Saunders, Philadelphia, 1991. With permission from Elsevier.)

where

$\overset{\circ}{V}_m$ is the maximum ventilation rate ($=50$ L/min)

$\overset{\circ}{V}_{O_2}$ (the O_2 consumption rate or absorption rate from the alveoli) $= Q(C_{O_2}^{AB} - C_{O_2}^{VB})$

Equation 6.5 indicates that as $\left(\overset{\circ}{V}/\overset{\circ}{V}_m\right)$ increases $P_{O_2}^{al}$ increases, and as $\overset{\circ}{V}_{O_2}$ increases $P_{O_2}^{al}$ decreases (as in Figure 6.6).

(2) Alveolar $P_{CO_2}^{al}$ can be expressed in terms of $\overset{\circ}{V}$ and $\overset{\circ}{V}_{CO_2}$ based on Figure 6.7 [2], as

$$P_{CO_2}^{al}(\text{in mmHg}) = k_3\, e^{-k_4\left[\left(\overset{\circ}{V}/\overset{\circ}{V}_m\right)/\overset{\circ}{V}_{CO_2}\right]} \tag{6.6}$$

where $\overset{\circ}{V}_{CO_2}$ (the CO_2 production rate or excretion rate from the blood) $= Q(C_{CO_2}^{VB} - C_{CO_2}^{AB})$.

This equation implies that as $\left(\overset{\circ}{V}/\overset{\circ}{V}_m\right)$ increases, $P_{CO_2}^{al}$ decreases. Also, as $\overset{\circ}{V}_{CO_2}$ increases, $P_{CO_2}^{al}$ increases, as per Figure 6.7.

(3) Blood P_{O_2} can be obtained in terms of blood C_{O_2}, from the O_2 disassociation curve (providing concentrations in arterial or venous blood, as represented in Figure 6.8) as

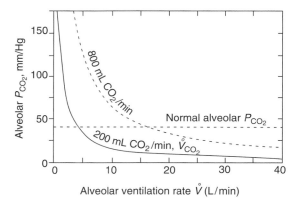

FIGURE 6.7
Effect on alveolar P_{CO_2} of Alveolar ventilation rate ($\overset{\circ}{V}$) and rate of Carbon dioxide excretion from the blood or CO_2 production rate (V_{CO_2}). This relationship is expressed by Equation 6.6. (Adopted from Guyton, A.C., *Text Book of Medical Physiology*, Saunders, Philadelphia, 1991. With permission from Elsevier.)

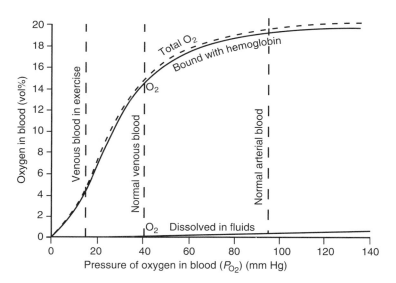

FIGURE 6.8
Oxygen dissociation curves showing the total oxygen vol% or mL of O_2 in each 100 mL of normal blood. We will adopt oxygen concentration units to be (mL of O_2/100 mL of blood), and hence we will divide the numbers on the *y*-axis by 100. (Adopted from Guyton, A.C., *Text Book of Medical Physiology*, Saunders, Philadelphia, 1991. With permission from Elsevier.)

$$C_{O_2}(\text{mL}/100\,\text{mL}) = C_{O_2}^m \left(1 - e^{-k_5 \left(P_{O_2}/P_{O_2}^m\right)}\right)$$

$$\text{or} \quad C_{O_2}^* = 1 - e^{-k_5 P_{O_2}^*} \tag{6.7}$$

wherein

- $C_{O_2}^m$ and $P_{O_2}^m$ are the maximum values of blood O_2 concentration and partial pressure, respectively.

- $C_{O_2}^* = \frac{C_{O_2}}{C_{O_2}^m}$.

- $P_{O_2}^* = \frac{P_{O_2}}{P_{O_2}^m}$.

(4) Blood P_{CO_2} can be obtained in terms of C_{CO_2}, from the CO_2 disassociation curve (providing CO_2 concentration in arterial or venous blood, as represented in Figure 6.9) as

$$C_{CO_2}(\text{mL}/100\,\text{mL}) = C_{CO_2}^m \left[\left(1 - e^{-k_6 \left(P_{CO_2}/P_{CO_2}^m\right)}\right)\right]$$

$$\text{or} \quad C_{CO_2}^* = 1 - e^{-k_6 \left(P_{CO_2}/P_{CO_2}^m\right)} = 1 - e^{-k_6 P_{CO_2}^*} \tag{6.8}$$

wherein

- $C_{CO_2}^m$ and $P_{CO_2}^m$ are the maximum values of blood CO_2 concentrations and partial pressure, respectively.

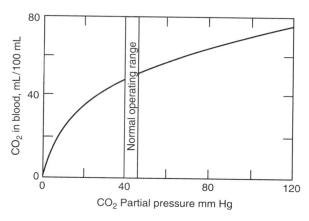

FIGURE 6.9
Carbon dioxide dissociation curve. We adopt carbon dioxide concentration units to be (mL of CO_2/100 mL of blood), and hence we divide the numbers on the *y*-axis by 100. (Adopted from Guyton, A.C., *Text Book of Medical Physiology*, Saunders, Philadelphia, 1991. With permission from Elsevier.)

- $C_{CO_2}^{*} = \frac{C_{CO_2}}{C_{CO_2}^{m}}$.
- $P_{CO_2}^{*} = \frac{P_{CO_2}}{P_{CO_2}^{m}}$.

6.5.2 Quantifying Alveolar O_2 and CO_2 Partial Pressure Expressions

We will now quantify the earlier mentioned empirical expressions of $P_{O_2}^{al}$ and $P_{CO_2}^{al}$ in Equations 6.5 and 6.6.

(1) Now, let us refer Equation 6.5 for the $P_{O_2}^{al}$ partial pressure curve (Figure 6.6), represented by the equation

$$P_{O_2}^{al} = k_1\left[1 - e^{-k_2\left[(\mathring{V}/\mathring{V}_m)/\mathring{V}_{O_2}\right]}\right] = k_1\left[1 - e^{-k_2\left[\mathring{V}^{*}/\mathring{V}_{O_2}\right]}\right]$$

$$\text{where } \mathring{V}^{*} = \mathring{V}/\mathring{V}_m \tag{6.9}$$

wherein

$\mathring{V}$ is the alveolar ventilation rate (in L/min)
$\mathring{V}_m$ is the maximum ventilation rate ($= 50$ L/min)
$\mathring{V}_{O_2}$ is the O_2 consumption rate (in L/min)

Herein, the coefficients k_1 and k_2 can be determined by having this equation match the Figure 6.6 data. Note, in this equation, when $\mathring{V} = 0$, $P_{O_2}^{al} = 0$ from the equation, which satisfies the data in Figure 6.6.

In Figure 6.6, for $\mathring{V}_{O_2} = 0.25$ L/min, when $\mathring{V}^{*} = \mathring{V}/\mathring{V}_m = 0.5$, $P_{O_2}^{al} = 140$ mmHg.
Hence,

$$140 = k_1\left[1 - e^{-k_2(0.5/0.25)}\right] = k_1(1 - e^{-2k_2}) \tag{6.10}$$

Also, when $\mathring{V}_{O_2} = 1$ L/min and $\mathring{V}^{*} = 0.3$ L/min, $P_{O_2}^{al} = 100$ mmHg.
Hence,

$$100 = k_1\left[1 - e^{-k_2(0.3/1)}\right] = k_1(1 - e^{-0.3k_2}) \tag{6.11}$$

From Equations 6.10 and 6.11, we get

$$\frac{140}{100} = \frac{k_1(1 - e^{-2k_2})}{k_1(1 - e^{-0.3k_2})} = \frac{1 - e^{-2k_2}}{1 - e^{-0.3k_2}}$$

$$\therefore 140 - 140e^{-0.3k_2} = 100 - 100e^{-2k_2}$$

$$\text{so that, } k_2 = 4.18 \text{ min/L} \tag{6.12}$$

Upon substituting $k_2 = 4.18$ min/L into Equation 6.10, we obtain

$$140 = k_1 \left(1 - e^{-(2 \times 4.18)}\right), \text{ so that } k_1 \approx 140 \text{ mmHg} \qquad (6.13)$$

Hence, the $P_{O_2}^{al}$ curve can be represented by

$$P_{O_2}^{al}(\text{mmHg}) = 140 \left[1 - e^{-4.18 \left[\overset{\circ}{V}{}^{*}/\overset{\circ}{V}_{O_2}\right]}\right] \qquad (6.14)$$

Wherein

$$\overset{\circ}{V}_{O_2} = Q\left(C_{O_2}^{AB} - C_{O_2}^{VB}\right)$$

$$\overset{\circ}{V}{}^{*} = \frac{\overset{\circ}{V}}{50 \text{ L/min}}$$

(2) Now, let us look at the $P_{CO_2}^{al}$ expression:

$$P_{CO_2}^{al} = k_3 e^{-k_4 \left[\left(\overset{\circ}{V}/\overset{\circ}{V}_m\right)/\overset{\circ}{V}_{CO_2}\right]} = k_3 e^{-k_4 \left[\overset{\circ}{V}{}^{*}/\overset{\circ}{V}_{CO_2}\right]}$$

We note, from Figure 6.7 that for $\overset{\circ}{V}_{CO_2} = 0.2$ L/min and $\overset{\circ}{V}{}^{*} = 0.2$, $P_{CO_2}^{al} = 12$. Hence, from the above equation, we get

$$12 = k_3 e^{-k_4} \qquad (6.15)$$

Also, for $\overset{\circ}{V}_{CO_2} = 0.8$ L/min and $\overset{\circ}{V}{}^{*} = 0.2$, $P_{CO_2}^{al} = 62$ mmHg. Hence,

$$62 = k_3 e^{-k_4(0.2/0.8)} = k_3 e^{-k_4/4} \qquad (6.16)$$

From Equations 6.15 and 6.16, we get

$$\frac{12}{62} = \frac{e^{-k_4}}{e^{\frac{-k_4}{4}}} = e^{-\frac{3}{4}k_4}$$

$$\therefore \ln\left(\frac{12}{62}\right) = -\frac{3}{4}k_4, \text{ so that } k_4 = 2.19 \qquad (6.17)$$

Substituting $k_4 = 2.19$ into Equation 6.16, we obtain

$$62 = k_3 e^{-(2.19/4)}, \quad \therefore k_3 = 107.18 \qquad (6.18)$$

Hence, the $P_{CO_2}^{al}$ curve can be represented as

$$P_{CO_2}^{al}(\text{mmHg}) = 107.18 e^{-2.19 \left[\left(\overset{\circ}{V}/\overset{\circ}{V}_m\right)/\overset{\circ}{V}_{CO_2}\right]} \qquad (6.19)$$

wherein
$$\overset{\circ}{V}{}^* = \overset{\circ}{V}/50 \text{ L/min}$$
$$\overset{\circ}{V}_{CO_2} = Q(C_{CO_2}^{VB} - C_{CO_2}^{AB})$$

6.5.3 Quantifying Arterial and Venous O_2 and CO_2 Partial Pressure Expressions

We will now quantify the previous expressions of $P_{O_2}^{AB}$ and $P_{CO_2}^{VB}$ in terms of $C_{O_2}^{AB}$ and $C_{CO_2}^{VB}$, as given by Equations 6.7 and 6.8.

(1) From the O_2 disassociation curve in Figure 6.8, we had put down

$$\text{Blood } C_{O_2} = C_{O_2}^m \left[1 - e^{-k_5 \left[P_{O_2}/P_{O_2}^m \right]} \right]$$

$$\text{or} \quad C_{O_2}^* = 1 - e^{-k_5 P_{O_2}^*} \tag{6.20}$$

wherein

$$C_{O_2}^* = \frac{C_{O_2}}{C_{O_2}^m}$$

$$P_{O_2}^* = \frac{P_{O_2}}{P_{O_2}^m}$$

$C_{O_2}^m = 0.2$ (or 20 mL of O_2/100 mL of blood)

$P_{O_2}^m = 140 \text{ mmHg}$

From Figure 6.8, at $P_{O_2}^* = (40\,\text{mmHg}/140\,\text{mmHg}) = 0.29$ (for normal venous blood),

$$C_{O_2}^* = \frac{15}{20} = 0.75$$

Hence, from Equation 6.20:

$$0.75 = 1 - e^{-0.29 k_5}$$
$$\therefore k_5 = 4.78 \tag{6.21}$$

Also, for

$$P_{O_2}^* = (95\,\text{mmHg}/140\,\text{mmHg}) = 0.68 \text{ (for normal blood)}$$
$$C_{O_2}^* = \frac{0.19}{0.20} = 0.95$$

Hence, from Equation 6.20:

$$0.95 = 1 - e^{-0.68k_5} \quad \text{or} \quad k_5 = 4.4 \qquad (6.22)$$

So we take the average value of k_5:

$$\therefore k_5 = \frac{(4.78 + 4.4)}{2} = 4.59 \qquad (6.23)$$

Then, the O_2 disassociation curve is given by

$$C_{O_2} = C_{O_2}^B = 0.2 \left[1 - e^{-4.59(P_{O_2}/140)} \right] \qquad (6.24)$$

Hence, from the above equation, the partial pressure of O_2 in blood $(P_{O_2}^B)$ can be expressed as

$$P_{O_2}^B = \frac{140}{4.59} \ln \left[\frac{0.2}{0.2 - C_{O_2}^B} \right] = 30.5 \ln \left[\frac{0.2}{0.2 - C_{O_2}^B} \right] \qquad (6.25)$$

(2) Finally, we look at CO_2 disassociation curve (in Figure 6.9), expressed as

$$C_{CO_2} = C_{CO_2/\max} \left(1 - e^{-k_6 \left(P_{CO_2}/P_{CO_2}^m \right)} \right)$$

$$\text{or} \quad C_{CO_2}^* = 1 - e^{-k_6 \left(P_{CO_2}/P_{CO_2}^m \right)} = 1 - e^{-k_6 P_{CO_2}^*} \qquad (6.26)$$

wherein

$C_{CO_2}^m = 0.8$ (or 80 mL of CO_2 in 100 mL of blood)

$P_{CO_2}^m = 140$ mmHg

On the basis of Figure 6.9, when $P_{CO_2}^* = (20 \text{ mmHg}/140 \text{ mmHg}) = 0.14$, $C_{CO_2}^* = \frac{0.38}{0.80} = 0.475$,

so that

$$0.475 = 1 - e^{-0.14k_6} \quad \text{and} \quad k_6 = 4.60 \qquad (6.27)$$

Also, when $P_{CO_2}^* = (70 \text{ mmHg}/140 \text{ mmHg}) = 0.5$, $C_{CO_2}^* = \frac{0.60}{0.80} = 0.75$, so that

$$0.75 = 1 - e^{-0.5k_6} \quad \text{and} \quad k_6 = 2.77 \qquad (6.28)$$

So we take the average value of k_6:

$$k_6 = \frac{(4.60 + 2.77)}{2} = 3.69 \tag{6.29}$$

Then, in the CO_2 disassociation curve, the CO_2 concentration is given (from Equations 6.26 through 6.30) by

$$C_{CO_2} = C_{CO_2}^B = 0.8\left[1 - e^{-3.69(P_{CO_2}/140)}\right] \tag{6.30}$$

so that the partial pressure of CO_2 in blood ($P_{CO_2}^B$) can be expressed as

$$P_{CO_2}^B(\text{mmHg}) = 37.94 \ln\left[\frac{0.8}{0.8 - C_{CO_2}^B}\right] \tag{6.31}$$

Now we have been able to express: (1) $P_{O_2}^{al}$ in terms $\overset{\circ}{V}_{O_2}$ and $\overset{\circ}{V}^*$, by means of Equation 6.14; (2) $P_{CO_2}^{al}$ in terms of $\overset{\circ}{V}_{CO_2}$ and $\overset{\circ}{V}^*$, by means of Equation 6.19; (3) $P_{O_2}^B$ (or partial pressure of blood O_2) in terms of O_2 blood concentration $C_{O_2}^B$, by means of Equation 6.25; and (4) $P_{CO_2}^B$ (or partial pressure of blood CO_2) in terms of CO_2 blood concentration $C_{CO_2}^B$, by means of Equation 6.31. We are now in a position to develop the expressions for $\Delta P_{av}^{O_2}$ and $\Delta P_{av}^{CO_2}$ in the pulmonary capillary bed.

6.5.4 Determining $\Delta P_{av}^{O_2}$ and $\Delta P_{av}^{CO_2}$

In order to determine D_{O_2} and D_{CO_2}, we also need to determine $\Delta P_{av}^{O_2}$ and $\Delta P_{av}^{CO_2}$ in Equations 6.3 and 6.4, respectively.

Figure 6.10 illustrates the variation of $\Delta P^{O_2}(=P_{O_2}^{al} - P_{O_2}^{cap})$ along the length (l) of the capillary bed.

$$\text{Let, } l^* = l/l_m \tag{6.32}$$

Now we can express ΔP^{O_2} as function l^*, as follows:

$$\Delta P^{O_2}(l^*) = \Delta P_{max}^{O_2} f_{O_2}(l^*) \tag{6.33}$$

where $f_{O_2}(l^*)$ is a function that varies from 1 (at $l^* = 0$) to 0 (at $l^* = 1$).
Then

$$\Delta P_{av}^{O_2} = \Delta P_{max}^{O_2}\left(\int_0^1 f_{O_2}(l^*)\, dl^*\right) = \Delta P_{max}^{O_2}(F_{O_2}) \tag{6.34}$$

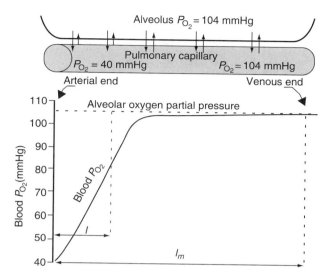

FIGURE 6.10
Uptake of oxygen by the pulmonary capillary blood. (Curve figure was constructed from Mihorn and Pulley, *Biophys. J.*, 8, 337, 1968; Guyton, A.C., *Text Book of Medical Physiology*, Saunders, Philadelphia, 1991. With permission from Elsevier.)

On the basis of the data in Chapter 10 of [3], since $\Delta P^{O_2}_{av} = 12$ mmHg for $\Delta P^{O_2}_{max} = 65$ mmHg, we have $F_{O_2} = 0.185$. We can hence put down

$$\Delta P^{O_2}_{av}(\text{mmHg}) = 0.185; \quad \Delta P^{O_2}_{max} = 0.185\left(P^{al}_{O_2} - P^{VB}_{O_2}\right) \tag{6.35}$$

We can similarly determine the average value of $\Delta P^{CO_2}_{av}$ from Figure 6.11, which shows the variation of $\Delta P^{CO_2}(=P^{cap}_{CO_2} - P^{al}_{CO_2})$ along the length l of the capillary bed.

$$\text{Let } l^* = l/l_m \tag{6.36}$$

Then, we can represent Figure 6.11 as

$$\Delta P^{CO_2}(l^*) = \Delta P^{CO_2}_{max} f_{CO_2}(l^*) \tag{6.37}$$

where f_{CO_2} is a function that varies from 1 (at $l^* = 0$) to 0 (at $l^* = 1$), so that,

$$\Delta P^{co_2}_{av} = \Delta P^{o_2}_{max}\left[\int_0^1 f_{co_2}(l^*)\,dl^*\right] = \Delta P^{co_2}_{max}(F_{co_2}) \tag{6.38}$$

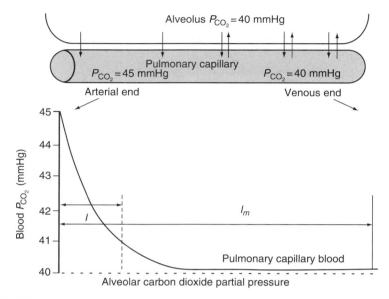

FIGURE 6.11
Diffusion of carbon dioxide from the pulmonary blood into the alveolus. (Curve figure was constructed from Mihorn and Pulley, *Biophys. J.*, 8, 337, 1968; Guyton, A.C., *Text Book of Medical Physiology*, Saunders, Philadelphia, 1991. With permission from Elsevier.)

On the basis of the data in Chapter 10 of [3], since $\Delta P_{av}^{CO_2} = 0.5$ mmHg for $\Delta P_{max}^{co_2} = 5$ mmHg; hence we have $F_{CO_2} = 0.1$.

We can now put down

$$\Delta P_{av}^{co_2}(\text{mmHg}) = 0.1\ \Delta P_{max}^{o_2} = 0.1\left(P_{co_2}^{VB} - P_{co_2}^{al}\right) \tag{6.39}$$

6.6   Sequential Procedure to Compute D_{O_2} and D_{CO_2}

On the basis of Equations 6.3 and 6.35, we have

$$D_{O_2} = \frac{\text{Total O}_2 \text{ consumed}}{\Delta P_{av}^{O_2}} = \frac{\overset{\circ}{V}_{O_2}}{\Delta P_{av}^{O_2}} = \frac{Q\left(C_{O_2}^{AB} - C_{O_2}^{VB}\right)}{0.185\left(P_{O_2}^{al} - P_{O_2}^{VB}\right)} \tag{6.40}$$

On the basis of Equations 6.4 and 6.39, we have

$$D_{CO_2} = \frac{\text{Total CO}_2 \text{ produced}}{\Delta P_{av}^{CO_2}} = \frac{\dot{V}_{CO_2}}{\Delta P_{av}^{CO_2}} = \frac{Q\left(C_{CO_2}^{VB} - C_{CO_2}^{AB}\right)}{0.1\left(P_{CO_2}^{VB} - P_{CO_2}^{al}\right)} \tag{6.41}$$

Note that in the denominators of these equations, we need to know $\overset{\circ}{V}_{O_2}(t)$ and $\overset{\circ}{V}_{CO_2}(t)$ for determining $P_{O_2}^{al}$ and $P_{CO_2}^{al}$, respectively, based on Equations 6.14 and 6.19.

(1) We first monitor: $V(t)$, $\overset{\circ}{V}(t)$, SV(stroke volume), EP(cardiac ejection period), $C_{O_2}^{VB}$, $C_{O_2}^{AB}$, $C_{CO_2}^{VB}$, and $C_{CO_2}^{AB}$ (O_2 and CO_2 concentrations in pre-oxygenated and post-oxygenated blood).

We can utilize the monitored values of $C_{O_2}^{AB}(=C_{O_2}^{VE})$ and $C_{O_2}^{VB}(=C_{O_2}^{AE})$ in Equation 6.40, and the values of $C_{CO_2}^{AB}(=C_{CO_2}^{VE})$ and $C_{CO_2}^{VB}(=C_{CO_2}^{AE})$ in Equation 6.41, as indicated below.

(2) We determine

$Q = SV/$ejection period, and therefrom

$$\overset{\circ}{V}_{O_2}(t) = Q(C_{O_2}^{AB} - C_{O_2}^{VB}), \quad \text{the } O_2 \text{ consumption rate} \qquad (6.42)$$

$$\overset{\circ}{V}_{CO_2}(t) = Q(C_{CO_2}^{VB} - C_{CO_2}^{AB}), \quad \text{the } CO_2 \text{ production rate} \qquad (6.43)$$

wherein we have utilized in the numerators of Equations 6.40 and 6.41, the monitored values of $C_{O_2}^{VB}$, $C_{O_2}^{AB}$, $C_{CO_2}^{VB}$, and $C_{CO_2}^{AB}$.

(3) We next obtain $P_{O_2}^{al}$ and $P_{CO_2}^{al}$ in the denominators of the above Equations 6.40 and 6.41, by substituting the expressions for $\overset{\circ}{V}_{O_2}(t)$ and $\overset{\circ}{V}_{CO_2}(t)$ as well as of $\overset{\circ}{V}^{*}(=\overset{\circ}{V}/50)$ into the above equations for $P_{O_2}^{al}$ (Equation 6.14) and $P_{CO_2}^{al}$ (Equation 6.19).

(4) We then obtain the values of $P_{O_2}^{VB}$(or $P_{O_2}^{AE}$) and $P_{CO_2}^{VB}$(or $P_{CO_2}^{AE}$) in the denominators of Equations 6.40 and 6.41, by substituting the monitored values of $C_{O_2}^{VB}(=C_{O_2}^{AE})$ and $C_{CO_2}^{VB}(=C_{CO_2}^{AE})$ as well as of $\overset{\circ}{V}^{*}(=\overset{\circ}{V}/50)$ into Equations 6.25 and 6.31.

With this information, we can evaluate D_{O_2} and D_{CO_2} in Equations 6.40 and 6.41.

(5) Alternately, in order to evaluate D_{O_2} and D_{CO_2}, we can also evaluate $\Delta P_{av}^{O_2}$ from Equation 6.35, by employing the values of $P_{O_2}^{al}$ in step (3) and $P_{O_2}^{VB}$ in step (4), for its utilization in Equation 6.40.

We can likewise evaluate $\Delta P_{av}^{CO_2}$ from Equation 6.39, by employing the values of $P_{CO_2}^{al}$ in step (3) and $P_{CO_2}^{VB}$ in step (4), for its utilization in Equation 6.41.

Then, in order to determine the values of the lung gas-exchange parameters (or indices) D_{O_2} and D_{CO_2}, we substitute into Equations 6.40 and 6.41 for (a) $\overset{\circ}{V}_{O_2}(t)$ and $\overset{\circ}{V}_{CO_2}(t)$ from step (2) and for (b) $\Delta P_{av}^{O_2}$ and $\Delta P_{av}^{CO_2}$ from step (5).

6.7 Case Studies

Let us now evaluate D_{O_2} and D_{CO_2} for a couple of typical cases' data values.

6.7.1 Case Study 1

The monitored data is as follows:

$$C_{O_2}^{AE} = C_{O_2}^{VB} = 0.13, \quad C_{O_2}^{VE} = C_{O_2}^{AB} = 0.18, \quad C_{CO_2}^{AE} = C_{CO_2}^{VB} = 0.525,$$

$C_{CO_2}^{VE} = C_{CO_2}^{AB} = 0.485$; alveolar ventilation rate $(\overset{\circ}{V}) = 5$ L/min, blood flow rate in the pulmonary vascular bed $(Q) = 5$ L/min.

From Equation 6.25, we obtain

$$P_{O_2}^{VB} = 30.5 \ln\left[\frac{0.2}{0.2 - C_{O_2}^{VB}}\right] = 30.5 \ln\left[\frac{0.2}{0.2 - 0.13}\right]$$

$$= 32.02 \text{ mmHg} \tag{6.44}$$

From Equation 6.31, we obtain

$$P_{CO_2}^{VB} = 37.94 \ln\left[\frac{0.8}{0.8 - C_{CO_2}^{VB}}\right] = 37.94 \ln\left[\frac{0.8}{0.8 - 0.525}\right]$$

$$= 40.51 \text{ mmHg} \tag{6.45}$$

We have monitored cardiac output $Q = 5$ L/min. From the ventilation rate $\overset{\circ}{V} = 5$ L/min, we obtain (from Equation 6.9):

$$\overset{\circ}{V}{}^{*} = \frac{5}{50} = 0.1 \tag{6.46}$$

Then, from Equation 6.3 or 6.42:

$$\overset{\circ}{V}_{O_2}(t) = Q\left(C_{O_2}^{AB} - C_{O_2}^{VB}\right), \quad \text{so that from the above data,}$$

$$\overset{\circ}{V}_{O_2}(t) = 5000 \times (0.05) = 250 \text{ ml of } O_2/\text{min consumption rate} \tag{6.47}$$

Likewise, from Equation 6.4 or 6.43 and the above data,

$$\overset{\circ}{V}_{CO_2}(t) = Q(C_{CO_2}^{VB} - C_{CO_2}^{AB}) = 5000(0.04)$$

$$= 200 \text{ ml of } CO_2/\text{min production rate} \tag{6.48}$$

Now, from Equation 6.14:

for $\overset{\circ}{V}{}^{*} = 0.1$ (Equation 6.46) and $\overset{\circ}{V}_{O_2} = 0.25$ L (Equation 6.47), we obtain $P_{O_2}^{al}$:

$$P_{O_2}^{al} = 140\left[1 - e^{-4.18\left[\overset{\circ}{V}{}^{*}/\overset{\circ}{V}_{O_2}\right]}\right]$$

$$= 140\left[1 - e^{-4.18[0.1/0.25]}\right] = 113.7 \text{ mmHg} \tag{6.49}$$

Likewise, from Equation 6.19, for $\overset{\circ}{V}{}^{*} = 0.1$ (Equation 6.46) and $\overset{\circ}{V}_{CO_2} = 0.20$ L (Equation 6.48), we obtain

$$P_{CO_2}^{al} = 107.18e^{-2.19\left(\overset{\circ}{V}{}^{*}/\overset{\circ}{V}_{CO_2}\right)} = 107.18e^{-2.19[0.1/0.2]}$$

$$= 35.86 \text{ mmHg} \tag{6.50}$$

Now, we can evaluate the diffusion coefficients as follows:

From Equations 6.40, 6.35, 6.44, 6.47, and 6.49, along with the given data, we get

$$D_{O_2} = \frac{Q\left(C_{O_2}^{AB} - C_{O_2}^{VB}\right)}{\Delta P_{av}^{O_2}} = \frac{Q\left(C_{O_2}^{AB} - C_{O_2}^{VB}\right)}{\Delta P_{max}^{O_2}(F_{O_2})} = \frac{\overset{\circ}{V}_{O_2}(\text{in mL/min})}{0.185\left(P_{O_2}^{al} - P_{O_2}^{VB}\right)\text{mmHg}}$$

$$= \frac{5000(0.05)}{(113.7 - 32.02) \times 0.185} = 16.84 \text{ (mL O}_2\text{/min)/mmHg} \tag{6.51}$$

Likewise, from Equations 6.41, 6.39, 6.45, 6.48, and 6.50, along with the given data, we get

$$D_{CO_2} = \frac{Q\left(C_{CO_2}^{CB} - C_{CO_2}^{AB}\right)}{\Delta P_{av}^{CO_2}} = \frac{\overset{\circ}{V}_{CO_2}(\text{in mL/min})}{0.1\left(P_{CO_2}^{VB} - P_{CO_2}^{al}\right)\text{mmHg}}$$

$$= \frac{5000(0.04)}{(40.51 - 35.86) \times 0.1} = 430.11(\text{mL CO}_2\text{/min)/mmHg} \tag{6.52}$$

6.7.2 Case Study 2

For this case study (to determine D_{O_2} and D_{CO_2}), we have the following data:

(1) From the inspired and expired air data analysis (such as that carried out in Section 6.4), we compute:

O$_2$ consumption rate = 283.2 mL/min (Section 6.4.1),

CO$_2$ production rate = 226.8 mL/min (Section 6.4.1), $\qquad$ (6.53)

$P_{O_2}^{al} = 103.21$ mmHg and $P_{CO_2}^{al} = 38.47$ mmHg (Table 6.2).

(2) From the venous blood–gas analysis:

$$C_{O_2}^{VB} = 0.13, \quad C_{CO_2}^{VB} = 0.548 \tag{6.54}$$

Then, as per Equation 6.25,

$$P_{O_2}^{VB} = 31.2 \text{ mmHg}, \tag{6.55}$$

corresponding to the monitored $C_{O_2}^{VB} = 0.13$.
Also, for the monitored $C_{CO_2}^{VB} = 0.548$, as per Equation 6.31, we get

$$P_{CO_2}^{VB} = 43.84 \text{ mmHg} \tag{6.56}$$

We have obtained, from air composition analysis,

$$\text{O}_2 \text{ consumption rate } \overset{\circ}{V}_{O_2}(t) = 283.3 \text{ mL/min} \tag{6.57}$$

$$\text{and CO}_2 \text{ production rate } \overset{\circ}{V}_{CO_2}(t) = 226.8 \text{ mL/min} \tag{6.58}$$

Hence, from Equation 6.40 with the calculated values of $\overset{\circ}{V}_{O_2} = 283.3$ mL/min (Equation 6.57), $P_{O_2}^{al}$ (Equation 6.53), and of $P_{O_2}^{VB}$ (Equation 6.55), we obtain

$$D_{O_2} = \frac{\overset{\circ}{V}_{O_2}}{\Delta P_{av}^{O_2}} = \frac{283.2}{(103.21 - 31.2) \times 0.185}$$

$$= 21.90 \text{ (mL O}_2/\text{min)}/\text{mmHg} \tag{6.59}$$

Likewise, from Equation 6.41, along with the calculated values of $\overset{\circ}{V}_{CO_2} = 226.8$ mL/min (Equation 6.58), $P_{CO_2}^{al}$ (Equation 6.53), and $P_{CO_2}^{VB}$ (Equation 6.56), we obtain

$$D_{CO_2} = \frac{\overset{\circ}{V}_{CO_2}}{\Delta P_{av}^{CO_2}} = \frac{226.8}{(43.84 - 38.47) \times 0.1}$$

$$= 417.68 \text{ (mL CO}_2/\text{min)}/\text{mmHg} \tag{6.60}$$

It is noted that both the methods (in Sections 6.7.1 and 6.7.2) yield similar values of D_{O_2} and D_{CO_2}, thereby providing validity to our analysis. However, the advantage of this method (in Section 6.7.2) over the previous method (in Section 6.7.1) is that it does not require monitoring the cardiac output, and is hence simpler to implement clinically.

6.8 Nondimensional Gas-Transfer Index (NDGTI)

We can even formulate a nondimensional gas-transfer index as

$$\text{NDGTI} = \frac{D_{CO_2}}{D_{O_2}} \times \frac{O_2 \text{ consumption rate}}{CO_2 \text{ production rate}} \tag{6.61}$$

Checking dimensions:

$$\begin{aligned}
\text{NDGTI} &= \frac{D_{CO_2}}{D_{O_2}} \times \frac{O_2 \text{ consumption rate}}{CO_2 \text{ production rate}} \\
&= \frac{(\text{mL } CO_2/\text{min})/\text{mmHg}}{(\text{mL } O_2/\text{min})/\text{mmHg}} \times \frac{\text{mL } O_2/\text{min}}{\text{mL } CO_2/\text{min}}
\end{aligned}$$

For our case 6.7.2, we have the value of NDGTI given by

$$\text{NDGTI} = \frac{417.68 \ (\text{mL } CO_2/\text{min})/\text{mmHg}}{21.90 \ (\text{mL } O_2/\text{min})/\text{mmHg}} \times \frac{283.2 \ \text{mL/min}}{226.8 \ \text{mL/min}} = 23.8 \tag{6.62}$$

In summary, we have derived expressions for diffusion coefficients D_{O_2} and D_{CO_2}, in terms of (1) evaluated cardiac-output CO, O_2, and CO_2 concentrations in arterial and venous blood as well as alveolar and blood O_2 and CO_2 partial-pressures or (2) inspired and expired air data analysis and O_2 and CO_2 concentrations of the venous blood. The coefficients D_{O_2} and D_{CO_2} represent the lung capability to oxygenate the blood.

The derived information of D_{O_2} and D_{CO_2} as well as of O_2 and CO_2 metabolic rates can be of considerable clinical use (including for SARS assessment). Finally, the NDGTI index can be employed by clinicians to assess the gas-transfer capacity of the lung-capillary bed system, in terms of just one nondimensional index.

References

1. Ghista, D.N., Loh, K.M., and Ng, D., Lung gas composition and transfer analysis: O_2 and CO_2 diffusion coefficients and metabolic rates, in *Human Respiration: Anatomy and Physiology, Mathematical Modeling, Numerical Simulation and Applications*, V. Kulish (Ed.), WIT Press, Southampton, U.K., 2005, Chapter 3, pp. 78–93.
2. Guyton A.C., *Text Book of Medical Physiology*, Saunders, Philadelphia, 1991.
3. Cooney, D.O., *Biomedical Engineering Principles. (An Introduction to Fluid, Heat and Mass Transport Processes)*, Marcel Dekker, New York, 1976, Chapter 10, pp. 341–397.

7

Indicators for Extubation of Mechanically Ventilated COPD Patients Using Lung Ventilation Modeling

Dhanjoo N. Ghista and Rohit Pasam

CONTENTS

7.1   Introduction

This chapter is about modeling of lung ventilation response of COPD patients on mechanical ventilation, and how we can develop a lung ventilatory index to enable us to assess their lung status and decide when they are ready to be weaned off the mechanical ventilator. In mechanically ventilated patients with chronic obstructive pulmonary disease (COPD), elevated airway resistance and decreased lung compliance (i.e., stiffer lung) are observed with rapid breathing. The need for accurate predictive indicators of lung status improvement is essential for ventilator discontinuation (or extubation) through stepwise reduction in mechanical support, as and when patients are increasingly able to support their own breathing, followed by trials of unassisted breathing preceding extubation and ending with extubation.

For determining if a patient is ready to be discontinued from mechanical ventilation after the clinician has chosen an appropriate indicator to assess lung status he or she will incorporate this indicator into three general approaches for ventilator discontinuation which are (1) synchronized intermittent mandatory ventilation (SIMV) where the number of breaths is supplied by the ventilator and lowering the ventilator breaths will initiate more spontaneous breaths in the patient; (2) pressure support ventilation (PSV) which provides inspiratory pressure assistance based on spontaneous efforts; and (3) spontaneous breathing trial (SBT). The intent of the ventilatory discontinuation process is to decrease the level of support provided by the ventilator, requiring the patient to assume a greater proportion of the ventilatory workload.

For stepwise reduction in mechanical ventilatory support, the most useful clinically employed indicators have been rapid shallow breathing index (RSBI) <65 breaths/min/L (measured using ventilatory settings) and respiratory rate or frequency (RF) <38 breaths/min. However, these are extrinsic empirical indices; currently, there is no known easy-to-use, reliable indicator that incorporates the intrinsic parameters governing the respiratory system mechanics for indicating lung status improvement or deterioration and eventual ventilator discontinuation. For this reason, we have developed an easy-to-employ lung ventilatory index (LVI), involving the intrinsic parameters of a lung ventilatory model, represented by a first-order differential equation in lung volume response to ventilator driving pressure. The LVI is then employed for evaluating lung status of COPD patients requiring mechanical ventilation because of acute respiratory failure.

7.2   Scope and Methodology

We recruited 13 mechanically ventilated patients with COPD in acute respiratory failure. All patients met the diagnostic criterion of COPD. The first attempt at discontinuation (or weaning off the ventilator) for every patient was made within a short duration (not exceeding 88 h). The patients in the study were between the ages of 54–83 years. All the patients were on SIMV mode with mandatory ventilation at initial intubation. Based on the physician's judgment, the modes were changed for eventual discontinuation of mechanical ventilation. The time period for recording observations was 1 h. For all purposes in this study, a successful ventilator discontinuation is defined as the toleration to extubation for 24 h or longer and a failed ventilator discontinuation is defined as either a distress when ventilator support is withdrawn or the need for reintubation. Our LVI was then employed to distinguish patients who could be successfully weaned off the mechanical ventilator.

Hence, the scope of our chapter is that we have developed an LVI, based on a lung model represented by a first-order differential equation in lung volume dynamics to assess lung function and efficiency in the case of COPD patients requiring mechanical ventilation because of acute respiratory failure. Herein, we have attempted to evaluate the efficacy of the LVI in identifying improving or deteriorating lung condition in such mechanically ventilated COPD patients, and consequently if LVI can be used as a potential indicator to predict ventilator discontinuation. In a bioengineering study of 13 COPD patients who were mechanically ventilated because of acute respiratory failure, when their LVI was evaluated, it provided clear separation between patients with improving and deteriorating lung condition. Finally, we formulated a lung improvement index (LII) representative of the overall lung response to treatment and medication, and a parameter m that corresponds to the rate of lung improvement and reflects the stability of lung status with time. This chapter is based on our Chapter 9 in Ref. [1] and the figures employed here are adopted from this chapter.* Other works on this subject (of COPD, mechanical ventilation, and weaning from mechanical ventilation) are given by Refs. [2–8].

7.3 Lung Ventilation Model

From a ventilatory mechanics viewpoint, the lungs can be considered analogous to a balloon, which can be inflated and deflated (passively). The gradient between the mouth pressure (P_m) and the alveolar pressure (P_{al}) causes respiration to occur. During inspiration, $P_m > P_{al}$ which causes air to enter the lungs. During expiration, P_{al} increases and is greater than P_m; this causes the air to be expelled out of the lungs passively. These pressure differentials provide a force driving the gas flow. The pressure difference between the alveolar pressure (P_{al}) and pleural pressure (P_p) counterbalances the elastic recoil. Thus, the assessment of respiratory mechanics involves the measurement of flows, volumes (flow integrated over time), and pressure-gradients. The lung ventilation model (shown in Figure 7.1) is based on the following dynamic equilibrium differential equation (Equation 7.1), expressing lung volume response to pressure across the lung:

$$R\mathring{V} + \frac{V}{C} = P_L(t) - P_e = P_N(t) \tag{7.1}$$

where
  P_L (total positive pressure across the lung) $= P_m - P_p$, where P_p is determined by intubating the patient, and assuming that the pressure in the relaxed esophageal tube equals the pressure in the pleural space surrounding it
  lung compliance (C) and airflow resistance (R) are the parameters of the governing Equation 7.1 with both R and C being instantaneous values
  $V = V(t) - V_e$ (where V_e is the end-expiratory lung volume)
  P_e is the end-expiratory pressure
  P_N (the net driving pressure) $= P_L - P_e$

---

* With the permission of the Publisher WIT Press, Southampton, U.K.

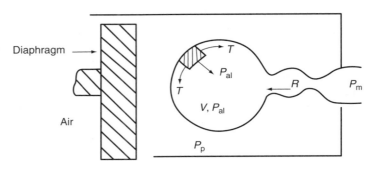

FIGURE 7.1
Lung model where the alveoli are lumped into one chamber (of volume V and pressure P_{al}) and the airway are lumped into one tube of resistance R to airflow rate. (Adopted from Ghista, D.N., Pasam, R., Vasudev, S.B., Bandi, P., and Kumar, R.V., in *Human Respiration: Anatomy and Physiology, Mathematical Modeling, Numerical Simulation and Applications*, V. Kulish (Ed), WIT press, Southampton, U.K., 2006.)

Let B be the amplitude of the net pressure waveform applied by the ventilator, C_a be the averaged dynamic lung compliance, R_a the averaged dynamic resistance to airflow, P_L (the driving pressure) $= P_e + B\sin(\omega t)$, and P_N the net pressure be given as $P_N = B\sin(\omega t)$. The governing Equation 7.1 then becomes

$$R_a \overset{\circ}{V} + \frac{V}{C_a} = P_N = B\sin(\omega t) \tag{7.2}$$

The volume response to P_N (the solution to Equation 7.2) is given by

$$V(t) = \frac{BC_a\{\sin(\omega t) - \omega k_a \cos(\omega t)\}}{1 + \omega^2 k_a^2} + He^{-(t/k_a)} \tag{7.3}$$

where
$k_a\ (= R_a C_a)$ is the averaged time constant
the integration constant H is determined from the initial conditions
the model parameters are C_a and k_a (i.e., C_a and R_a)
ω is the frequency of the oscillating pressure profile applied by the ventilator

An essential condition is that the flow rate is zero at the beginning of inspiration and end of expiration. Hence, the flow rate $dV/dt = 0$ at $t = 0$. Applying this initial condition to our differential Equation 7.3, the constant H is obtained as

$$H = \frac{BC_a \omega k_a}{1 + \omega^2 k_a^2} \tag{7.4}$$

When this expression for H is substituted in Equation 7.3, we also get $V(t = 0) = 0$. Then, from Equations 7.3 and 7.4, we obtain

$$V(t) = \frac{BC_a\{\sin(\omega t) - \omega k_a \cos(\omega t)\}}{1 + \omega^2 k_a^2} + \frac{BC_a \omega k_a}{1 + \omega^2 k_a^2} e^{-(t/k_a)} \tag{7.5a}$$

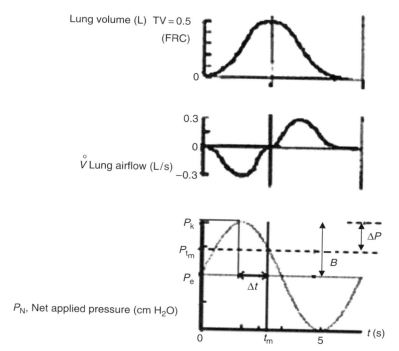

FIGURE 7.2 (See color insert following page 266.)
Lung ventilatory model data showing airflow ($\dot{V}$), volume (V), and net pressure (P_N). Pause pressure (P_{t_m}) occurs at t_m, at which the volume is maximum (TV = tidal volume). Δt is the phase difference between the time of maximum volume and peak pressure (P_k). It is also the time lag between the peak and pause pressures. B is the amplitude of the net pressure waveform P_N applied by the ventilator. This P_N oscillates about P_e with amplitude of B. The difference between peak pressure P_k and pause pressure P_{t_m} is ΔP. (Adopted from Ghista, D.N., Pasam, R., Vasudev, S.B., Bandi, P., and Kumar, R.V., in *Human Respiration: Anatomy and Physiology, Mathematical Modeling, Numerical Simulation and Applications*, V. Kulish (Ed), WIT press, Southampton, U.K., 2006.)

For $t = t_m$, $V(t)$ is maximum and equal to the tidal volume (TV). In a normal person, k_a is of the order of 0.1 and 0.5 in ventilated patients with respiratory disorders, which is relevant to our study of COPD patients. At $t = t_m$ at which the lung volume is maximum, we note from Figure 7.2 that t_m is of the order of 2 s. Hence t_m/k_a is of the order 20 to 4, so that e^{-t/k_a} is of the order of e^{-20} to e^{-4}, which is very small and hence negligible. Hence, in Equation 7.5, we can neglect the exponential term, so that

$$V(t) = \frac{BC_a\{\sin(\omega t) - \omega k_a \cos(\omega t)\}}{1 + \omega^2 k_a^2} \tag{7.5b}$$

Figure 7.2 illustrates a typical data of V, $\dot{V}$, and P_N (in which P_k is of the order of 10 cm H_2O for COPD patients). For evaluating the parameter k_a, we will determine the time at which $V(t)$ is maximum and equal to the TV. Hence putting $dV/dt = 0$ in Equation 7.5b, we obtain

$$\cos(\omega t) + \omega k_a \sin(\omega t) = e^{-(t/k_a)} \tag{7.6}$$

Hence, from Equation 7.6, we obtain the following expression for k_a:

$$\tan(\omega t) = 1/\omega k_a \tag{7.7a}$$

or

$$k_a = -1/\omega \tan(\omega t_m) \tag{7.7b}$$

where

$$\text{since } \tan(\omega t_m) = -1/\omega k_a, \text{ then} \tag{7.8a}$$

$$\tan(\pi - \omega t_m) = 1/\omega k_a \quad \text{or} \quad \pi - \omega t_m = \tan^{-1}(1/\omega k_a) \tag{7.8b}$$

Hence

$$t_m = \frac{\pi - \tan^{-1}(1/\omega k_a)}{\omega} \tag{7.8c}$$

or

$$t_m = \frac{\pi - \theta}{\omega}, \quad \text{where} \quad \theta = \tan^{-1}(1/\omega k_a) = \pi - \omega t_m, \text{ and } k_a = 1/\omega \tan\theta \tag{7.8d}$$

Now, since both ω and t_m are known, we can evaluate k_a from Equation 7.7b. Also, from the above equations and as illustrated in Figure 7.2, the phase difference (Δt) between $t = \pi/2\omega$ at which pressure P_N ($= B \sin \omega t$) is maximum (and equal to P_k) and t_m (at which $P_N = P_m$) is

$$\Delta t = t_m - t = \frac{\pi - \tan^{-1}(1/\omega k_a)}{\omega} - \frac{\pi}{2\omega} \tag{7.9a}$$

or

$$\Delta t = \frac{(\pi/2) - \tan^{-1}(1/\omega k_a)}{\omega} \tag{7.9b}$$

From Equations 7.8c and 7.8d

$$\tan^{-1}(1/\omega k_a) = \theta = \pi - \omega t_m \tag{7.10}$$

Hence

$$\sin\theta = \frac{1}{\sqrt{1 + \omega^2 k_a^2}} \tag{7.11a}$$

$$\cos\theta = \frac{\omega k_a}{\sqrt{1 + \omega^2 k_a^2}} \tag{7.11b}$$

Figure 7.2 shows the clinical lung volume response dynamics in response to the net ventilatory driving pressure (P_N). Referring to Figure 7.2, t_m denotes the time at which the lung volume is maximum. Since ω (the frequency of the oscillatory pressure profile applied by the ventilator) and t_m (the time at which $V(t) = TV$) are known, we can also evaluate the model parameter k_a, from Equations 7.10 and 7.11. Now, from Equation 7.5b, we obtain

$$V(t = t_m) = TV = \frac{BC_a\{\sin(\omega t_m) - \omega k_a \cos(\omega t_m)\}}{1 + \omega^2 k_a^2} \qquad (7.12)$$

In Equation 7.12, if we put

$$N = \sin(\omega t_m) - \omega k_a \cos(\omega t_m) \qquad (7.13a)$$

then, from Equation 7.8a, Equation 7.13a becomes

$$N = \frac{1}{\sqrt{1 + \omega^2 k_a^2}} + \frac{\omega^2 k_a^2}{\sqrt{1 + \omega^2 k_a^2}} \qquad (7.13b)$$

or

$$N = \sqrt{1 + \omega^2 k_a^2} \qquad (7.13c)$$

Then, based on Equations 7.13a and 7.13c, Equation 7.12 becomes

$$V(t = t_m) = TV = \frac{BC_a}{\sqrt{1 + \omega^2 k_a^2}} \qquad (7.14)$$

Based on this equation, we can evaluate the compliance C_a, if B (the amplitude of P_N) is known, since we have already evaluated k_a (from Equation 7.7b).

7.4 Determining Lung Compliance (C_a) and Airflow Resistance (R_a)

As shown in Figure 7.2, the peak pressure (P_k) is the maximum pressure in the lungs during inspiration. The pause pressure (P_{t_m}) is defined to occur when $V(t)$ is maximum, i.e., at $t = t_m$ at the end of inspiration. The peak pressure occurs when the driving pressure is maximum at $t = \pi/2\omega$, while the pause pressure occurs when the lung volume is maximum, i.e., at end of inspiration, when $t_m = (\pi - \theta)/\omega$, as indicated by Equation 7.8d. It can be interpreted that there is a phase lag of Δt between pause and peak pressures, which is described by Equation 7.9. It is known that the driving pressure (P_L) is given as $P_L = P_e + B \sin(\omega t)$, which leads us to

$$\text{Peak pressure } (P_k) = P_L \text{ (at } t = \pi/2\omega) = P_e + B \qquad (7.15)$$

$$\text{Pause pressure } (P_{t_m}) = P_L \text{ (at } t = t_m = (\pi - \theta)/\omega)$$

$$= P_e + B \sin\left\{\omega\left(\frac{\pi - \theta}{\omega}\right)\right\} \qquad (7.16)$$

$$\text{i.e., pause pressure } (P_{t_m}) = P_e + B \sin\theta \qquad (7.17)$$

Now, based on Equation 7.11a, Equation 7.17 becomes

$$\text{Pause pressure } (P_{t_m}) = P_e + \frac{B}{\sqrt{1 + \omega^2 k_a^2}} \qquad (7.18)$$

Then, from Equations 7.15 and 7.17, we get (as shown in Figure 7.2)

$$P_k - P_{t_m} = \Delta P = B - B \sin\theta = B(1 - \sin\theta) \qquad (7.19)$$

$$\text{or} \quad B = \frac{P_k - P_{t_m}}{(1 - \sin\theta)} = \frac{\Delta P}{(1 - \sin\theta)} \qquad (7.20)$$

where $\sin\theta$ is given by Equation 7.11b.

In Equation 7.20, since P_k, and P_{t_m} as well as θ $(= \pi - \omega t_m)$ are measurable, the amplitude B (of P_N) can be evaluated. Then from Equation 7.14, we can evaluate the averaged lung compliance (C_a), because k_a $(= R_a C_a)$ has already been evaluated by Equation 7.7b. Hence, based on Equations 7.14, 7.11a, and 7.20, we obtain

$$C_a = \frac{TV\sqrt{1 + \omega^2 k_a^2}}{B} = \frac{TV}{B \sin\theta} = \frac{TV(1 - \sin\theta)}{\Delta P \sin\theta} \qquad (7.21)$$

Then, from Equations 7.21 and 7.8d, the average value of airflow resistance (R_a) can be evaluated as

$$R_a = k_a/C_a = \frac{\Delta P \sin\theta(1/\omega \tan\theta)}{TV(1 - \sin\theta)} = \frac{\Delta P \cos\theta}{TV\omega(1 - \sin\theta)} \qquad (7.22)$$

For our patients, the ranges of the computed values of these parameters are

$$R_a = 9 - 43 \text{ cm H}_2\text{O s/L}$$
$$C_a = 0.020 - 0.080 \text{ L/cm H}_2\text{O} \qquad (7.23)$$

Now that we have determined the expressions for the parameters R_a and C_a, the next step is to develop an integrated index lung ventilatory incorporating these parameters.

7.5 Formulating a Lung Ventilatory Index (LVI) Incorporating R_a and C_a

Correlations between average airflow resistance (R_a), average lung compliance (C_a), tidal volume (TV), respiratory rate (RF), and maximum inspiratory pressure or peak pressure (P_k) could be used as a possible indicator for determining lung status in a mechanically ventilated COPD patient with acute respiratory failure. We hence propose that a composite index (LVI), incorporating these isolated parameters, can have a higher predictive power for assessing lung status and determining when a patient should be put on a mechanical ventilator.

For this purpose, we note that COPD patients have higher R_a, lower C_a, lower TV, higher P_k, and higher respiratory rate (or frequency) RF. If we want the LVI to have a high value for a COPD patient, further increasing LVI for deteriorating lung status and decreasing LVI for improving lung status in a mechanically ventilated COPD patient in acute respiratory failure, then the LVI can be expressed as

$$LVI = \frac{R_a(RF)P_k}{C_a(TV)} \tag{7.24}$$

where RF is the respiratory rate frequency.

Let us obtain the order of magnitude values of this LVI expression for a mechanically ventilated COPD patient in acute respiratory failure (by using representative computed values of the parameters R_a, C_a, RF, TV, and P_k), in order to verify that the formula for LVI (given by Equation 7.24) can enable distinct separation of COPD patients in acute respiratory failure from patients ready to be weaned off the respirator. For an intubated COPD patient, we have

for $R_a = 15$ cm H_2O s/L, $C_a = 0.035$ L/cm H_2O, RF $= 0.33$ s^{-1}
TV $= 0.5$ L and $P_k = 20$ cm H_2O,

$$LVI \text{ (intubated COPD)} = \frac{[15 \text{ cm } H_2O \text{ s/L}][0.33\text{s}^{-1}][20 \text{ cm } H_2O]}{[0.035 \text{ L/cm } H_2O][0.5 \text{ L}]}$$

$$= 5654 \text{ (cm } H_2O/L)^3 \tag{7.25}$$

Now let us obtain the order of magnitude of LVI (by using representative computed values of R_a, C_a, RF, TV, and P_k) for a COPD patient with improving lung status just before successful discontinuation or extubation. For a successfully weaned COPD patient (examined in an outpatient clinic), we have

$$\text{for } R_a = 10\, \text{cm H}_2\text{O s/L}, \quad C_a = 0.050 \text{ L/cm H}_2\text{O}, \quad RF = 0.33 \text{ s}^{-1}$$
$$TV = 0.35 \text{ L} \quad \text{and} \quad P_k = 12 \text{ cm H}_2\text{O},$$

$$
\begin{aligned}
\text{LVI (outpatient COPD)} &= \frac{[10 \text{ cm H}_2\text{O s/L}][0.33 \text{ s}^{-1}][12 \text{ cm H}_2\text{O}]}{[0.05 \text{ L/cm H}_2\text{O}][0.35 \text{ L}]} \\
&= 2263 \text{ (cm H}_2\text{O/L)}^3
\end{aligned}
\tag{7.26}
$$

Hence, for LVI to reflect lung status improvement in a mechanically ventilated COPD patient in acute respiratory failure, it has to decrease from LVI_0 at the time of admission to the range of LVI for an outpatient COPD patient at the time of discontinuation. Hence, we now put down for lung ventilatory index variation (LVIV):

$$LVIV = LVI(t) = LVI_0\, e^{-mt} \tag{7.27}$$

where
  LVI_0 represents the value of LVI at the time of admission of the patient
    to the respiratory care unit
  Coefficient m, represents the rate of improvement (or deterioration) of
    the patients lung status; $m = 0$ implies no change in lung condition
  Coefficient m (the rate of decrease of LVI or improvement in lung
    status) will be positive with deteriorating lung condition and nega-
    tive for improving lung condition

We can also formulate a lung improvement an index LII (or ΔLVI%), as a measure for overall lung status improvement or deterioration, as

$$LII = \Delta LVI(\%) = \frac{\text{LVI (at entry or intubation)} - \text{LVI (at discharge or extubation)}}{\text{LVI (at entry)}} \times 100 \tag{7.28}$$

and employ it for justifying discontinuation of mechanical ventilation.

7.6 Evaluating Lung Ventilatory Index

7.6.1 LVI Characteristics

Now that we have formulated LVI (Equation 7.24), let us verify for select patient data that LVI is indicative of lung status improvement or deterioration. For this purpose, we have categorized the intubated patients into two categories: (1) patients who were successfully discontinued and (2) patients who failed discontinuation. Table 7.1 provides the range of LVI values for the two categories, where for all successful discontinuations the LVI is close to the value for an outpatient COPD patient.

TABLE 7.1

Range of LVI Values at Intubation and Outcome

Outcome	Number	Age (years)	Sex (M/F)	Time of Intubation (h)	LVI at Intubation (cm H_2O/L)^3	LVI at Outcome (cm H_2O/L)^3
Successful discontinuation	6	54–74	6/0	11–55	3,959–13,568	1,194–4,589
Failed discontinuation	7	64–83	5/2	29–88	3,350–21,152	7,144–15,658

In Table 7.1, all patients who were successfully discontinued have LVI at discontinuation in the range of 1,194–4,589 cm H_2O/L^3. Similarly, patients with failed discontinuation have LVI at discontinuation in the range of 7,144–15,658 cm H_2O/L^3. Thus LVI is indicating a clear separation between successful and failed discontinuation of mechanical ventilation. In Figure 7.3, we have provided the distribution of LVI as a measure of outcome for successful and failed discontinuation of mechanically ventilated COPD patients in acute respiratory failure (approximated as a normal distribution). In Figure 7.4, we have provided the LVI characteristics for four patients, indicating their lung status. Among these patients, patient 4 had failed extubation while the other three patients were successfully extubated.

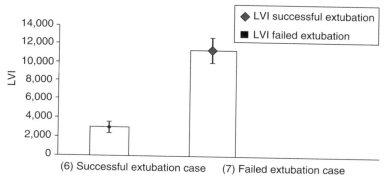

FIGURE 7.3

Distribution of lung ventilatory index (LVI) (as a measure of outcome) at discontinuation for patients with failed and successful weaning off the mechanical ventilator. For the six successfully discontinued cases, the LVI was 2,900 ± 567 cm H_2O/L^3; for the seven failed discontinuation cases the LVI was 11,400 ± 1,433 cm H_2O/L^3. It is seen that LVI indicates clear separation between failed and successful discontinuation. (Adopted from Ghista, D.N., Pasam, R., Vasudev, S.B., Bandi, P., and Kumar, R.V., in *Human Respiration: Anatomy and Physiology, Mathematical Modeling, Numerical Simulation and Applications*, V. Kulish (Ed), WIT press, Southampton, U.K., 2006.)

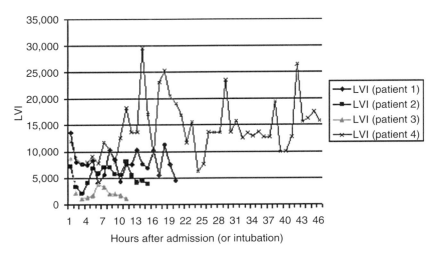

FIGURE 7.4 (See color insert following page 266.)
Lung ventilatory index (LVI) (showing lung status) in four mechanically ventilated COPD patients in acute respiratory failure. Note that patients 1, 2, and 3 were successfully discontinued, and patient 4 had failed discontinuation. (Adopted from Ghista, D.N., Pasam, R., Vasudev, S.B., Bandi, P., and Kumar, R.V., in *Human Respiration: Anatomy and Physiology, Mathematical Modeling, Numerical Simulation and Applications*, V. Kulish (Ed), WIT press, Southampton, U.K., 2006.)

7.6.2   Comparing the Efficacies of R_a and C_a with LVI

Now let us evaluate the significance of R_a and C_a with lung status. In Table 7.2, we have provided information on the values of R_a and C_a at intubation as well as for successful and failed discontinuation for the two classes of patients. In Figures 7.5 and 7.7, we have provided R_a and C_a characteristics for the same four patients discussed in Figure 7.4. We observe that patients with successful discontinuation had (1) R_a at discontinuation in the range of 9–14 cm H_2O s/L compared to 17–23 cm H_2O s/L in failed discontinuation cases and (2) C_a for successful discontinuation in the range of 0.03–0.08 L/cm H_2O compared to 0.028–0.042 L/cm H_2O in failed discontinuation cases.

TABLE 7.2

Ranges of R_a and C_a at Intubation and Associated with Outcomes for Successful and Failed Ventilator Weaning

Outcome	R_a at Intubation (cm H_2O s/L)	R_a at Outcome (cm H_2O s/L)	C_a at Intubation (L/cm H_2O)	C_a at Outcome (L/cm H_2O)	k_a at Intubation (s)	k_a at Outcome (s)
Successful discontinuation	14–32	9–14	0.03–0.047	0.030–0.080	0.42–1.50	0.27–1.12
Failed discontinuation	14–25	17–23	0.03–0.037	0.028–0.042	0.42–0.925	0.47–0.99

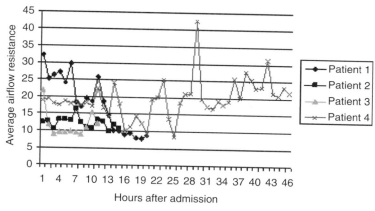

FIGURE 7.5 (See color insert following page 266.)
The variations in R_a for four mechanically ventilated patients, showing that patients 1, 2, and 3 are all discontinued at lower airflow resistance (R_a) values of 8–10 cm H_2O s/L, i.e., closer to outpatient COPD values of R_a; for these patients, the R_a values decreased with mechanical ventilation. For patient 4 (who could not be weaned off), the R_a values remained at a high level. (Adopted from Ghista, D.N., Pasam, R., Vasudev, S.B., Bandi, P., and Kumar, R.V., in *Human Respiration: Anatomy and Physiology, Mathematical Modeling, Numerical Simulation and Applications*, V. Kulish (Ed), WIT press, Southampton, U.K., 2006.)

Figure 7.5 shows the time variation in the parameter R_a of successfully extubated patients (SEPs) and the unsuccessfully extubated patients (UEPs). It is noted that R_a decreases steadily in SEPs, but continues to be high in UEPs. The distribution of R_a (in Figure 7.6), graphically illustrates distinct

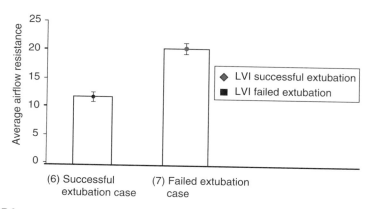

FIGURE 7.6
Distribution of R_a at discontinuation for successful and failed extubation. $R_a = 11.5 \pm 0.83$ cm H_2O s/L for the six successful discontinuation cases, and 20 ± 1 cm H_2O s/L for the seven failed discontinuation cases. Hence, R_a indicates a clear separation between failed and successful discontinuations. (Adopted from Ghista, D.N., Pasam, R., Vasudev, S.B., Bandi, P., and Kumar, R.V., in *Human Respiration: Anatomy and Physiology, Mathematical Modeling, Numerical Simulation and Applications*, V. Kulish (Ed), WIT press, Southampton, U.K., 2006.)

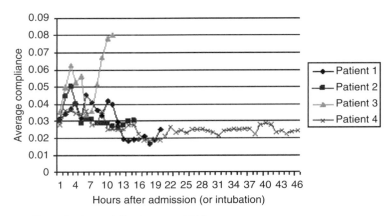

FIGURE 7.7 (See color insert following page 266.)
The variations in C_a for four mechanically ventilated patients, indicating that the C_a values for patients 1, 2, and 4 were all in the lower values and did not change significantly from the time of intubation; incidentally, the lung status for patients 1 and 2 improved and they were successfully discontinued, while patient 4 failed discontinuation. (Adopted from Ghista, D.N., Pasam, R., Vasudev, S.B., Bandi, P., and Kumar, R.V., in *Human Respiration: Anatomy and Physiology, Mathematical Modeling, Numerical Simulation and Applications*, V. Kulish (Ed), WIT press, Southampton, U.K., 2006.)

separation of UEPs and SEPs. Yet in the case of patient 2 in Figure 7.5, the resistance at intubation was 12.38 cm H_2O s/L and decreased only slightly at discontinuation to 10.82 cm H_2O s/L within the period of 15 h of intubation to a value corresponding to that of an outpatient COPD patient. However, for this patient 2, the LVI changed from 7300 to 4500 cm H_2O/L^3 as shown in Figure 7.4, and is significantly closer to the outpatient COPD patient LVI at successful discontinuation as shown in Figure 7.3 and Table 7.1. This indicates that R_a by itself is less sensitive to the change in lung status compared to LVI.

Now let us observe the dynamics of C_a, to see if it can by itself provide proper separation between normal and COPD patients. Figure 7.7, illustrating the C_a dynamics of the same four patients, does not show a definitive trend of increasing C_a for SEPs; in fact, the C_a of even patient 1 kept decreasing (although, as shown in Figure 7.4, its k_a decreased). In fact, from both Table 7.2 and Figure 7.8, it is seen that C_a (by itself) cannot provide clear separation between patients with improving and deteriorating lung status.

Hence, it is seen that both R_a and C_a (by themselves) cannot be reliable indicators for assessing lung status in mechanically ventilated COPD patients in acute respiratory failure, compared to the LVI.

7.6.3   LVI as a Reliable Predictor of Ventilator Discontinuation

We have noted that only LVI can be a reliable indicator of lung status in a mechanically ventilated COPD patient with acute respiratory failure. But

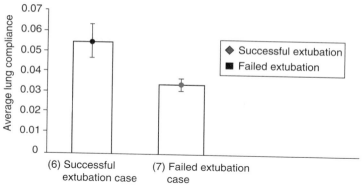

FIGURE 7.8

Distribution of C_a at extubation for successful and failed discontinuation. $C_a = 0.055 \pm 0.0083$ L/cm H_2O for the six successful discontinuation cases, and 0.034 ± 0.0027 L/cm H_2O for the seven failed discontinuation cases. Hence, C_a (by itself) does not provide a clear separation between failed and successful discontinuations. (Adopted from Ghista, D.N., Pasam, R., Vasudev, S.B., Bandi, P., and Kumar, R.V., in *Human Respiration: Anatomy and Physiology, Mathematical Modeling, Numerical Simulation and Applications*, V. Kulish (Ed), WIT press, Southampton, U.K., 2006.)

can it be used as a predictor for ventilator discontinuation? Now, an ideal predictor of ventilator discontinuation would safely and easily distinguish patients needing discontinuation and continued ventilatory support, which we have successfully indicated with our patient data. Also, it has been speculated that as much as 42% of the time spent by patients on mechanical ventilation are instances when they could have been extubated, and in many cases unnecessary delays in the discontinuation process is associated with further complications [2–8].

In this context, we observe that LVI has adequately addressed the issue by identifying patients whose LVI values have stayed consistently in the LVI range for successful extubation. For example, in Figure 7.4 there are four instances when the LVI value for patient 4 is lower than 5000 cm H_2O/L^3, and closer to the value for an ideal COPD outpatient; yet this patient was not weaned off the mechanical ventilator. This leads us to conclude that LVI can be a reliable factor in the clinician's judgment to identify patients needing discontinuation. Thus, we can state that LVI is a reliable index for discontinuation.

7.7 Assessing Lung Improvement Index (LII) and Rate of Lung Improvement (*m*)

The rationale behind designating LII (Equation 7.28) as an index is not as an indicator for lung status per se, because there are instances when the patient

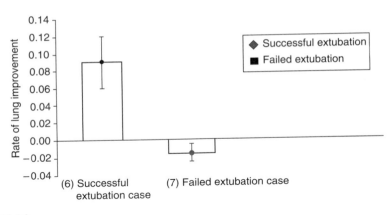

FIGURE 7.9
The distribution of m for patients with successful and failed discontinuation or extubation. The parameter $m = 0.09 \pm 0.03$ for the six successful discontinuation cases, and -0.015 ± 0.01 for the seven failed discontinuation cases. Hence, m indicates appreciable separation between successful and failed discontinuations (or weaning off the ventilator). (Adopted from Ghista, D.N., Pasam, R., Vasudev, S.B., Bandi, P., and Kumar, R.V., in *Human Respiration: Anatomy and Physiology, Mathematical Modeling, Numerical Simulation and Applications*, V. Kulish (Ed), WIT press, Southampton, U.K., 2006.)

lung had improved from the time of initial intubation but could not be sustained upon discontinuation. However, LII could be representative of the overall lung response to treatment and medication. As regards use of LVIV for assessment of continuing improvement of mechanically ventilated COPD patients, the coefficient m in Equation 7.27 corresponds to the rate of improvement or deterioration in lung status. It is to be noted that for a patient with improving lung condition, m will be positive and vice versa.

We have observed that in most patients, m decreased immediately after the first few hours of mechanical ventilation, and later on varied somewhat and stabilized before successful discontinuation. Hence, we propose that m provides a measurement of lung status change at any point in time, and hence provide a clear separation between patients with improving and deteriorating lung status, as indicated by Figure 7.9. Table 7.3 provides information on both LII and m for patients with successful and failed

TABLE 7.3

Range of LII and m at Intubation and Outcome

Outcome	Number	Age (years)	Sex (M/F)	Time of Intubation (h)	LII (%) at Outcome	m at Outcome
Successful discontinuation	6	54–74	6/0	11–55	26% to 86%	0 to 0.18
Failed discontinuation	7	64–83	5/2	29–88	−101% to 49%	−0.045 to 0.015

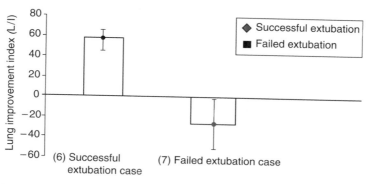

FIGURE 7.10

The distribution of the lung improvement index (*LLI*) for patients with successful and failed discontinuations. $LLI = 56\% \pm 10\%$ for the 6 successful-discontinuation cases, and $-26\% \pm 25\%$ for the 7 failed-discontinuation cases. (Adopted from Ghista, D.N., Pasam, R., Vasudev, S.B., Bandi, P., and Kumar, R.V., in *Human Respiration: Anatomy and Physiology, Mathematical Modeling, Numerical Simulation and Applications*, V. Kulish (Ed), WIT press, Southampton, U.K., 2006.)

discontinuation, while Figures 7.9 and 7.10 illustrate their distributions. Based on Table 7.3 and Figure 7.10, we can say that both LII and *m* show distinct outcomes for patients with successful extubation.

7.8 Conclusion

We have shown LVI, LII, and *m* to be reliable indicators for mechanically ventilated COPD patients in acute respiratory failure. We have also indicated how they can be collectively employed to decide extubating a patient. In conclusion, the way in which we could employ LVI, LII, and *m* in combination is as follows. Starting with evaluation of LVI at the time of intubation, we can employ LII and *m* to designate signs of lung improvement. Then, when LVI value persists being less than 3000 cm H_2O/L^3 for 2–3 h, we could decide to extubate the patient.

References

1. Ghista DN, Pasam R, Vasudev SB, Bandi P, and Kumar RV, Indicator for lung-status in mechanically-ventilated COPD patients using lung ventilation modeling and assessment, *Human Respiration: Anatomy and Physiology, Mathematical Modeling, Numerical Simulation and Application*, edited by Kulish V, WIT Press, Southampton, U.K., 2006, Chapter 9, pp. 169–185.
2. MacIntyre NR, Evidence-based guidelines for weaning and discontinuing ventilatory support by a collective task force of the American College of Chest Physicians, *Respir. Care*, 47(1), 69–90, 2002.

3. Menzies R, Gibbons W, and Goldberg P, Determinants of weaning and survival among patients with COPD who require mechanical ventilation for acute respiratory failure, *Chest*, 95(2), 398–405, 1989.
4. Yang KL, Tobin MJ, and Engl N, A prospective study of indexes predicting the outcome of trials of weaning from mechanical ventilation, *J. Med.*, 324(21), 1445–1450, 1991.
5. Alvisi R, Volta CA, Righini ER, Capuzzo M, Ragazzi R, Verri M, Candini G, Gritti G, and Milic-Emili J, Predictors of weaning outcome in chronic obstructive pulmonary disease patients, *Eur. Respir. J.*, 15(4), 656–662, 2000.
6. Pauwels RA, Buist AS, Calverley PM, Jenkins CR, and Hurd SS, Global strategy for the diagnosis, management, and prevention of chronic obstructive pulmonary disease, NHLBI/WHO Global Initiative for Chronic Obstructive Lung Disease (GOLD) Workshop summary, *Am. J. Respir. Crit. Care Med.*, 163(5), 1256–1276, 2001.
7. Meade M, Guyatt G, Cook D, Griffith L, Sinuff T, Kergl C, Mancebo J, Esteban A, and Epstein S, Predicting success in weaning from mechanical ventilation, *Chest*, 120(Suppl. 6), 400S–424S, 2001.
8. Ubran A, Van de Graaff WB, and Tobin MJ, Variability of patient-ventilator interaction with pressure support ventilation in patients with chronic obstructive pulmonary disease, *Am. J. Respir. Crit. Care Med.*, 152(1), 129–136, 1995.

Section III

Glucose–Insulin Regulation (in Diabetes) Engineering Mechanics

8

Glucose–Insulin Dynamics Modeling

Dhanjoo N. Ghista and Dittakavi Sarma

CONTENTS

8.1   Background

8.1.1   History and Prevalence of Diabetes

Albert Einstein once said, "The most incomprehensible thing about the universe is that it is comprehensible." Human or more specifically the human body is a part of this still largely incomprehensible universe. There has been and will always be constant effort to unravel the mystery of the human body.

Diseases are part of this mystery. How do they happen? How do we cure them? Why some diseases are more prevalent in one ethnic group than the others? Can we use the power of science and technology to know more about them? Diabetes is one of the diseases that has been around for a very long time and yet has not been fully understood amid progress in many fields of science and technology.

There was a prescription for frequent urination, the most common symptom of diabetes, on an Egyptian papyrus dating back to 1500 BC. Much later in 100 AD, the Greek physician Aretaeus of Cappadocia first named the condition "diabetes," which is Greek for "siphon," since people with diabetes urinated often. In 1889, two European scientists discovered that removing pancreases from dogs caused diabetes. Until this century, the only way doctors had to treat diabetes was through diet. Then the first major progress happened in 1922, with the successful purification and subsequent injection of insulin taken from an animal into a boy with diabetes. It greatly improved his condition and life expectancy, very uncommon at that time.

Diabetes is now one of the commonest noninfectious medical conditions in the world and the numbers are rising. It currently affects about 151 million people worldwide and the number will double by 2010 according to professor Paul Zimmet of the World Health Organization. Diabetes is a leading cause of blindness, renal failure, and limb amputation all over the world. The need to detect diabetic risk factors and treat organ disorders and complications associated with diabetes has provided the motivation for the technology for more precise categorization of at-risk subjects, confirmed diabetics, and severely diabetics. To gain understanding about diabetes, let us now take a look at blood glucose regulation of the human body.

8.1.2   Blood Glucose Regulation

Glucose enters the blood from two sources: from the intestine where carbohydrates are absorbed from digested food and from liver. Figure 8.1 shows

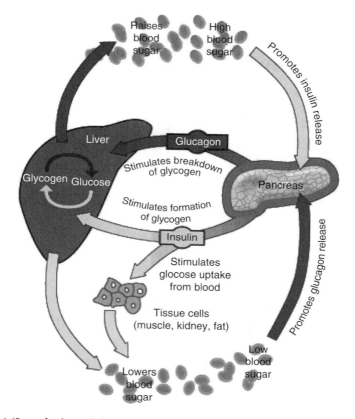

FIGURE 8.1 (See color insert following page 266.)
Effects of insulin and glucagons released by the pancreas in achieving normal blood glucose level.

how the body relies on two hormones, which have two opposite actions: insulin and glucagon, produced in the pancreas to maintain a constant blood glucose level.

Insulin is made and secreted by the beta cells of the pancreatic islets, small islands of endocrine cells in the pancreas. Glucagon is made and secreted by the alpha cells of the pancreatic islets, shown in Figure 8.2. The liver stores glucose in the form of glycogen, converts it back into glucose, and releases it into the blood stream. The pancreas assists cells to assimilate blood glucose by secreting insulin. This lowers the blood glucose level.

Since glucose is the essential fuel for the cells, normal blood glucose level is necessary for the proper functioning of the organs. Diabetes is a metabolic disorder in which blood glucose levels are abnormally high, because the body does not produce enough insulin or becomes resistant to its effects. The abnormally high blood glucose level found in diabetes can be detected using two commonly employed tests: intravenous glucose tolerance test (IVGTT) and oral glucose tolerance test (OGTT).

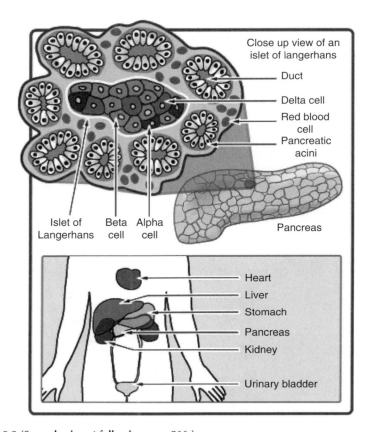

FIGURE 8.2 (See color insert following page 266.)
The pancreas has many islets that contain insulin-producing beta cells and glucagons-producing alpha cells.

8.1.3 Classification of Diabetes

There are two major types of diabetes:

(1) Type 1 diabetes (previously known as juvenile-onset diabetes or insulin-dependent diabetes mellitus [IDDM]). Type 1 diabetes accounts for 5% to 10% of diabetes cases, and most often occurring in children and young adults. People with type 1 diabetes must take daily insulin injections to stay alive, since their bodies do not produce any insulin. Without insulin, glucose cannot get to the cells for energy or to the liver for storage. Further, excess glucose in the blood causes organ and tissue damage over time.

(2) Type 2 diabetes (previously known as adult-onset diabetes or noninsulin-dependent diabetes mellitus [NIDDM]). Type 2 diabetes accounts for 90% to 95% of diabetes cases, and usually develops in middle age or later. Type 2 diabetes is associated with insulin resistance, rather than due to lack of insulin as seen

in type 1 diabetes. Insulin levels in these cases are usually normal or higher than average, but the body's cells are rather sluggish or insensitive in responding to it. This results in the build up of unused glucose in the blood stream, causing higher than normal blood glucose levels, and organ and tissue damage over time.

8.1.4   Objective

Our objective, in the Chapters 8, 9, and 10 of this Section III is to (1) model glucose–insulin dynamics by means of appropriate differential equations, (2) plot the glucose and insulin response solutions to these model equations, (3) conduct clinical simulation of the model differential equations' solutions by means of parameter-identification of intravenous and oral glucose tolerance test models, and (4) develop a nondimensional index for diabetes (made up of the model parameters), and demonstrate its sensitivity to differentiate diabetics from normals as well as identify so-called normals who are at risk for getting diabetes. In this chapter, we will concern ourselves with objectives 1 and 2, and also discuss the model's characteristics and physiological significance.

The conventional clinical diagnosis of diabetes entails obtaining the fasting blood glucose concentration values from OGTT. However, this evaluation based on just one reading does not provide insight into the glucose-regulatory dysfunctional etiology of diabetes disease–disorder and effectiveness of antidiabetic drugs.

Therefore, it is deemed necessary to monitor blood glucose and insulin concentration response dynamics to administered inputs of glucose during the glucose tolerance tests. Then, model analysis of this clinical data needs to be carried out by making the model solutions (or expressions for blood glucose and insulin concentrations, in terms of the input glucose bolus amount) match the blood glucose and insulin concentrations data, and evaluating the model parameters.

Finally, the model parameters are grouped together to formulate a nondimensional index. A distribution of this index values demonstrates demarcation of nondiabetics, patients at risk for getting diabetes, and distinctly diabetic patients.

8.2   Introduction to Glucose-Regulatory Modeling

Diabetes mellitus is a heterogeneous clinical syndrome characterized by hyperglycemia and long-term specific complications: retinopathy, neuropathy, nephropathy, and cardiomyopathy. Automatic neuropathy leads to visceral denervation, producing a variety of clinical abnormalities: cardiac and respiratory dysrhythmias, gastrointestinal motility disorders, urinary bladder dysfunction, and impotence.

Kinks in the body's glucose factory: Glucose is an essential fuel for the cells. Normal blood glucose level is necessary for the functioning of the organs. Blood sugar level rises in the case of damaged pancreas or when the receptors on the cell walls (supposed to absorb the sugar) are defective. In that situation, when the liver fails to detect the high blood sugar level, excessive glucose is released into the blood. When the blood glucose level exceeds 180–200 mg/dL, it starts spilling into urine.

The currently employed intravenous and oral glucose tolerance tests merely monitor readings of blood glucose concentration, and do not provide insight into the dynamics of blood glucose and insulin concentrations to reliably diagnose risk-to-diabetes, resistance to insulin, and severity index. For that purpose, it is necessary to model the glucose–insulin regulatory (GIR) process, and employ the solution to simulate the clinical glucose concentration curves for intravenous and oral glucose test inputs. The parameters of the glucose–insulin regulatory system (GIRS) models can then be employed as diagnostic indices.

For GIRS modeling, to simulate IVGTT and OGTT, we require to formulate the nature of glucose-input function and the GIRS transfer function to simulate monitored glucose and insulin blood concentration response curves. The objective and scope of this chapter is to provide the foundations for the GIRS model, so that it can be employed (in the succeeding chapters) to simulate the GIR process (and its dysfunction in the case of diabetes) by means of IVGTT and OGTT.

In Chapter 9, we will deal with:

- modeling of blood glucose regulation and tolerance testing,
- elucidating how the glucose-regulatory system model's transfer-function can explain the blood glucose response data in intravenous and oral glucose tolerance tests, and
- demonstrating patient simulation of the blood glucose-regulatory models, by means of which the model parameters can be evaluated and related to physiological parameters.

8.3 Modeling of Glucose–Insulin Regulation

We adopt the linearized biomathematical model of Bolie [1], as the basis of our modeling, because it is simple but still compatible with the known physiological mechanisms. This model characterizes the glucose–insulin system by means of differential equations (given below as Equations 8.1 and 8.2) with four parameters: α, β, γ, and δ, representing pancreatic insulin sensitivity to insulin and glucose blood concentrations, tissue glycogen storage, and tissue glucose utilization to elevated blood glucose concentrations.

8.3.1 Differential-Equation Model of the Glucose–Insulin System

The compartmental block diagram of the blood glucose and insulin regulatory system (BGIRS) is illustrated in Figure 8.3. The glucose-input rate into

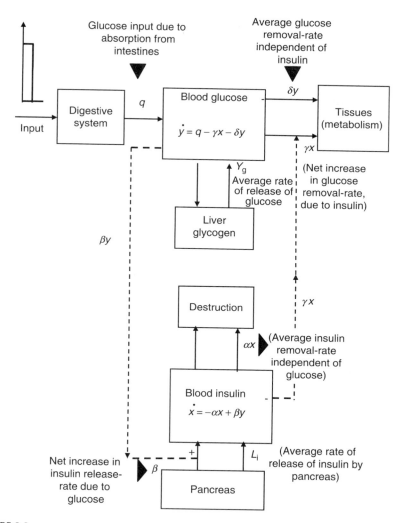

FIGURE 8.3
Block diagram of how (1) insulin level and rate of change of insulin $\dot{x}$ governs blood glucose concentration $y(t)$, and (2) rate of change of glucose $\dot{y}$ is influenced by insulin concentration $x(t)$ and ingested glucose-input rate $q(t)$. The parameter β corresponds to insulin release factor and the parameter γ represents sensitivity or resistance to insulin (representative of risk-to-diabetes). The term γx is the insulin-dependent glucose transport and the term δy is the insulin-independent glucose transport.

the blood pool is represented by q in the figure. From the blood pool, glucose is metabolized into the tissues, as represented by the two terms δy (removal-rate of glucose from the blood pool independent of insulin) and γx (removal-rate of glucose under the influence of insulin). In return, the glucose influences the release rate of insulin into the blood pool by the pancreas (as represented by the βy term). The insulin is also removed independently of glucose (as per the term αx).

By considering the conservation rate of glucose and insulin in their respective compartments, we obtain the basic equations, governing BGIRS: With reference to the blood glucose–insulin system (depicted in Figure 8.3), the corresponding first-order differential equations of the insulin and glucose-regulatory subsystems are given by [1,2]:

$$x' = p(t) - \alpha x + \beta y \qquad (8.1)$$

$$y' = q(t) - \gamma x - \delta y \qquad (8.2)$$

where
  x is the blood insulin concentration (from its fasting level)
  y is the blood glucose concentration (from its fasting level)
  p is the insulin input rate
  q is the glucose input rate, for unit blood glucose compartment volume (V)
  x', y' denote the first-derivatives of x and y with respect to time

In these equations, the glucose–insulin model system parameters (regulatory coefficients) are α, β, γ, and δ (Tables 8.1 through 8.3). These coefficients, when multiplied by the blood glucose compartment volume V (which is proportional to the body mass) denote, respectively (as depicted in Figure 8.4):

- Sensitivity of insulinase activity to elevated insulin concentration (αV)
- Sensitivity of pancreatic insulin output to elevated glucose concentration (βV)
- Combined sensitivity of liver glycogen storage and tissue glucose utilization to elevated insulin concentration (γV)
- Combined sensitivity of liver glycogen storage and tissue glucose utilization to elevated glucose concentration (δV)

TABLE 8.1

Nomenclature

Symbol	Meaning	Dimension
T	Time	Hour
V	Extracellular fluid volume	Liter
p	Rate of insulin injection	Units $(\text{liter})^{-1}$ $(\text{hour})^{-1}$
q	Rate of glucose injection	Grams $(\text{liter})^{-1}$ $(\text{hour})^{-1}$
x	Extracellular insulin concentration	Units/liter
y	Extracellular glucose concentration	Grams/liter

TABLE 8.2

Abbreviations and Meanings

IVGTT	Intravenous glucose tolerance test
OGTT	Oral glucose tolerance test
GI	Gastrointestinal
BGCS	Blood glucose control system
GIRS	Glucose–insulin regulatory system
αV	Sensitivity of insulinase activity to elevated insulin concentration
βV	Sensitivity of pancreatic insulin output to elevated glucose concentration
γV	Combined sensitivity of liver glycogen storage and tissue glucose utilization to elevated insulin concentration
δV	Combined sensitivity of liver glycogen storage and tissue glucose utilization to elevated glucose concentration

8.3.2 Differential-Equation Model in Glucose Concentration [y], for Insulin Infusion Rate [p = 0], and Glucose Inflow Rate [q]

In Equations 8.1 and 8.2, let the insulin infusion rate $p = 0$ and the glucose infusion rate is q. Then, on differentiating Equation 8.2 on either side with respect to t, we get the differential equation of blood glucose response (y) to q input [2], as

$$y'' = q' - \gamma x' - \delta y'$$
$$= q' - \gamma(-\alpha x + \beta y) - \delta y', \text{ on substituting for } x' \text{ from Equation 8.1}$$
$$= q' + \alpha(\gamma x) - \beta \gamma y - \delta y'$$
$$= q' + \alpha(q - \delta y - y') - \beta \gamma y - \delta y', \text{ on substituting for } \gamma x \text{ from Equation 8.2}$$
$$= q' + \alpha q - y'(\alpha + \delta) - y(\alpha \delta + \beta \gamma)$$

TABLE 8.3

Units of Terms

Symbol	Dimension
p	Units $(L)^{-1} (h)^{-1}$
q	Grams $(L)^{-1} (h)^{-1}$
α	$(h)^{-1}$
β	(units) $(h)^{-1} (g)^{-1}$
γ	(g) $(h)^{-1}$ (units)$^{-1}$
δ	h^{-1}
A	h^{-1}
ω_n	rad/h
ω	rad/h
T_d	h
$\lambda (= \omega_n^2)$	h^{-2}

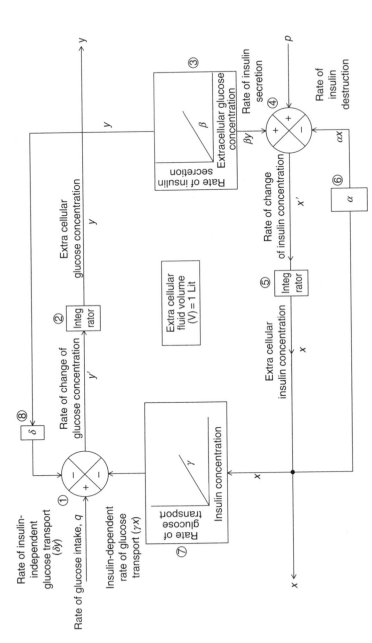

Note:　$\alpha, \beta, \gamma, \delta$ are the slopes of linearized curves of blocks 6,3,7,8. It is implied that the input–ouput relation of these blocks is proportional with zero initial conditions.

FIGURE 8.4
Physiological model of Blood glucose–insulin regulatory control system. Adopted from Dittakavi, S.S. and Ghista, D.N., *J. Mech. Med. Bio.,* 1, 193, 2001.

Rearranging, we get the differential equation:

$$y'' + y'(\alpha + \delta) + y(\alpha\delta + \beta\gamma) = q' + \alpha q \tag{8.3}$$

where y' and y'' denote first and second time-derivatives of y.

8.3.3 Differential-Equation Model in Insulin Concentration (x), for $p = 0$ and q Input

Differentiating Equation 8.1 on both sides with respect to t, we get [2]

$$x'' = -\alpha x' + \beta y', \text{ for } p = 0$$
$$= -\alpha x' + \beta(q - \gamma x - \delta y), \text{ upon substituting for } y' \text{ from Equation 8.2}$$
$$= -\alpha x' + \beta q - \beta\gamma x - \delta(\beta y)$$
$$= -\alpha x' + \beta q - \beta\gamma x - \delta(x' + \alpha x), \text{ upon substituting for } \beta y \text{ from Equation 8.1}$$

Rearranging, we get the differential equation for blood insulin response (x) to q input, as

$$x'' + x'(\alpha + \delta) + x(\alpha\delta + \beta\gamma) = \beta q \tag{8.4}$$

where x' and x'' denote first and second time-derivatives of x.

The solution analyses of the governing Equations 8.3 and 8.4 are given in the next Section 8.4.

8.3.4 Laplace Transform Representation of Governing Equations 8.3 and 8.4 [2]

1. Appendices A.8.1–8.4 provides a brief tutorial elaboration of the Laplace transform (LPT) application to the solution of differential equations. The transfer function (TF) corresponding to Equation 8.3 is obtained by taking LPT on both sides (assuming the initial conditions to be zero), as indicated in Appendix A.8.5.

$$s^2 Y(s) + sY(s)(\alpha + \delta) + Y(s)(\alpha\delta + \beta\gamma) = Q(s)(s + \alpha) \tag{8.5}$$

Thereby, we obtain (for glucose response):

$$G(s) = \frac{Y(s)}{Q(s)} = \frac{(s + \alpha)}{s^2 + s(\alpha + \delta) + (\alpha\delta + \beta\gamma)} = G(s) \tag{8.5a}$$

This transfer function, $G(s)$ can be expressed in the form

$$G(s) = \frac{Y(s)}{Q(s)} = \frac{(s + \alpha)}{(s + p_1)(s + p_2)} \tag{8.5b}$$

where
  $G(s)$ will have poles at $s = -p_1$ and $s = -p_2$, and a "zero" at $s = -\alpha$
  $p_1 + p_2 = \alpha + \delta$
  $p_1 p_2 = \alpha\delta + \beta\gamma$

Thus, p_1 and p_2 are the roots of the characteristic (quadratic) equation:

$$p^2 - (\alpha + \delta)p + (\alpha\delta + \beta\gamma) = 0, \tag{8.6a}$$

and are given by

$$\begin{aligned}
p_1 \text{ and } p_2 &= \frac{(\alpha + \delta) \pm \left[(\alpha + \delta)^2 - 4(\alpha\delta + \beta\gamma)\right]^{1/2}}{2} \\
&= \frac{(\alpha + \delta) \pm \left[(\alpha - \delta)^2 - 4\beta\gamma\right]^{1/2}}{2}
\end{aligned} \tag{8.6b}$$

Now, let us make the left-hand-side (LHS) of Equation 8.3 (of blood glucose response),
  $y'' + y'(\alpha + \delta) + y(\alpha\delta + \beta\gamma)$, correspond to

$$y'' + 2Ay + \omega_n^2 y = y'' + \lambda T_d y + \lambda y = q' + \alpha q \tag{8.6c}$$

where
  A is the attenuation or damping constant of the system (in h^{-1})
  ω_n is the natural frequency of the system (in rad/h)
  T_d is the derivative-time constant (in h)

For this purpose, we put

$$\alpha + \delta = 2A = \lambda T_d, \quad \alpha\delta + \beta\gamma = \omega_n^2 = \lambda$$

and

$$\begin{aligned}
\omega &= (\omega_n^2 - A^2)^{1/2} \quad \text{if } A < \omega_n \\
&= (A^2 - \omega_n^2)^{1/2} \quad \text{if } A > \omega_n
\end{aligned} \tag{8.6d}$$

so that λy is the proportional-control term and $(\lambda T_d y')$ is the derivative feedback control term with derivative time T_d. Also, ω_n is the natural frequency of y response, and ω is its damped frequency (in rad/h).
  Then, from Equations 8.6b and 8.6c, we get

$$(A^2 - \omega_n^2)^{1/2} = \frac{[(\alpha + \delta)^2 - 4(\alpha\delta + \beta\gamma)]^{1/2}}{2} = \frac{[(\alpha - \delta)^2 - 4\beta\gamma]^{1/2}}{2}$$

So, from Equations 8.6b and 8.6c, the roots of the characteristic equation are given by

$$p_1, p_2$$
$$= A \pm (A^2 - \omega_n^2)^{1/2} = A \pm i(\omega_n^2 - A^2)^{1/2}$$
$$= A \pm \omega \quad \text{if } A > \omega_n \text{ (overdamaged response)}$$
$$= A \pm i\omega \quad \text{if } A < \omega_n \text{ (underdamaged response)} \tag{8.6e}$$

Then, for $A > \omega_n$ (i.e., for overdamped and aperiodic response)

$$p_1 p_2 = A^2 - \omega^2, \quad p_1 + p_2 = 2A,$$
$$p_1 - p_2 = 2\omega, \quad \frac{A}{\omega} = \frac{p_1 + p_2}{p_1 - p_2} = \frac{p_1}{p_2} \tag{8.6f}$$

and

$$(\alpha - \delta) = \alpha + \alpha - \delta - \alpha = \alpha + \alpha - 2A = 2(\alpha - A)$$

On the other hand, for $A < \omega_n$ (i.e., for underdamped response)

$$p_1 p_2 = A^2 + \omega^2 = \omega_n^2, \quad p_1 + p_2 = 2A, \quad p_1 - p_2 = 2i\omega \tag{8.6g}$$

The response $y(t)$ of the system, i.e., the solution of the Deq (8.3), is obtained by taking the inverse transform of

$$Y(s) = \frac{Q(s)(s + \alpha)}{s^2 + s(\alpha + \delta) + (\alpha\delta + \beta\gamma)} = \frac{Q(s)(s + \alpha)}{(s + p_1)(s + p_2)} \tag{8.7}$$

given by Equation 8.5.

2. Similarly from Equation 8.4, we get for blood insulin response:

$$X(s) = \frac{\beta Q(s)}{s^2 + s(\alpha + \delta) + (\alpha\delta + \beta\gamma)} \tag{8.8}$$

In Section 8.5, we determine $y(t)$ and $x(t)$ responses for different forms of glucose-input function $q(t)$.

8.4 Block Diagrams of Blood Glucose Control System

8.4.1 Blood Glucose–Insulin Regulatory Control System (BGCS) Model

In the block diagram of blood glucose–insulin regulatory (BGIR) system of Figure 8.4, which is essentially a modification of Guyton's BGIR system [3], where the glucose intake q (either injected into blood or by natural absorption

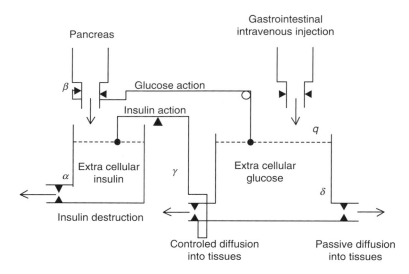

FIGURE 8.5
Hydraulic analog of insulin–glucose regulation. (Adopted from Dittakavi, S.S. and Ghista, D.N., *J. Mech. Med. Bio.*, 1, 193, 2001.)

from intestines) is the input and the blood glucose concentration (y) is the output of the blood glucose control system; the functions of the eight blocks are indicated therein.

In the block diagram, if the extracellular fluid volume (V) is considered equal to unity (1 L), it is readily noted that the summing points No. 4 and No. 1 of the block diagram, respectively represent the governing pair of differential Equations 8.1 and 8.2:

$$x' = p - \alpha x + \beta y \quad \text{and} \quad y' = q - \gamma x - \delta y$$

A hydraulic analog of the insulin–glucose regulation is shown in Figure 8.5.

8.4.2 Control-System Derivation and Representation of Equations 8.1 and 8.2 to Obtain Equations 8.5a and 8.8

Figure 8.6 illustrates the steps involved in the process of obtaining the transfer functions for glucose and insulin response. With reference to Figure 8.6b, we can

(1) Rewrite Equation 8.1, in LPT representation, as

$$sX(s) = -\alpha X(s) + \beta Y(s) \quad \text{or} \quad X(s) = \frac{\beta Y}{(s + \alpha)}$$

(2) Rewrite Equation 8.2, in LPT format, as

$$sY(s) = Q(s) - \gamma X(s) \quad \text{or} \quad Y(s) = \frac{Q - \gamma X}{s + \delta}$$

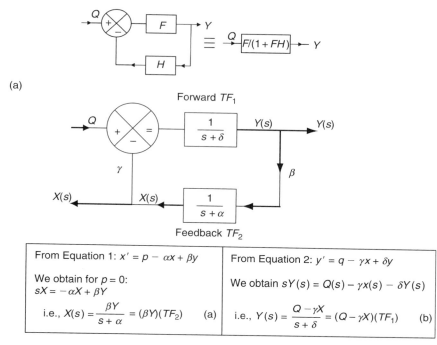

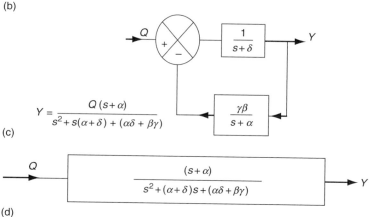

FIGURE 8.6

Control-system derivation and representation of Equation 8.5a. (a) General rule for getting the transfer function of a closed-loop system represented in s-domain. (b) Representation of governing Equations 8.1 and 8.2. Note that the $X(s)$ and $Y(s)$ expressions can be combined to obtain $X(s)$ given by Equation 8.8 and $Y(s)$ given by Equation 8.5. (c) Closed-loop system of s-domain, obtained from (b). (d) Transfer function.

(3) Combine $X(s)$ and $Y(s)$, to obtain

$$X(s) = \frac{\beta Q(s)}{s^2 + s(\alpha + \delta) + (\alpha\delta + \beta\gamma)}, \text{ which is Equation 8.8}$$

and

$$Y(s) = \frac{Q(s)(s + \alpha)}{s^2 + s(\alpha + \delta) + (\alpha\delta + \beta\gamma)}, \text{ which is Equation 8.5a}$$

8.5 Analyses of Glucose and Insulin Responses to Clinically Representative Glucose-Input Functions [2]

8.5.1 Glucose Response Characterization to Different Forms of Input Functions $q(t)$ into the Blood Pool

1. Glucose $y(t)$ response to step glucose input $q(t) = S\,u(t)$, where $S =$ 1 g of glucose (kg of body weight)$^{-1}$ (h)$^{-1}$, depicted in Figure 8.7a:

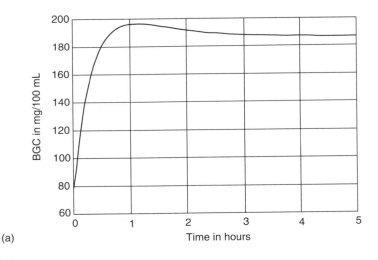

(a)

FIGURE 8.7
(a) Simulated blood glucose response $y(t)$ to unit-step input 1.0 g(kg)$^{-1}$ (h)$^{-1}$ (or 400 mg [100 mL]$^{-1}$ (h)$^{-1}$) of glucose infusion, given by Equation 8.11 as the solution of Equation 8.3, for $\alpha = 0.916$, $\beta = 0.198$, $\gamma = 3.23$, and $\delta = 3.04$ (units provided in Table 8.3), these being the values adopted by Bolie [1]. According to Equations 8.6c and 8.6d, these parametric values correspond to ω_n (the natural frequency of the system) $= (\alpha\delta + \beta\gamma)^{1/2} = 1.85$ rad/h, and the damping coefficient A $(= \lambda T_d/2) = \alpha + \delta = 3.956/2 = 1.978$ h^{-1}. Also, because $A > \omega_n$, these values correspond to an over-damped system, associated with a diabetic patient. However, because A is only slightly greater than ω_n, this case may be deemed to represent critically damped response associated with the subject being at risk to become diabetic. (Adopted from Dittakavi, S.S. and Ghista, D.N., *J. Mech. Med. Bio.*, 1, 193, 2001.)

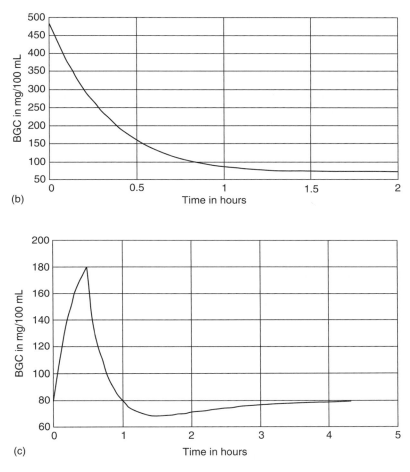

FIGURE 8.7 (continued)
(b) Simulated glucose concentration response to impulse input of glucose, of the blood glucose control system $Y(s)/Q(s)$, given by Equation 8.13 as the solution of Equation 8.3, for $\alpha = 0.916$, $\beta = 0.198$, $\gamma = 3.23$, $\delta = 3.04$. (c) Simulated glucose response $y(t)$ to glucose infusion as a rectangular-pulse input of magnitude 1.0 $g(kg)^{-1}(h)^{-1}$ ($=400$ mg $[100 \text{ mL}]^{-1} (h)^{-1}$) for a period of 30 min. The response $y(t)$ is given by Equation 8.14b as the solution of Equation 8.3, for $\alpha = 0.916$, $\beta = 0.198$, $\gamma = 3.23$, $\delta = 3.04$; their units provided in Table 8.3. The $y(t)$ response (and hence even the peak value) depends on the values of the parameters. In all these glucose response curves, $y(0)$ is taken to be 80 mg/100 mL (assumed to be the fasting glucose concentration). (Adopted from Dittakavi, S.S. and Ghista, D.N., *J. Mech. Med. Bio.*, 1, 193, 2001.)

If $q(t) = u(t)$, a unit-step function (as with the case of IVGTT), $= 1$ for $t > 1$(and 0 for $t < 0$) then, $Q(s) = 1/s$, and Equation 8.5b becomes

$$Y(s) = Q(s) \times G(s) = (s + \alpha)/s(s + p_1)(s + p_2) \qquad (8.9)$$

where p_1 and p_2 are the poles defined in Equation 8.6b.

Performing partial-fraction expansion, we get

$$Y(s) = \frac{A_0}{s} + \frac{A_1}{s + p_1} + \frac{A_2}{s + p_2} \tag{8.10}$$

For the above expression to be equal to $Y(s)$ given by Equation 8.9, we obtain

$$A_0 = \frac{\alpha}{p_1 p_2}$$

$$A_1 = \frac{(\alpha - p_1)}{(p_1^2 - p_1 p_2)}$$

$$A_2 = \frac{(\alpha - p_2)}{(p_2^2 - p_1 p_2)}$$

(note that the dimension of each term of $Y(s)$ is t^{-1})

Now the subject's mass is taken to be 70 kg, which corresponds to a blood volume of 17.5 L. Hence 1 g of glucose/kg is equivalent to 70 kg/17.5 L = 400 mg/100 mL. Hence the glucose-input magnitude = 400 mg (100 mL)$^{-1}$ h^{-1} or 4 g L^{-1} h^{-1}.

Now taking the inverse LPT, and noting that the glucose-input magnitude,

$$S = 1 \text{ g of glucose kg}^{-1} \text{ h}^{-1} = 4 \text{ g L}^{-1} \text{ h}^{-1} = 400 \text{ mg } 100 \text{ mL}^{-1} \text{ h}^{-1},$$

we get, for input

$$q(t) = S \, u(t)$$

$$y(t) = 4(A_0 + A_1 e^{-p_1 t} + A_2 e^{-p_2 t}) \text{g/L} \tag{8.11}$$

Although, based on Equation 8.11, $y(0) = 0$, Equation 8.11 is depicted by the graph in Figure 8.7a for $y(0) = 80$ mg/100 mL (assumed to be the fasting glucose concentration value), by displacing the origin to (0, 80).

2. Glucose $y(t)$ response to unit-impulse glucose input: $q(t) = I \, \delta(t)$, where $I = 1$ g of glucose kg^{-1}, depicted in Figure 8.7b.

   If $q(t) = \delta(t)$ (as in the case of IVGTT), then $Q(s) = 1$, and Equation 8.5b becomes

$$Y(s) = \frac{(s + \alpha)}{[(s + p_1)(s + p_2)]} \tag{8.12}$$

We can employ this model to simulate IVGTT. In this test, for the input glucose dose, $I = 1$ g per kg of body weight. If we adopt a blood volume of 17.5 L for a 70 kg person, then $y(0) = (70 \text{ g}/17.5) = 400$ mg/100 mL.

Hence, for the input, we note that $I = 1$ g of glucose input per 1 kg of body weight or $I = 4$ g of glucose/L of blood, i.e., 1 g of glucose $kg^{-1} = 4$ g of glucose $L^{-1} = 400$ mg of glucose/100 mL.

Performing partial-fraction expansion, we get (from Equation 8.12):

$$Y(s) = \frac{A_3}{(s + p_1)} + \frac{B_3}{(s + p_2)} = \frac{(s + \alpha)}{(s + p_1)(s + p_2)}$$

where

$$A_3 = \frac{s + \alpha}{s + p_2} \text{ (at } s = -p_1) = \frac{(\alpha - p_1)}{(p_2 - p_1)}$$

$$B_3 = \frac{s + \alpha}{s + p_1} \text{ (at } s = -p_2) = \frac{(\alpha - p_2)}{(p_1 - p_2)}$$

Then, taking inverse LPT, we get for $q(t) = I\,\delta(t)$, with $I = 4$ g of glucose/L of blood:

$$y(t) = I(A_3 e^{-p_1 t} + B_3 e^{-p_2 t})\text{g/L}$$
$$= \frac{4[(\alpha - p_1)e^{-p_2 t} - (\alpha - p_2)e^{-p_2 t}]}{(p_2 - p_1)}\text{g/L} \qquad (8.13)$$

Therefore,

$$y(0) = 4 \text{ g/L} \qquad (8.13a)$$

Note that at $t = 0$, $y = 4$ g L^{-1}, or 400 mg/dL. However, when this solution (Equation 8.13) is depicted in Figure 8.7b to simulate IVGTT for an injected glucose bolus of 4 g/L or 400 mg/100 mL, then $y(0) = 400$ mg/100 mL + 80 mg/100 mL (of fasting blood glucose concentration) = 480 mg/100 mL.

Equation 8.13 can also be expressed in terms of α, β, γ, and δ. Let us recall that we had put

$$\alpha\delta + \beta\gamma = \omega_n^2, \quad \text{and} \quad \alpha + \delta = 2A, \text{ in Equations 8.6c and 8.6d}$$

Then, as per Equation 8.6e, we had obtained: p_1 and $p_2 = A \pm (A^2 - \omega_n^2)^{1/2} = A \pm \omega$ for $A > \omega_n$ (for overdamped response), and $= A \pm i\omega$ for $A < \omega_n$ (for underdamped response)

Then

$$p_2 - p_1 = -2i\omega \quad \text{for} \quad A < \omega_n$$
$$= -2\omega \quad \text{for} \quad A > \omega_n$$

Again, it is noted that for the values of the parameters (α, β, γ, and δ) as indicated in Figure 8.7b, $A[= (\alpha + \delta)/2 = 1.978 \text{ h}^{-1}] > \omega_n[= (\alpha\delta + \beta\gamma)^{1/2} = 1.85 \text{ h}^{-1}]$. Hence this case corresponds to an overdamped system, corresponding to a diabetic subject. However, we could say that $A \cong \omega_n$, and hence this could represent a critically damped case, which would imply that this subject is at risk of becoming diabetic.

We also note that

$$\dot{y}(t) = \frac{4}{(p_2 - p_1)} \left[-p_1(\alpha - p_1)e^{-P_1 t} + p_2(\alpha - p_2)e^{-P_2 t} \right] \qquad (8.13b)$$

and

$$\dot{y}(0) = 4[\alpha - (p_1 + p_2)] = -4\delta \qquad (8.13c)$$

It can be noted that the negative slope of the curve at $t = 0$ corresponds to the model analysis (Equation 8.13c) $y'(0) = 4 \text{ g L}^{-1}\delta = 4 \text{ gL}^{-1}(3.04 \text{ h})^{-1} = 12.16 \text{ g L}^{-1} \text{ h}^{-1} = 1216 \text{ mg}(100 \text{ mL})^{-1}(\text{h})^{-1} = 3.04 \text{ gkg}^{-1} \text{ h}^{-1}$.

3. Glucose Response $y(t)$ to Glucose input $q(t) = $ Rectangular-Pulse function $h(t)$, where $h = 4 \text{ g (L)}^{-1} \text{ (h)}^{-1}$ or $400 \text{ mg} (100 \text{ mL})^{-1} \text{ (h)}^{-1}$, shown in Figure 8.7c

Here, the input function

$$h(t) = h \text{ grams of glucose } (\text{kg})^{-1}(\text{h})^{-1} \quad \text{for} \quad 0 < t \le t_0$$
$$= 0 \quad \text{for} \quad t \ge t_0$$

In other words, a rectangular pulse consists of $h(t) = h\, u(t) - h\, u(t - t_0)$.

The LPT of a rectangular pulse is

$$L[h(t)] = H(s) = (h/s)(1 - e^{-t_0 s}) = (h/s) - (he^{-t_0 s})/s \qquad (8.14)$$

where the first part of the equation is the LPT of the positive step and the second part is that of delayed negative step.

The transfer function $G(s)$ of the blood glucose control system is given by Equation 8.5b. Therefore, from Equations 8.14 and 8.5, we get the LPT of the system response:

$$Y(s) = H(s) * G(s) = \frac{h(s + \alpha)}{s(s + p_1)(s + p_2)} - \frac{h(s + \alpha)e^{-t_0 s}}{s(s + p_1)(s + p_2)} \qquad (8.14a)$$

where the poles are defined by Equation 8.6b, and p_1 and p_2 are the roots of the characteristic Equation 8.6a.

Now taking the inverse LPT of $Y(s)$ in Equation 8.14a gives the theoretical curve $y(t)$, which is graphically depicted in Figure 8.7c (for $y(0) = 80$ mg/100 mL of fasting glucose concentration):

$$y(t) = h\left(A_o + A_1 e^{-p_1 t} + A_2 e^{-p_2 t}\right)$$
$$- h\left(A_o + A_1 e^{-p_1(t - t_0)} + A_2 e^{-p_2(t - t_0)}\right) U(t - t_0) \qquad (8.14b)$$

where the unit-step function is

$$U(t - t_0) = 0 \quad \text{for} \quad t < t_0$$
$$= 1 \quad \text{for} \quad t > t_0$$

$$A_0 = \frac{\alpha}{p_1 p_2}, \quad A_1 = \frac{(\alpha - p_1)}{(p_1^2 - p_1 p_2)}, \quad \text{and} \quad A_2 = \frac{(\alpha - p_2)}{(p_2^2 - p_1 p_2)} \qquad (8.14c)$$

In Equation 8.14b, the first term represents the response to positive step starting at $t = 0$, while the second term represents the response to the delayed negative step starting at $t = t_0$. Also, Equation 8.14b can be written as two separate equations to represent $y(t)$, for different ranges of "t"

$$y(t) = h(A_0 + A_1 e^{-p_1 t} + A_2 e^{-p_2 t}) \quad \text{for} \quad 0 < t \le t_0 \qquad (8.15)$$
$$y(t) = h[A_1\{e^{-p_1 t} - e^{-p_1(t - t_0)}\} + A_2\{e^{-p_2 t} - e^{-p_2(t - t_0)}\}] \quad \text{for} \quad t \ge t_0 \qquad (8.16)$$

When $t = t_0$, we have:

$$y(t) = h\left[-(A_1 + A_2) + A_1 e^{-p_1 t_0} + A_2 e^{-p_2 t_0}\right]$$
$$= h\left[A_0 + A_1 e^{-p_1 t_0} + A_2 e^{-p_2 t_0}\right] \qquad (8.17)$$

since the sum of the residues $A_0 + A_1 + A_2 = 0$, as per Hazony and Riley rules. It may be noted that Equation 8.17 is essentially Equation 8.15 for $t = t_0$. Both the equations give the same $y[t]$ for $t = t_0$, since in this case the response $y[t]$ is continuous at $t = t_0$.

As regards the implication and application of this rectangular pulse, we can consider that when glucose bolus is administered to a patient, as an impulse input to the gastrointestinal (GI) compartment, the GI compartment converts this impulse input of glucose into a rectangular-pulse response of glucose, which becomes the input into the blood-pool compartment (BPC).

Hence $y(t)$, given by Equation 8.14b and simulated in Figure 8.7c (based on the previously indicated values of α, β, γ, and δ) could be deemed to simulate the OGTT blood glucose response data depicted in Figure 9.5 of the next chapter (except for the pointed and high peak value, as depicted in Figure 8.7c).

Note that in the above equations we can put $h =$ either 1 g $(kg)^{-1}$ $(h)^{-1}$ or 4 g $(L)^{-1}$ $(h)^{-1}$ or 400 mg $(100 \text{ mL})^{-1}$ $(h)^{-1}$, depending on the dimensional requirement of measuring $y(t)$ being in grams of glucose/liter of blood or in milligram of glucose/liter of blood.

At the rate of 1 g $kg^{-1} h^{-1}$ ($= 400$ mg $[100 \text{ mL}]^{-1} h^{-1}$) over a 30 min period, a total of 0.5 g kg^{-1} dose would have been delivered. So for a 70 kg person, the total dose would be 35 g delivered in 30 min.

8.5.2 Insulin $x(t)$ Responses of the System: $x(t)$ Output to Glucose $q(t)$ Input

The insulin response $x(t)$ is obtained by taking the inverse LPT of Equation 8.8 on both sides, for various inputs of $q(t)$, as considered above for $y(t)$.

1. Insulin response to $q(t) = $ step $[U(t)]$ glucose input, as depicted in Figure 8.8a
   The insulin response, for unit-step glucose input, is given by

   $$x(t) = k_0 + k_1 e^{-p_1 t} + k_2 e^{-p_2 t}$$

   where

   $$k_0 = \frac{\beta}{p_1 p_2}, \quad k_1 = \frac{\beta}{(p_1^2 - p_1 p_2)}, \quad \text{and} \quad k_2 = \frac{\beta}{(p_2^2 - p_1 p_2)} \qquad (8.18)$$

   If the input glucose mass-flow rate, $q(t) = S\ u(t)$, with $S = 1$ g of glucose (kg body mass^{-1} h^{-1} or 4 g L^{-1} h^{-1}), then

   $$x(t) = 4(k_0 + k_1 e^{-p_1 t} + k_2 e^{-p_2 t}) \text{ units/L} \qquad (8.18a)$$

   as depicted in Figure 8.8a, where the resting insulin concentration is taken to be 4 μ/100 mL (where the symbol μ denotes milli units).

   Here, the resting insulin concentration is assumed to be 4 μ/100 mL. At $t = \infty$, Equation 8.18b yields $x(t = \infty) = 4\ k_0 = 4\beta/p_1 p_2$, which (for the parameters values given in Figure 8.8a) = $4(0.198)/\omega_n^2 = 0.792/1.85^2 = 0.228$ units/L or 22.8 μ/100 mL. Now since the resting insulin concentration is taken to be 4 μ/100 mL, hence $x(t = \infty) = 26.8$ μ/100 mL, as shown by the graph.

2. Insulin response to $q(t) = $ impulse $[I\ \delta(t)]$ glucose input, as depicted in Figure 8.8b

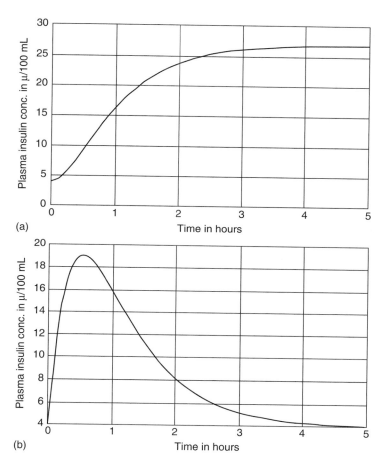

FIGURE 8.8
(a) Plasma-insulin simulated response $x(t)$, to unit-step glucose input of 1.0 g (kg mass)$^{-1}$ (h)$^{-1}$ of glucose infusion, given by Equation 8.18 as the solution of Equation 8.4, for $\alpha = 0.916$, $\beta = 0.198$, $\gamma = 3.23$, $\delta = 3.04$; their units are provided in Table 8.3. (b) Simulated insulin-concentration response $x(t)$ to Impulse input of glucose for the system $X(s)/Q(s)$, as given by Equation 8.19, as the solution of Equation 8.4, with $\alpha = 0.916$, $\beta = 0.198$, $\gamma = 3.23$, $\delta = 3.04$; their units are provided in Table 8.3. The resting value of insulin concentration is taken to be 4 μ/100 mL of blood, where μ denotes milli units. (Adopted from Dittakavi, S.S. and Ghista, D.N., *J. Mech. Med. Bio.*, 1, 193, 2001.)

The insulin response is given by

$$x(t) = I\beta \frac{(e^{-p_1 t} - e^{-p_2 t})}{(p_2 - p_1)} \tag{8.19a}$$

which is also the derivative of Equation 8.18
  If $q(t) = I\,\delta(t)$, with $I = 1$ g (kg body weight)$^{-1}$ or 4 g (liter of blood volume)$^{-1}$,

then,

$$x(t) = 4\beta(e^{-p_1 t} - e^{-p_2 t})/(p_2 - p_1) \text{ units/L} \qquad (8.19b)$$

This function is graphically depicted in Figure 8.8b, for $x(0) = 4$ $\mu/100$ mL of blood.

We can also express $x(t)$ in terms of the basic parameters of the model using the same expressions (as in Equation 8.6d and the equation following it), namely:

$$\alpha\delta + \beta\gamma = \omega_n^2, \ \alpha + \delta = 2A, \ \omega = (\omega_n^2 - A^2)^{1/2} = \frac{[4\beta\gamma - (\alpha - \delta)^2]^{1/2}}{2}$$

with p_1, p_2 given, as in Equation 8.6e, by

$$\begin{aligned} &= A \pm i(\omega_n^2 - A^2)^{1/2} \\ &= A \pm (A^2 - \omega_n^2)^{1/2} \\ &= A \pm \omega \ (\text{for} \quad A > \omega_n) \\ &= A \pm i\omega \ (\text{for} \quad A < \omega_n) \end{aligned}$$

Then, we obtain

$$\begin{aligned} x(t) &= 4\beta e^{-At}\left(\frac{e^{-i\omega t} - e^{-i\omega t}}{2i\omega}\right) \ \text{for} \ A < \omega_n \text{ or underdamped response} \\ &= \frac{4\beta}{\omega} e^{-At} \sin\omega t \, (\text{units/L}) \end{aligned} \qquad (8.19c)$$

and

$$\begin{aligned} x(t) &= 4\beta e^{-At}\left(\frac{e^{-\omega t} - e^{\omega t}}{-2\omega}\right) \ \text{for} \quad A > \omega_n \text{ or overdamped response} \\ &= \frac{4\beta}{\omega} e^{-At} \sinh\omega t \, (\text{units/L}) \end{aligned} \qquad (8.19d)$$

Let us refer to Equation 8.19d for insulin response to impulse glucose input:

$$x(t) = (4\beta/\omega)e^{-At} \sinh \omega t, \text{ in units } L^{-1}$$

so that

$$x'(t) = (4\beta/\omega)[\omega \cosh \omega t - A \sinh \omega t] \, e^{-At} = 0$$

Then, to obtain the time at which $x(t)$ is maximum, we put $x'(t) = 0$, and get

$$\omega t = \tanh^{-1} \omega/A = \tanh^{-1}(0.7/1.98) = \tanh^{-1}(0.35) = 0.38$$

$$\text{i.e., } t = 0.38/\omega = 0.38/0.7 = 0.54 \text{ h}$$

Hence, as per model analysis, $x(t)$ is maximum at $t = 0.54$ h. The maximum value of insulin concentration is (as depicted in Figure 8.8b) 19 $\mu/100$ mL, for the resting value of 4 $\mu/100$ mL.

3. Insulin response to rectangular-pulse [$h(t)$] glucose input

The insulin response $x(t)$ is given by

$$\begin{aligned}
x(t) &= h(k_0 + k_1 e^{-p_1 t} + k_2 e^{-p_2 t}) \quad \text{for} \quad t < t_0 \\
&= h(k_0 + k_1 e^{-p_1 t} + k_2 e^{-p_2 t}) - h\left(k_0 + k_1 e^{-p_1(t-t_0)} + k_2 e^{-p_2(t-t_0)}\right) U(t - t_0) \\
&\quad \text{for} \quad t > t_0 \\
&= h[k_1\{e^{-p_1 t} + e^{-p_1(t-t_0)}\} + k_2\{e^{-p_2 t} - e^{-p_2(t-t_0)}\}] \, U(t - t_0) \\
&\quad \text{for} \quad t > t_0
\end{aligned} \tag{8.20}$$

We put $h = 400$ mg 100 mL^{-1} h^{-1} in Equation 8.20, if $x(t)$ is measured in milli units (μ)/100 mL of blood-pool volume.

8.6 Model Characteristics and Physiological Significance [2]

8.6.1 Model Characteristics

The solutions to Equations 8.5 and 8.8, to different modalities of glucose input, have been provided in Section 8.5 and graphically illustrated in Figures 8.7 and 8.8, to simulate glucose and insulin responses.

The Equations 8.11, 8.13, and 8.14b constitute the solutions of governing Equation 8.3, employing $Y(s)$ given by Equation 8.7, as glucose concentration (y) responses (in gram/liter) to $q(t)$, for unit-step, unit-impulse, and rectangular-pulse glucose inputs. Likewise, solutions for Equation 8.4 employing $X(s)$ given by Equation 8.8, are given by Equations 8.18 through Equations 8.20 for insulin-concentration response (x) in units/liter, for the same three types of glucose input. The input glucose function $q(t)$ is expressed in gram per liter per hour (g L^{-1} h^{-1}) for impulse, step, and rectangular-pulse input signals.

As regards the data adopted by us to obtain these response curves, we have assumed that a 70 kg normal person's body contains 17.5 L of blood, i.e., 4 kg body mass ≈ 1 L of blood or 1 g of glucose input/kg of body mass ≈ 4 g of glucose input/L of blood. This relation requires that all the unit-input response functions are to be multiplied by 4, for obtaining the response solutions to the governing Equations 8.3 and 8.4, because of glucose input of 1 g/kg body mass being ≈ 4 g of glucose/L of blood; moreover 1 g (or 1 unit)/L $= 100$ mg (or 100 μ) per dL.

All of this data is employed (in the MATLAB program) for developing the glucose and insulin output solutions to Equations 8.7 and 8.8, for the above indicated forms of glucose-input functions. The resulting (computer-simulated) response graphs are depicted in Figure 8.7a through c and Figure 8.8a and b. For the sake of convenience, we have employed the same values of parameters employed by Bolie [1]: $\alpha = 0.916$, $\beta = 0.198$, $\gamma = 3.23$, and $\delta = 3.04$ (units given in Table 8.3) for all the types of response curves depicted by Figures 8.7 and 8.8. By associating Equation 8.3 with Equation 8.6c, corresponding to the values of these parameters, we can also obtain values of the natural frequency of the system $\omega_n (= \sqrt{\alpha\delta + \beta\gamma})$ and of the damping coefficient $A = (\alpha + \delta)/2$.

The discriminant of the quadratic Equation 8.6a, which is the under-root term in Equation 8.6b, dictates the damping category into which the system falls, namely

- For overdamped system, $(\alpha - \delta)^2 > 4\beta\gamma$, i.e., $A > \omega_n$
- For critically damped system, $(\alpha - \delta)^2 = 4\beta\gamma$, i.e., $A\ =\ > \omega_n$
- For underdamped system, $(\alpha - \delta)^2 < 4\beta\gamma$, i.e., $A < \omega_n$

For the above given values of $(\alpha, \beta, \gamma, \delta)$ based on Equations 8.6c and 8.6d, we obtain the damping coefficient, $A = (\alpha + \delta)/2 = 1.98$ h^{-1} the system's natural frequency $\omega_n = (\alpha\delta + \beta\gamma)^{1/2} = (2.78 + 0.64)^{1/2} = 1.85$ rad/h and

$$T = \frac{2\pi}{\omega_n} = 3.396$$
$$\lambda(= \alpha + \delta) = 3.96 \text{ h}^{-1}$$
$$\lambda T_d(= \alpha\delta + \beta\gamma)$$
$$T_d = 1.16 \text{ h}$$

Hence the system's natural frequency, $\omega_n < A$, and the system is over-damped.

Also, as a check, $\dfrac{[(\alpha - \delta)^2 - 4\beta\gamma]^{1/2}}{2} = (A^2 - \omega_n^2)^{1/2} = 0.7$

Hence, because $(\alpha - \delta)^2 > 4\beta\gamma$, the adopted parameters' values correspond to the case of overdamped system associated with a diabetic patient. However, $A \cong \omega_n$ and also $T(= 2\pi/\omega_n)$ is less than the critical value of 4 h. Hence it is more appropriate to associate this data with a critically damped system, characterizing a subject at risk of being diabetic.

Although the values of the parameters (adopted by us) yield overdamped system response curves of Figures 8.7 and 8.8, the glucose and insulin response transfer-functions Equations 8.7 and 8.8 are infact also valid to generate the response curves for critically damped and underdamped systems.

8.6.2 Explanation of the Glucose- and Insulin-Simulated Response Curves in Terms of the Model Characteristics

1. Transfer-function models are generally used to describe a system output versus input. Hence in Equation 8.1, p is made to be zero so as to make the system have only one input "q." The outputs, $y(t)$ of Figure 8.7a and $x(t)$ of Figure 8.8a, to unit-step glucose input of 4 g of glucose/L of blood, are respectively obtained by taking the inverse LPT of the corresponding s-domain equations.

2. The curves in Figures 8.7a and 8.8a represent the glucose and insulin outputs $y(t)$ and $x(t)$ of the system, in response to the unit-step input of glucose-intake q(with $p = 0$). The characteristics of the regulatory system, represented by Equation 8.3, can be explained in terms of the roots of the characteristic equation, obtained by putting the denominator of $G(s)$ (in Equation 8.5a) equal to zero, as:

$$s^2 + s(\alpha + \delta) + (\alpha\delta + \beta\gamma) = (s + p_1)(s + p_2) = 0 \qquad (8.21)$$

where p_1 and p_2 are the roots of: $p^2 - p(\alpha + \delta) + (\alpha\delta + \beta\gamma) = 0$

3. The zero(s) of the system as well as the nature of the Input function are responsible for the system response and behavior. Hence, in Equations 8.5a and 8.5b, the zero $(s + \alpha)$ of the system also influences the system response.

   The presence of a zero in the closed-loop transfer function decreases the rise-time and increases the maximum overshoot of the step-response [4]. This concept is illustrated by the early peaking of computer-simulated glucose response $y(t)$ in Figure 8.7a, in response to a step-input $q(t)$ of 1.0 g $(\text{kg})^{-1}$ $(\text{h})^{-1}$. It is contended that although the characteristic equation's roots are generally used to study the relative damping and stability of linear control systems, the zero(s) of the transfer function should not be overlooked in their effects on the overshoot and rise-time of the step-response [4].

4. Figure 8.7b shows the impulse-input response of glucose, as the solution of Equation 8.3, with $q = I\delta(t)$. We can infact employ

this model response curve to simulate IVGTT response, to a rapidly injected glucose bolus into the blood. By matching this analytical response to the IVGTT clinical data (curve) of glucose concentration, we can obtain the values of the parameters α, β, γ, and δ.

In the simulated IVGTT data in Figure 8.7b, the glucose dose $= 1.0$ g/kg, and the blood volume $(V) = 17.5$ L for a 70 kg person; then $y(0) = 70$ g/17.5 L $= 400$ mg/100 mL. However, because the fasting BGC value is 80 mg/100 mL, hence the glucose-response curves (to glucose impulse input, simulating IVGTT), starts at 480 mg/dL on the y-axis (at time $t = 0$) and ends at $y(\infty) = 0$ (at $t = \infty$), as per Equation 8.13.

5. Because of the "zero" in Equations 8.5a and 8.5b, not only is the glucose step-response of the system (i.e., response to step-function input) of Figure 8.7a affected, but also the glucose impulse–response depicted in Figure 8.7b.

   However, when the second-order system consists of only poles but no zeros, as in the case of Equation 8.8 representing insulin-concentration output, the impulse–response curve of the system starts from the origin [4,5], as shown in Figure 8.8b, for resting insulin concentration of 4 μ/100 mL.

6. So, the effect of zero in the glucose response transfer function is to alter the initial value of glucose concentration $y(0)$ from 0 to 400 mg/100 mL, i.e., from $Y(0) = 80$ (the assumed fasting glucose concentration) to $y(0) = 480$, as depicted in Figure 8.7b. In other words, this glucose-tolerance model curve would have started from the origin instead of 480 mg/100 mL, if there were no zero.

   Interestingly, in Section 9.3 of the next chapter, we will see that the zero of BGCS is compensated (canceled) by introducing a pole due to the gastrointestinal tract, resulting in the OGTT response equation not containing zero.

8.6.3   Physiological Significance of the Model Simulation

1. Blood glucose-regulatory system model of Equation 8.2 yields the response to a step-velocity glucose input (q), with zero steady-state error. This means that in the steady-state, the rate of glucose injected $(q) =$ total rate of glucose metabolism $(\gamma x + \delta y)$; this condition is obtained by putting $y' = 0$ in Equation 8.2. Hence when $x' = y' = 0$, $x = x$ (ss or steady-state) and $y = y$(ss). Note that the term (γx) is the insulin-dependent glucose transport, and the term (δy) is the insulin-independent (or glucose concentration-dependent) glucose transport.

2. On the basis of Equation 8.2, we note that the y' and y responses are damped by the term γx (representing the glucose utilization by the tissue due to insulin concentration). Hence, the value of γ can characterize insulin-resistance (Figure 8.3), indicative of "diabetic risk factor."

3. In Equations 8.5a and 8.5b, it is the zero of the system, represented by $(s+\alpha)$, which introduces the proportional-derivative effect, resulting in a high rate of initial response to a step-input of glucose, as depicted in Figure 8.7a. This feature of the system, i.e., the early peaking phenomena (rapid rise of blood glucose level), is due to the zero $(s+\alpha)$ of the system transfer function $[Y(s)/Qs)]$ of Equations 8.5a and 8.5b.

Appendix A: Tutorial on Laplace Transform Methodology to Solve DEQ(S)

A.8.1 Types of Inputs (Forcing Functions)

The strength of a pulse is given by the product of pulse height and pulse width i.e., its area. When the pulse width or duration is negligibly small (when compared to the time-constants of the system and the pulse height is infinitely large, the pulse input is called "impulse input." Thus the impulse function is a special limiting case of a pulse function. The unit-impulse function $\delta(t)$ has zero value everywhere except at $t=0$, where its impulse-strength is unity.

The step function whose height is unity is called "unit-step function." It is defined as follows:

$$U(t) = 0 \quad \text{for} \quad t < 0$$
$$= 1 \quad \text{for} \quad t > 0$$

A.8.2 Laplace Transform

$$F(s) = \int_0^\infty f(t)e^{-st}\,dt$$

$F(s) =$ Laplace transform of $f(t)$
$F(t) =$ Inverse LPT of $F(s)$
"s" is the Laplace operator, whose dimensions is t^{-1}

Some LPT of variables (with assumed zero initial conditions) are given below:

$y{:}Y(s)$ $y^{|}{:}sY(s)$ $y^{||}{:}s^2\,Y(s)$

$x{:}X(s)$ $x^{|}{:}sX(s)$ $x^{||}{:}s^2\,X(s)$

$q{:}Q(s)$ $q^{|}{:}sQ(s)$

if $q(t) = \delta(t)$, $Q(s) = 1$, and if $q(t) = U(t)$, $Q(s) = 1/s$

A.8.3 Application of Laplace Transform to the Solution of Linear Differential Equation

The procedure is outlined as follows:

1. Transform the differential equation to the s-domain, using LPT tables.
2. Manipulate the transformed algebraic equation and solve for the output variable.
3. Perform partial-fraction expansion, so that the inverse LPT can be obtained from the LPT table.
4. Perform the inverse LPT operation.

A.8.4 Transfer Function of a Linear Time Invariant System

The TF is defined to be the ratio of the LPT of the output (response function) to the LPT of input (driving function) under the assumption that all initial conditions are zero. This statement can be represented by the equation $G(s) = Y(s)/Q(s)$

$$G(s) = Y(s) \text{ if } Q(s) = 1, \text{ i.e., if } q(t) = \delta(t)$$

A.8.5 Transfer Function of the Glucose–Insulin System Equation 8.3 Given by Equation 8.5a

For the system given by Equation 8.3:

$$y'' + y'(\alpha + \delta) + y(\alpha\delta + \beta\gamma) = q + \alpha q$$

$$L[y(t)] = Y(s)$$

$$L[y'(t)] = sY(s) - y(0) = sY(s)$$

$$L[y''(t)] = s^2 Y(s) - sY(0) - y'(0) = s^2 Y(s)$$

assuming all initial conditions to be zero.

Similarly, $L[q(t)] = Q(s)$

$$L[\dot{q}(t)] = sQ(s) - q(0) = sQ(s)$$

Therefore,

$$s^2Y(s) + (\alpha + \delta)(sY(s)) + (\alpha\delta + \beta\gamma)Y(s) = sQ(s) + \alpha Q(s)$$
$$Y(s)[s^2 + s(\alpha + \delta) + (\alpha\delta + \beta\gamma)] = Q(s)[s + \alpha]$$
$$\frac{Y(s)}{Q(s)} = \frac{(s + \alpha)}{s^2 + s(\alpha + \delta) + (\alpha\delta + \beta\gamma)} = G(s). \tag{8.A1}$$

A.8.6 Transfer Function of the Governing Glucose–Insulin System Equation 8.6c

The TF of a general second-order system (with a zero at $s = -z$) is represented by

$$(s + z)/(s^2 + 2As + \omega_n^2) \tag{8.A2}$$

where

ω_n = natural frequency of oscillation of system
A is the attenuation constant

The system is said to be under damped if $A < \omega_n$, critically damped if $A = \omega_n$, and over damped if $A > \omega_n$.

Comparing the above equation with Equation 8.A1 and the system transfer function of Equation 8.5a, we have the following relations:

$$z = \alpha, \quad (\alpha\delta + \beta\gamma) = \omega_n^2, \quad (\alpha + \delta) = 2A$$

Hence, for the system represented by Equations 8.3 and 8.6c, we can put their transfer-function denominators as

$$s^2 + s(\alpha + \delta) + (\alpha\delta + \beta\gamma) = (s + p_1)(s + p_2) = s^2 + s(p_1 + p_2) + p_1p_2$$

so that

$$(\alpha + \delta) = p_1 + p_2 \quad \text{and} \quad (\alpha\delta + \beta\gamma) = p_1p_2 \tag{8.A3}$$

We can solve this pair of equations for p_1 and p_2, by substituting

$$p_1 = \frac{\alpha\delta + \beta\gamma}{p_2} \quad \text{and} \quad p_2 = \frac{\alpha\delta + \beta\gamma}{p_1}$$

into equation: $\alpha + \delta = p_1 + p_2$, from which we get

$$p_2^2 - p_2(\alpha + \delta) + (\alpha\delta + \beta\gamma) = 0$$

and

$$p_1^2 - p_1(\alpha + \delta) + (\alpha\delta + \beta\gamma) = 0$$

Hence p_1 and p_2 are the roots of the characteristic-quadratic equation:

$$p^2 - p(\alpha + \delta) + (\alpha\delta + \beta\gamma) = 0$$

from which we obtain

$$
\begin{aligned}
p_1, p_2 &= \frac{(\alpha + \delta) \pm [(\alpha + \delta)^2 - 4(\alpha\delta + \beta\gamma)]^{1/2}}{2} \\
&= \frac{(\alpha + \delta) \pm [(\alpha - \delta)^2 - 4\beta\gamma]^{1/2}}{2}
\end{aligned} \tag{8.A4}
$$

Hence, both the transfer functions (8.A1) and (8.A2) of system's Equations 8.3 and 8.6c can be represented by

$$G(s) = \frac{(s + \alpha)}{(s + p_1)(s + p_2)} \tag{8.A5}$$

where p_1 and p_2 are given by Equation 8.A4.

The singular points at which $G(s)$ (or its derivatives) approach infinity are called poles. The points at which the function $G(s)$ is equal to zero are called zeros. Hence, the system represented by Equation 8.A1 has two poles at $s = -p_1$ and $s = -p_2$ and one zero at $s = -\alpha$.

The inverse transform of the system transfer function is

$$
L^{-1}\left[\frac{(s + \alpha)}{s^2 + s(\alpha + \delta) + (\alpha\delta + \beta\gamma)}\right] = L^{-1}\left[\frac{(s + \alpha)}{(s + p_1)(s + p_2)}\right]
$$

$$
= \frac{(\alpha - p_1)e^{-p_1 t} - (\alpha - p_2)e^{-p_2 t}}{(p_2 - p_1)} \tag{8.A6}
$$

References

1. Bolie, V.W., Coefficients of normal blood glucose regulation, *Journal of Applied Physiology*, 16(5):783–788, 1961.
2. Dittakavi, S.S. and Ghista, D.N., Glucose Tolerance Tests Modeling & Patient-Simulation for Diagnosis, *Journal of Mechanics in Medicine and Biology*, 1(2): 193–223, 2001.
3. Guyton, A.C. and Hall, J.E., *Text Book of Medical Physiology*, W.B. Saunders, Philadelphia, 1991.
4. Kuo, B.C., *Automatic Control Systems*, Prentice-Hall of India Pvt. Ltd., New Delhi, India, 1993, pp. 352–355.
5. Rama Rao, A.V.S.S., *Text Book of Biochemistry*, 4th ed. L.K. & S. Publishers, Tirupati, India, 1983, pp. 247–248.

9

Glucose Tolerance Tests Modeling

Dhanjoo N. Ghista and Dittakavi Sarma

CONTENTS

9.1 Scope

In this chapter, we apply our blood glucose–insulin control system (BGCS) model developed in the previous chapter to simulate glucose tolerance tests, both intravenous glucose tolerance test (IVGTT) and oral glucose tolerance test (OGTT).

9.2 Intravenous Glucose Tolerance Test (IVGTT)

There are three methods of performing this test: (1) constant infusion of glucose [at the rate of $g(kg)^{-1}(h)^{-1}$] into blood, until the steady state of $y(t)$ curve is reached; (2) single dose (bolus) of glucose (1 g/kg mass of body) rapidly injected into blood; (3) constant infusion at the same rate for 30 min.

In these three methods, blood glucose concentration is monitored at uniform intervals, to obtain corresponding $y(t)$ curves, as step response, impulse response, and rectangular-pulse response. The corresponding time-domain equations simulating these glucose response curves are represented by Equations 8.11, 8.13, and 8.14b of Chapter 8. These responses are respectively illustrated in Figure 8.7a through c.

9.2.1 Glucose Regulation Model

Most of the glucose regulatory models (IVGTT) involve similar differential equations with different notations. Here, we have decided to adopt Bolie's model [1], namely Equation 8.3 of Chapter 8, i.e.,

$$y'' + (\alpha + \delta)y' + (\alpha\delta + \beta\gamma)y = q' + \alpha q \tag{9.1}$$

9.2.2 Clinical Support for Model Justification

Among the system-response curves, those illustrated by Figure 8.7a through c are clinically useful for diagnosis of diabetes, by the three methods indicated above. Guyton and Hall [2] have demonstrated that the overshoot of the response curve is caused by the initial buildup of glucose in the extracellular fluid, before the insulin function has had time to become fully activated. In the previous chapter, we have explained the same overshoot by means of (1) the control-engineering concept of zero in the system and (2) the system's response property to a step input.

The clinical support for Figure 8.7c is given by Ref. [3], that in a normal person when glucose is infused intravenously into the body at the rate of $1 \text{ g}(kg)^{-1}(h)^{-1}$ for a period of half an hour, the concentration of blood glucose does not rise more than 250 mg/100 mL at the end of infusion and falls below the fasting level in 2 h. These two conditions are satisfied in our simulated response curve of Figure 8.7c.

9.2.3 BGCS Model Parametric Determination from Simulation of IVGTT Patient Data

Many authors, as for instance Insel et al. [3], Bergman et al. [4], Ferrannini et al. [5], and Galvin et al. [6], have given definitions of increasing complexity for the insulin sensitivity parameter (γ, the reciprocal of insulin resistance) and for other parameters α, β, and δ. However, this can only be ascertained after we have evaluated the parameters for a broad range of normal and

TABLE 9.1

Intravenous Kinetics Data: Data from a Normal Subject
(Weight 63.8 kg; Dose = 31.9 g)

Time in Minutes	0	5	10	20	30	40	50	60
Glucose concentration (y) in mg/100 mL	74.8	278.8	245.9	176.0	124.6	84.8	64.7	53.2
Insulin concentration (x) in μ/100 mL	4.4	102.7	107.0	69.6	49.1	20.3	17.6	12.4

Note: Glucose dose of 0.5 g/kg body weight is given as a bolus injection in 5 min (2.5% solution). Readings of glucose and insulin at 0 min indicate fasting levels, and those at 5 min indicate the levels reached after the bolus injection. The symbol μ denotes milli units.

diabetic subjects, and tracked some borderline subjects in order to determine the sensitivity of the (insulin resistance) parameter γ to "risk of becoming diabetic."

1. Protocol suggested by us to evaluate the parameters α, β, γ, and δ involves collection of glucose kinetics data from IVGTT, following a single glucose bolus of 0.5 g/kg weight of the subject, representing a unit-impulse input of glucose. For example, if the subject weighs 60 kg, the prescribed intravenous glucose dose is 30 g (or 30,000 mg). Thereafter, at uniform time intervals of 10 min, blood samples are collected for a total experimental period of 1 h to determine both glucose and insulin concentrations; these data are provided in Tables 9.1 and 9.2.

   For the administered glucose input, the model-response curves $y(t)$ and $x(t)$ of the model Equations 9.3 and 9.4, given by Equations 8.13 and 8.19, can be made to match the glucose test data, and the model system parameters (α, β, γ, δ) can thereby be evaluated. Alternatively (but not as accurately), we can employ the basic Equations 8.1 and 8.2 for data analysis, to evaluate the four parameters α, β, γ, and δ.

   Let the glucose and insulin levels at time t, $t+10$, $t+20$, and $t+30$ min (after subtracting their corresponding fasting levels) be respectively represented by y_t, y_{t+10}, y_{t+20}, y_{t+30}, and x_t, x_{t+10}, x_{t+20}, x_{t+30}. Let y'_t be the slope of the glucose curve at t min, and

TABLE 9.2

Intravenous Kinetics Data: Data from a Hypertensive Patient
(Weight = 55.2 kg; Dose 27.6 g)

Time in Minutes	0	5	10	20	30	40	50	60
Glucose concentration in mg/100 mL	92.8	330.8	286.3	232.4	187.2	171.1	150.0	127.8
Insulin concentration in μ/100 mL	4.7	62.9	120.9	146.0	166.4	130.0	139.8	81.6

Note: Glucose dose of 0.5 g/kg body weight is given as a bolus injection in 5 min (2.5% solution). Readings of glucose and insulin at 0 min indicate fasting levels, and those at 5 min indicate the levels reached after the bolus injection. The symbol μ denotes milli units.

y'_{t+10} the slope at $t+10$ min. Similarly, the slopes x'_t and x'_{t+20} can also be found from $x(t)$.

Thereafter, the values of model parameters α and β in Equation 8.1 can be found by determining their "best" values to satisfy the four simultaneous equations corresponding to times t, $t+10$, $t+20$, and $t+30$:

$$-x'_t = \alpha x_t - \beta y_t; \quad -x'_{t+10} = \alpha x_{t+10} - \beta y_{t+10}$$
$$-x'_{t+20} = \alpha x_{t+20} - \beta y_{t+20}; \quad -x'_{t+30} = \alpha x_{t+30} - \beta y_{t+30} \tag{9.2}$$

arising from Equation 8.1, assuming that α and β have the same values during the 30 min monitoring period.

The values of model parameters γ and δ in Equation 8.2 can similarly be determined by solving the four simultaneous equations (corresponding to the four time intervals):

$$-y'_t = \gamma x_t + \delta y_t; \quad -y'_{t+10} = \gamma x_{t+10} + \delta y_{t+10}; \quad -y'_{t+20} = \gamma x_{t+20} + \delta y_{t+20}$$
$$-y'_{t+30} = \gamma x_{t+30} + \delta y_{t+30} \tag{9.3}$$

assuming that γ and δ have the same values during the 30 min time period.

This methodology is employed to determine the parameters for one typical normal subject (A) and for one typical atherosclerotic hypertensive (ath-hyp) nondiabetic subject (B), both of whose intravenous kinetics test data are provided in Tables 9.1 and 9.2. The parametric values of these subjects are obtained as

$$\alpha = 0.05 \quad \beta = 0.96 \quad \gamma = 6.48 \quad \delta = 1.76 \text{ for the normal subject (A)}$$
$$\alpha = 2.25 \quad \beta = 15.12 \quad \gamma = 0.93 \quad \delta = 0.12 \text{ for the atherosclerotic} \quad (9.4)$$
$$\text{hypertensive patient (B)}$$

The dimensions of these parameters $(\alpha, \beta, \gamma, \delta)$ are given in Table 8.3. The following are the values for the parameters: (1) the damping coefficient $A(=\alpha+\delta)=1.8\,\text{h}^{-1}$ for subject A and $2.37\,\text{h}^{-1}$ for subject B; and (2) the natural frequency $\omega_n(=\sqrt{(\alpha\delta+\beta\gamma)})=2.74\,\text{rad/h}$ for subject A and $3.79\,\text{rad/h}$ for subject B. It is seen that for both the subjects $A<\omega_n$, indicative of underdamped model response, characteristic of normal nondiabetic subjects.

2. There are points concerning data in Tables 9.1 and 9.2 that are worth noting. For the normal subject, the dose of 31.9 g (at 0.5 g/kg mass of the patient) is administered in 5 min. Hence,

$$y'(0) = \frac{278.8 - 74.4}{5 \text{ min}} = \frac{204 \text{ mg}/100 \text{ mL}}{5 \text{ min}} = \frac{204/100}{1/12 \text{ h}} \text{ g L}^{-1}\text{h}^{-1}$$
$$= 24 \text{ g L}^{-1}\text{h}^{-1}$$

Now, from the model Equation 8.2, $q = y'(0)$, since normalized $x(0)$ and $y(0) = 0$. Correspondingly, the administered dosage input rate $q(t) = I\ \delta(t) = $ (dose 0.5 g kg^{-1})/(1/12) h = 6 g kg^{-1} h^{-1} = 24 g L^{-1} h^{-1} = $y'(0)$.

Also, in the data, the value of y (the blood glucose concentration) 5 min after the dose is administered is equal to (278.8 – 74.4 =) 204 mg/100 mL and this is for a dose of 0.5 g kg^{-1}h^{-1}. Correspondingly, our model solution in Equation 8.13a gives $y(0) = 400$ mg/100 mL for a dose of 1 g kg^{-1} h^{-1}. After the first 5 min of glucose administration, the blood glucose concentration becomes 204 mg/100 mL. Thereafter, the blood glucose system response to this glucose input starts. Note that the blood glucose concentration drops from 278.8 to 245.9 mg/100 mL within the next 5 min, giving us a negative slope $y'(5)$ of 3.95 g L^{-1} h^{-1}.

The parameters' values for both the subjects yield $(\alpha - \delta)^2 < 4\beta\gamma$, corresponding to an underdamped glucose regulatory system. However, as noted in Equation 9.4, for the atherosclerotic hypertensive subject (relative to the normal subject), we obtain a reduced value of γ (i.e., reduced sensitivity or increased resistance to insulin for tissue utilization of glucose) and an elevated value of β (i.e., the insulin release factor in response to elevated glucose concentration due to increased resistance to insulin). This could imply that in an atherosclerotic hypertensive patient, there is overworking of the pancreas due to reduced tissue utilization of glucose. Although this is a preliminary study, this model is shown to sensitively bring out these features, to diagnose atherosclerotic hypertensive patients as having a high diabetes risk factor.

3. Even more interesting aspect is the phase comparison between blood glucose and blood insulin concentration data, plotted in Figure 9.1a and b. These plots show that in the case of the

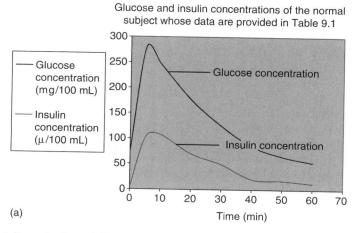

FIGURE 9.1 (See color insert following page 266.)
Blood glucose and insulin concentration dynamics. The symbol μ denotes milli units.

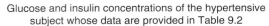

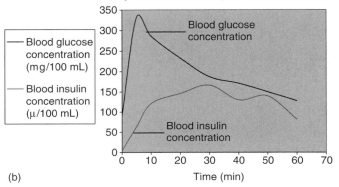

(b)

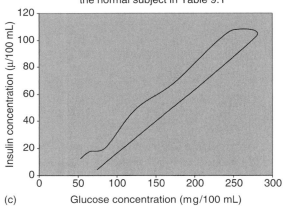

(c)

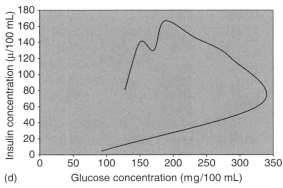

(d)

FIGURE 9.1 (continued)
(See color insert following page 266.)

atherosclerotic hypertensive subject (B), the phase lag between the blood glucose and the blood insulin concentration curves is quite pronounced compared with that for the normal subject (A). The situation appears even more fascinating when blood insulin concentration is plotted versus blood glucose concentration for the two patients in Figure 9.1c and d. A sort of hysteresis-type loop becomes manifest, and the loop area is more pronounced for subject B than for subject A. The loop area could be deemed to correspond to an imbalance between the blood glucose amount and the insulin release rate. Perhaps then, this loop area could constitute another means for designating risk to diabetes and glucose for clearly detecting diabetes. This phenomenon and concept of characterizing it as a diagnostic index needs to be explored further.

9.3 Oral Glucose Tolerance Test

9.3.1 OGTT Process

This OGTT test differs from IVGTT in that the glucose is administered in a single dose orally (instead of being injected into blood). In this test, a fasting person is given an oral glucose dose of 1 g/kg, for the purpose of diagnosing diabetes. If the subject is normal and free from diabetes, the blood glucose level rises from the fasting value of, say, 80 to 120–140 mg/dL, and then falls back to below normal in about 2 h.

The physiology of the GI tract suggests that the intestinal glucose absorption rate is constant for a limited time duration. Hence, the glucose rectangular-pulse input $h(t)$ into the blood pool can be deemed to be representative of this phenomena. This in fact is made physiologically possible by means of the combined effect of two mechanisms: (1) due to the pyloric sphincter valve resistance which controls the transfer of glucose from the stomach to the intestines in inverse proportion to the stomach distension, and (2) due to the active transport of glucose from the intestines into the blood (across the intestinal wall) at its maximum rate, according to the Michaelis–Menten equation graphically depicted in Figure 9.2.

Based on the similarity between Figure 8.7c (simulating glucose response to a rectangular-pulse input of glucose) and the clinical data curve of Figure 9.5, we could say that the OGTT response curve can be characterized as glucose response to a rectangular-pulse input of glucose into the blood-pool compartment (BPC) from the gastrointestinal compartment (GIC). In this integrated system made up of the GIC and the BPC, the input to GIC is an impulse (ingested) input of glucose. The GIC converts the impulse input of glucose into a rectangular-pulse response of glucose.

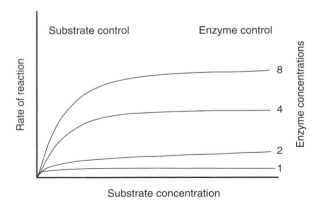

FIGURE 9.2

Effect of substrate and enzyme concentrations on the rate of enzyme-catalyzed reaction. *Note*: The following is the general form of Michaelis–Menten equation for active transport: Rate of reaction $= k_1$ (enzyme concentration) (substrate concentration)/$(k_2 +$ substrate concentration). When there is no glucose in blood, during hypoglycemia (i.e., when "y" is much below the fasting level), the active transport is carried out at its maximum rate. Thus, the passive and active transports work together to maximize the intestinal glucose transport, so as to tide over the crisis of hypoglycemia in diabetics.

Thereafter, the BPC converts the rectangular-pulse input into the OGTT response curve.

As such, Equations 8.14b and 8.14c can supposedly simulate the OGTT data. However, for this glucose rectangular-pulse input $h(t)$ into the blood pool (from the GI tract), the model-response curve $y(t)$, shown in Figure 8.7c and described by Equations 8.14b and 8.14c, has a sharp corner at its peak. On the other hand, the clinically monitored OGTT data (illustrated in Figure 9.5) are obtained as smooth curves. So we need a more representative model for simulation of OGTT data.

This concept gets further support from Davenport [7], who indicates that "variable rates-of-gastric emptying balance the variable-rates-of absorption, to give a spuriously constant rate-of-absorption for the entire GI tract. The introduction of hypertonic sugar solutions into the stomach causes gastric distention and delays gastric emptying, and this in turn determines the rate at which sugar is delivered to the small intestines for absorption."

9.3.2 Model Analysis to Simulate OGTT Response Curve

The OGTT model-simulation response curve is considered to be the result of giving impulse glucose dose (of 4 g of glucose/L) of blood-pool volume to the combined system consisting of GI tract and blood-pool systems.

Based on the analysis carried out by Fisher and Teo in Equations 2.5 through 2.7 of Ref. [8], we can put down the transfer function (TF) of the GI tract to be

$$GI\ TF = G_1(s) = 1/(s + \alpha) \tag{9.5}$$

because the intestinal glucose concentration variation is an exponential decay, and its exponential parameter value is close to that of the parameter α.

In other words, we are indicating that the rate at which the ingested glucose enters the blood pool is given by

$$q(t) = G \exp(-\alpha t) \tag{9.6}$$

it being the solution of the DEq:

$$q' + \alpha q = G\delta(t) \tag{9.7}$$

representing the process occurring in the GI wall, by which the glucose impulse input [$G\delta(t)$] is converted into the exponential decay function of Equation 9.6. This yields the GI tract TF to be (as given by Equation 9.5)

$$GI\ TF = G_1(s) = 1/(s + \alpha) \tag{9.8}$$

When we multiply this GI tract TF $G_1(s)$ by the TF $G_2(s)$ $[=(s+\alpha)/s^2 + s(\alpha+\delta)+(\alpha\delta+\beta\gamma)]$ of the blood-pool glucose metabolism given by Equation 8.5a to get the overall TF of the combined digestive tract and blood pool system as $GS = 1/[s^2 + s(\alpha + \delta) + (\alpha\delta + \beta\gamma)]$, and put $Q(S) = L$ $[G\delta(t)] = G$ gram of glucose per liter of blood-pool volume per hour, we get for the digestive tract and blood-pool conglomerate,

$$Y(s) = G/\{s^2 + s(\alpha + \delta) + (\alpha\delta + \beta\gamma)\} \tag{9.9}$$

corresponding to impulse injected glucose-bolus input of $q(t) = G\delta(t)$.

The corresponding OGTT response curve can then be represented by the inverse Laplace transform of Equation 9.9, as follows:

$$y(t) = G(e^{-p_1 t} - e^{-p_2 t})/(p_2 - p_1) \quad g/L \tag{9.10}$$

where p_1 and p_2 are given by Equations 8.6b and 8.6e. The graphical representation of this Equation 9.10 is illustrated in Figure 9.3.

9.3.3 Damped Response Model for OGTT Simulation

1. OGTT response curve needs to satisfy the following features [2,9] of a normal clinical case, with a test glucose bolus q [$= G\delta(t)$] of 1 g of glucose per kilogram of patient mass (i.e., for $G = 1$ g of glucose per kilogram of patient per hour):

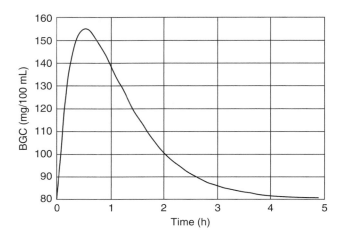

FIGURE 9.3
Impulse-input response of the combined system, consisting of the GI tract and blood pool, as represented by Equation 9.10. *Note*: This curve is generated for parametric values of $\alpha = 0.196$, $\beta = 0.198$, $\gamma = 3.23$, $\delta = 3.04$, whose units are provided in Table 8.3.

- Peak value does not exceed 200 mg/dL.
- Peak is reached in approximately 0.5–1.5 h from the time $t = 0$ (i.e., from time of glucose ingestion), at least in normal subjects.
- There should not be a sharp turning point at the peak of the curve, as obtained in Figure 8.7c for a rectangular-pulse input.
- Curve may be oscillatory in certain subjects, whose system is underdamped.

The glucose-bolus impulse input at the mouth is after its transport through the GIC deemed to be manifested as a rectangular-pulse input to the glucose BPC, as depicted by Hobbie [9]. Even though the TF of the GI tract can be represented by a definite time integrator $(1/s)$, the precise curve shape of the glucose input to the blood pool cannot correspond to the clinical data. In order to circumvent this problem, we have considered that the gut and the blood pool together form one single compartment, with even the glucose-bolus input (G) into this integrated compartment to be regarded as an unknown (and to be determined) magnitude of the glucose input impulse function $G\delta(t)$ of the clinically monitored OGTT data curve (representing the system response).

2. The rate-control first-order model was first suggested by Hobbie [9], but it could not explain the oscillatory features of the OGTT curve. Earlier, we have indicated that the glucose-bolus impulse

input at the mouth becomes manifest as an exponential decay input into the BPC, based on the TF $G_1(s)$ of the GI tract as $1/(s + \alpha)$. When this exponential decay function becomes the input into the BPC, represented by the TF $G_2(s) = (s + \alpha)/[s^2 + s(\alpha + \delta) + (\alpha\delta + \beta\gamma)]$, we get the overall TF $= 1/[s^2 + s(\alpha + \delta) + (\alpha\delta + \beta\gamma)]$, whereby

$$Y(s) = \frac{G}{s^2 + s(\alpha + \delta) + (\alpha\delta + \beta\gamma)} \quad (9.11)$$

for the combined digestive tract and blood-pool system, wherein the glucose mass input-rate $G(\text{gmL}^{-1}\text{h}^{-1})$ is regarded as an unknown (and to be determined) magnitude of the glucose-bolus impulse-input $G\delta(t)$ into the combined system (whose clinically monitored OGTT data represents the system response).

We can hence satisfactorily and conveniently represent the OGTT response curve by means of an oral glucose-regulatory second-order system model (involving a proportional-control term plus a derivative feedback-control term), represented (in Laplace transform) by another version of Equation 9.11 as follows [10]:

$$Y(s) = G/[s^2 + \lambda(1 + T_d s)] = G/(s^2 + \lambda T_d s + \lambda)$$
$$\lambda = \alpha\delta + \beta\gamma, \quad \lambda T_d = \alpha + \delta \quad (9.12)$$

where the glucose-bolus input rate into the combined gut and BPC is $G(\text{g L}^{-1}\text{h}^{-1})$, λy is the proportional control term, and $\lambda T_d(dy/dt)$ is the derivative feedback control term having derivative time T_d [with λT_d replacing $(\alpha + \delta)$ of Equations 9.9 and 9.11].

Equation 9.12 is now adopted to represent the response of the blood glucose (proportional + derivative) feedback control system model for simulating glucose metabolism during OGTT, as illustrated in Figure 9.4.

Equation 9.12 can also be written as [10]

$$Y(s) = \frac{G}{(s^2 + 2As + \omega_n^2)} \quad (9.13)$$

where

G is in gram of glucose per liter of blood-pool volume per hour
$\omega_n \ (= \lambda^{1/2})$ is the natural frequency of the system
A is the attenuation or damping constant of the system (in h^{-1})
$\lambda = 2A/T_d = \omega_n^2$ (in h^{-2})
ω_d (or ω) $= (\omega_n^2 - A^2)^{1/2}$ is the angular frequency (in rad/h) of damped oscillation of the system

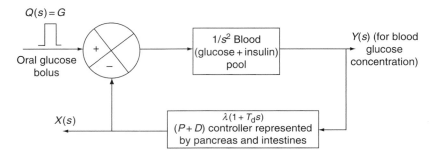

F = Feedforward transfer function = $1/s^2$
H = Feedback transfer function = $\lambda(1 + T_d s)$ where $\lambda = \alpha\delta + \beta\gamma$ and $\lambda T_d = \alpha + \delta$
G = Closed loop transfer function = $F/(1 + FH) = [G/(s^2 + \lambda T_d s + \lambda)]$

FIGURE 9.4
Pancreas in combination with intestines depicted as a proportional derivative feedback controller $\lambda(1 + T_d s)$. It is seen that the blood insulin concentration is given by: $x = \lambda y + \lambda T_d y'$.

The corresponding governing differential equations are [10]

$$y'' + 2Ay' + \omega_n^2 y = G\delta(t); \quad \text{damped frequency } \omega_d(\text{or } \omega) = (\omega_n^2 - A^2)^{1/2}$$

$$\tag{9.14a}$$

$$y'' + \lambda T_d y' + \lambda y = G\delta(t); \quad \lambda(= 2A/T_d) = \omega_n^2, \quad T_d = 2A/\omega_n^2 \tag{9.14b}$$

3. It is seen that the solution of Equation 9.14 (detailed in Appendix A) for an underdamped response, i.e., for $A^2 < \omega_n^2$ (corresponding to that of normal subjects, represented by the lower curve in Figure 9.5), is given by

$$y(t) = (G/\omega)e^{-At}\sin\omega t \tag{9.15}$$

where ω (or ω_d) $= (\omega_n^2 - A^2)^{1/2}$.
Then, the solution for overdamped response (in which $A^2 > \omega_n^2$, corresponding to the upper curve of Figure 9.5 of a diabetic subject) is given (as derived in Appendix A) by

$$y(t) = (G/\omega)e^{-At}\sinh\omega t \tag{9.16}$$

where ω (or ω_d) $= (A^2 - \omega_n^2)^{1/2}$.
The solution for a critically damped response (in which $A = \omega_n$) is given (as derived in Appendix A) by

$$y(t) = Gte^{-At} \tag{9.17}$$

for $\omega_n^2 = A^2 = \lambda$, and derivative time period $T_d = 2A/\lambda = 2A/\omega_n^2$.
These solutions will be employed to simulate the clinical glucose concentration data, and to therefrom evaluate the model system

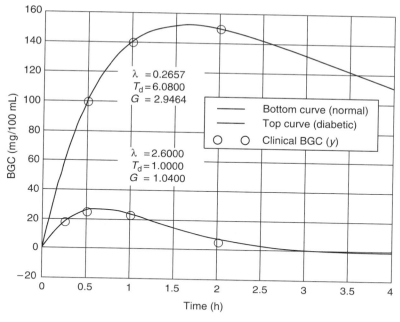

FIGURE 9.5
OGTT clinical data of "normal" and "diabetic" subjects, and their model-simulation response curves. *Note*: For the bottom curve (representing a normal subject), $\lambda = 2.6$ h^{-2}, $T_d = 1$ h, and $G = 1.04$ g L^{-1} h^{-1}. For the top curve (representing a diabetic subject), $\lambda = 0.2657$ h^{-2}, $T_d = 6.08$ h, and $G = 2.9464$ g L^{-1} h^{-1}. (Adopted from Dittakavi, S.S. and Ghista, D.N., *J. Mech. Med. Biol.*, 1, 193, 2001).

parameters A and ω_n (or λ and T_d), to not only differentially diagnose diabetic subjects but also to characterize resistance to insulin.

The corresponding expressions for blood insulin concentrations are obtained (on the basis of Figure 9.4) from

$$x = \lambda y + \lambda T_d y' = (\alpha\delta + \beta\gamma)y + (\alpha + \delta)y' \qquad (9.18)$$

9.3.4 Model Parameter Identification: Simulation with Patient Data and Evaluation of Model Parameters

1. For Normal Subjects (Tables 9.3 and 9.4 and Figure 9.5 Bottom Curve)
The subjects were orally administered a bolus glucose dose of 4 g/L of blood volume or 1 g of glucose/kg weight. Normalized BGC data (of Table 9.3) are provided in Table 9.4. The averaged values of the last row are employed in Figure 9.5. The average clinically obtained readings (above the fasting level, i.e., normalized with respect to the fasting glucose values) of the five normal male subjects (in the age group of 18–20 years) are at 1/4 h = 18.4 mg/dL, at 1/2 h = 25.4 mg/dL, and at 1 h = 23.2 mg/dL.

TABLE 9.3

Sample OGTT Clinical Data for a Normal Subject Observed BGC

No. of Subject	Fasting BGC (mg/dL)	0 h (mg/dL)	1/4 h (mg/dL)	1/2 h (mg/dL)	1 h (mg/dL)	2 h (mg/dL)
1	77	77	90	98	83	55
2	78	78	91	93	109	80
3	75	75	85	97	100	102
4	68	68	99	100	96	58
5	74	74	99	111	100	105

From this set of averaged values (given in the last row of Table 9.4) of monitored OGTT blood glucose data in Figure 9.5 (bottom curve), the response appears to be underdamped.

The solution of Equation 9.14 for an underdamped system (simulating the normal subjects data in Figure 9.5) is given by Equation 9.15, as

$$y(t) = (G/\omega)e^{-At} \sin \omega t$$

where G is the (to be determined) value of gram of glucose (administered to the system) per liter of blood-pool volume per hour, A is the attenuation constant, ω (or ω_d) $= (\omega_n^2 - A^2)^{1/2}$ is the damped frequency of the system, ω_n is the natural frequency $= (\lambda)^{1/2}$, and $T_d = 2A/\omega_n^2$.

We apply Equation 9.15 to the data points of the clinically monitored patient's blood glucose response. The resulting three simultaneous equations (with parameters A, ω, and G, obtained by using the three data points) can be solved to yield: $A = 1.4 \text{ h}^{-1}$, $\omega = 0.78 \text{ rad/h}$, and $G = 1.04 \text{ g L}^{-1} \text{ h}^{-1}$. Note that, $A^2 (= 1.96) < \omega_n^2 (= 2.6)$. The detailed procedure and related analysis are given in Appendix B.

Knowing the values of ω and A, the parametric relationships of Equation 9.14 are then employed to evaluate the model parameters λ and T_d, as: $\lambda = 2.6 \text{ h}^{-2}$ and $T_d = 1 \text{ h}$, for normal subjects, represented by the bottom curve of Figure 9.5.

TABLE 9.4

Observed BGC—Fasting BGC for the Normal Subject Whose Data Is Given in Table 9.3

No. of Subject	0 h (mg/dL)	1/4 h (mg/dL)	1/2 h (mg/dL)	1 h (mg/dL)	2 h (mg/dL)
1	0	13	21	06	−22
2	0	13	15	31	02
3	0	10	22	25	27
4	0	31	32	28	−10
5	0	25	37	26	31
Average	0	18.4	25.4	23.2	5.6

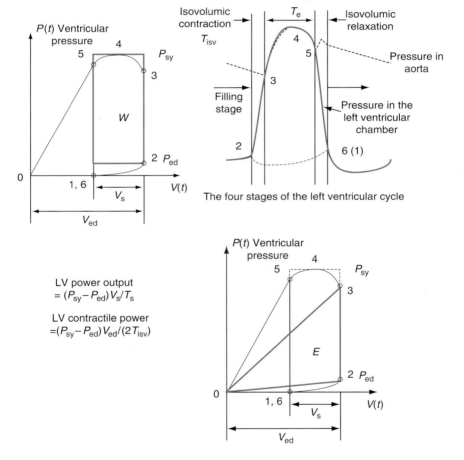

FIGURE 1.2
Left-ventricular work (W), energy input (E), power output and input, and efficiency (ε). (Adopted from Ghista, D.N., *J. Mech. Med. Biol.*, 4, 401, 2004.)

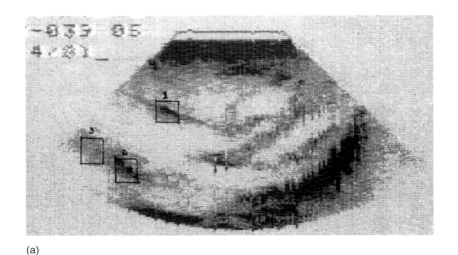

(a)

yfx	99	100	101	102	103	104	105	106	107	108	109	110	111	112	113	114
98	79	78	88	90	99	96	102	108	91	77	92	86	135	122	73	55
99	114	115	101	114	126	128	114	116	119	126	82	68	84	103	78	57
100	151	137	125	128	136	135	133	134	149	137	91	75	74	73	82	83
101	175	177	171	151	144	143	154	147	138	142	139	139	126	64	76	71
102	202	196	174	125	192	193	183	164	131	131	125	132	92	89	81	116
103	139	143	183	193	206	217	233	248	209	146	116	102	111	113	117	116
104	147	136	143	178	203	251	250	255	229	201	75	71	92	82	88	95
105	108	110	132	151	210	223	227	249	255	255	230	210	104	87	81	112
106	84	104	88	121	147	184	227	239	255	255	252	247	220	125	76	70
107	83	110	108	122	135	175	194	183	206	228	211	255	255	184	141	131
108	68	92	122	131	145	147	149	151	217	181	189	222	241	178	190	167
109	56	76	81	122	132	137	145	143	154	150	156	156	195	190	206	190
110	76	63	96	96	82	83	103	120	142	128	133	141	153	181	192	194
111	59	57	63	66	70	103	106	118	96	94	86	110	129	150	95	66
112	58	60	59	57	58	61	71	77	106	89	91	92	110	147	97	85
113	74	71	78	60	56	58	57	62	71	70	79	83	78	92	67	76
114	57	57	65	63	57	56	63	56	51	56	58	80	85	78	67	55
115	51	60	63	63	58	57	56	57	54	59	57	58	59	76	68	81

(b)

FIGURE 2.6
(a) Long axis 2-D ultrasonic view of a pediatric patient's heart, showing highly echoreflectile regions 1 and 2 and a healthy region 3. (b) Echocardiographic texture analysis, showing echo-intensity levels from myocardial region 1. (Adopted from Figure 2 of Kamath, M.V., Way, R.C., Ghista, D.N., Srinivasan, T.M., Wu, C., Smeenk, S., Marning, C., and Cannon, J., *Eng. in Med.*, 15, 137, 1986.)

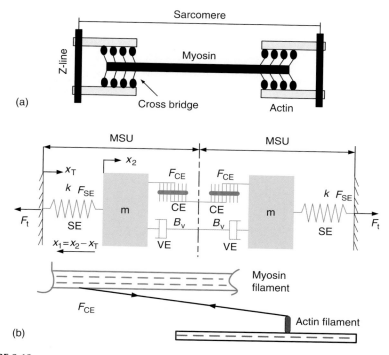

(a)

(b)

FIGURE 3.12

(a) The actin and myosin filaments constituting the contractile components of the myocardial fibril; (b) Myocardial fibril model composed of two symmetrical myocardial structural units (MSUs), which are mirror images of each other. Each MSU is composed of (i) an effective mass (m) that is accelerated; (ii) connective-tissue series element having parameter k (elastic modulus of the series element) and the force F_{SE}; (iii) the parallel viscous element of the sarcolemma having viscous damping parameter B_v and force F_{VE}; (iv) the contractile element (CE), which generates contractile force F_{CE} between the myosin (thick) and actin (thin) filaments. When the contractile element shortens (by amount x_2), the series element lengthens (i.e., x_1 increases). During ejection, the MSU x_T decreases, and during filling the MSU x_T increases. (Adopted from Ghista, D.N., Zhong, L., Eddie, Y.K.Ng., Lim, S.T., Tan, R.S., and Chua, T., *Mol. Cell. Biomech.*, 2, 217, 2005.)

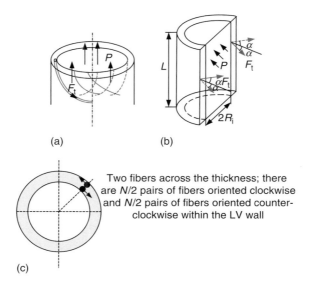

(a)   (b)

(c)

FIGURE 3.14
(a) Equilibrium of fiber force and LV pressure on the top circular plane of the LV cylindrical model. (b) Equilibrium of fiber force and LV pressure in the circumferential direction. (c) Location of two sets of fibers across the LV wall thickness. (Adopted from Ghista, D.N., Zhong, L., Eddie, Y.K.Ng., Lim, S.T., Tan, R.S., and Chua, T., *Mol. Cell. Biomech.*, 2, 217, 2005.)

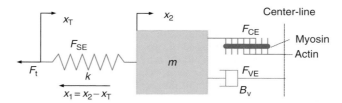

FIGURE 3.15
Dynamic model of MSU having effective mass m; k is the elastic modulus of series element; B_v is the viscous-damping parameter of parallel viscous element; F_t denotes the total generated force caused by the contractile stress F_{CE}; F_{SE} is the force in the series element $[=k(x_1 + x_{1ed})]$, where x_{1ed} is the deformation of the SE at end diastole; F_{VE} is the force in the viscous element $(=B_v \dot{x}_2)$; x_1 then represents the added deformation of the SE during systole (over and above its deformation during the filling phase) due to the development of F_{CE}. (Adopted from Ghista, D.N., Zhong, L., Eddie, Y.K.Ng., Lim, S.T., Tan, R.S., and Chua, T., *Mol. Cell. Biomech.*, 2, 217, 2005.)

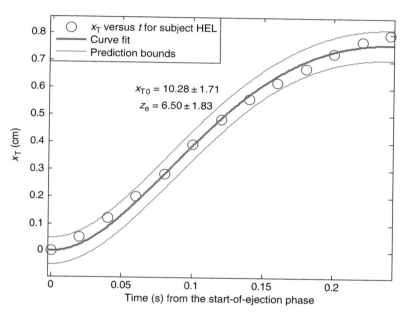

FIGURE 3.18

Computed $x_T(t)$ during the ejection phase ($t = 0$ corresponds to start-of-ejection): From the data shown in Figure 3.17, we calculate the model x_T during the ejection phase by using Equation 3.39, as shown by the round points in the figure. This data is now fitted with Equation 3.41. The resulting values of the parameters (x_{T0} and z_e) are shown in the figure and also listed in Table 3.4. Here $t = 0$ corresponds to the start-of-ejection. (Adopted from Ghista, D.N., Zhong, L., Eddie, Y.K.Ng., Lim, S.T., Tan, R.S., and Chua, T., *Mol. Cell. Biomech.*, 2, 217, 2005.)

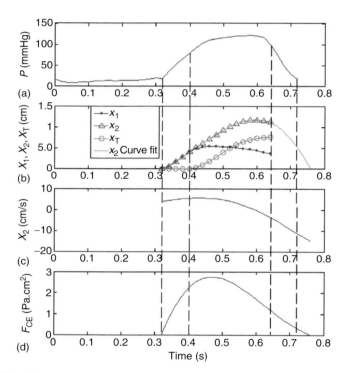

FIGURE 3.20
Computed results of MSU model-dynamics terms x_1, x_2, x_T, $\dot{x}_2$, and F_{CE}, for subject HEL. Diastolic phase: 0–0.32 s; isovolumic contraction phase: 0.32–0.4 s; ejection phase: 0.4–0.64 s; isovolumic relaxation phase: 0.64–0.72 s. Here $t = 0$ corresponds to the start-of-filling. Note that F_{CE} and x_2 extend into the filling phase; $t_0 = 0.04$ s. (Adopted from Ghista, D.N., Zhong, L., Eddie, Y.K.Ng., Lim, S.T., Tan, R.S., and Chua, T., *Mol. Cell. Biomech.*, 2, 217, 2005.)

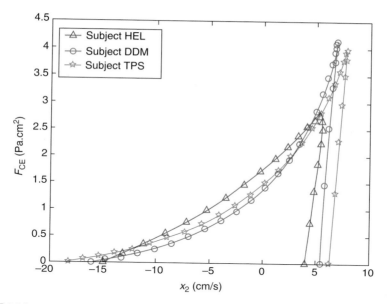

FIGURE 3.24

CE force (F_{CE})–velocity ($\dot{x}_2$) relationships for subjects HEL, DDM, and TPS. Of the three subjects, the subject TPS has the biggest area encircled within the F_{CE} versus $\dot{x}_2$ curve, and hence has the bigger contractile power input. (Adopted from Ghista, D.N., Zhong, L., Eddie, Y.K.Ng., Lim, S.T., Tan, R.S., and Chua, T., *Mol. Cell. Biomech.*, 2, 217, 2005.)

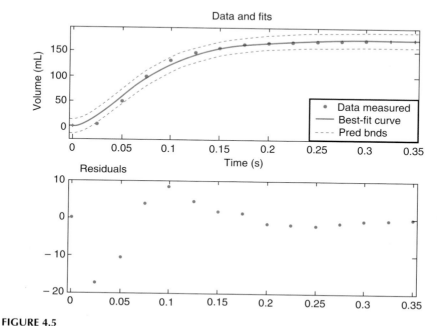

FIGURE 4.5

Plot of computed aortic volume versus time during the systolic phase of the aorta. The round points represent the measured data, while the solid line represents the filled computed volume curve for values of a and b given in Table 4.1. The prediction bounds define the width of the interval with a level of 95%. The values of these parameters and the RMS 1 (root-mean-square error) are given in Table 4.1.

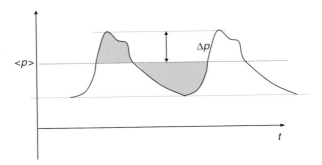

FIGURE 4.9
Schematic of a typical arterial pressure waveform.

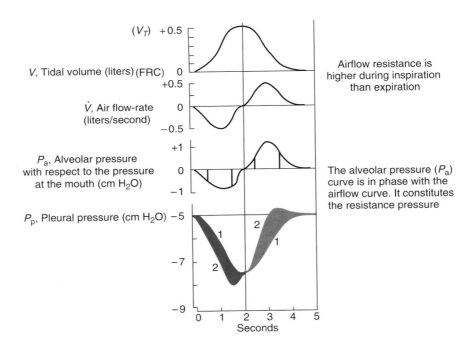

FIGURE 5.2
Lung ventilatory model and lung-volume and pleural-pressure data. In the bottom figure, graph 1 represents $(P_p - P_a) = -P_{el}$ (the pressure required to overcome lung elastance plus lung elastic recoil pressure at the end of expiration $= V/C + P_{el0}$); graph 2 represents P_p, obtained by adding $(P_a - P_m)$ to graph 1. The driving pressure $P_N(t)$ in Equation 5.1 equals P_p minus P_{el0} at the end of expiration. (Adopted from Ghista, D.N., Loh, K.M., Damodaran, M., in *Human Respiration; Anatomy and Physiology, Mathematical Modeling, Numerical Simulation and Applications*, V. Kulish (Ed), WIT Press, Southampton, U.K., 2006.)

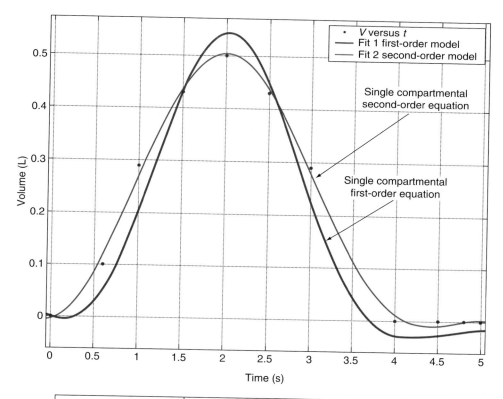

FIGURE 5.5
Results of Second-order Single-compartment model (based on differential equation formulation), compared with the First-order model. (Adopted from Ghista, D.N., Loh, K.M., and Damodaran, M., in *Human Respiration; Anatomy and Physiology, Mathematical Modeling, Numerical Simulation and Applications*, V. Kulish (Ed), WIT Press, Southampton, U.K., 2006.)

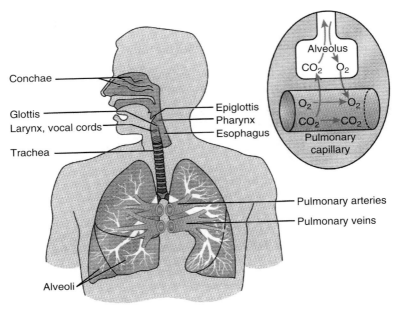

Conchae

Glottis
Larynx, vocal cords

Trachea

Epiglottis
Pharynx
Esophagus

Alveolus

CO_2 O_2

O_2 O_2
CO_2 CO_2

Pulmonary
capillary

Pulmonary arteries
Pulmonary veins

Alveoli

FIGURE 6.1
Respiratory passages. (Adopted from Guyton, A.C., *Text Book of Medical Physiology*, Saunders, Philadelphia, 1991. With permission from Elsevier.)

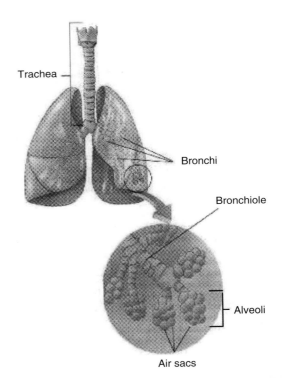

Trachea

Bronchi

Bronchiole

Alveoli

Air sacs

FIGURE 6.2
Trachea, bronchi, and alveoli.

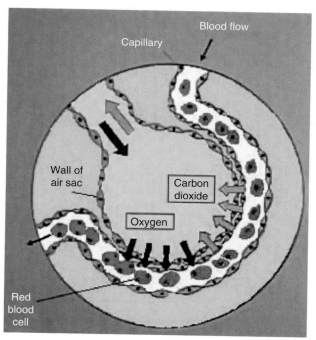

FIGURE 6.3
Exchange of carbon dioxide and oxygen.

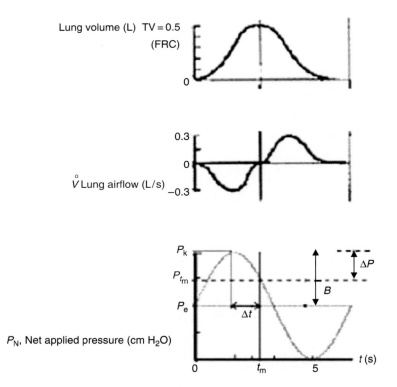

FIGURE 7.2
Lung ventilatory model data showing airflow ($\overset{\circ}{V}$), volume (V), and net pressure (P_N). Pause pressure (P_{t_m}) occurs at t_m, at which the volume is maximum (TV = tidal volume). Δt is the phase difference between the time of maximum volume and peak pressure (P_k). It is also the time lag between the peak and pause pressures. B is the amplitude of the net pressure waveform P_N applied by the ventilator. This P_N oscillates about P_e with amplitude of B. The difference between peak pressure P_k and pause pressure P_{t_m} is ΔP. (Adopted from Ghista, D.N., Pasam, R., Vasudev, S.B., Bandi, P., and Kumar, R.V., in *Human Respiration: Anatomy and Physiology, Mathematical Modeling, Numerical Simulation and Applications*, V. Kulish (Ed), WIT press, Southampton, U.K., 2006.)

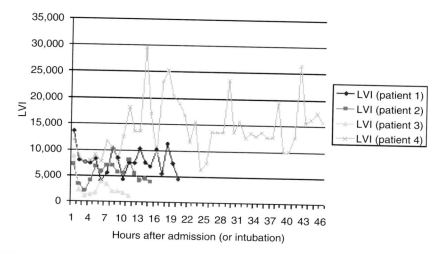

FIGURE 7.4

Lung ventilatory index (LVI) (showing lung status) in four mechanically ventilated COPD patients in acute respiratory failure. Note that patients 1, 2, and 3 were successfully discontinued, and patient 4 had failed discontinuation. (Adopted from Ghista, D.N., Pasam, R., Vasudev, S.B., Bandi, P., and Kumar, R.V., in *Human Respiration: Anatomy and Physiology, Mathematical Modeling, Numerical Simulation and Applications*, V. Kulish (Ed), WIT press, Southampton, U.K., 2006.)

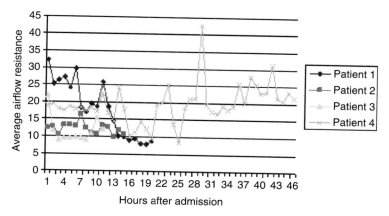

FIGURE 7.5

The variations in R_a for four mechanically ventilated patients, showing that patients 1, 2, and 3 are all discontinued at lower airflow resistance (R_a) values of 8–10 cm H_2O s/L, i.e., closer to outpatient COPD values of R_a; for these patients, the R_a values decreased with mechanical ventilation. For patient 4 (who could not be weaned off), the R_a values remained at a high level. (Adopted from Ghista, D.N., Pasam, R., Vasudev, S.B., Bandi, P., and Kumar, R.V., in *Human Respiration: Anatomy and Physiology, Mathematical Modeling, Numerical Simulation and Applications*, V. Kulish (Ed), WIT press, Southampton, U.K., 2006.)

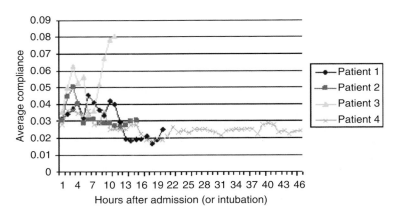

FIGURE 7.7
The variations in C_a for four mechanically ventilated patients, indicating that the C_a values for patients 1, 2, and 4 were all in the lower values and did not change significantly from the time of intubation; incidentally, the lung status for patients 1 and 2 improved and they were successfully discontinued, while patient 4 failed discontinuation. (Adopted from Ghista, D.N., Pasam, R., Vasudev, S.B., Bandi, P., and Kumar, R.V., in *Human Respiration: Anatomy and Physiology, Mathematical Modeling, Numerical Simulation and Applications*, V. Kulish (Ed), WIT press, Southampton, U.K., 2006.)

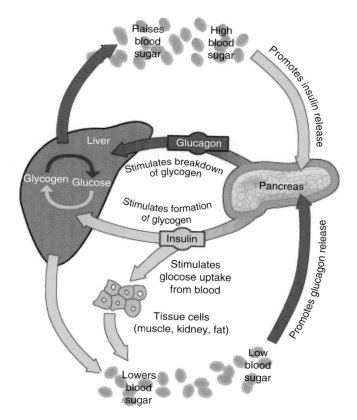

FIGURE 8.1
Effects of insulin and glucagons released by the pancreas in achieving normal blood glucose level.

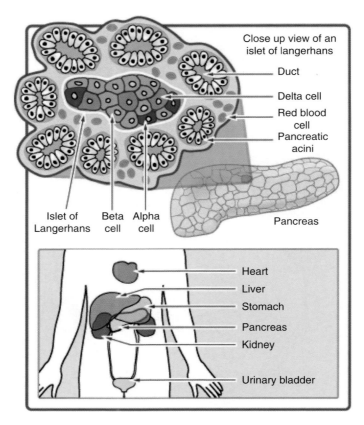

FIGURE 8.2
The pancreas has many islets that contain insulin-producing beta cells and glucagons-producing alpha cells.

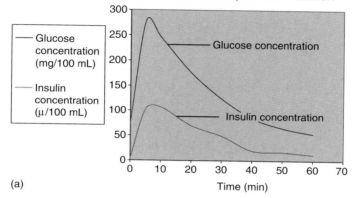

(a)

Glucose and insulin concentrations of the normal subject whose data are provided in Table 9.1

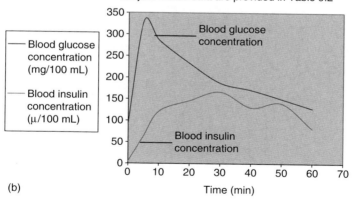

(b)

Glucose and insulin concentrations of the hypertensive subject whose data are provided in Table 9.2

FIGURE 9.1

Blood glucose and insulin concentration dynamics. The symbol μ denotes milli units.

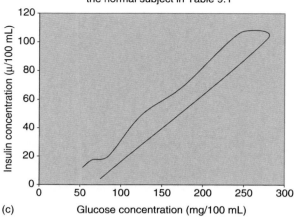

(c)

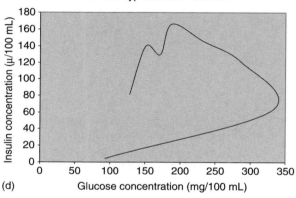

(d)

FIGURE 9.1 (continued)

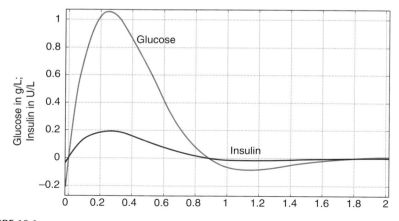

FIGURE 10.1
When glucose bolus is administered to a normal person, a typical response of blood glucose and insulin correlation (normalized) with respect to their fasting or initial concentration values. Blood glucose concentration is measured in g/L and insulin concentration is measured in U/L.

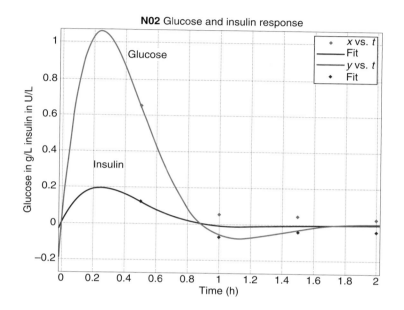

Parameters	Values	Fitting			Remarks
A	2.863	Glucose fit	SSE	0.001478	Normal, just as the clinician diagnosed
G	9.9				
ω	3.558		R-square	0.9961	
α	1.1219	Insulin fit	SSE	0.001394	
β	0.04619				
γ	4.6041		R-square	0.8254	
δ	18.4				

FIGURE 10.5
The glucose–insulin response of a sample normal (nondiabetic) subject response.

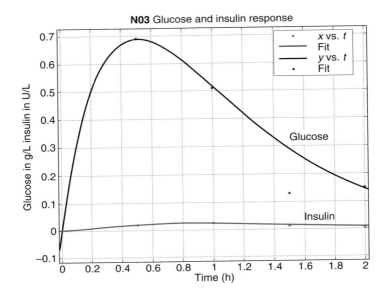

Parameters	Values	Fitting			Remarks
A	1.9650	Glucose fit	SSE	4.06e-005	N03 is at risk of becoming diabetic. This insulin response is inadequate
G	3.6740				
ω	0.079		R-square	0.9999	
α	3.0000	Insulin fit	SSE	9.172e-006	
β	0.1194				
γ	8.9717		R-square	0.9719	
δ	0.9300				

FIGURE 10.7
The glucose–insulin response of **N03** is a good example of a "missed" clinical diagnosis. Even though the subject was diagnosed as normal, clinically (i.e., pronounced to be nondiabetic), the subject in fact is at-risk of becoming diabetic, based on this subject's data being best fitted by "critically damped response" Equation 10.18, as well as based on the value of the nondimensional diabetes index (Equation 10.28) and as indicated by Table 10.6.

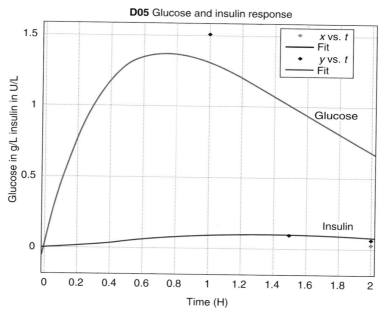

Parameters	Values	Fitting			Remarks
A	1.3530	Glucose fit	SSE	0.001906	Subject D05 has been diagnosed as diabetic, but (as per our model) is only at risk of being diabetic
G	5.0750				
ω	0.001507		R-square	0.9031	
α	2.2400	Insulin fit	SSE	0.0004872	
β	0.1862				
γ	4.2254		R-square	0.9238	
δ	0.4660				

FIGURE 10.8

The glucose–insulin response of **D05** is another example of a missed clinical diagnosis. This subject was diagnosed to be diabetic; however, based on our diabetes index DNDI, this subject is at risk of being diabetic, as indicated by Table 10.6.

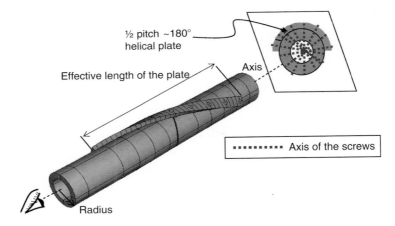

½ pitch ~180°
helical plate

Effective length of the plate

Axis

Axis of the screws

Radius

FIGURE 11.22
Terminology for a helical plate. Inset shows the side view of the helical plate, depicting the half pitch of the plate.

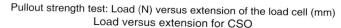

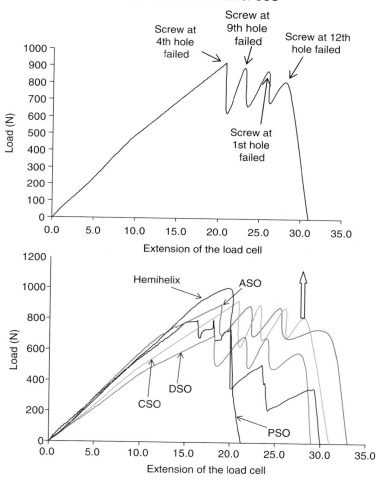

FIGURE 11.25

Pullout tests: load versus extension curves for all the configurations of fixations shown in Figure 11.22. An inset depicts the four peaks in the curve that represent the screw pullout for the CSO (convergent) configuration. Similar screw pullout patterns were observed for PSO (perpendicular), ASO (alternating), and DSO (divergent) configurations. The stiffness of the assembly is the slope of the load versus extension curve and the area under the load versus extension curve till the initiation of pullout represents the energy to pullout.

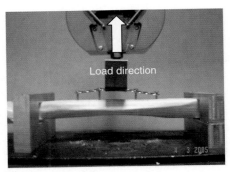

(a)

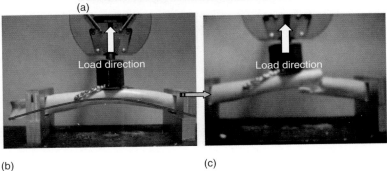

(b)   (c)

FIGURE 11.26

(a) Sequential pullout of straight plate PSO (perpendicular), CSO (convergent), DSO (divergent), and ASO (alternating) configurations. (b) The holding power of hemihelical plate (HHP) is high, such that sequential screw pullout is not observed. (c) However, the bone failed before screw pullout, indicating that the fixation is stiff enough so that screw loosening does not occur.

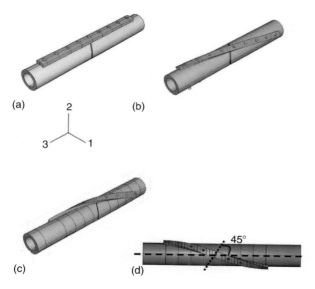

(a)

2
3   1

(b)

(c)

(d)

45°

FIGURE 11.27
Finite element models in ABA-
QUS: (a) straight plate model,
(b) 90° helical plate model, (c)
180° helical plate model, and
(d) oblique fracture fixed by a
helical plate. The bone axis
(depicted in the figure) is
along coordinate 3.

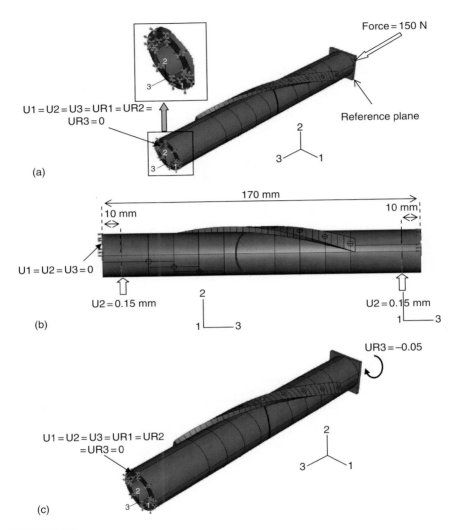

FIGURE 11.28
Loading and boundary conditions applied on the 180° helical plate fixation on the simulated fractured bone (with 45° oblique fracture): (a) compressive load, (b) bending load, and (c) torsional load. Similar loading conditions are applied on the straight plate and the 90° helical plate fixations. Here U1, U2, and U3 represent displacements along axes 1, 2, and 3, respectively; UR1, UR2, and UR3 represent rotations about axes 1, 2, and 3, respectively.

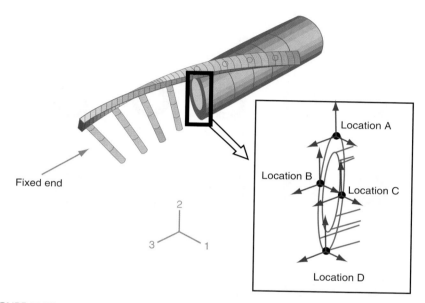

FIGURE 11.30

Different locations considered in the finite element analyses (FEA), for computing fracture gap movement. Location A is on the bone on the fracture gap and underneath the plate. Location D is on the bone at fracture gap but on the opposite side of the plate. Locations B and C are on the bone on the fracture gap and between locations A and D. *Note*: For better presentation of the fracture gap, the left bone fragment was made invisible.

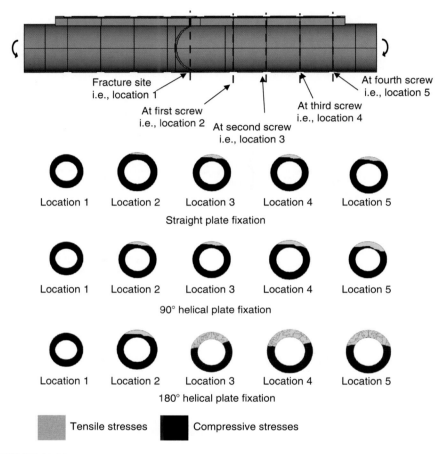

FIGURE 11.34
Locations of the neutral axis (NA) on the bone cross section at different places along the length of the bone for straight plate, 90° helical plate, and 180° helical plate fixations subjected to bending moment. The change of color from grey to black represents the NA. (a) Locations considered for the NA along the length of the bone; (b) cross sections of bone showing NA at different locations along the length of the bone fixed by the straight plate; (c) cross sections of bone showing NA at different locations along the length of the bone fixed by the 90° helical plate; and (d) cross sections of bone showing NA at different locations along the length of the bone fixed by the 180° helical plate.

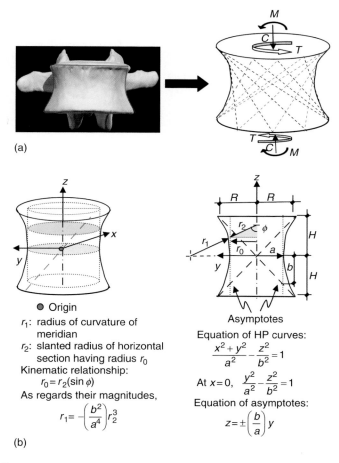

(a)

(b)

● Origin

r_1: radius of curvature of meridian

r_2: slanted radius of horizontal section having radius r_0

Kinematic relationship:

$$r_0 = r_2(\sin \phi)$$

As regards their magnitudes,

$$r_1 = -\left(\frac{b^2}{a^4}\right)r_2^3$$

Asymptotes

Equation of HP curves:

$$\frac{x^2 + y^2}{a^2} - \frac{z^2}{b^2} = 1$$

At $x = 0$, $\dfrac{y^2}{a^2} - \dfrac{z^2}{b^2} = 1$

Equation of asymptotes:

$$z = \pm\left(\frac{b}{a}\right)y$$

FIGURE 12.2

(a) Shows a typical vertebral body (VB) where the cortical VB is shaped as a hyperboloid (HP) shell formed of two sets of generators. The height of the HP can be expanded or reduced by the inclination of the generators. (b) Depicts geometry of HP shells. (Adopted from Ghista, D.N., Fan, S.C., Ramakrishna, K., and Sridhar, I., *Int. J. Des. Nat.*, 1, 34, 2006.)

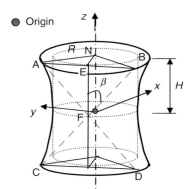

Origin

ABCD constitutes a plane tangent to the waist circle at F.

$$\tan \beta = \frac{EB}{FE} = \frac{\sqrt{R^2 - a^2}}{H} \tag{I}$$

Coordinates of point A are $[-a, \sqrt{R^2 - a^2}, H]$

Substituting into the HP equation: $\dfrac{x^2 + y^2}{a^2} - \dfrac{z^2}{b^2} = 1$

We get $\dfrac{a^2 + R^2 - a^2}{a^2} - \dfrac{H^2}{b^2} = 1$

i.e., $\dfrac{R^2}{a^2} - \dfrac{H^2}{b^2} = 1$ or $b = \dfrac{aH}{\sqrt{R^2 - a^2}}$ \hfill (II)

FIGURE 12.3
Geometry of cortical vertebral body (VB) hyperboloid (HP) shell: The generators AD and BC form the basis of the construction of the HP shell. (Adopted from Ghista, D.N., Fan, S.C., Ramakrishna, K., and Sridhar, I., *Int. J. Des. Nat.*, 1, 34, 2006.)

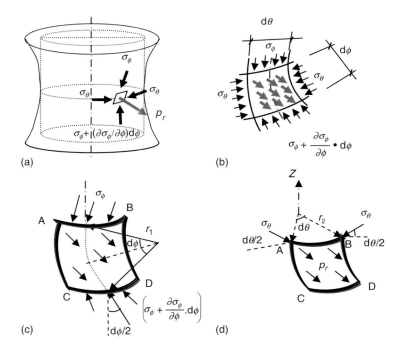

FIGURE 12.4
Stresses acting on an element of the vertebral body (VB) hyperboloid (HP) shell: (a) HP shell element in equilibrium, (b) σ_ϕ and σ_θ equilibrating internal pressure p_r, (c) orientation of σ_ϕ, and (d) orientation of σ_θ. (Adopted from Ghista, D.N., Fan, S.C., Ramakrishna, K., and Sridhar, I., *Int. J. Des. Nat.*, 1, 34, 2006.)

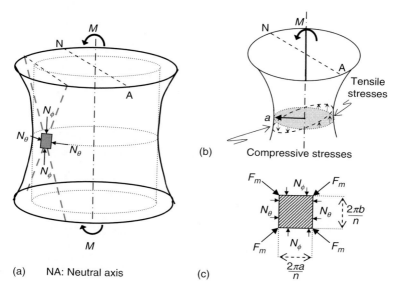

(a)   NA: Neutral axis   (c)

FIGURE 12.8

(a) Stress resultants at the waist section of vertebral body (VB) hyperboloid (HP) shell under bending, (b) equilibrium of forces on a shell segment, and (c) equivalent diagonal forces in the intersecting bars to take up the stresses around a shell element on the compression side of the VB HP shell. (Adopted from Ghista, D.N., Fan, S.C., Ramakrishna, K., and Sridhar, I., *Int. J. Des. Nat.*, 1, 34, 2006.)

FIGURE 12.16
Rezaian spinal fixator is placed along the loading axis of the spine. (From Rezaian, S.M. and Ghista, D.N., *IEEE Eng. Med. Biol.*, 13, 525, 1994. Copyright 1994.)

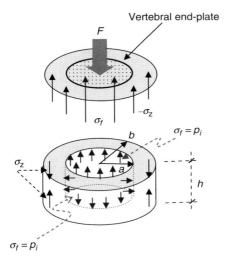

FIGURE 13.3
Normal stresses σ_f and σ_z under the applied force compressive F. (Adopted from Ghista, D.N., Fan, S.C., Sridhar, I., and Ramakrishna, K., *Int. J. Des. Nat.*, 1, 146, 2007.)

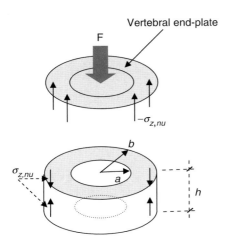

FIGURE 13.5
Normal stress $\sigma_{z,nu}$ equilibrating the applied force F in a nucleotomized disc. (Adopted from Ghista, D.N., Fan, S.C., Sridhar, I., and Ramakrishna, K., *Int. J. Des. Nat.*, 1, 146, 2007.)

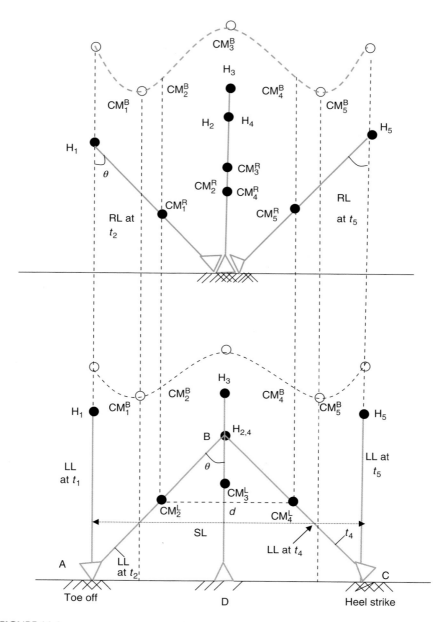

FIGURE 14.4
Depiction of the orientation of left and right legs, and their centers of mass (CM_i^L, CM_i^R) at times t_1 to t_5; H_1: hip joint locations. The right leg is in stance phase from t_2 to t_4; the left leg is in swing phase from t_2 to t_4. Note, that SL (stride length) $= 2d = 2l \sin \theta$; H_2 is at $l \cos \theta$ above the ground, while H_3 is at l above the ground; hence CM^B is raised by $l(1 - \cos \theta)$ from t_2 to t_3.

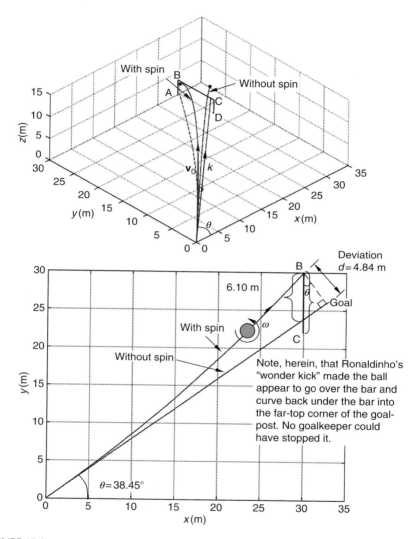

FIGURE 15.2
Analytical simulation of the trajectory of Ronaldinho's famous free kick in the quarter-final match against England in the 2002 World Cup (won by Brazil). The top figure shows the 3-D trajectories of the ball, with and without spin. The bottom figure shows the top view (or the horizontal projection) of the ball trajectory to its final location B into the goal. In doing so, to the goalkeeper Seaman, the ball must have actually appeared to be sailing over the bar, only to see it curve back to dip below the bar into the goal. In the figure, BC represents the goal bar, k is the unit vector making an angle θ with the x-axis, and β is the angle that the initial velocity vector ($\mathbf{v_0}$) makes with Ok (in the xOy plane). The initial velocity vector $\mathbf{v_0}$ lies in the zOk plane. The lateral deviation of the ball along the goal bar is 6.10 m, as shown in the figure.

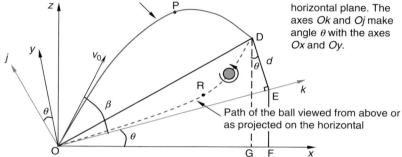

The ball is kicked in the zOk plane at an angle β to the horizontal plane or to the axis Ok in the horizontal plane. The axes Ok and Oj make angle θ with the axes Ox and Oy.

OPD is the ball trajectory; this trajectory plane (*OPD*) slopes with respect to the vertical zOk plane

Path of the ball viewed from above or as projected on the horizontal

FIGURE 15.3

Notations for soccer-ball kick-velocity and trajectory. The orthogonal lines (or axes) Ok and Oj are in the horizontal plane xOy, and make angles θ with the Ox and Oy axes, respectively. ED is the total horizontal deviation (*d*) of the ball when it lands on the ground at D. The curve ORD is the horizontal projection of the trajectory of OPD.

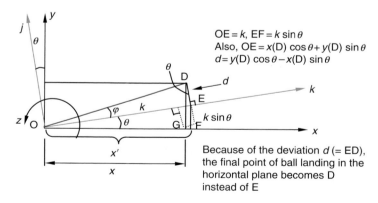

$OE = k$, $EF = k \sin\theta$
Also, $OE = x(D)\cos\theta + y(D)\sin\theta$
$d = y(D)\cos\theta - x(D)\sin\theta$

Because of the deviation d (= ED), the final point of ball landing in the horizontal plane becomes D instead of E

FIGURE 15.4

Ball displacements in the horizontal plane. The ball is kicked in the zOk vertical plane. However, because of the counter-clockwise angular velocity (ω) imparted to it, it has deviated by an amount "d" (=ED) perpendicular to Ok (i.e., parallel to Oj axis) when it lands on the ground.

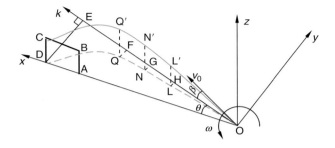

FIGURE 15.5
Corner kick by a right footer, straight into the goal. The player kicks the ball in the zOk plane, with a counter-clockwise angular velocity ω. The ball curves along OL′N′Q′ to C into the far-top corner. The deviations of the ball trajectory projected on the horizontal plane are HL, GN, FQ, and ED.

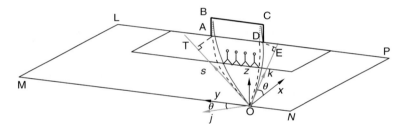

FIGURE 15.7
Left-footer kick (OB) and right-footer kick (OC) around the players' wall into the goalpost. The ball is kicked in the vertical planes zOs and zOk.

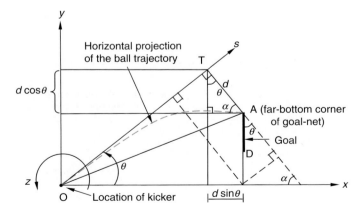

FIGURE 15.8
Geometry of the left-footer kick in the horizontal xOy plane: zOs is the vertical plane in which the ball is kicked, TA (d) is the horizontal deviation of the ball trajectory, A is the horizontal projection of the ball-location B (into the net).

Brazil won a free kick when Scholes tackled Kleberson from
behind, 30 m out on the right flank. Five Brazilians
lined up across the edge of the penalty area, seemingly
ready for the ball to be crossed towards the far-post.

Goalkeeper David Seaman obviously expected this too.
He was only 3 m off his line and took a small step
forward when Ronaldinho struck the free kick.

But Ronaldinho's shot was aimed at the far-top corner and dipped just under
the bar with Seaman flapping helplessly. In a few short minutes, Ronaldinho had
turned the game on its head.

FIGURE 15.9

Ronaldinho's wonder goal, the famous right-foot free kick, that made the ball curve into the far-
top corner of the goalpost and won the game for Brazil in the 2002 World-cup quarter-finals.

FIGURE 15.10

Seaman tried to reach the ball but failed, and the ball just dipped below the bar into the top
corner of the goal-net.

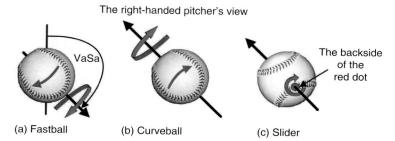

The right-handed pitcher's view

(a) Fastball (b) Curveball (c) Slider

The backside of the red dot

VaSa

FIGURE 16.4

The direction of spin (circular arrows) and the spin axes (straight arrows) of a three-quarter arm (a) fastball, (b) curveball, and (c) slider from the perspective of a right-handed pitcher, meaning the ball is moving into the page. VaSa is the angle between the vertical axis and the spin axis. (From Bahill, A.T., http://www.sie.arizona.edu/sysengr/slides. With permission. Copyright 2005.)

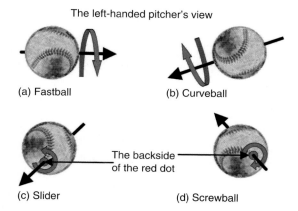

The left-handed pitcher's view

(a) Fastball (b) Curveball

(c) Slider (d) Screwball

The backside of the red dot

FIGURE 16.5

The direction of spin (circular arrows) and the spin axes (straight arrows) of an overhand (a) fastball, (b) curveball, (c) slider, and (d) screwball from the perspective of a left-handed pitcher, meaning the ball is moving into the page. (From Bahill, A.T., http://www.sie.arizona.edu/sysengr/slides. With permission. Copyright 2004.)

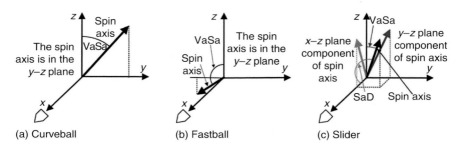

(a) Curveball   (b) Fastball   (c) Slider

FIGURE 16.8

Rectangular coordinate system and illustration of the angles VaSa and SaD for (a) curveball, (b) three-quarter arm fastball, and (c) slider all thrown by a right-handed pitcher. The origin is the pitcher's release point. For the curveball, the spin axis is in the y–z plane. For the fastball, the spin axis is also in the y–z plane, but it is below the y-axis. For the slider, the spin axis has components in both the y–z and x–z planes. (From Bahill, A.T., http://www.sie.arizona. edu/sysengr/slides. With permission. Copyright 2006.)

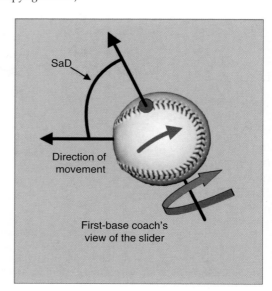

FIGURE 16.9

The first-base coach's view of a slider thrown by a right-handed pitcher. This illustrates the definition of the angle SaD. (From Bahill, A.T., http://www. sie.arizona.edu/sysengr/slides. With permission. Copyright 2007.)

TABLE 9.5

OGTT Data for a Typical Diabetic Subject

Time of Observation	0 h (mg/dL)	1/2 h (mg/dL)	1 h (mg/dL)	2 h (mg/dL)
Observed BGC	175	275	315	325
Observed BGC – fasting BGC	0	100	140	150

Note: Fasting BGC = 175 mg/dL.

2. For a Typical Diabetic Subject (Table 9.5 and Figure 9.5 Top Curve)

The clinically obtained readings (above the fasting level) of the diabetic subject are at $1/2\,h = 100\,mg/dL$, at $1\,h = 140\,mg/dL$, and at $2\,h = 150\,mg/dL$.

The OGTT data of the diabetic subject (with the fasting glucose level of 175 mg/dL) depicted in Figure 9.5 (top) appear to be overdamped.

The corresponding solution, for an overdamped system, is given by Equation 9.16, as

$$y(t) = (G/\omega)\,e^{-At}\sinh\,\omega t$$

where $\omega\,(\text{or}\ \omega_d) = (A^2 - \omega_n^2)^{1/2} = $ damped frequency of the system, using the earlier given notations for A and ω_n in Equations 9.13 and 9.14.

By simulating the diabetic subject's clinical data (in Table 9.5) by Equation 9.16, we get simultaneous equations in A, ω, and G. These equations can be solved to get their best-fit values as $A = 0.81\ h^{-1}$, $\omega = 0.622$ rad/h, and $G = 2.9464\ g\ L^{-1}h^{-1}$.

By employing the relations in Equation 9.14, we then get $\lambda = 0.2657\ h^{-2}$, $T_d = 6.08$ h as the parametric values associated with the upper curve of Figure 9.5, for injected glucose input. The analytical details are provided in Appendix C.

3. Interpretation of Results

The clinical data as well as its curve simulation by Equation 9.14 are depicted in Figure 9.5. A comparison of the model's parameter values for the normal and diabetic patients is provided by Table 9.6. It is noted that the two parameters G and T_d are higher in the diabetic subject compared to the normal subject, while $\lambda\ (= \omega_n^2)$ is very low in the diabetic subject. Alternatively, it can be noted that the damping constant A is $1.4\ h^{-1}$ for the sample normal subject compared to $0.81\ h^{-1}$ for the sample diabetic subject, and

TABLE 9.6

Comparison of the Normal and Diabetic Subjects' Parameters of the Model Given by Equation 9.13

	$G\ [g\,(L)^{-1}(h)^{-1}]$	$\lambda(h^{-2}) = \omega_n^2$	T_d (h)	$T = 2\pi/\omega_n$ (h)	$A\ (h^{-1})$
Normal	1.04	2.6	1.0	3.9	1.4
Diabetic	2.9464	0.2657	6.08	12.2	0.81

$T(2\pi/\omega_n)$ is 3.9 h for the sample normal subject compared to 12.2 h for the sample diabetic subject.

The time period T of the OGTT curve for the sample normal subject is equal to 3.9 h. According to Ackerman, the value of T should be less than 4 h. Thus, our analysis also agrees with Ackerman et al.'s theory [11]. If we reiterate Cramp and Carson [12], concerning the glucose dynamics OGTT, it is noteworthy that in many physiological and clinical situations, the normal or abnormal patient response is characterized by the presence or absence of features such as overshoots, pure delays, and specific frequencies of oscillation. It is these features, which are of prime significance to the clinician.

Appendix A: Nature of Solution for Equation 9.14

The OGTT model equation to be solved is

$$y'' + 2Ay' + \omega_n^2 y = G\,\delta(t) \tag{9.A1}$$

or

$$y'' + 2Ay' + \omega_n^2 y = 0$$

where $y_0 = 0$, $y_0' = G$, and the damped oscillation frequency $\omega = (\omega_n^2 - A^2)^{1/2}$. We can represent the solution as

$$y = c\,e^r t \tag{9.A2}$$

Upon substituting Equation 9.A2 into Equation 9.A1, we obtain

$$r^2 + 2Ar + \omega_n^2 = 0 \tag{9.A3}$$

from which

$$r_1, r_2 = -A \pm \left(A^2 - \omega_n^2\right)^{1/2} \tag{9.A4}$$

Now the evaluation of r_1 and r_2 depends on the square-root term $(A^2 - \omega_n^2)$.

Case 1: For $A^2 < \omega_n^2$ or $A^2 - \omega_n^2 < 0$, i.e., for $\omega_n^2 - A^2(= \omega^2) > 0$, we have

$$A^2 - \omega_n^2 = i^2(\omega_n^2 - A^2) = i^2\omega^2\left[\omega^2 = \omega_n^2 - A^2 > 0\right] \tag{9.A5}$$

From Equations 9.A4 and 9.A5, we have

$$r_1, r_2 = -A \pm \left(i^2\omega^2\right)^{1/2} = -A \pm i\omega \tag{9.A6}$$

The general solution of Equation 9.A1 is

$$y = \frac{c_1}{2}\left(e^{r_1 t} + e^{r_2 t}\right) + \frac{c_2}{2i}\left(e^{r_1 t} - e^{r_2 t}\right) \tag{9.A7}$$

Noting that $\cos \omega t = (e^{-i\omega t} + e^{-i\omega t})/2$ and $\sin \omega t = (e^{-i\omega t} - e^{-i\omega t})/2i$,

$$y = e^{-At}(c_1 \cos \omega t + c_2 \sin \omega t) \tag{9.A8}$$

Since we have $y(0) = 0$, $c_1 = 0$, and

$$y = c_2 e^{-At} \sin \omega t \tag{9.A9}$$

and

$$y' = c_2\left(e^{-At}\omega \cos \omega t + \sin \omega t\, e^{-At}(-A)\right) \tag{9.A10}$$

Since

$$y'(0) = G = c_2\,\omega, \quad c_2 = G/\omega \tag{9.A11}$$

and hence

$$y = (G/\omega)e^{-At} \sin \omega t \tag{9.A12}$$

Case 2: $A^2 > \omega_n^2$, or $A^2 - \omega_n^2 > 0$
Then $A^2 - \omega_n^2 = i^2(\omega_n^2 - A^2) = \omega^2$
and

$$r_1, r_2 = -A \pm (A^2 - \omega_n^2)^{1/2} = -A \pm \omega \tag{9.A13}$$

The general solution is then

$$y = c_1 e^{(-A+\omega)t} + c_2 e^{(-A-\omega)t}; \quad r_1 - r_2 = 2\omega \tag{9.A14}$$

with

$$y(0) = 0 = c_1 + c_2 \tag{9.A15}$$

and

$$y_1(0) = G = r_1 c_1 + r_2 c_2 \tag{9.A16}$$

Therefore,

$$c_1 = \frac{G}{(r_1 - r_2)} = \frac{G}{2\omega} \quad \text{and} \quad c_2 = -\frac{G}{(r_1 - r_2)} = -\frac{G}{2\omega} \tag{9.A17}$$

Hence, the solution is obtained as

$$\begin{aligned} y &= \frac{G}{2\omega} \left[e^{(-A+\omega)t} - e^{(-A-\omega)t} \right] \\ &= (G/\omega)e^{-At} \sinh(\omega t) \end{aligned} \tag{9.A18}$$

$$\text{because } \sinh \omega t = \frac{e^{\omega t} - e^{-\omega t}}{2}, \text{ and } \cosh \omega t = \frac{e^{\omega t} + e^{-\omega t}}{2} \tag{9.A19}$$

Case 3: $A = \omega_n$ and $r_1 = r_2 = -A$
Hence, in the case of repeated roots

$$y = c_1 e^{-At} + c_2 t e^{-At} \tag{9.A20}$$

where

$$c_1 = y_0 = 0, \quad c_2 = y'(0) = G \tag{9.A21}$$

$$y = Gte^{-At} \text{ (for } A = \omega_n, \, \omega = 0) \tag{9.A22}$$

Now

$$y' = G(-Ae^{-At}t + e^{-At}t) = Gt(1 - At)e^{-At} = 0, \text{ when } t = 1/A = T \tag{9.A23}$$

Hence,

$$y = y_{\max} \text{ (when } t = 1/A = T) = (G/A)e^{-AT} \tag{9.A24}$$

From Equations 9.A22 and 9.A24, we obtain

$$y = y_{\max}(t/T) \exp A(T - t) \tag{9.A25}$$

Appendix B: Model Simulation for the Normal Patient's OGTT Glucose Response

The impulse response of an underdamped (nearly critical) glucose regulatory system of normal subjects, whose fasting glucose level is taken to be zero (lower curve of Figure 9.5) is (Equation 9.15)

$$y(t) = (G/\omega)e^{-At} \sin \omega t \tag{9.B1}$$

where

G (in gram of glucose per liter of blood-pool volume per hour) is the initial rate of glucose transport from GI tract to blood pool across the intestinal membrane

A is the attenuation constant

$\omega = (\omega_n^2 - A^2)^{1/2}$ is the damped frequency

ω_n is the natural frequency of the system.

The means of the five blood glucose response values of the normal subject, presented in Table 9.4 (obtained by subtracting the fasting glucose level from clinically observed data of OGTT), are

at $1/4$ h = 18.4 mg/dL = 0.184 g/L,
at $1/2$ h = 25.4 mg/dL = 0.254 g/L, and
at 1 h = 23.2 mg/dL = 0.232 g/L.

From Equation 9.B1, we get

$$y(1/4) = (G/\omega)e^{-A/4} \sin \omega/4 = 0.184 \tag{9.B2}$$

$$y(1/2) = (G/\omega)e^{-A/2} \sin \omega/2 = 0.254 \tag{9.B3}$$

$$y(1) = (G/\omega)e^{-A} \sin \omega = 0.232 \tag{9.B4}$$

Using the trigonometric identity $\sin 2\theta = 2 \sin \theta \cos \theta$ we obtain

from Equations 9.B3/9.B2 : $2e^{-A/4} \cos \omega/4 = (0.254/0.184) = 1.380$ (9.B5)

from Equations 9.B4/9.B3 : $2e^{-A/2} \cos \omega/2 = (0.232/0.254) = 0.913$ (9.B6)

Squaring Equation 9.B5 : $4e^{-A/2} \cos^2 \omega/4 = 1.9$ (9.B7)

and employing trigonometric identity $2\cos^2 \omega/4 = 1 + \cos \omega/2$, we get from Equations 9.B7 and 9.B6:

$$(2\cos^2 \omega/4)/(\cos \omega/2) = (1 + \cos \omega/2)/(\cos \omega/2) = 2.08 \tag{9.B8}$$

$$\cos \omega/2 = 0.926; \quad \text{hence } \omega/2 = 0.388 \text{ rad/h and } \omega = 0.775 \text{ rad/h} \tag{9.B9}$$

Putting $\cos \omega/2 = 0.926$ in Equation 9.B6, we get

$$2e^{-A/2} \, 0.926 = 0.913, \text{ i.e., } e^{-A/2} = 0.493 \tag{9.B10}$$

Hence, $A = 1.4 \text{ h}^{-1}$.

Now $\omega^2 = \omega_n^2 - A^2$, or $\omega_n^2 = \omega^2 + A^2 = (0.775)^2 + (1.4)^2 = 2.6 = \lambda$ (note, $\omega_n^2 > A^2$),

$A = (\lambda T_d/2) = (2.6 \times T_d)/2 = 1.4 \text{ h}^{-1}$,
$T_d = (2 \times 1.4)/2.6 = 1.08 \text{ h, and}$
$\lambda = 2.6 \text{ h}^{-2}$.

Upon substituting the values of ω and A in Equation 9.B4, we get
$G = 1.04 \text{ g L}^{-1} \text{ h}^{-1}$

Appendix C: Model Simulation for the Diabetic Person's OGTT Data

The impulse response of an overdamped system of the diabetic patient, normalized with respect to fasting glucose level (upper curve of Figure 9.5), is the solution of Equation 9.14 for an overdamped system, given by (Equation 9.16)

$$y(t) = (G/\omega)e^{-At} \sinh \omega t \qquad (9.C1)$$

where
    A is the attenuation constant
    $\omega = (A^2 - \omega_n^2)^{1/2}$ is the damped frequency
    ω_n is the natural frequency of the system

The clinical glucose concentration values for the diabetic subject (obtained by subtracting the fasting glucose level of the diabetic from the clinically observed data of OGTT) are (Table 9.5)

    at $1/2$ h $= 100 \text{ mg/dL} = 1.0 \text{ g/L}$,
    at 1 h $= 140 \text{ mg/dL} = 1.4 \text{ g/L}$,
    at 2 h $= 150 \text{ mg/dL} = 1.5 \text{ g/L}$.

From Equation 9.C1, we get

$$y(1/2) = (G/\omega)e^{-A/2} \sinh \omega/2 = 1.0 \qquad (9.C2)$$

$$y(1) = (G/\omega)e^{-A} \sinh \omega = 1.4 \qquad (9.C3)$$

$$y(2) = (G/\omega)e^{-2A} \sinh 2\omega = 1.5 \qquad (9.C4)$$

Using the trigonometric identity, $\sinh 2\theta = 2 \sinh \theta \cosh \theta$, we obtain

from Equations 9.C3 and 9.C2 : $2e^{-A/2} \cosh \omega/2 = (1.4/1.0) = 1.4 \quad (9.C5)$

from Equations 9.C4 and 9.C3 : $2e^{-A}\cosh\omega = (1.5/1.4) = 1.07$ (9.C6)

Squaring Equation 9.C5 : $4e^{-A}\cosh^2\omega/2 = 1.96$ (9.C7)

Now employing the trigonometric identity : $2\cosh^2\omega/2 = 1 + \cosh\omega$, we get from Equations 9.C7 and 9.C6

$$(2\cosh^2\omega/2)/(\cosh\omega) = (1 + \cosh\omega)/(\cosh\omega) = (1.96/1.07) = 1.83$$

(9.C8)

$$(1/\cosh\omega) + 1 = 1.83, \text{ and hence } \omega = 0.6222 \text{ rad/h}$$ (9.C9)

Putting $\cosh\omega = 1.2$ in Equation 9.C6, we get
$1.07 = 2 \times e^{-A} \times 1.2$

$$\text{i.e., } e^{-A} = 0.446 \text{ and } A = 0.808 \text{ h}^{-1}$$ (9.C10)

Now, $\omega^2 = (A^2 - \omega_n^2)$, or $\omega_n^2 = A^2 - \omega^2 = (0.808)^2 - (0.622)^2 = 0.266$;

$\lambda = \omega_n^2 = 0.266 \text{ h}^{-2}$,
$A = (\lambda\, T_d/2) = (0.266 \times T_d)/2 = 0.808 \text{ h}^{-1}$
i.e., $T_d = 6.08 \text{ h}$

Upon substituting the above values of λ and T_d in Equation 9.C3, we get $G = 2.946 \text{ g L}^{-1}\text{ h}^{-1}$ as the third parameter.

References

1. Bolie, V.W., Coefficients of normal blood glucose regulation, *J. Appl. Physiol.* 16(5):783–788, 1961.
2. Guyton, A.C. and J.E. Hall, *Text Book of Medical Physiology*, W.B. Saunders, Philadelphia, 1991.
3. Insel, P.A., J.E. Llljenquist, J.D. Tobin, R.S. Sherwin, P. Watkins, R. Andres, and M. Berman, Insulin control of glucose metabolism in man, *J. Clin. Invest.* 55:1057–1066, 1975.
4. Bergman, R.N., Y.Z. Ider, C.R. Bouden, and C. Cobeill, Quantitative estimation of insulin sensitivity, *Am. J. Physiol.* 236(6):E667–E677, 1979.
5. Ferrannini, E., J.D. Smith, C. Cobell, G. Toffolo, A. Pilo, and R.A. De Fronzo, Effect of insulin on the distribution and disposition of glucose in man, *J. Clin. Invest.* 76:357–364, 1985.
6. Galvin, P., G. Ward, J. Walters, R. Pestell, M. Koschmann, A. Vaag, I. Martin, J.D. Best, and F. Alford, A simple method for quantitation of insulin sensitivity and insulin release from an intravenous glucose tolerance test, *Diabetic Med.* 9:921–928, 1992.
7. Davenport, M.H.W., *Physiology of Digestive Tract*, Year Book Medical Publishers Inc., Chicago, 1961, pp. 175–176.

8. Fisher, M.E. and K.L. Teo, Optimal infusion resulting from a mathematical model of blood glucose dynamics, *IEEE Trans. Biomed. Eng.* 36:479–486, 1989.

9. Hobbie, R.K., *Intermediate Physics for Medicine and Biology*, John Wiley & Sons, New York, second edition, 1988, pp. 295–299.

10. Dittakavi, S.S. and D.N. Ghista, Glucose tolerance test modeling & patient simulation for diagnosis, *J. Mech. Med. Biol.* 1(2):193–223, 2001.

11. Ackerman, E., J.W. Rosevear, and W.F. McGuckin, A mathematical model of glucose tolerance test, *Phys. Med. Biol.* 9:203–213, 1964.

12. Cramp, D.G. and E.R. Carson, Dynamics of blood glucose and its regulating hormones, In: *Biological Systems, Modeling and Control*, IEEE Control Engineering Series I, edited by D.A. Linkens, London: Peter Peregrinus Ltd., 1979, pp. 171–201.

10

Modeling of OGTT Blood Glucose and Insulin Responses and Diagnostic Indices

Dhanjoo N. Ghista, Kah Meng Loh, and Meena Sankaranarayanan

CONTENTS

In this chapter, we again provide the derivation of the basic model to represent blood glucose and insulin responses to glucose ingestion in the oral glucose tolerance test (OGTT). We then provide solutions, to the basic Blood glucose–insulin regulatory system (BGIRS) model equations, for insulin response $x(t)$ to glucose bolus injestion in OGTT. Finally, we develop a nondimensional index (NDI) based on the model parameters to facilitate characterization of subjects as normal or diabetic or at risk of becoming diabetic.

10.1 Oral Glucose Tolerance Test (OGTT) Protocol

The test subjects need to fast for 12 h before the test and during the 2 h test. A blood sample of the subject is taken before the beginning of the test. After the subject drinks a 75 g of glucose solution dissolved in 250–300 mL of water, the subject's blood glucose and insulin concentrations are measured at specified intervals of 30, 60, 90, and 120 min.

Qualitative interpretation of the results, for preliminary categorization of the patients:

1. Blood glucose normal values [1,2]
   Fasting: 70–115 mg/dL
   30 min: less than 200 mg/dL
   1 h: less than 200 mg/dL
   2 h: less than 140 mg/dL
   Normal insulin level (reference range): 1–30 mU/L [1,2]; mU denotes milliunits

2. Impaired glucose regulation

When a person has a fasting glucose equal to or greater than 110 mg/dL and less than 126 mg/dL, it is considered as impaired fasting glucose. This is considered as a risk factor for diabetes and will likely trigger another test in the future, but by itself does not provide sufficient evidence for the diagnosis of diabetes.

A person is said to have impaired glucose tolerance when the 2 h glucose results from the oral glucose tolerance test are greater than or equal to 140 mg/dL but less than 200 mg/dL. This is also considered a risk factor for future diabetes. A person is deemed to be diabetic when the oral glucose tolerance tests show that the blood glucose level at 2 h is equal to or more than 200 mg/dL. This must be confirmed by a second test on another day.

10.2 Representing Glucose and Insulin Responses by Their Concentration–Time Profiles

When a glucose bolus is administered to a normal person, typical blood glucose and insulin concentration–time profiles are illustrated in Figure 10.1, and can be regarded to be underdamped responses [1]. However, when a glucose bolus is administered to a typical diabetic patient, the blood glucose and insulin concentration–time profiles (illustrated in Figure 10.2) appear to be overdamped.

Comparing Figures 10.1 and 10.2 [1–3], we note that a normal person's glucose response is such that the blood glucose concentration (normalized with respect to the fasting or initial value) peaks at levels of up to 1.06 g/L (or 106 mg/dL), and is back to 0 g/L by the end of 2 h. Likewise, the response curve for blood insulin (normalized with respect to fasting blood insulin concentration) peaks up to 0.2 U/L (or 20 mu/dL) and returns to 0 U/L at the end of 2 h. On the other hand, for this typically diabetic patient, the blood glucose concentration is peaking at 2.28 g/L (or 228 mg/dL), and falls to only 2.20 g/L by the end of 2 h. The blood insulin response is quite meager. It barely peaks at 0.04 U/L (or 4 mu/dL) at the end of 2 h, and remains at this level.

The advantage in plotting the responses in the form of curves is that it enables us to quantify the clinical criteria, by means of parametric modeling of the response curve (i.e., underdamped or overdamped), as demonstrated in the previous chapters. There are patients who do not fall into either clinical category. Their response curves can place them into a critically damped domain, whereby they are neither normal nor diabetic but at risk of becoming

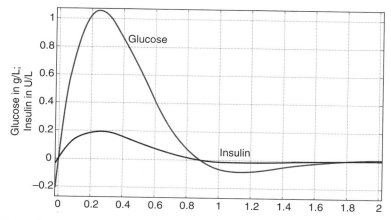

FIGURE 10.1 (See color insert following page 266.)
When glucose bolus is administered to a normal person, a typical response of blood glucose and insulin correlation (normalized) with respect to their fasting or initial concentration values. Blood glucose concentration is measured in g/L and insulin concentration is measured in U/L.

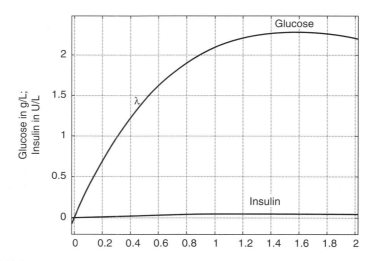

FIGURE 10.2
When glucose bolus is administered to a diabetic patient, a typical response of blood glucose and insulin correlation (normalized) with respect to their fasting or initial concentration values.

diabetic. Even more relevantly, we demonstrate that we can combine the parameters into an NDI, which is more conveniently clinically implementable.

There are many works on modeling of glucose–insulin dynamic and regulation [1–9]. However, this chapter is specifically oriented to modeling the glucose and insulin responses to ingested glucose bolus for OGTT. Herein, the glucose concentration–time data are simulated by an appropriate type of solution (under- or over- or critically damped) of the governing differential equations for glucose and insulin responses to glucose bolus ingestion, presented in Chapter 8. The diagnosis of the patient (as normal, at risk of being diabetic, as borderline diabetic, as diabetic) depends on the solution category that the clinical data fall into.

In this chapter, we employ three sets of equations for the above mentioned three categories of glucose and insulin responses. These three sets of equations are given by (i) Equations 10.10–10.12 for blood glucose response and (ii) by Equations 10.14, 10.16, and 10.18 for insulin response. Now, each set of clinical data is analyzed by the above three types of model response equations, to determine the class of system response based on the best-fit results. All the clinical data sets, presented herein, have been collected and a priori analyzed by a qualified clinician. Also, the clinical data are normalized by the respective fasting values. Normalization helps to facilitate the accuracy of model computational simulation. Since both the blood glucose and insulin responses are analyzed simultaneously, it is necessary to know both the glucose and insulin responses simultaneously to ascertain that the patient has been diagnosed correctly.

The subjects studied were of age range from 19 to 67 years old. The physiological system responses to the orally ingested bolus of glucose were monitored closely for 2 h at 30 min intervals, and their data were simulated by our model response expressions.

10.3 Differential Equation Model of the Glucose–Insulin System

The compartmental block diagram of the blood glucose and insulin regulatory system (BGIRS) [1–3] is illustrated in Figure 10.3 (which is essentially Figure 8.3 of Chapter 8), repeated here in the interests of maintaining continuity. The glucose input-rate into the blood-pool is represented by "q" in the figure. From the blood-pool, glucose is metabolized into the tissues in two ways, as represented by the two terms δy (removal-rate of glucose from the blood-pool independent of insulin) and γx (removal-rate of glucose under the influence of insulin). In return, the glucose influences the release-rate of insulin into the blood-pool by the pancreas, as represented

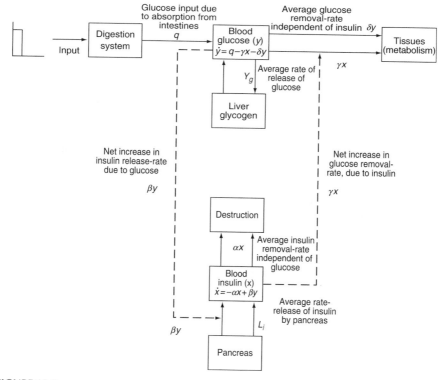

FIGURE 10.3

Blood glucose–insulin control system (BGIRS). Block diagram of how (1) insulin level and rate-of-change of insulin $\mathring{x}(t)$ govern blood glucose concentration $y(t)$, and (2) the rate-of-change of glucose $\mathring{y}(t)$ is influenced by insulin concentration $x(t)$ and ingested glucose input rate $q(t)$.

by the term βy. The insulin is also removed independently of glucose, as per the term αx.

10.3.1 Modeling of Glucose–Insulin Regulation of Oral Glucose Tolerance Test (OGTT)

We have adopted the linearized biomathematical model of Bolie [1], as the basis of our modeling, because it is simple but still compatible with the known physiological mechanisms. This model characterizes the glucose–insulin system by means of the differential equations (given later as Equations 10.1 and 10.2) with four parameters: α and β, γ and δ, representing (i) pancreatic insulin sensitivity to insulin and glucose blood concentrations, and (ii) combined sensitivity of liver glycogen storage and tissue glucose utilization to elevated insulin and glucose concentrations.

The above block diagram of Figure 10.3 [1–3] illustrates how (1) the insulin release-rate is governed by blood glucose concentration $y(t)$ and insulin level $x(t)$, and (2) the rate-of-change of glucose is influenced by insulin concentration $x(t)$, glucose level $y(t)$, and injested glucose input rate $q(t)$. In Figure 10.3

> α represents pancreatic insulin sensitivity to elevated insulin concentration, in $(h)^{-1}$
>
> β is the pancreatic insulin sensitivity to elevated glucose blood concentration, in $(units) (h)^{-1} (g)^{-1}$
>
> γ is the combined sensitivity liver glycogen storage and tissue glucose utilization to elevated blood insulin concentration, in $(g) (h)^{-1} (units)^{-1}$
>
> δ is the combined sensitivity of liver glycogen storage and tissue glucose utilization to elevated blood glucose concentration, in $(h)^{-1}$

10.3.2 The Governing Differential Equations for Glucose and Insulin Systems

By considering the conservation-rates of glucose and insulin in their respective compartments, we obtain the basic differential equations, governing the BGIRS (depicted in Figure 10.3), as (given by Equations 10.1 and 10.2 of Chapter 8)

$$\dot{x} = p(t) - \alpha x + \beta y \tag{10.1}$$

$$\dot{y} = q(t) - \gamma x - \delta y \tag{10.2}$$

where
 x is the blood insulin concentration (from its fasting level) in U/L
 y is the blood glucose concentration (from its fasting level) in g/L
 p is the insulin input-rate in $(U) (L)^{-1} (h)^{-1}$
 q is the glucose input-rate in $g(L)^{-1}(h)^{-1}$
 $\dot{x}$ and $\dot{y}$ denote the first-derivatives of x and y with respect to time

In these equations, the glucose–insulin model system parameters (regulatory coefficients) are α, β, γ, δ.

From Equations 10.1 and 10.2, we obtain the differential equation model for glucose concentration (y) and insulin concentration (x), for insulin infusion rate ($p = 0$) and glucose inflow rate (q). For glucose response, we obtain (by differentiating both sides of Equation 10.2 with respect to "t")

$$\ddot{y} = q' - \gamma \dot{x} - \delta \dot{y}$$
$$= \dot{q} + \alpha q - \dot{y}(\alpha + \delta) - y(\alpha \delta + \beta \gamma)$$

Rearranging, we get the differential equation:

$$\ddot{y} + \dot{y}(\alpha + \delta) + y(\alpha \delta + \beta \gamma) = \dot{q} + \alpha q \tag{10.3}$$

where $\dot{y}$ and $\ddot{y}$ denote first and second time derivatives of y.

Similarly, differentiating both sides of Equation 10.1 with respect to t, we get

$$\ddot{x} = -\alpha \dot{x} + \beta \dot{y}, \quad \text{assuming } p = 0$$
$$= -\alpha \dot{x} + \beta q - \beta \gamma x - \delta(\dot{x} + \alpha x),$$

or,

$$\ddot{x} + \dot{x}(\alpha + \delta) + x(\alpha \delta + \beta \gamma) = \beta q \tag{10.4}$$

where $\dot{x}$ and $\ddot{x}$ denote first and second time derivatives of x.

10.3.3 Laplace Transform Representation of the Governing Equations 10.3 and 10.4

1. Glucose response solution

   The transfer-function (TF) corresponding to Equation 10.3 is obtained by taking Laplace transforms of both sides (assuming the initial conditions to be zero), to obtain the transform function $G(s)$ (as we showed in Chapter 8)

$$G(s) = \frac{Y(s)}{Q(s)} = \frac{(s + \alpha)}{(s + p_1)(s + p_2)} \tag{10.5}$$

   Then the $y(t)$ response of the system (i.e., the solution of the differential Equation 10.3) is obtained by taking the inverse transform of

$$Y(s) = \frac{Q(s)(s + \alpha)}{s^2 + s(\alpha + \delta) + (\alpha \delta + \beta \gamma)} = \frac{Q(s)(s + \alpha)}{(s + p_1)(s + p_2)} \tag{10.6}$$

2. Similarly from Equation 10.4, we get the transfer function for insulin response as

$$\frac{X(s)}{Q(s)} = \frac{\beta}{s^2 + s(\alpha + \delta) + (\alpha \delta + \beta \gamma)} \tag{10.7}$$

In OGTT test, a fasting person is given an oral glucose dose of 1 g/kg, for diabetes diagnostic purpose. If the subject is normal and free from diabetes, the blood glucose level rises from the fasting value of, say, 80 mg/dL to 120–140 mg/dL, and then falls back to below normal in about 2 h.

The physiology of the GI tract suggests that the intestinal glucose absorption rate is constant for a limited time duration. Hence, the glucose rectangular-pulse input $h(t)$ into the blood-pool could be representative of this phenomena. This is due to the combined effect of two mechanisms: (1) the pyloric sphincter valve-resistance which controls the transfer of glucose from the stomach to the intestines in inverse proportion to the stomach distension, and (2) the active transport of glucose from the intestines into the blood (across the intestinal wall) at its maximum rate, according to Michaelis–Menten equation, graphically depicted in Figure 9.2 of Chapter 9. We combine these two mechanisms into a single equation representing blood glucose concentration response due to oral ingestion of glucose, as represented by Figure 10.4.

Now for OGTT simulation, we note that the GI compartment has the transfer function $G_1(s) = 1/(s + \alpha)$ [2,4], which is tantamount to a decay in glucose concentration (at the rate of α) during its transmission through the GI tract, in response to an impulse input $G\delta(t)$, wherein G is the amount of glucose per liter of blood-pool volume per hour. This is depicted in Figure 10.4. The blood-pool compartment has the transfer function $G_2(s)[=(s + \alpha)/s^2 + s(\alpha + \delta) + (\alpha\delta + \beta\gamma)]$, as depicted in Figure 10.4, and given by Equation 10.5.

Hence, by multiplying $G_1(s)$ and $G_2(s)$, we get the transfer function of the digestive tract and blood-pool conglomerate as

$$Y(s) = \frac{(s + \alpha)}{s^2 + s(\alpha + \delta) + (\alpha\delta + \beta\gamma)} \times \frac{1}{(s + \alpha)} \times G$$

$$= \frac{G}{s^2 + s(\alpha + \delta) + (\alpha\delta + \beta\gamma)} = \frac{G}{s^2 + \lambda T_d s + \lambda} \quad (10.8a)$$

where

$$\lambda = (\alpha\delta + \beta\gamma)$$

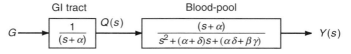

GI tract $\qquad$ Blood-pool

$$G \longrightarrow \boxed{\frac{1}{(s+\alpha)}} \xrightarrow{Q(s)} \boxed{\frac{(s+\alpha)}{s^2+(\alpha+\delta)s+(\alpha\delta+\beta\gamma)}} \longrightarrow Y(s)$$

FIGURE 10.4

Laplace transform format of the GI tract and blood-pool system of governing Equation 10.5, to simulate the monitored oral glucose tolerance test (OGTT) glucose response curve.

and

$$\lambda T_{\mathrm{d}} = (\alpha + \delta)$$

This Equation 10.7 can be adopted to represent the response of the blood glucose (proportional + derivative feedback) control system model, for simulating glucose metabolism during OGTT, as illustrated by Figure 10.4. Equation 10.8a can also be written as

$$Y(s) = \frac{G\,(\text{in g L}^{-1}\,\text{h}^{-1})}{s^2 + 2As + \omega_n^2} \qquad (10.8b)$$

where

$\omega_n =$ the natural frequency of the system
$A \; =$ the attenuation or damping constant of the system in h^{-1}
$\lambda \;\; = 2A/T_{\mathrm{d}} = \omega_n^2$ in h^{-2}
$\omega \;\; = \sqrt{(\omega_n^2 - A^2)}$ is the angular frequency of damped oscillation of the system in rad/h
$G \; =$ grams of glucose per liter of blood pool volume per hour: $\text{g(L)}^{-1}\,\text{(h)}^{-1}$

10.4 Solutions to the OGTT Model Governing Differential Equation, for Glucose Response (*y*) to Glucose Bolus Ingestion (in OGTT)

On the basis of Equation 10.8, the governing differential equation of glucose response (of the combined GI and blood-pool compartment) to glucose bolus ingestion represented by the impulse function $G\delta(t)$ is given by

$$\ddot{y} + 2A\dot{y} + \omega_n^2 y = G\delta(t) \qquad (10.9a)$$

where the damped oscillation frequency

$$\omega_{\mathrm{d}}\,(\text{or}\;\omega) = (\omega_n^2 - A^2)^{1/2}; \quad 2A = \alpha + \delta, \text{ and } \omega_n^2 = \alpha\delta + \beta\gamma$$

or,

$$\ddot{y} + \lambda T_{\mathrm{d}}\,\dot{y} + \lambda y = 0; \quad \lambda(= 2AT_{\mathrm{d}}) = \omega_n^2, \quad T_{\mathrm{d}} = 2A/\omega_n^2 \qquad (10.9b)$$

wherein $y_0 = 0$ and $y' = G$.

10.4.1 Solution for Glucose Response *y(t)*

Solution for underdamped case:
For $A^2 < \omega_n^2$ or $A^2 - \omega_n^2 < 0$, i.e., for $\omega_n^2 - A^2 (= \omega^2) > 0$, we have the solution to Equation 10.9 as

$$y(t) = (G/\omega)e^{-At}\sin/\omega t \tag{10.10}$$

Solution for overdamped case:
  For $A^2 > \omega_n^2$, or $A^2 - \omega_n^2 > 0$, we have

$$y(t) = (G/\omega)e^{-At}\sinh \omega t \tag{10.11}$$

Solution for critically damped case:
  For $\omega = 0$, we have

$$y(t) = Gte^{-At} \tag{10.12}$$

From the above equations, we can determine the model equation parameters G, ω, A, for each clinical case, by making the corresponding solution (for underdamped or overdamped, or critically damped) response match the OGTT clinical data, by Matlab-based parameter identification procedure. In Chapter 9, we showed that we can also analytically evaluate these parameters.

10.4.2   Solutions for Insulin Response $x(t)$ to Glucose Bolus Ingestion (OGTT)

Our aim is to formulate the blood insulin concentration expressions involving the original model parameters (α, β, γ, δ), so that we can simulate the OGTT insulin concentration data, and evaluate these parameters. For this purpose, we could employ the OGTT model expression for insulin concentration, given by Equation 9.18. However, in this expression, the parameters are coupled. Instead, we can simply consider that the above expressions (10.10–10.12) best represent the OGTT glucose concentration (y) data, whose parameters (G, ω, A) can be determined by parametric simulation of the OGTT blood glucose concentration data. We can then proceed to solve for the original parameters α, β, γ and δ, of the model by referring to the basic model Equations 10.1 and 10.2.

For underdamped response of normal subjects, we obtain from Equations 10.1 and 10.10:

$$\dot{x} = -\alpha x + \beta y = -\alpha x + \beta\frac{G}{\omega}e^{-At}\sin \omega t \tag{10.13}$$

Solving Equation 10.13, we get the corresponding insulin response (derived in Appendix A) for $x(0) = 0$, as

$$x(t) = \frac{-[Ae^{-At}\sin \omega t - \alpha e^{-At}\sin \omega t - \omega e^{-\alpha t} + \omega e^{-At}\cos \omega t](\beta G/\omega)}{A^2 - 2A\alpha + \alpha^2 + \omega^2} \tag{10.14}$$

For overdamped response of diabetic subjects, we obtain from Equations 10.1 and 10.11:

$$\dot{x} = -\alpha x + \beta y = -\alpha x + \beta\frac{G}{\omega}e^{-At}\sinh \omega t \tag{10.15}$$

Solving Equation 10.15, we get the insulin response (derived in Appendix B) for $x(0) = 0$ as follows:

$$x(t) = \frac{\begin{bmatrix} \frac{1}{2}[A\cosh(\omega t + At - \alpha t) + A\sinh(-\omega t + At - \alpha t) - A\cosh(-\omega t + At - \alpha t) - \\ A\sinh(\omega t + At - \alpha t) + \omega\sinh(\omega t + At - \alpha t) - \omega\cosh(-\omega t + At - \alpha t) - \\ \alpha\sinh(-\omega t + At - \alpha t) + \alpha\sinh(\omega t + At - \alpha t) - \alpha\cosh(\omega t + At - \alpha t) - \\ \omega\cosh(\omega t + At - \alpha t) + \alpha\cosh(-\omega t + At - \alpha t) + 2\omega + \omega\sinh(-\omega t + At - \alpha t)]e^{-(\alpha t)}(\beta G/\omega) \end{bmatrix}}{(-\omega^2 + A^2 - 2A\alpha + \alpha^2)}$$

(10.16)

For a critically damped response (of borderline subjects), we obtain the insulin response from Equations 10.1 and 10.12 (derived in Appendix C) for $x(0) = 0$, as

$$\dot{x} = -\alpha x + \beta y = -\alpha x + \beta G t e^{-At} \tag{10.17}$$

The solution of Equation 10.17 is given by

$$x(t) = \frac{-\beta G\left[tAe^{-At} - t\alpha e^{-At} + e^{-At} - e^{-\alpha t}\right]}{(A - \alpha)^2} \tag{10.18}$$

10.5   Evaluating the Model Parameters (α, β, γ, δ)

By simulating the glucose response with our $y(t)$ solutions (of Equations 10.10, 10.11, or 10.12), we can evaluate the parameters G, A, and ω. We will now proceed to solve for the parameters: β, γ, and δ.

From Equation 10.9, we have the following relationships:

$$\omega_n^2 = \lambda = (\alpha\delta + \beta\gamma) \tag{10.19}$$

$$\lambda T_d = 2A = (\alpha + \delta) \tag{10.20}$$

Then, after evaluating the parameter A (by matching the $y(t)$ solution to the glucose response data), we can obtain the value of $(\alpha + \delta)$ from Equation 10.20, as

$$\alpha + \delta = 2A \tag{10.21}$$

Now, by matching the $x(t)$ solution to the insulin response data, we can evaluate the parameters α and β. Hence, from Equation 10.21, by substituting this evaluated value of α, we can obtain the value of δ. Now, we have solved for α, β, and δ. But we still need to determine the parameter γ.

For instance, in the case of underdamped response (of normal subjects),

$$\omega_n^2 = \omega^2 - A^2 \tag{10.22}$$

Also, from Equation 10.19, we obtain

$$\omega_n^2 = \lambda = (\alpha\delta + \beta\gamma)$$

and hence

$$\omega^2 - A^2 = \lambda = (\alpha\delta + \beta\gamma) \tag{10.23}$$

Now, in this equation, since only γ is unknown, we can evaluate it. Hence, we can determine all the four model parameters α, β, γ, and δ for normal subjects.

In the case of overdamped response characteristic (of diabetic subjects),

$$\omega^2 = A^2 - \omega_n^2 \tag{10.24}$$

Also, from Equation 10.19

$$\omega_n^2 = \lambda = (\alpha\delta + \beta\gamma)$$
$$A^2 - \omega^2 = (\alpha\delta + \beta\gamma) \tag{10.25}$$

Hence, again in this above equation, since only γ is as yet unknown, we can evaluate it. Hence all the model parameters (α, β, γ, and δ) can be determined.

10.6   Clinical Applications of the Model

We have decided that for a patient to be classified in any one of the three (underdamped, overdamped, and critically damped) response categories, the model solution equation should fit the data with a very high degree of correlation coefficients: R-square ≥ 0.90 and SSE ≤ 0.1. In case a patient data fit all the three (underdamped, overdamped, and critically damped) response categories, we will designate the patient to the category for which the R-square value is the highest.

10.6.1   Underdamped Category for Patients Clinically Designated to be Normal (Table 10.1)

Clinically designated normal patients were assessed for their blood glucose and insulin responses, based on the criteria presented in Section 10.1 of this chapter. If both their blood glucose and insulin concentrations data demonstrated underdamped features (by being best fitted by Equations 10.13 and 10.14), they were characterized as normal.

The underdamped features of a sample subject (that met the underdamped criterion with a high R-square value) are illustrated in Figure 10.5.

TABLE 10.1

Subjects That Were Well Fitted
by Underdamped Model Equation
Solutions 10.13 and 10.14.

Best Fitted (R-Square $\geq 90\%$)
N11
N12
N13
N16
N18

Note: These subjects are placed in Table 10.4, based
on their DNDI index (Equation 10.28) value.

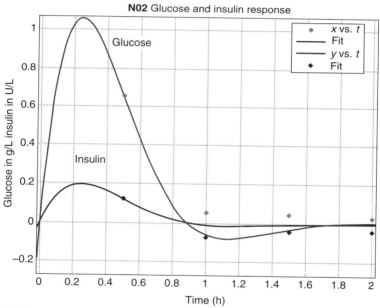

Parameters	Values	Fitting			Remarks
A	2.863	Glucose fit	SSE	0.001478	Normal, just as the clinician diagnosed
G	9.9				
ω	3.558		R-square	0.9961	
α	1.1219	Insulin fit	SSE	0.001394	
β	0.04619				
γ	4.6041		R-square	0.8954	
δ	18.4				

FIGURE 10.5 (See color insert following page 266.)
The glucose–insulin response of a sample normal (nondiabetic) subject response.

Equation 10.9 parameters (A, G, ω) as well as the computed Equations 10.1 and 10.2 model parameters (α, β, γ, δ) are tabulated in the table of Figure 10.5 for this subject.

10.6.2   Overdamped Category of Patients Clinically Designated as Diabetic (Table 10.2)

Now, we designate those patients whose blood glucose and insulin concentrations data were well fitted by overdamped response expressions (10.15 and 10.16) of the model equation, and designate them to be diabetic, as per our model response category.

It is to be noted that subject N17 (clinically diagnosed to be normal) falls into the category of overdamped model response. The response of the patient D01 (who was also clinically diagnosed to be diabetic) is illustrated in Figure 10.6. Additionally the Equation 10.9 parameters (A, G, ω) and Equations 10.1 and 10.2 parameters (α, β, γ, δ) are tabulated in the table of Figure 10.6, for this subject.

10.6.3   Critically Damped Category of Patients (Table 10.3)

There are some patients (shown in Table 10.3) who were clinically diagnosed to be normal, for whom the critically damped solution Equations 10.17 and 10.18 give a better fit of their glucose and insulin response data (and a higher value of R-square) than the underdamped solution. One such patient is N03, who was clinically diagnosed to be normal but whose response curves (shown in Figure 10.7) best fit the critically damped model response. Hence, this subject is only at risk of becoming diabetic, based on our model solution.

Likewise, patient D05 was identified as diabetic, but he is only at risk of becoming diabetic because his data is best fitted by the critically damped response equation. His response curve is illustrated in Figure 10.8. The tables

TABLE 10.2

Subjects That Were Well Fitted by Overdamped
Model Equation Solutions 10.15 and 10.16

Best Fitted (R-Square $\geq 90\%$)
N17
D01
D04
D07
D11
D13
D15
D16

Note: These subjects are placed in Table 10.5, based on their values of the DNDI index (Equation 10.28).

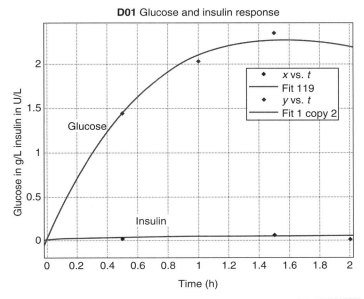

Parameters	Values	Fitting			Remarks
A	0.6417	Glucose fit	SSE	0.005257	Note that the blood glucose level comes down very slowly. D01 is a diabetic subject, and his model result confirms the clinical diagnosis
G	3.969				
ω	0.0002767		R-square	0.9986	
α	0.3084	Insulin fit	SSE	0.0002635	
β	0.004821				
γ	23.0383		R-square	0.8995	
δ	0.9750				

FIGURE 10.6
The glucose–insulin response of **D01** is a good example of an overdamped response, characteristic of a diabetic subject.

TABLE 10.3

Subjects That Were Well Fitted by Critically Damped Characteristic Model Equations 10.17 and 10.18

Best Fitted (R-Square $\geq$ 90%)

N03
N04 (almost normal, based on DNDI value)
N10 (almost normal, based on DNDI value)
N15
D05
D12

Note: These subjects are placed in Table 10.6, based on their values of the DNDI index (Equation 10.28).

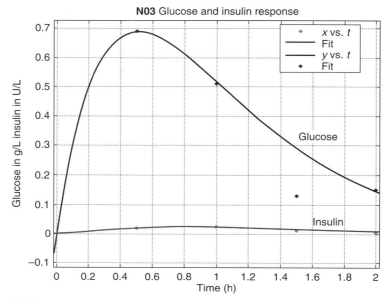

Parameters	Values	Fitting			Remarks
A	1.9650	Glucose fit	SSE	4.06e−005	N03 is at risk of becoming diabetic. This insulin response is inadequate
G	3.6740				
ω	0.079		R-square	0.9999	
α	3.0000	Insulin fit	SSE	9.172e−006	
β	0.1194				
γ	8.9717		R-square	0.9719	
δ	0.9300				

FIGURE 10.7 (See color insert following page 266.)
The glucose–insulin response of **N03** is a good example of a "missed" clinical diagnosis. Even though the subject was diagnosed as normal, clinically (i.e., pronounced to be nondiabetic), the subject in fact is at-risk of becoming diabetic, based on this subject's data being best fitted by the "critically damped response " Equations 10.17 and 10.18, as well as based on the value of the nondimensional diabetes index (Equation 10.28) and as indicated by Table 10.6.

of Figures 10.7 and 10.8 provide the values of (1) the parameters (A, G, and ω) of Equation 10.9 and (2) of the model Equations 10.1 and 10.2 parameters (α, β, γ, δ), for these subjects.

10.7   Combined Nondimensional Index Representing Both Glucose and Insulin Model Responses

There have been some works [5–9] dealing with insulin sensitivity to elevated blood glucose. However, we have come up with a nondimensional

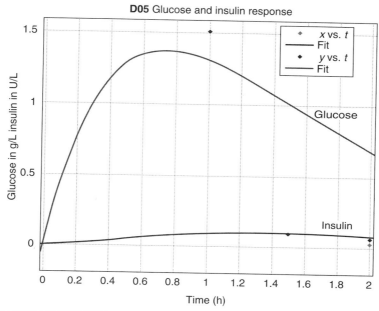

Parameters	Values	Fitting			Remarks
A	1.3530	Glucose fit	SSE	0.001906	Subject D05 has been diagnosed as diabetic, but (as per our model) is only at risk of being diabetic
G	5.0750				
ω	0.001507		R-square	0.9031	
α	2.2400	Insulin fit	SSE	0.0004872	
β	0.1862				
γ	4.2254		R-square	0.9238	
δ	0.4660				

FIGURE 10.8 (See color insert following page 266.)
The glucose–insulin response of **D05** is another example of a missed clinical diagnosis. This subject was diagnosed to be diabetic; however, based on our diabetes index DNDI, this subject is at risk of being diabetic, as indicated by Table 10.6.

index, based on glucose and insulin responses, which is based on a combination of model parameters and some important values of the glucose-tolerance test. Herein, we will develop this index, and show (by means of Tables 10.4 through 10.6) how it can be employed to categorize patients as normal or diabetic or at risk of becoming diabetic.

10.7.1 System Parameters Identification

The OGTT glucose and insulin data of individuals will be used for system parameter identification, and employed to evaluate the indices. The non-dimension index for blood glucose response (GNDI) is formulated as

Applied Biomedical Engineering Mechanics

TABLE 10.4

Results of Patients Who Were Classified as Normal (Nondiabetic) and Having Underdamped Response in the Oral Glucose Tolerance Test (OGTT)

Subject	A	G	ω	T_d	$\text{GNDI} = \frac{y_{max} \times \psi_2}{G^2} \times \frac{T_d}{A} \times \frac{T_{max}}{T_2} \times 10^6$	α	β	γ	δ	$\text{INDI} = \beta\gamma/\alpha\delta$	GNDI/INDI
N11	1.01	1.46	3.64	0.17	4.63	0.04	0.07	176.47	1.97	156.76	0.03
N12	3.92	5.53	1.41	0.51	0.18	1.11	2.47	3.19	6.73	1.05	0.17
N13	2.02	3.45	1.62	0.99	12.21	4.00	1.30	3.04	0.03	32.9	0.37
N16	1.66	3.44	3.61	0.32	6.71	0.42	0.08	118.17	2.89	7.79	0.86
N18	2.72	5.77	1.57	0.74	3.56	5.40	3.20	2.25	0.04	33.33	0.18

TABLE 10.5

Results of Patients Who Were Classified as Diabetic, Based on Their Having Overdamped Response in the Oral Glucose Tolerance Test (OGTT)

Subject	A	G	ω	T_d	$\text{GNDI} = \frac{y_{max} \times y_{1/2}}{G^2} \times \frac{T_d}{A} \times \frac{T_{max}}{T_2^3} \times 10^6$	α	β	γ	δ	$\text{INDI} = \beta\gamma/\alpha\delta$	GNDI/INDI
N17	1.06	2.64	0.00	1.89	183.04	1.17	0.04	0.27	0.95	0.01	19177
D01	0.64	3.97	0.00	3.12	75254.31	0.31	0.005	23.04	0.97	0.38	198037
D04	0.63	2.55	0.00	3.20	83360.10	0.29	0.003	105.38	0.96	0.87	95816
D07	0.98	4.90	0.00	2.01	6937.83	1.64	0.11	3.86	0.32	0.82	8460
D11	0.61	2.34	0.02	3.30	98639.12	0.18	0.004	56.64	1.03	0.99	99289
D13	1.05	3.29	0.00	1.88	3654.79	0.32	0.02	30.19	1.81	0.02	188740
D15	0.86	1.70	0.00	2.33	17795.68	0.07	0.02	36.48	1.65	6.32	2816
D16	0.93	2.02	0.00	2.15	8254.73	0.14	0.04	15.22	1.72	2.53	3263

Note: In the 'ω' column, the values of ω are <0.01 and hence put as '0.00'.

TABLE 10.6

Results of Patients Who Were Classified at "Being at Risk" of Becoming Diabetic, Based on Their Having Critically Damped Response in the Oral Glucose Tolerance Test (OGTT)

Subject	A	G	ω	T_d	$GNDI = \frac{y_{max} \times y_2}{G^2} \times \frac{T_d}{A} \times \frac{T_{max}}{T_2^2} \times 10^6$	α	β	γ	δ	$INDI = \beta\gamma/\alpha\delta$	$GNDI/INDI$
N03	1.97	3.67	0.00	1.02	63.09	3.00	0.12	8.97	0.93	0.36	177.7
N04	2.31	3.05	0.00	0.87	19.46	3.90	2.37	1.07	0.72	0.90	21.6
N10	2.38	2.37	0.00	0.84	33.31	4.70	1.82	2.95	0.06	0.19	1.75
N15	1.78	3.02	0.00	1.13	152.38	1.01	0.21	2.89	2.55	0.24	634
D05	1.35	5.08	0.00	1.48	81.21	2.24	0.19	4.23	0.47	0.76	106.92
D12	1.32	2.36	0.00	1.52	1324.10	0.27	0.29	3.79	2.37	1.72	769.8

Note: In the 'ω' column, the values of ω are <0.01 and hence put as '0.00'. Based on the values of the DNDI ($=GNDI/INDI$) index, we can say that subjects N04 and N10 are almost normal.

$$\text{GNDI} = \frac{(y_{\max} \times y_2)}{G^2} \times \frac{T_d}{A} \times \frac{T_{\max}}{T_2^5} \times 10^6 \quad . \tag{10.26}$$

where

$y_{\max}$ is the maximum blood glucose value in g/L
y_2 is the blood glucose value at 2 h
G is the glucose administered to the system in $gL^{-1}h^{-1}$
T_d is the derivative-time in h
A is the attenuation constant in h^{-1}
$T_{\max}$ is the time at which $y_{\max}$ occurs
T_2 is 2 h

The nondimensionless index for insulin response (INDI) is formulated as

$$\text{INDI} = \frac{\beta\gamma}{\alpha\delta} \tag{10.27}$$

where

α is the pancreatic insulin sensitivity to insulin in $(h)^{-1}$
β is the pancreatic insulin sensitivity to elevated glucose blood concentrations in $(units)\ (h)^{-1}(g)^{-1}$
γ is the liver glycogen storage and tissue glucose utilization due to elevated insulin concentration in $g(h)^{-1}(units)^{-1}$
δ is the tissue glucose utilization to elevated blood glucose concentrations in $(h)^{-1}$

The final combined diabetes nondimensionless index, combining both the above indices and involving all the four model parameters (α, β, γ, δ), is then given by

$$\text{DNDI} = \frac{\text{GNDI}}{\text{INDI}}$$

$$= \frac{\left[(y_{\max} \times y_2)/G^2\right](T_d/A)(T_{\max}/T_2^5) \times 10^6}{\beta\gamma/\alpha\delta} \tag{10.28}$$

wherein

α is the increase, means insulin removed
β is the increase, means insulin responsive to glucose concentration
γ is the decrease, means blood glucose increases and not enough glucose absorbed by tissues
δ is the decrease, means blood glucose increase, and inadequate tissue glucose utilization

The computed values of GNDI, INDI, and the integrated DNDI index, for the undamped subjects, overdamped subjects, and critically damped subjects, are provided in Tables 10.4 through 10.6, respectively. It is seen that undamped (and hence normal, as per our model) subjects have the value of DNDI in less than 1.0 range, while the overdamped (and hence diabetic, as per our model) subjects have DNDI in the 1000–200,000 range. Most critically damped (and hence at-risk, as per our model) subjects have DNDI in the intermediate 50–1000 range; it is seen, from Table 10.6, that subjects N04 and N10 are almost normal. Thus, this nondimensional index DNDI can be seen to provide effective separation of underdamped (normal), overdamped (diabetic), and critically damped (at-risk of becoming diabetic) subjects.

10.8   Conclusion

We have shown that we can obtain more accurate discrimination of normal diabetic subjects, by means of their being characterized by our model as being underdamped, overdamped, and critically damped. Some patients (clinically diagnosed to be normal) were designated by us to be in the borderline category. However, some patients who were clinically declared to be diabetic turned out to be only borderline. With more data, we can obtain more assured DNDI ranges of underdamped, overdamped, and critically damped subjects.

Appendix A: Solution of Equation 10.13

$$\text{To solve: } \dot{x} = -\alpha x + \beta \frac{G}{\omega} e^{-At} \sin \omega t \tag{A.1}$$

The auxiliary equation is:

$$m + \alpha = 0$$

Hence, the complimentary function (CF) is given by:

$$\text{CF} = x(t) = k e^{-\alpha t} \tag{A.2}$$

where k is the constant of integration.
   The particular integral (PI) is given by:

$$\text{PI} = \frac{1}{(D + \alpha)} \beta \frac{G}{\omega} e^{-At} \sin \omega t \tag{A.3}$$

$$\text{PI} = \beta \frac{G}{\omega} e^{-At} \frac{1}{(D - A + \alpha)} \sin \omega t = \beta \frac{G}{\omega} e^{-At} \frac{D - (\alpha - A)}{(D^2 - (\alpha - A)^2)} \sin \omega t$$

$$= \beta \frac{G}{\omega} e^{-At} \frac{D - (\alpha - A)}{-\omega^2 - (\alpha - A)^2} \sin \omega t = -\beta \frac{G}{\omega} e^{-At} \frac{\omega \cos \omega t - \alpha \sin \omega t + A \sin \omega t}{\omega^2 + \alpha^2 - 2A\alpha + A^2}$$

$$\tag{A.4}$$

Hence the general solution (GS) of Equation A.1 is given by:

$$GS = CF + PI \tag{A.5}$$

$$x(t) = ke^{-\alpha t} - \beta \frac{G}{\omega} e^{-At} \frac{\omega \cos \omega t - \alpha \sin \omega t + A \sin \omega t}{\omega^2 + \alpha^2 - 2A\alpha + A^2} \tag{A.6}$$

Applying the initial condition: at $t = 0$, $x(t) = 0$ to Equation A.6, we get

$$0 = k - \beta \frac{G}{\omega} \left[\frac{\omega}{\omega^2 + \alpha^2 - 2A\alpha + A^2} \right] \tag{A.7}$$

Substituting the value of k from Equation A.7 into Equation A.6, we get

$$x(t) = -\left(\beta \frac{G}{\omega} \right) \frac{Ae^{-At} \sin \omega t - \alpha e^{-At} \sin \omega t - \omega e^{-\alpha t} + \omega e^{-At} \cos \omega t}{A^2 - 2A\alpha + \alpha^2 + \omega^2} \tag{A.8}$$

which is the solution of Equation 10.13.

Appendix B: Solution of Equation 10.15

$$\text{To solve: } \dot{x} = -\alpha x + \beta \frac{G}{\omega} e^{-At} \sinh \omega t \tag{B.1}$$

The auxiliary equation is:

$$m + \alpha = 0 \tag{B.2}$$

Hence, the complimentary function (CF) is given by:

$$CF = x(t) = ke^{-\alpha t} \tag{B.3}$$

where k is the constant of integration.
The particular integral (PI) is given by:

$$PI = \frac{1}{(D + \alpha)} \beta \frac{G}{\omega} e^{-At} \sinh \omega t = \frac{1}{(D + \alpha)} \beta \frac{G}{\omega} e^{-At} \left(\frac{e^{\omega t} - e^{-\omega t}}{2} \right)$$

$$= \frac{1}{(D + \alpha)} \beta \frac{G}{\omega} \left(\frac{e^{(-A + \omega)t} - e^{-(A + \omega)t}}{2} \right) = \beta \frac{G}{2\omega} \left(\frac{e^{(-A + \omega)t}}{-A + \omega + \alpha} - \frac{e^{-(A + \omega)t}}{-A - \omega + \alpha} \right)$$

$$= \beta \frac{G}{2\omega} \left(\frac{(\alpha - A - \omega)e^{(-A + \omega)t} - (\alpha - A + \omega)e^{-(A + \omega)t}}{(\alpha - A - \omega)(\alpha - A + \omega)} \right) \tag{B.4}$$

Hence, general solution of Equation B.1 is:

$$x(t) = ke^{-\alpha t} + \beta \frac{G}{2\omega} \left(\frac{(\alpha - A - \omega)e^{(-A+\omega)t} - (\alpha - A + \omega)e^{-(A+\omega)t}}{(\alpha - A - \omega)(\alpha - A + \omega)} \right) \tag{B.5}$$

Applying the initial condition: at $t = 0$, $x(t) = 0$ to Equation B.5, we get

$$0 = k + \beta \frac{G}{2\omega} \left(\frac{-\omega - \omega}{(\alpha - A)^2 - \omega^2} \right) \tag{B.6}$$

$$\text{Hence, } k = \frac{\beta G}{(\alpha - A)^2 - \omega^2} \tag{B.7}$$

Substituting the value of k into Equation B.5, we get

$$x(t) = \frac{\beta G e^{-\alpha t}}{2\omega[(\alpha - A)^2 - \omega^2]} \left(2\omega + (\alpha - A - \omega)e^{(-A+\omega+\alpha)t} - (\alpha - A + \omega)e^{-(A+\omega-\alpha)t} \right) \tag{B.8}$$

Now in order to express the solution in hyperbolic form, we simplify Equation B.8 as follows:

$$x(t) = E \left(2\omega + (\alpha - A - \omega)e^{(-A+\omega+\alpha)t} - (\alpha - A + \omega)e^{-(A+\omega-\alpha)t} \right) \tag{B.9}$$

where $E = \frac{\beta G e^{-\alpha t}}{2\omega[(\alpha-A)^2 - \omega^2]}$

$$
\begin{aligned}
x(t) &= E \left(
\begin{aligned}
&2\omega + (\alpha - A - \omega)\left[\frac{e^{(\alpha-A+\omega)t} - e^{-(\alpha-A+\omega)t}}{2} \right] + (\alpha - A - \omega)\frac{e^{-(\alpha-A+\omega)t}}{2} \\
&+ (\alpha - A - \omega)\frac{e^{(\alpha-A+\omega)t}}{2} - (\alpha - A + \omega)\left[\frac{e^{(\alpha-A-\omega)t} + e^{-(\alpha-A-\omega)t}}{2} \right] \\
&+ (\alpha - A + \omega)\frac{e^{-(\alpha-A-\omega)t}}{2} - (\alpha - A + \omega)\frac{e^{(\alpha-A-\omega)t}}{2}
\end{aligned}
\right) \\[2mm]
&= E \left(
\begin{aligned}
&2\omega + (\alpha - A - \omega)\sinh(\alpha - A + \omega)t - (\alpha - A + \omega)\cosh(\alpha - A - \omega)t \\
&+ (\alpha - A - \omega)\cosh(\alpha - A + \omega)t - (\alpha - A + \omega)\sinh(\alpha - A - \omega)t
\end{aligned}
\right) \\[2mm]
&= E \left(
\begin{aligned}
&2\omega - \alpha \sinh(-\omega t + At - \alpha t) + A\sinh(-\omega t + At - \alpha t) + \omega \sinh(-\omega t + At - \alpha t) \\
&- \alpha \cosh(\omega t + At - \alpha t) + A\cosh(\omega t + At - \alpha t) - \omega \cosh(\omega t + At - \alpha t) \\
&+ \alpha \cosh(-\omega t + At - \alpha t) - A\cosh(-\omega t + At - \alpha t) - \omega \cosh(-\omega t + At - \alpha t) \\
&+ \alpha \sinh(\omega t + At - \alpha t) - A\sinh(\omega t + At - \alpha t) + \omega \sinh(\omega t + At - \alpha t)
\end{aligned}
\right)
\end{aligned}
\tag{B.10}
$$

Hence, the solution is given by:

$$x(t) = \frac{\beta G e^{-\alpha t}}{2\omega[(\alpha - A)^2 - \omega^2]}$$

$$\times \begin{pmatrix} 2\omega - \alpha \sinh(-\omega t + At - \alpha t) + A \sinh(-\omega t + At - \alpha t) + \omega \sinh(-\omega t + At - \alpha t) \\ -\alpha \cosh(\omega t + At - \alpha t) + A \cosh(\omega t + At - \alpha t) - \omega \cosh(\omega t + At - \alpha t) \\ +\alpha \cosh(-\omega t + At - \alpha t) - A \cosh(-\omega t + At - \alpha t) - \omega \cosh(-\omega t + At - \alpha t) \\ +\alpha \sinh(\omega t + At - \alpha t) - A \sinh(\omega t + At - \alpha t) + \omega \sinh(\omega t + At - \alpha t) \end{pmatrix}$$

(B.11)

which is the solution to Equation 10.15.

Appendix C: Solution of Equation 10.17

$$\text{To solve: } \dot{x} = -\alpha x + \beta G t e^{-At} \tag{C.1}$$

The auxiliary equation is:

$$m + \alpha = 0 \tag{C.2}$$

Hence, the complimentary function (CF) is given by:

$$CF = x(t) = k e^{-\alpha t}$$

where k is the constant of integration.

The particular integral (PI) is given by:

$$PI = \frac{1}{(D + \alpha)} \beta G t e^{-At} \tag{C.3}$$

$$PI = \beta G e^{-At} \frac{1}{(D - A + \alpha)} t = \frac{-\beta G e^{-At}}{(A - \alpha)} \times \frac{1}{(1 - \frac{D}{A-\alpha})} t$$

$$= \frac{-\beta G e^{-At}}{(A - \alpha)} \times \left[1 - \frac{D}{A - \alpha}\right]^{-1} t = \frac{-\beta G e^{-At}}{(A - \alpha)} \times \left[1 + \frac{D}{A - \alpha} - \frac{D^2}{(A - \alpha)^2} + \dots \right] t$$

$$= \frac{-\beta G e^{-At}}{(A - \alpha)} \left(t + \frac{1}{A - \alpha}\right) \tag{C.4}$$

$$\therefore PI = \frac{-\beta G e^{-At}}{(A - \alpha)^2} (t(A - \alpha) + 1) \tag{C.5}$$

Hence the general solution (GS) of Equation C.1 is given by:

$$GS = CF + PI \tag{C.6}$$

$$x(t) = ke^{-\alpha t} - \frac{\beta G e^{-At}}{(A - \alpha)^2}(t(A - \alpha) + 1) \tag{C.7}$$

Applying the initial condition: at $t = 0$, $x(t) = 0$ to Equation C.7, we get

$$0 = k - \frac{\beta G}{(A - \alpha)^2}$$

$$\therefore k = \frac{\beta G}{(A - \alpha)^2} \tag{C.8}$$

Substituting the value of k from Equation C.8 into Equation C.7, we get

$$x(t) = \frac{\beta G}{(A - \alpha)^2}e^{-\alpha t} - \frac{\beta G e^{-At}}{(A - \alpha)^2}(t(A - \alpha) + 1) \tag{C.9}$$

$$\therefore x(t) = \frac{-\beta G}{(A - \alpha)^2}\left[tAe^{-At} - t\alpha e^{-At} + e^{-At} - e^{-\alpha t}\right] \tag{C.10}$$

which is the solution of Equation 10.17.

References

1. Bolie, V.W., Coefficients of normal blood glucose regulation, *Journal of Applied Physiology*, 16(5): 783–788, 1961.
2. Dittakavi, S.S. and D.N. Ghista, Glucose tolerance test modeling and patients-simulation for diagnosis, *Journal of Mechanics in Medicare and Biology*, 1(2): 193–223, 2001.
3. Insel, P.A., J.E. Llljenquist, J.D. Tobin, R.S. Sherwin, P. Watkins, R. Andres, and M. Berman, Insulin control of glucose metabolism in man, *The Journal of Clinical Investigation*, 55: 1057–1066, 1975.
4. Fisher, M.E. and K.L. Teo, Optimal Insulin infusion resulting from a mathematical model of blood glucose dynamics, *IEEE Transactions on Biomedical Engineering*, 36: 479–486, 1989.
5. Bergman, R.N., Y.Z. Ider, C.R. Bouden, and C. Cobeill, Quantitative estimation of insulin sensitivity, *American Journal of Physiology*, 236(6): E667–E677, 1979.
6. Ferrannini, E., J.D. Smith, C. Cobell, G. Toffolo, A. Pilo, and R.A. De Fronzo, Effect of insulin on the distribution and disposition of glucose in man, *The Journal of Clinical Investigation*, 76:357–364, 1985.

7. Galvin, P., G. Ward, J. Walters, R. Pestell, M. Koschmann, A. Vaag, I. Martin, J.D. Best, and F. Alford, A Simple method for quantitation of insulin sensitivity and insulin release from an intravenous glucose tolerance test, *Diabetic Medicine*, 9:921–928, 1992.

8. Naoki, S., et al., Dynamics of blood glucose and its regulating hormones, In: Biological Systems, Modelling and Control, An Adaptive Interface Based on Physiological Index, *IEEE Transactions of Biomedical Engineering*, 2793–2798, 1996.

9. Cramp, D.G. and E.R. Carson, Dynamics of blood glucose and its regulating hormones, In: *Biological systems, Modelling and Control, IEEE Control Engineering Series I 1*, D.A. Linkins (Ed.), Peter Peregrinus Ltd (London), 1979, pp. 171–201.

Section IV

Orthopedic Engineering Mechanics

11

Analysis for Internal Fixation of a Fractured Bone by Means of Bone–Fixator Plate

Dhanjoo N. Ghista, Sridhar Idapalapati, and Ramakrishna Kotlanka

CONTENTS

11.1   Introduction and Scope

Axial compressive load is more prominent in fracture-fixed long bones and hence internal fracture fixation by interfragmentary compression at the fracture site needs to be achieved. However, due to the applied load eccentricity (with respect to the central axis of the bone–plate assembly), bending moment is also applied to the fracture-fixed bone. Bending moment will induce both tensile and compressive stresses across the fracture site, and open up the fracture, leading to the reduction in the stability of the fixation. From an engineering perspective, fracture-fixed bone–plate assembly is weakest in bending though it gets subjected to axial compressive loading. Hence, in this chapter, we will analyze the plate-fixed fractured bone under compression as well as in bending loading.

In this chapter, firstly using the composite beam theory based on the mechanics of material approach, an analytical model is developed to calculate the forces in the screws used in bone fracture fixation by the plate. Based on the forces in the screws and stresses in the plate and the bone, an optimal selection criterion of the fixation plate is proposed to ensure minimal deformation of the fractured bone as well as for necessary and sufficient stress shielding. Secondly, employing the finite element method, the use of stiffness-graded plate as a potential substitute to the homogeneous stainless steel bone–plate is analyzed to determine the extent of increased stress shielding. Thirdly, we demonstrate a novel concept of osteosynthesis using hemihelical plates for fixation of oblique bone fractures. Parts of this chapter are based on our work [1,2] concerning the biomechanical analysis of bone–plate assemblies.

11.2 Analysis of Forces in the Plate Screws of an Internally Fixed Bone under Axial Loading

When a plate is fixed (with the help of screws) to a fractured long bone for fracture healing, the bone–plate assembly can be analyzed as a composite beam sharing the load between the plate and the bone. In case of a bone–plate assembly, remodeling removes most of the stress concentration effects of the screw holes. However, the transfer of load between bone fragments is through screws and plates, until the fracture is healed completely. Only axial load is considered in this section, for the sake of simplicity (although the stress state of bone is in practice also due to bending and torsional loadings). In Figure 11.1, the axial load "P" is applied on the fracture-fixed bone, where the distance between screws is "s." The elastic calculation shows that the load transferred by screws "a" and "b" in axial loading condition is based on the geometrical and material properties of the bone and the plate. The free body diagram of the load transfers among the bone, plate, and screw is shown in Figure 11.2.

The plate and the bone are assumed to have the same amount of uniform axial strain (based on compatibility consideration), and it is assumed that the screws do not bend. In Figure 11.2, it is shown that Q_2 denotes that part of the applied load P diverted by screw a through the bone, and the remaining load Q_1 is the load transmitted to the remainder of the plate through screw b into the bone. Thus, $Q_1 + Q_2 = P$, and this is represented in Figure 11.2b.

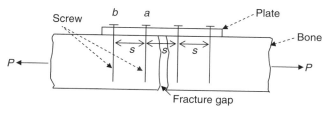

FIGURE 11.1
Free body diagram of bone and plate fixation under uniaxial tensile force. (From Ramakrishna, K., Sridhar, I., Sivashanker, S., Ganesh, V.K., and Ghista, D.N., *J. Mech. Med. Biol.*, 5, 89, 2005.)

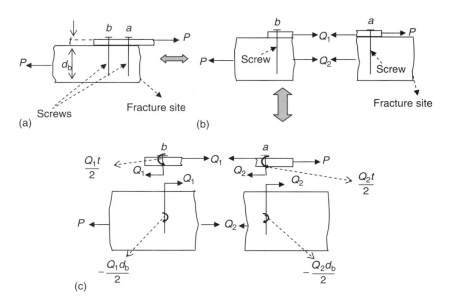

FIGURE 11.2
Free body diagram with detailed representation of forces in the fixation plate and bone. In the figure "t" is the thickness of the plate and "d_b" is the outer diameter of the bone. (From Ramakrishna, K., Sridhar, I., Sivashanker, S., Ganesh, V.K., and Ghista, D.N., *J. Mech. Med. Biol.*, 5, 89, 2005.)

The internal forces applied by the screws on bone are shown in Figure 11.2c. The plate and the bone are deemed to be firmly held together by the screws. In other words, it is assumed that the force P applied to the plate–screw–bone assembly gets distributed into forces Q_1 and Q_2 in the plate and the bone (through the screws), as illustrated in Figure 11.2a. Hence, if the axial strain in the plate and the bone segments is such that the screw is not deformed by the bending moments exerted on it (as illustrated in Figure 11.2c), then we have the condition:

$$\frac{Q_1}{A_p E_p} = \frac{Q_2}{A_b E_b} \quad \text{or} \quad \frac{Q_1}{btE_p} = \frac{4(P - Q_1)}{\pi d_b^2 E_b}$$

Therefore,

$$\frac{Q_1}{P} = \frac{A_p E_p}{A_b E_b + A_p E_p} \quad \text{or} \quad \frac{Q_1}{P} = \frac{\overline{A}\,\overline{E}}{1 + \overline{A}\,\overline{E}} \tag{11.1}$$

where $\overline{A}$ is the nondimensional cross-sectional area equal to A_p/A_b (i.e., ratio of cross-sectional area of plate to cross-sectional area of bone) and $\overline{E}$ is

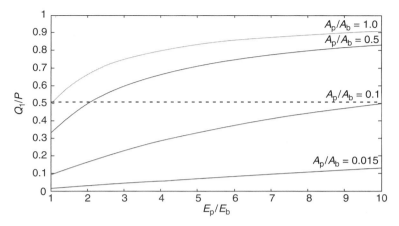

FIGURE 11.3
This figure represents the influence for $\bar{E}$ and $\bar{A}$ on the shear force on the outer screw in axial loading condition. For example, for $t = 2$ mm and width $= 10$ mm and outer diameter of the bone $= 40$ mm, we get $A_p = 20$ mm^2, $A_b = 1256$ mm^2, and $\bar{A} = 0.0156$. (From Ramakrishna, K., Sridhar, I., Sivashanker, S., Ganesh, V.K., and Ghista, D.N., *J. Mech. Med. Biol.*, 5, 89, 2005.)

the dimensionless modulus equal to E_p/E_b (i.e., ratio of plate modulus to bone modulus). Figure 11.3 provides a graphical representation of Equation 11.1. The proportion of force P diverted by screw a into the bone is substantial for a normal plate dimension (of thickness 2 mm and width 10 mm), and of the order of 90% for $\bar{E} = 10$.

From Equation 11.1, it is observed that for a bone–plate assembly under uniaxial tensile/compressive load, the load transferred by the screws is independent of their separation distance. Upon substituting representative values of $t = 2$ mm, $b = 10$ mm, and $d_b = 40$ mm in Equation 11.1, Q_1 is $P/7$. Since Q_1 is less than half the value of force P, it indicates that the transfer of loads is not the same in both the screws, and screw b transfers less of plate force P than screw a proximal to the fracture site. The amount of load transferred by the screw depends on the ratio of the cross-sectional area and material properties of the bone and the plate, as depicted in Figure 11.3.

It is shown analytically that the two screws cannot equally share the applied load. This is because, as seen in Figure 11.3, if Q_1/P were to be 0.5, $\bar{A}$ would be 0.1 for a representative value of $\bar{E} = 10$. That is, for $E_p = 10E_b$, A_p should be $0.1A_b$; hence A_p would be 125.6 mm^2 for a representative value of $A_b = 1256$ mm^2, and for a plate of width 10 mm, t would be 12.5 mm, which is not practical. It is to be noted that the presented analysis is applicable for only small strains.

11.3   Structural Analysis of Plate-Reinforced Fractured Bone under Bending (to Determine the Forces Applied by the Screws)

After fracture fixation, the plate (attached to the bone on the side which is in tension) shields the bone from tensile stresses at the fracture interface (site 1) and away from the fracture interface (site 2), as shown in Figure 11.4a. While the tensile stress shielding at fracture interface is necessary to promote healing, that away from the fracture can cause osteoporosis and reduction in bone strength. This problem may be resolved by satisfying two objectives: (1) designing for the neutral axis (NA) at the fracture interface to be at the plate–bone interface, to ensure that no tensile stresses are transferred to the callus while it is being formed and (2) designing for the NA away from the fracture interface to be as far into the bone as possible, so that the bone is subjected to the normally prevalent tensile stresses. These two requirements can be satisfied by careful tailor-made design of the plate modulus and geometry.

11.3.1   Bending Analysis of the Bone–Plate Assembly for Stresses in Bone and Plate, Using Composite Beam Theory of Perfect Bond between the Bone and the Plate

When a bone–plate assembly beam is subjected to bending moment (as shown in Figure 11.4a), the material above the NA is subjected to tensile stresses and the material below the NA bears compressive stresses. For a fractured bone, the plate is preferably fixed on the tension side of the bone, as per the tension band principle. The compatibility criterion for the bending analysis is that the curvature or bending deformation should be the same in the plate and the bone, along the contact (bone and plate) interface. The plate is assumed to be attached to the bone by means of two screws on

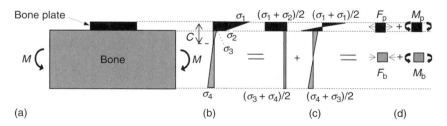

FIGURE 11.4
(a) Fracture fixation by plate under bending moment; (b) normal stress distribution for bone–plate assembly under bending moment; (c) the stresses in Figure 11.4b are a combination of the stresses due to axial forces and bending moment; and (d) representation of the axial and bending moment acting on the fracture fixation. (From Ramakrishna, K., Sridhar, I., Sivashanker, S., Ganesh, V.K., and Ghista, D.N., *J. Mech. Med. Biol.*, 5, 89, 2005.)

either side of the fracture site. However, the bone–plate assembly (in internal fixation) is assumed to behave as a composite beam during and after healing, because the plate is assumed to be perfectly bonded to the bone. In order to identify the role of the screws under bending, the total moment taken up by the plate and the forces in the screws need to be calculated.

The normal stress distribution on the plate–bone transverse section for the applied moment (M) is shown in Figure 11.4b, where σ_1, σ_2, σ_3, and σ_4 are the stresses at the top surface of the plate, bottom surface of the plate, top surface of the bone, and bottom surface of the bone, respectively. As shown in Figure 11.4c and d, the stress in the plate can be regarded to consist of (1) an axial stress $(\sigma_1 + \sigma_2)/2$ due to an axial tensile force, F_p and (2) a bending stress $(\sigma_1 - \sigma_2)/2$ due to bending moment, M_p. Similarly, the stress in the bone can be regarded to consist of (1) an axial compressive force, F_b and (2) a moment, M_b.

At a cross section through the bone and the plate, the total normal stresses are hence given by the following equations [1]:

$$\sigma_1 = \frac{ME_pC}{S} \tag{11.2}$$

$$\sigma_2 = \frac{ME_p(C - t)}{S} \tag{11.3}$$

$$\sigma_3 = \frac{ME_b(C - t)}{S} \tag{11.4}$$

$$\sigma_4 = -\frac{ME_b(d_b + t - C)}{S} \tag{11.5}$$

where
$S = E_p\, I'_p + E_b\, I'_b$
$I'_p = I_p + A_p\, y_1^2$
$I'_b = I_b + A_b\, y_2^2$
S is the equivalent flexural stiffness of the plate–bone assembly
C is the distance between NA and the top surface of the plate
y_1 is the distance between NA and center of plate
y_2 is the distance between NA and center of bone

The moments and the axial forces that are shared by the plate and the bone are given by

$$M_p = \frac{MI'_p E_p}{S} \tag{11.6}$$

$$M_b = \frac{MI'_b E_b}{S} \tag{11.7}$$

$$F_p = \frac{ME_p(2C - t)(bt)}{2S} \tag{11.8}$$

$$F_b = \frac{\pi(ME_b)(2C - 2t - d_b)d_b^2}{8S} \tag{11.9}$$

11.3.2 Bending Analysis of Bone–Plate Assembly, in which the Plate Is Attached to the Fractured Bone by Means of Screws

11.3.2.1 Analysis for Four Screws (Two on Each Side of Bone Callus at the Fracture Site): Determination of Forces in Screws and Stresses at the Top of Bone Surface

Now, the bone and the fixator plate are assumed to be held together by means of four screws, two on either side from the midspan. We now carry out analysis of the plate and the bone under the action of the moments M_p and M_b (Equations 11.6 and 11.7) and the axial forces F_p and F_b (Equations 11.8 and 11.9), due to forces (W) exerted on the plate and the bone by the inner screws and the forces (R) exerted on them by the outer screws, as is carried out in Figure 11.5. Further, it is assumed that the two extreme

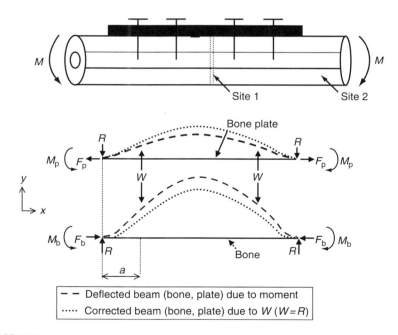

FIGURE 11.5
Bone fracture fixation by bone plate fixed to the bone by means of two screws on either side of fracture site 1. The two figures below illustrate the free body diagram of bone and plate as clamped boundary condition. The bone and the plate are regarded as two independent structures, held together by (1) forces $R = (W)$ applied by the two outer screws and (2) forces W applied by the two inner screws. (From Ramakrishna, K., Sridhar, I., Sivashanker, S., Ganesh, V.K., and Ghista, D.N., *J. Mech. Med. Biol.*, 5, 89, 2005.)

screws rigidly fix the bone–plate assembly by applying forces R, as shown in Figure 11.5.

With moments M_p and M_b sustained by the plate and the bone, the deflections in plate and bone will be different. Under this loading, the compliant bone would deform more than the plate. Hence, in order for the bone and the plate to deform by the same amount, forces have to be applied through screws on the plate and the bone. Let the forces through the screws be represented as "W," as illustrated in Figure 11.5. The screw forces W on the plate and the bone are equal in magnitude and opposite in direction, and are assumed to make the plate and the bone have the same deformation. The free body diagrams of the plate and the bone are shown in Figure 11.5. As illustrated therein, W is the force acting on plate through the screw to increase the plate deflection; similarly, W is the force imposed on the bone by the screw to restrict its deflection.

In order to determine the force (W) exerted by the screws and stresses on the top surface of the bone at site 1 (at the midspan of the plate) and at site 2, the following analysis is performed:

1. Express the flexure-bending moment equations for the plate and the bone, under the action of end forces (F_p and F_b) and end moments (M_b and M_p), and the screw forces (W), as depicted in Figure 11.5.

2. Impose the boundary condition that both plate and bone have zero deflection and slope at the ends.

3. Equate the deflections in the bone to plate at the location of the screws as well as at the midspans of the bone and the plate.

The parameters assigned for the calculations are as follows: Young's modulus (E) of the bone is assumed to be 21 GPa, and Young's modulus of plate is as varied as 200, 110, and 50 GPa. Further just after fracture fixation we assume the presence of callus modulus taken to be $E_c = 0.05E_b$. Additional data for the analysis are as follows: bone outer radius of 12 mm, bone inner radius of 6 mm, plate of 15 mm width and 60 mm length, distance of 40 mm between the screws, and applied bending moment is 1 Nm. With the applied bending moment on the fracture-fixed bone–plate assembly (along with the above-mentioned bone and plate geometrical dimensions and moduli), Table 11.1 summarizes the results.

From Table 11.1, it is noticed that in case 1, to have the NA inside the plate at site 1 and maximum tensile stress in the top layer of the bone at site 2, the thickness of the plate should be at least 3 mm. Similarly in case 2, the plate thickness should be 4 mm. However, for the modulus of plate to be less than 50 GPa (in case 3, using carbon fiber reinforced polymer, CFRP), the NA is shifted into the bone and thus causes some tensile stress in the upper layer of bone (i.e., at site 1) leading to delayed healing. Hence, the optimal plate should have a modulus of 210 GPa and thickness of 3 mm for the cases considered.

TABLE 11.1

Design Parameters of Plate and Calculated Stress Values in the Bone Callus
at Site 1 and in the Bone at Site 2

Modulus of Callus (GPa)		Modulus of Plate (GPa)	Thickness of Plate (mm)	Magnitude of Forces in Each Screw (N). Outer Screws Are in Tension and Inner Screws in Compression.	Stress at Site 1 on the Top Surface of the Bone Callus (N/mm²)	Stress at Site 2 on the Bone Top Surface (N/mm²)
Case 1	0.1	210 (316L	2	25.12	0.20	0.64
		Stainless	3	26.72	−0.19	0.43
		steel)	4	29.07	−0.30	0.27
			5	30.2	−0.32	0.17
Case 2	0.1	110 (Titanium	2	18.86	0.38	0.72
		alloy:	3	24.81	0.08	0.56
		Ti–6Al–4V)	4	28.44	−0.12	0.40
			5	29.42	−0.39	0.29
Case 3	0.1	50 (CFRP)	2	12.2	0.41	0.77
			3	18.21	0.36	0.68
			4	25.72	0.26	0.55
			5	26.32	0.11	0.43

Ideally, it is preferable that the NA be located at the plate–bone interface and the stress in the bone underneath the plate be zero (so that there is no tensile stress) at the callus site; at the same time, the plate should not be overly stiff, so that there is no tensile stress in the bone away from the callus site (due to the NA being inside the plate). However, since the plate thicknesses are standardized, what is being implied from Table 11.1 is that the stainless steel plate of 3 mm thickness yields the best option for (1) the callus to be totally in compression, with the stress underneath the plate being minimal and (2) the bone away from the callus fracture site to have the maximal tensile stress.

11.3.2.2 Analysis of Six Screws (Three on Each Side of Callus at the Fracture Site): Determination of Forces in the Screws and Stresses at the Top of Bone Surface

Now, stress analysis is performed for the bone and the stainless steel plate assembly with six screws (three on either side of the fracture), with the following additional data: moment of inertia of plate $(I_p) = 156$ mm^4, the dimensions $b = 40$ mm and $a = 20$ mm (see Figure 11.6). The stresses in the bone at fracture interface and away from it, for E (plate) $= 210$ GPa are

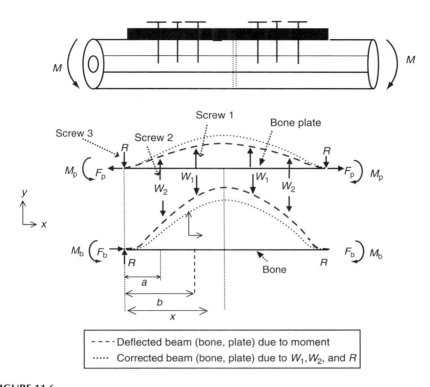

FIGURE 11.6
Free body diagram of bone–plate assembly with six screws holding the bone and the plate together, three on either side of the fracture site. From the calculations, it is noted that screw 1 (W_1) is in compression, while screws 2 (W_2) and 3 (R) are in tension.

tabulated in Table 11.2; and the forces in the screws are given in Table 11.3. The optimal plate thickness is still 3 mm, even with the increase in the number of screws from four to six.

11.4 Finite Element Analysis of Bone Fracture Fixation

11.4.1 Representation of Constitutive Properties of Callus and Plate for the Finite Element Analysis

The objective of this section is to address the role of functionally graded plate stiffness for optimal fracture fixation, using the ABAQUS finite element package. The methodology will be illustrated only for bending moment (as bending is deemed to be the predominant load that opens the crack), applied onto a fractured long cylindrical bone–plate assembly. The purpose of stiffness-graded plate is to explore the extent of stress shielding. In this

TABLE 11.2

Design Parameters of the Plate and Calculated Stress Values in the Bone Callus at the Fracture Site 1 and in the Bone at Site 2

Modulus of Callus (GPa)	Modulus of SS Plate (GPa)	Plate Thickness (mm)	Stresses at Site 1 on the Top Surface of Callus (N/mm²)	Stresses at Site 2 on the Bone Top Surface (N/mm²)
0.1	210	2	0.20	0.64
		3	−0.17	0.43
		4	−0.30	0.27
		5	0.31	0.17

model, the callus at fracture site is considered to be homogeneous and isotropic, with its modulus varying from 1% of bone modulus (grown after fracture) to 100% of bone modulus at full healing. The bone and the plate modulus assigned in the model are 21 and 210 GPa, respectively.

11.4.2 Two-Dimensional Analysis of Internally Fixed Fractured Bone under Bending, Using Stiffness-Graded Plate in Comparison with Stainless Steel Plate, for Perfect Bonding of the Plate to the Bone

Functionally graded materials (FGMs) are currently being used for a range of mechanical and structural applications. FGMs (made from a mixture of ceramics and metals) are characterized by a smooth and continuous change of the mechanical properties from one characteristic surface to the other. Finite element analysis (FEA-ABAQUS) is used to determine the stress distribution for the fracture-fixed bone–plate assembly, as illustrated in Figure 11.7a through d, where the geometry and loading conditions (represented by four-point bending corresponding to a bending moment of 1 N mm) are shown in Figure 11.7a. Figure 11.7b represents the case of a 316L stainless steel (SS) of uniform stiffness; Figure 11.7c is for the case of a

TABLE 11.3

Forces in the Screws for Six-Screw Bone–Plate Fixation for the Plate Modulus of 210 GPa

Thickness of Plate (mm)	Force in Screw 1 (W_1) in N (Compressive)	Force in Screw 2 (W_2) in N (Tensile)	Force in Screw 3 (R) in N (Tensile)
2	23.17	1.38	21.78
3	25.73	2.41	23.31
4	28.24	2.61	25.63
5	29.96	3.16	26.80

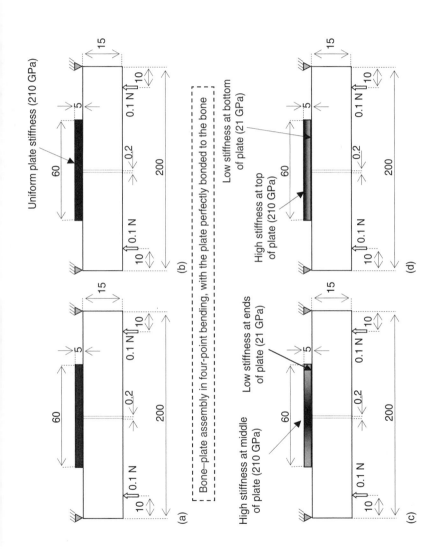

FIGURE 11.7

(a) Schematic representation of dimensions (mm) and loading conditions used for FEA; (b) uniform stiffness plate; (c) stiffness-graded plate in length; and (d) stiffness-graded plate in thickness. (From Ramakrishna, K., Sridhar, I., Sivashankar, S., Ganesh, V.K., and Ghista, D.N., *J. Mech. Med. Biol.*, 5, 89, 2005.)

lengthwise stiffness-graded plate (SGL); and Figure 11.7d is for the case of thickness-wise stiffness-graded plate (SGT). In all the cases, it is assumed that the plate is bonded to the bone, the primary purpose of the analysis being to investigate the reduction in stress shielding provided by stiffness-graded plates.

In order to make the model continuous at the fracture interface, callus (of length 1% of the beam length) is assumed at the fracture site. Young's modulus of the callus is assumed to be 1% of that of bone at the initial stages of healing, while its value is equal to that of bone modulus at the final stages of healing. In Figure 11.7b, Young's modulus of SS is 210 GPa. In Figure 11.7c and d, Young's modulus of SGL and SGT are varied linearly from 210 to 21 GPa along the length and thickness directions of the plate, respectively. The transfer of load between the bone and the plate (bypassing the crack) enables the plate-reinforced bone to bear the loading. At the same time, the plate prevents the crack from opening up, and helps to induce compressive stress in the lower portion of the crack interface.

Herein, the bone and the plate are assumed to be completely bonded at the interface (using CONTACT PAIR option in ABAQUS, i.e., the nodes on the contact surfaces are glued to each other, to thereby develop uniform strain across the interface). Four noded (quadratic) plane-strain elements are used to discretize the geometry. With the use of finite element method (FEM), it is illustrated (in Figure 11.8) how the stresses at the fracture interface vary, with an increase in callus stiffness due to fracture healing (in Figure 11.8). In the early stages of healing, when the callus modulus is 1% of the bone modulus (where bone is modeled, it is assigned with Young's modulus of 20 GPa), the NA is located inside the plate and the callus interface is in compression. As healing proceeds (the callus modulus increases), the NA shifts down into the bone, allowing the callus to also bear some tensile stress. In this way, the total tensile stress borne by the plate decreases with increase in the callus modulus.

The computed variations of normal stresses along the plate–bone interface at initial and final stages of healing are shown in Figure 11.9a and b, respectively. It is observed that initially after fracture, all the three types of plates shield the bone, by not allowing any tensile stress in the upper bone layers (underneath the plate) close to the fracture interface. In other words, at cross sections close to the fracture site, the NA is located within the plate. However, about 10 mm away from the fracture site, the NA becomes located into the bone region, and the bone starts bearing the tensile stress. It is also observed that the stress-shielding zone size for the SS plate is at least 50% greater than that for the SGL and SGT plates. Qualitatively, it can be seen that with SGT, a higher compressive stress is developed in the top layer at the fracture interface. It is noted that the stress distribution of SGL follows that of SS plate at fracture site, and that of SGT plate away from the fracture site.

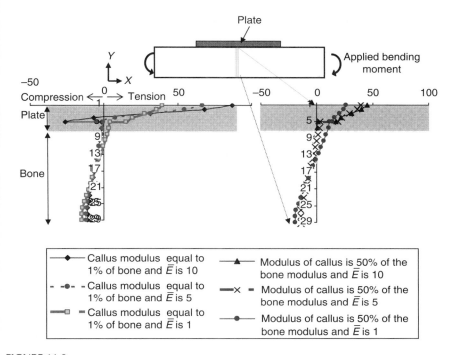

FIGURE 11.8

Normal stress distribution at the callus (crack interface) and NA location for different callus and plate properties, as healing progresses (calculated using FE simulations). (From Ramakrishna, K., Sridhar, I., Sivashanker, S., Ganesh, V.K., and Ghista, D.N., *J. Mech. Med. Biol.*, 5, 89, 2005.)

11.4.3 Two-Dimensional Analysis of Fracture-Fixed Bone with Stiffness-Graded Plate for Different Screw Locations

In the previous case study, the plate was assumed to be perfectly bonded to the bone. The same four-point bending problem is now analyzed by fixing the plate with a maximum of three screws and a minimum of one screw on either side of the fractured callus surface, as shown in Figure 11.10. In the finite element (FE) simulations, the contact surface between the plate and the bone is assumed to have a coefficient of friction of 0.37. The transfer of load between the plate and the bone is through the screws; to simulate this function, a finite length of 1 mm of plate is held (through a tie option) to the bone.

Figures 11.11 through 11.16 depict the computed variations of normal stresses on the bone and across the bone–plate interface for all the six modalities of fixation at the initial and final stages of healing for the six-screw fixation (S6), four-screw mode 1 fixation (S4-1), four-screw mode 2 fixation (S4-2), four-screw mode 3 fixation (S4-3), two screws near the fracture site (S2N), and two screws away from the fracture site (S2E).

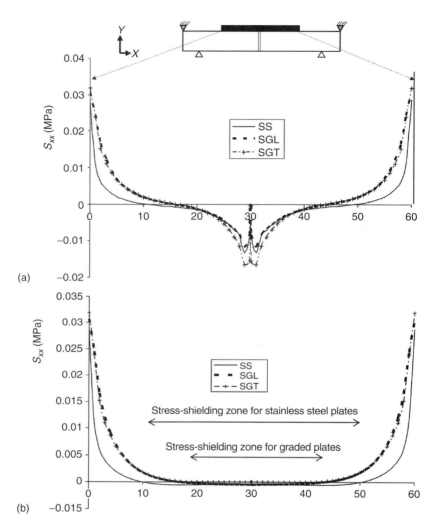

FIGURE 11.9
Results for stiffness-graded plate perfectly bonded to the bone. Comparison of the stresses along the length of bone–plate interface for stainless steel (SS), lengthwise stiffness-graded plate (SGL), and thickness-wise stiffness-graded plate (SGT); (a) stresses on bone versus distance at initial stage of healing and (b) stresses on bone versus distance at final stage of healing. These comparisons show that SGL and SGT offer less stress shielding compared to SS. (From Ramakrishna, K., Sridhar, I., Sivashanker, S., Ganesh, V.K., and Ghista, D.N., *J. Mech. Med. Biol.*, 5, 89, 2005.)

The variation of normal stresses in top layer of the bone for the S6 is shown in Figure 11.11, for both initial and final stages of the healing. It can be seen that in the early stages of healing, the bone is almost de-stressed at the fracture. Also, it is noted that, because the screws are functioning as elements holding the plate and the bone together, large variations of normal

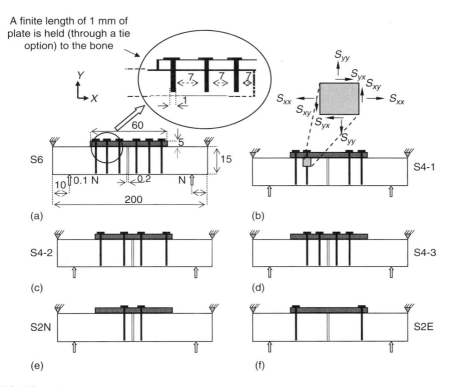

FIGURE 11.10
Different screw locations for which finite element analysis (FEA) is carried out: (a) six screws (S6), (b) four screws, mode 1 (S4-1), (c) four screws, mode 2 (S4-2), (d) four screws, mode 3 (S4-3), (e) two screws near the fracture site (S2N), and (f) two screws away from the fracture site (S2E). The distance (in mm) between screws is shown in the inset of (a). In (b) inset, an element underneath the plate and on the bone is enlarged to depict the stress notation used in Table 11.4. (From Ramakrishna, K., Sridhar, I., Sivashanker, S., Ganesh, V.K., and Ghista, D.N., *J. Mech. Med. Biol.*, 5, 89, 2005.)

stresses are seen at the screw sites of the bone. This is understandable because in the initial stage of healing, the bone–plate assembly is almost equivalent to that of two separate bone fragments held together by the plate through the screws.

In the final stages of healing, the NA becomes relocated from the plate to within the bone. At this stage, the screws are not playing such a major role in maintaining the integrity of the bone, as in the early stages of healing. Hence (as seen in Figure 11.11), the magnitudes of the stresses at the screws near the fracture site are considerably lesser than those at the initial stage of healing. From Figures 11.11 through 11.16, it is noted that the stress distributions associated with fixation by modes S6, S4-1, S4-2, and S2E give maximal stress-shield zones. Model S4-3 provides a smaller stress-shield zone, while fixation by S2N provides minimal stress-shield zone.

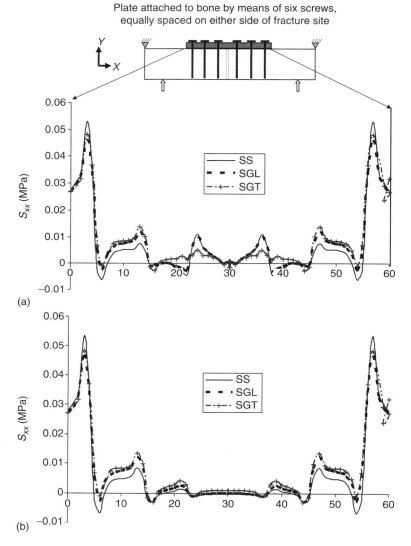

FIGURE 11.11

Fixation mode S6. The figure shows normal stresses at the top face of bone in contact with the plate for S6 fixation mode (a) at initial stages of healing and (b) at final stages of healing. It can be seen that the bone is under compression stress at the fracture site underneath the plate. Upon complete healing, the bone bears some tensile stress at the fracture site. (From Ramakrishna, K., Sridhar, I., Sivashanker, S., Ganesh, V.K., and Ghista, D.N., *J. Mech. Med. Biol.*, 5, 89, 2005.)

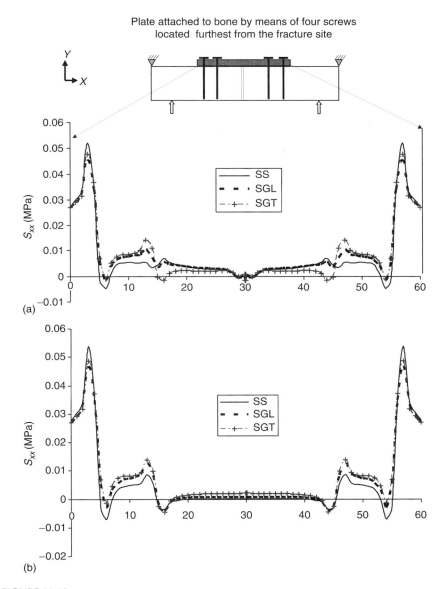

FIGURE 11.12
Fixation mode S4-1. The figure shows normal stresses at the top face of bone in contact with the plate for S4-1 (a) at initial stages of healing and (b) at final stages of healing. Even in this fixation configuration, upon complete healing the bone bears some tensile stresses at the fracture site. (From Ramakrishna, K., Sridhar, I., Sivashanker, S., Ganesh, V.K., and Ghista, D.N., *J. Mech. Med. Biol.*, 5, 89, 2005.)

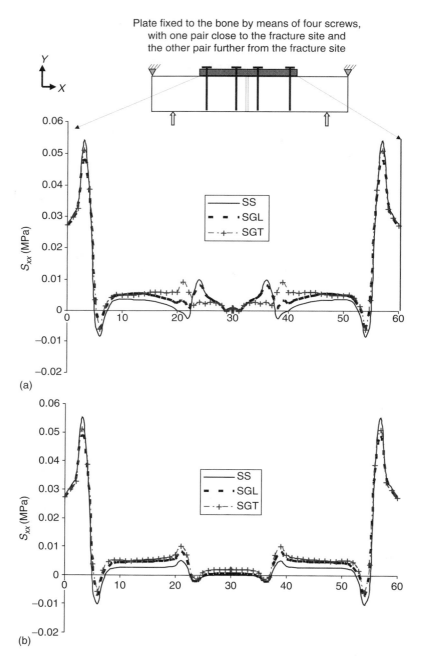

FIGURE 11.13

Fixation mode S4-2. The figure shows normal stresses at the top face of bone in contact with the plate for S4-2 (a) at initial stages of healing and (b) at final stages of healing. Here too, upon complete healing the bone bears some tensile stresses at the fracture site. (From Ramakrishna, K., Sridhar, I., Sivashanker, S., Ganesh, V.K., and Ghista, D.N., *J. Mech. Med. Biol.*, 5, 89, 2005.)

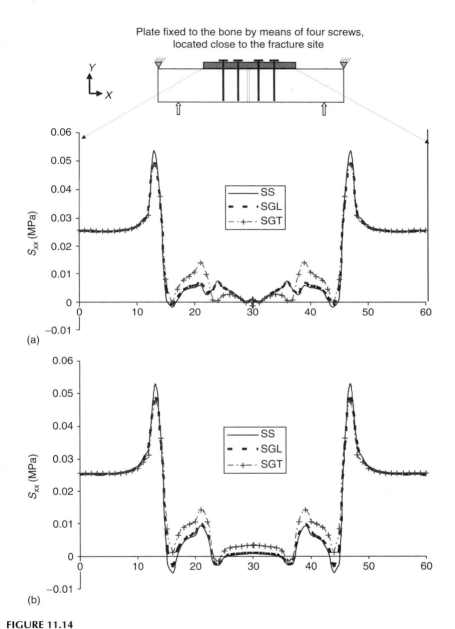

FIGURE 11.14
Fixation mode S4-3. The figure shows normal stresses at the top face of bone in contact with the plate for S4-3 (a) at initial stages of healing and (b) at final stages of healing. It is to be noted that the stress-shielding zone is reduced in this fixation configuration. The stress-shielding zone is dependent on the working length of the plate, which depends on the location of the screw. (From Ramakrishna, K., Sridhar, I., Sivashanker, S., Ganesh, V.K., and Ghista, D.N., *J. Mech. Med. Biol.*, 5, 89, 2005.)

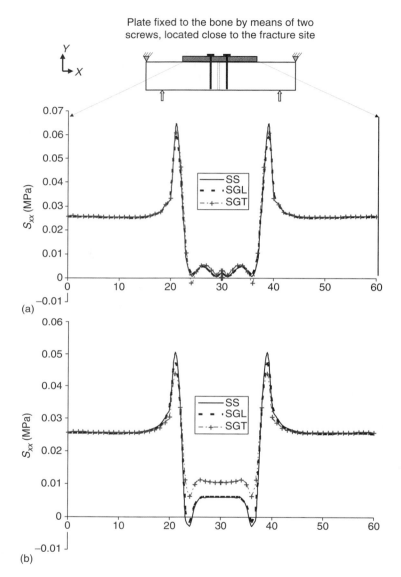

FIGURE 11.15

Fixation mode S2N. The figure shows normal stresses at the top face of bone in contact with the plate for S2N (a) at initial stages of healing and (b) at final stages of healing. (From Ramakrishna, K., Sridhar, I., Sivashanker, S., Ganesh, V.K., and Ghista, D.N., *J. Mech. Med. Biol.*, 5, 89, 2005.)

The influence of length of the plate in bearing the bending moment is governed by the extreme screws.

The stress-shield zone is dependent on the working length of the plate, and on the location of the screws. In fixation modes S4-3 and S2N,

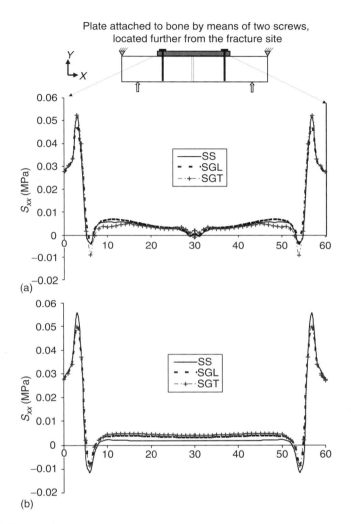

FIGURE 11.16
Fixation mode S2E. The figure shows normal stresses at the top face of bone in contact with the plate for S2E (a) at initial stages of healing and (b) at final stages of healing. In this fixation configuration, bone is shielded from stresses for the entire length of the plate due to the usage of the extreme screws. The extreme screws dictate the stress-shielding zone of the bone. (From Ramakrishna, K., Sridhar, I., Sivashanker, S., Ganesh, V.K., and Ghista, D.N., *J. Mech. Med. Biol.*, 5, 89, 2005.)

the entire length of the plates (S4-3 and S2N) is not utilized, as the extreme screws are not used. If the extreme screws are not used, the ends of the plates are not held to the bone, and will provide loose fixation. Hence, fixation modes S4-3 and S2N are avoided in clinical practice. Fixation by just two screws at the fracture site may not provide enough stability;

hence, S2E is not clinically applicable. However, fixation modes S4-1 and S4-2, of plate with two screws on either side of the fracture, are clinically relevant.

Now, considering the effect of plate stiffness on the stress distribution, it is noticed that the SGL plate simulates the SS plate stress distribution closer to the fracture zone, but simulates the SGT plate further away from the callus. The level of stress shielding is not significantly affected by the variation in plate modulus. This implies that the screw location on the plate (rather than the stiffness of the plate) has a more dominating role in minimizing the stress-shield zone. The shear stress near the screw and the maximum compressive stress in the callus for various plate–screw fixation modes are summarized in Table 11.4. Also from this table, it can

TABLE 11.4

Central Deflection, Shear Stresses at the Screw (at Bone–Plate Interface), and the Maximum Compressive Normal Stress at the Fracture Site for Two-, Four-, and Six-Screw Fixation Illustrated in Figure 11.10

Mode of Fixation	Central Deflection (mm) at Fracture Site (10^{-5})			Screw Number from Fracture Site	S_{xy}-Shear Stress at Screw (MPa)			Maximum Compressive Stress in Callus (10^{-4} MPa)		
	SS	SGL	SGT		SS	SGL	SGT	SS	SGL	SGT
S6				1	−0.0037	−0.0025	−0.0025			
	68	70	72	2	0.0038	0.00557	0.007	−56	−59	−69.2
				3	0.0263	0.0178	0.0196			
S4-1				1	—	—	—			
	69	71	73	2	0.009	0.003	0.0068	−57	−58	−69
				3	0.0254	0.0172	0.0193			
S4-2				1	0.002	0.00007	0.00009			
	69	72	74	2	—	—	—	−56	−59	−70.1
				3	0.0274	0.0193	0.0221			
S4-3				1	0.00075	0.00098	0.0033			
	80	86	87	2	0.026	0.0213	0.0200	−58	−61	−70.6
				3	—	—	—			
S2N				1	0.0314	0.0267	0.0123			
	96	98	100	2	—	—	—	−86	−83.4	−90
				3	—	—	—			
S2E				1	—	—	—			
	69	71	75	2	—	—	—	−57.3	−61.2	−71
				3	0.025	0.018	0.023			

Note: SS: stainless steel, SGL: stiffness graded along length, SGT: stiffness graded along thickness.

be observed that the S6 gives minimal central deflection at fracture site and minimal shear stresses in the screws. For S4-1 design, the recommendation (based on Table 11.4) for minimal deflection is that one screw should be close to the fracture site and the other should be as far as possible from the fracture site.

11.5 Mechanics of Osteosynthesis Using Hemihelical Plates

Fixation of fractured bone or for the healed bone to regain its prefracture stiffness and strength, particularly in the case of oblique and helical cracks, continues to pose an immense osteosynthesis challenge. This section addresses the unique features of hemihelical plate (HHP) internal bone fracture fixation, where screws are used to anchor an HHP onto the fractured bone. This is a relatively new concept in internal fracture fixation. The screws help in transferring forces between the plate and the fractured bone. Consequently, these screws may be subjected to relatively large shear stresses during movement of the fractured limb. In addition, the bone surface adjacent to the screw could be subjected to the trauma of insertion, which may cause some temporary bone necrosis. These factors can lead to loosening and subsequent "pulling out" of the screws, thereby destabilizing the entire assembly of bone fracture fixation [3–7].

Such a phenomenon of screw pullout is commonly observed in straight plate fixation, where the screws are oriented normal to the plate in the same plane. Therein, inclination (i.e., angling) of the screws is considered to be a possible solution to mitigate the loosening of bone fracture fixation (as shown in Figure 11.17) [8]. Apart from the possibility of screw loosening, the straight plate also induces undue stress shielding of the fractured bone. This is because the straight plate fixture is fastened onto the tensile surface of the fractured bone (thereby de-stressing the bone beneath the plate), and the plate material Young's modulus (which is ~200 GPa for 316L stainless steel) is typically an order of magnitude higher than that of bone, which is about 20 GPa. In order to reduce the mismatch of the material properties of the plate and the bone, CFRP composite materials and stiffness-graded plates have been recommended [1,9,10].

Herein, firstly four-point bending experiments on a straight plate fixation assembly, for four different screw configurations—convergent (CSO), divergent (DSO), alternating (ASO), and perpendicular (PSO) are discussed to determine (1) the forces present in the screws, (2) the stiffness of the fixation assembly, and (3) the energy to failure. Secondly, the concept of HHP fixation is delineated by providing the results of pullout experiments with the aim of comparing the holding strengths between a straight plate and a hemihelical plate for the various screw configurations

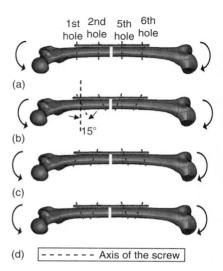

FIGURE 11.17

Schematics of straight plate and fractured bone assembly models (for bending experiments), for four different screw configurations: (a) PSO (perpendicular), (b) CSO (convergent), (c) DSO (divergent), and (d) ASO (alternating). All the inclinations are 15° normal to the axis of the bone.

mentioned above. Thirdly, FEA for three fracture fixation configurations (i.e., straight plate, 90° helical plate, and 180° helical plate), under uniaxial compression as well as bending moment and torsional loadings, are performed to compare the overall stiffness of the various assemblies and the fracture gap movement.

11.5.1 Bending Experiments of Straight Plate and Fractured Bone Fixation Assembly with Different Screw Orientations

11.5.1.1 Materials and Methods

A six-hole dynamic compression plate (DCP, 110 mm long) was attached onto the tensile surface of a transversely fractured Synthes* femur bone segment, using two 4.5 mm cortical screws on each side of the fracture location. The fracture gap was about 2 mm. As depicted in Figure 11.17, four different screw orientations were employed, namely PSO, CSO, DSO, and ASO. In the PSO setup, the screws were arranged such that plate axis is perpendicular to the screw axis (as shown in Figure 11.17a). In the CSO setup, the screws were inclined at 15° (to the normal of the bone axis) toward the fracture location (as shown in Figure 11.17b). In the DSO setup, the screws were inclined at 15° (to the normal of the bone axis) away from the fracture (see Figure 11.17c). In the ASO setup, the inner screws were inclined at 15° away from the fracture, while the outer screws were inclined at 15° toward the fracture interface (see Figure 11.17d). In all of the above configurations, the screws were placed in the first, second, fifth, and sixth screw holes (counted from the left-hand side of the bone plate) of the DCP.

---

* Synthes bones from Mathys Ltd.

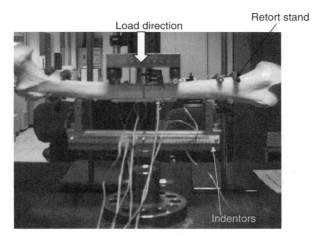

FIGURE 11.18
Bending experiments setup of the fracture-fixed bone before loading.

Figure 11.18 shows the experimental arrangement of the (dynamic compression) straight plate bone fracture fixation assembly employed for four-point bending experiments (before loading). For each test specimen, two bolt gauges were placed along the axes of the screws corresponding to the first and second holes of the DCP (as shown in Figure 11.19), to determine the forces in each screw due to its bending. In addition, three strain gauges (from Tokyo Sokki Kenkyulo Co. Ltd.) were used to monitor the nature of the stresses at various locations in the fracture fixation assembly: one strain gauge was near the sixth hole of the DCP, a second strain gauge was placed on the lateral side of the Synthes femur bone, and a third strain gauge was attached to the compression side of the femur bone. All the bolt and strain gauges (from Tokyo Sokki Kenkyulo Co. Ltd.) were connected to

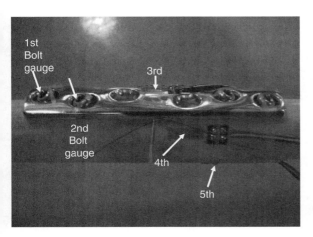

FIGURE 11.19
Position of strain gauges on synbone and plate. First and second are bolt gauges; third is strain gauge for stainless steel; and fourth and fifth are strain gauges for plastics.

a data logger. Retort stands were used to grip the bone and prevent it from rotating during loading due to the irregular geometry of the bone. Four-point bending tests were conducted on all four configurations (i.e., PSO, CSO, DSO, and ASO), using an Instron testing machine under displacement control at a loading rate of 5 mm/min.

11.5.1.2 Results and Discussion

Muscle forces applied to the bone and compressive loading on the fracture interface often give rise to bending due to the natural curvature of bones. Hence, bending tests were carried out (on CSO, DSO, ASO, and PSO configurations) to determine the relative (1) strengths, (2) bending stiffness, and (3) energy to failure of the straight plate bone fracture fixation assemblies. Figure 11.20 shows the load–displacement curves for PSO, CSO, DSO, and ASO configurations of fracture fixation assemblies. The slopes of these load–displacement curves indicate that ASO and PSO configurations offer, respectively, the greatest and lowest bending stiffness for the bone fracture fixation assembly. The CSO and DSO configurations result in similar but intermediate levels of bending stiffness, as listed in Table 11.5.

    The areas under the respective load–displacement curves indicate that the CSO and DSO configurations result, respectively, in the greatest and lowest energy to fracture or toughness for the bone fracture fixation assembly. The PSO and ASO configurations result in similar but intermediate levels of energy to fracture or toughness. This suggests that the ASO and CSO configurations appear to provide the best compromise between bending stiffness and toughness for the bone fracture fixation assembly, compared to those of PSO and DSO configurations (from the values depicted in

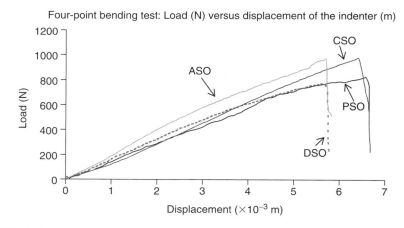

FIGURE 11.20
Bending tests: load–displacement graphs for PSO, CSO, DSO, and ASO specimen configurations. The slope of the load–displacement curve represents the stiffness of the assemblies and the area of the load–displacement curve represents energy to refracture.

TABLE 11.5

Stiffness and Energy to Refracture for Perpendicular
Screw Orientation (PSO), Convergent Screw Orientation
(CSO), Divergent Screw Orientation (DSO), and
Alternating Screw Orientation (ASO) Configurations

Screw Orientation	Stiffness Fracture Fixation (kN/m)	Energy to Refracture of Fracture Fixation (J)
PSO	126	2.66
DSO	135	2.10
CSO	140	2.92
ASO	201	2.69

Table 11.5). Strain gauge measurements provide verification that the fractured bone and the DCP assembly behave as a composite beam in bending.

Figure 11.21a and b shows, respectively, plots of the "measured load transfer versus the applied moment" in the innermost and outermost screws in the PSO bone–plate fixation assembly. One of the primary roles

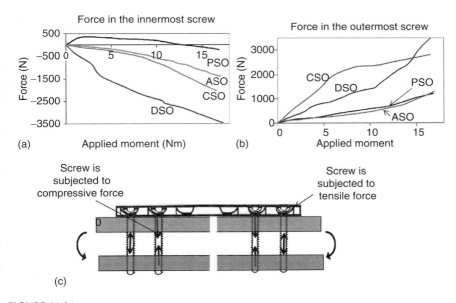

FIGURE 11.21
Force in each screw versus moment applied for PSO (perpendicular), DSO (divergent), CSO (convergent), and ASO (alternating) configurations. (a) For the screw near to the fracture interface, (b) for the screw furthest from the fracture interface, and (c) schematic of the forces in the screws. Extreme screw holds the plate onto the bone while subjected to bending. Hence, the extreme screws are subjected to tensile force, while inner screws maintain the bone and the plate to deform the same amount (i.e., deflection at the midspan of the plate equal). Thus, inner screws are subjected to compressive force.

of the screws is to ensure that the bone–plate fracture fixation assembly (comprising of the bone, plate, and screws) deforms in an integrated fashion when subjected to loading. As depicted in Figure 11.21c, the outermost screws at the fracture interface experience predominantly tensile forces in the process of keeping the plate fixed onto the bone, while the innermost screws experience predominantly compressive forces in the process of ensuring that the plate and the bone deform in an integrated manner.

At any given load, the magnitude of the force transfer by the innermost screws is the greatest in the DSO configuration and lowest in the PSO configuration. The CSO and ASO configurations provide intermediate levels of force transfer in the innermost screws. In contrast, the magnitude of force transfer by the outermost screws is greatest in the CSO configuration and lowest in the PSO configuration. The DSO and ASO configurations provide intermediate levels of force transfer in the innermost screws.

Overall, the experimental results suggest that angling the screws improves the load transfer between the plate and the fractured bone during bending. The DSO and CSO configurations appear to provide the best compromise for improved load transferred between the plate and the fractured bone for the outer and inner screws.

11.5.2   Hemihelical Plate versus Straight Plate Bone Fracture Fixation: Experimental Observations

11.5.2.1   Materials and Methods

In this section, the use of HHP fixation is analyzed experimentally by conducting pullout experiments with the aim of comparing the holding strengths between a "straight plate fractured bone" assembly (with various screw orientations) and a "hemihelical plate fractured bone" assembly. Figure 11.22 illustrates the nomenclature of a helical plate. The "axis" is

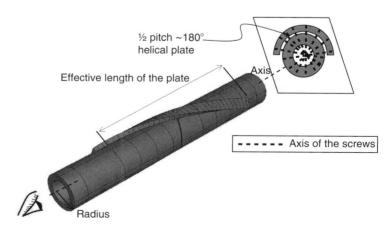

FIGURE 11.22 (See color insert following page 266.)
Terminology for a helical plate. Inset shows the side view of the helical plate, depicting the half pitch of the plate.

defined as a straight line along the length of the bone, while the "radius" is defined as the distance from the axis of the bone to the outer surface of the bone (at midshaft). The "effective length" is the distance between the ends of the plate along the axis of the bone, and the "pitch" is the degree of rotation (or twist) of the plate. For example, 1 pitch represents a 360° twist of the plate, $\frac{1}{2}$ pitch represents a 180° twist of the plate, and $\frac{1}{4}$ pitch represents a 90° twist of the plate.

Fernandez [11] was the first to clinically moot the idea of fracture fixation using helical bone plates. In particular, his anatomical study postulated the advantage of helical plates for treating fractures on the humerus. Further applications of helical plate have been illustrated by Gardner et al. [12], Apivatthakakul et al. [13], and Yang [14], as well as by us [15]. Bone fracture fixation using a 360° or 270° helical plate is not considered herein due to the clinical application constraints. Hence, only 180° and 90° helical plates are considered in our investigation. In these experiments, a 12-hole Zimmer dual compression contourable plate (DCCP), with 4.5 mm cortical screws, was fixed onto a Synthes femur bone specimen by screws located at the first, fourth, ninth, and twelfth hole positions.

Experiments were conducted to determine the axial pullout strengths of both (1) hemihelical and (2) straight bone plates with PSO, CSO, DSO, and ASO configurations, as indicated in Figure 11.23, in accordance with ASTM

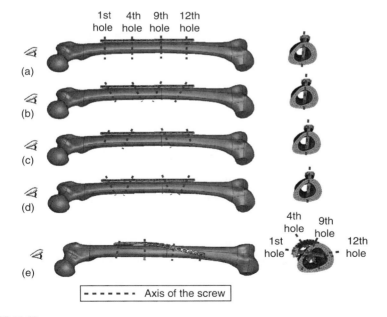

FIGURE 11.23
Pullout test models: inclined screw orientations; Straight plate fixation with (a) PSO (perpendicular), (b) CSO (convergent), (c) DSO (divergent), and (d) ASO (alternating) configurations; Helical plating (e) hemihelix ~180° Helical plate. In the hemihelical plate fixation setup, it is to be noted that the screws are oriented in different planes, compared to all the screws being in the same longitudinal plane in straight plate fixation.

FIGURE 11.24
Setup for the pullout test of 180° helical plate and bone fixation held by four cortical screws with
newly designed grippers holding the assembly.

F1691-96. The experimental setup for the pullout experiments is shown in
Figure 11.24. Special grippers were designed and built to hold the ends of
the femur bone specimens, and comply with ASTM F1691-96 test techniques. Each bone–plate fracture fixation assembly was loaded by a 5 kN
Instron testing machine, under displacement control, at a loading rate of
5 mm/min. The experiments were confined to one sample of each configuration (i.e., PSO, ASO, CSO, DSO, and hemihelical fixation assembly) due
to limited bone specimens; this is because the saw bones are not reusable as
they fracture after the test even though the plates remain in the elastic state.

11.5.2.2 Results and Discussion

Figure 11.25 shows the load–displacement responses for the hemihelical
bone plate and straight bone plate (with four different screw orientations)
during pullout testing. These pullout tests enable assessment of the
holding capacity of the plate and screws onto the fractured bone. As
shown in Figure 11.25, four distinct peaks were observed in the load-
displacement curves of straight bone plates (with various screw orientations). These peaks reflect sequential screw pullout, where the screws
located at the fourth hole loosens initially, followed by progressive loosening of the ninth, first, and twelfth holes in the bone plate (see Figures 11.24
and 11.26a). In contrast, no sequential screw pullout was observed in the
hemihelical bone–plate fixation (refer Figure 11.26b and c). It is also noteworthy that the holding strength of the hemihelical bone–plate fixation is
higher than that of the straight bone plate with PSO, CSO, DSO, and ASO
configurations. Moreover, no screw loosening was observed in the hemihelical bone–plate fixation (as noted in Figure 11.26b and c).

The gradients of these load–displacement curves (Figure 11.25)
indicate that the straight bone plate with ASO and DSO configurations

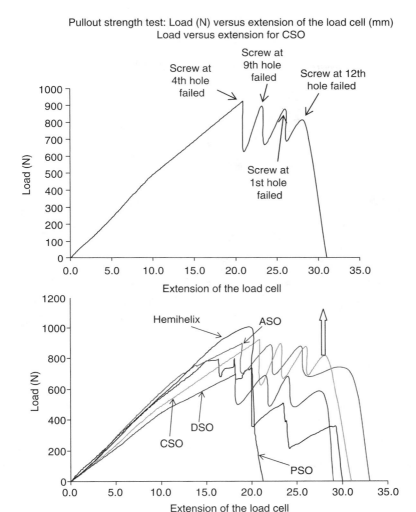

FIGURE 11.25 (See color insert following page 266.)
Pullout tests: load versus extension curves for all the configurations of fixations shown in Figure 11.22. An inset depicts the four peaks in the curve that represent the screw pullout for the CSO (convergent) configuration. Similar screw pullout patterns were observed for PSO (perpendicular), ASO (alternating), and DSO (divergent) configurations. The stiffness of the assembly is the slope of the load versus extension curve and the area under the load versus extension curve till the initiation of pullout represents the energy to pullout.

offer, respectively, the greatest and lowest stiffness for the bone fracture fixation assembly for bending loading. Convergent and perpendicular straight plate screw configurations as well as hemihelical bone plates result in similar intermediate levels of stiffness (as listed in Table 11.6). The areas under the respective load–displacement curves indicate that the hemihelical and straight bone plates (with DSO configuration) result, respectively, in the

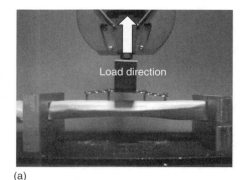

(a)

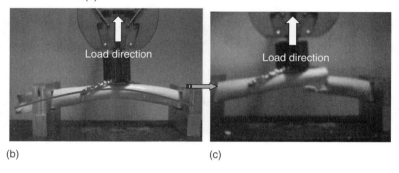

(b)   (c)

FIGURE 11.26 (See color insert following page 266.)
(a) Sequential pullout of straight plate PSO (perpendicular), CSO (convergent), DSO (divergent), and ASO (alternating) configurations. (b) The holding power of hemihelical plate (HHP) is high, such that sequential screw pullout is not observed. (c) However, the bone failed before screw pullout, indicating that the fixation is stiff enough so that screw loosening does not occur.

greatest and lowest energy to fracture (or toughness) for the bone fracture fixation assembly. The PSO, ASO, and CSO configurations for straight plate fixation result in similar intermediate levels of energy to fracture (as shown in Table 11.6).

TABLE 11.6

Stiffness and Energy to Initiate Pullout and Peak Pullout for PSO, CSO, DSO, ASO, and Hemihelix Configurations

Screw Orientation	Stiffness of Plate Fixation (kN/m)	Energy to Pullout of Plate Fixation (J)	Peak Pullout Load of Plate Fixation (N)
PSO	55.1	7.17	796
ASO	58.0	9.67	900
CSO	49.3	10.12	924
DSO	44.7	6.88	679
Hemihelix	55.4	10.83	1000

The HHP also confers the highest pullout strength in the bone fracture fixation assembly, compared to the straight plate configurations with four different screw orientations. Overall, these results suggest that hemihelical bone–plate fixation provides the optimal combination of strength, stiffness, and toughness compared to straight bone plates with different screw orientations. In addition, there is the risk of loosening and sequential screw pullout in straight bone plates, since the axes of the screws are all in the same plane (as seen from cross-sectional views along the longitudinal plane of the bone fracture fixation assembly, see Figure 11.23a through d). In contrast, the axes of the screws in the hemihelical bone fracture fixation assembly are all in different planes (intersecting one another as shown in Figure 11.23e), thus eliminating or minimizing the problem of loosening and progressive sequential pullout of screws.

The primary aim of fixation is to promote swift healing of the fractured bone and restore the strength to that of the intact bone. The results from this study show that the properties of the fractured bone–plate assembly stiffness and the energy required for bone refracture or screw pullout are dependent on the orientation of the plate and thereby on the inclinations of the screws. They indicate that not all types of inclined screw configurations (as defined in Figure 11.17) are beneficial; only CSO and ASO configurations show significant improvement over PSO and DSO configurations in stiffness and energy for refracture; the CSO and ASO configurations allow the fractured bone–plate assembly to fail at higher load (1.16 and 1.13 times, respectively) and absorb more energy (1.41 and 1.35 times, respectively) than the PSO configuration. It is also apparent that inclining the screws enhances the load transferring capability between the plate and the bone.

The pullout strength test has also shown the superior capability of angled (inclined) screw fixation over the usual perpendicular screw fixation. Similarly, CSO and ASO configurations give more holding strength between the plate and the bone. However, there is still a chance of sequential pullout (which is undesirable for the progress of fracture healing), as the axis of each screw is still in the longitudinal plane (Figure 11.23a through d). On the other hand, the 180° helical plate performs better (in terms of pullout strength or load-holding capacity) than any inclined screw configuration (because of the multiplanar planes of screw fixations), and thus it solves the problem of sequential pullout, because the axis of each screw is oriented in different planes (Figure 11.23e). Thus, the helical plate provides much higher holding strength than any straight plate fixation designs.

11.5.3 Hemihelical Plate versus Straight Plate Bone Fracture Fixation—Finite Element Analysis

It can be observed (from Sections 5.1 and 5.2) that helical plating has an improved load-holding capacity, which is necessary to avoid implant loosening (which is more prominent in the osteoporotic bones or less quality bones). It is also necessary to analyze the fracture gap movement characteristics and the assembly (fracture-fixed bone and plate) stiffness of the helical plating with respect to the straight plate. For this purpose, FEA are carried out for fixations of an oblique fracture (angled 45° to the axis of the bone) fixed with a straight plate, a 90° helical plate, and a 180° helical plate under compression, bending, and torsion loadings using ABAQUS finite element program.

11.5.3.1 Modeling of the Fractured Bone Fixation by a Helical Plate

FEA are carried out on the bone–plate–screw assembly, for (1) bone with screw holes modeled as hollow cylinders (of outer diameter 24 mm, inner diameter 16 mm, length 170 mm, and eight screw holes), (2) the eight screws considered to be cylinders of diameter 3.5 mm (in order to reduce the number of elements required for screws meshing, as screw threads require finer elements for meshing), and (3) different plate configurations (i.e., straight plate, 90° helical plate, and 180° helical plate). The plate dimensions are 12 mm width, 4 mm thick, 140 mm effective length, while the radius of the helix is 12 mm. In order to "produce" a helical plate, the straight plate is curved, such that it fits exactly the outer diameter of the bone. The distance between the screw holes is equal in all the plate configurations.

All three configurations of fixation (i.e., straight plate, 90° helical plate, and 180° helical plate) are modeled by commercial computer aided design (CAD) software UNIGRAPHICS. The corresponding FE models are depicted in Figure 11.27a through c. The modeled parts (i.e., plate, bone, and screws) are imported into the commercial FEA software ABAQUS from UNIGRAPHICS through a standard CAD translator format STEP. After

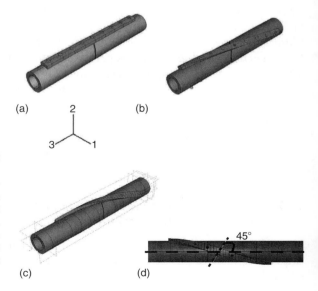

(a) 2 (b)

3 1

FIGURE 11.27 (See color insert following page 266.)
Finite element models in ABAQUS: (a) straight plate model, (b) 90° helical plate model, (c) 180° helical plate model, and (d) oblique fracture fixed by a helical plate. The bone axis (depicted in the figure) is along coordinate 3.

45°

(c) (d)

importing these parts into ABAQUS, the bone is modified to incorporate the oblique fracture gap of 2 mm at its midspan, angled 45° to the axis of bone (as shown in Figure 11.27d). The screws are positioned such that they are always perpendicular to the top surface of the plate; in this way, the screws are in different planes incorporating the bone axis (as previously illustrated by Figure 11.23).

The bone is modeled as a transversely isotropic material, with moduli and Poisson's ratios of $E_1 = 14.5$ GPa, $E_2 = 14.5$ GPa, $E_3 = 19.7$ GPa, $G_{12} = 7.0$ GPa, $G_{13} = 7.0$ GPa, $G_{23} = 5.28$ GPa, $v_{12} = 0.285$ GPa, $v_{13} = 0.285$ GPa, and $v_{23} = 0.26515$ GPa (the directions 1, 2, 3 are shown in Figure 11.28) [16]. The plate and screw materials are assumed to be in elastic state for the applied loading, and assigned 200 GPa Young's modulus (316L stainless steel) and 0.3 Poisson's ratio values. For the locking-screw mechanism during analysis, the screw head and the contoured surface of the plate are tied together by means of a contact option available in ABAQUS. Similar contact conditions are assigned for the cylindrical screw and bone, so that the screw holds the bone during loading. A coefficient of friction of 0.37 is assigned between the bone and the plate contact, and a value of 1 between the broken bone fragments (i.e., at fracture interface). The analysis is carried out using finite strain–displacement relation.

11.5.3.2 Loading and Boundary Conditions Imposed on Fracture-Fixed Bone

Compressive, torsion, and bending loadings are applied on obliquely fractured bones fixed by straight plate, 90° helical plate, and 180° helical plate. In order to apply a compression load of 150 N, one end of the fracture-fixed bone is fully constrained and a compressive force of 150 N in the axial (U3) direction is applied on the reference plane (as the loading condition) that is fixed to the free end of the bone (as shown in Figure 11.28a). A four-point bending loading state has been simulated, by applying a normal displacement (U2) of 0.15 mm at the inner supports, located at 10 mm of the outer simple supported ends, as shown in Figure 11.28b. For the torsional loading the condition on the reference plane is modified to make the displacement UR3 = −0.05 rad to simulate the torsional load, while the boundary conditions (of the fixed end) are maintained (Figure 11.28c).

11.5.3.3 Results and Discussion

11.5.3.3.1 Stiffness of the Assembly

In both compression and bending, it is observed that the stiffness of the fracture fixation (slope of the load deflection curves for the fracture-fixed assembly) is lowest for 180° helical plate, second lowest for 90° helical plate, and highest for straight plate, as shown in Figure 11.29a and b. In other words, it is minimum for 180° helical plate and maximum for straight plate. In torsion, the stiffness is lowest for straight plate, second

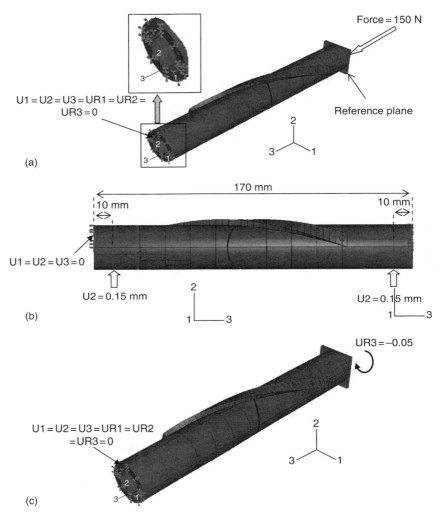

FIGURE 11.28 (See color insert following page 266.)
Loading and boundary conditions applied on the 180° helical plate fixation on the simulated fractured bone (with 45° oblique fracture): (a) compressive load, (b) bending load, and (c) torsional load. Similar loading conditions are applied on the straight plate and the 90° helical plate fixations. Here U1, U2, and U3 represent displacements along axes 1, 2, and 3, respectively; UR1, UR2, and UR3 represent rotations about axes 1, 2, and 3, respectively.

lowest for 90° helical plate, and highest for 180° helical plate, as indicated in Figure 11.29c. This means that with the increase in the degree of contouring in helical plating, the compression and flexural stiffnesses reduce, whereas the torsional stiffness increases. Hence, the helical plate fixation makes the assembly flexible in compression and bending loading conditions, while providing maximum torsional stiffness.

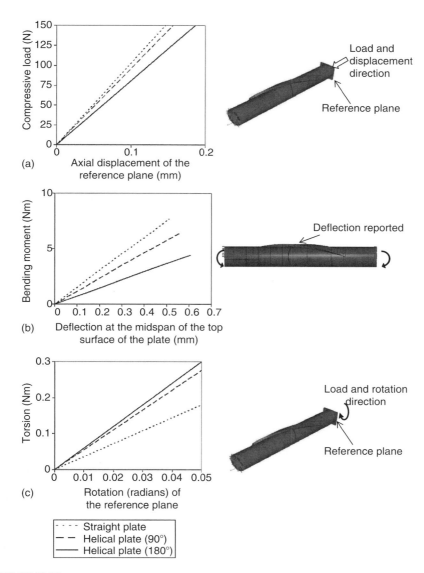

FIGURE 11.29

Comparison of stiffness of bone–plate assembly: (a) compressive load versus deflection of the reference plane (depicted in the figure); (b) bending moment versus deflection at the midspan of the top surface of the plate; and (c) torsion versus rotation of the reference plane (depicted in the figure). Helical plate fixations offer less stiffness than the straight plate fixation in compression and bending loadings. In torsional loading helical plate fixation provides maximum stiffness. Oblique fractures produced by torsion have been a big concern for fixation by straight plates. Our helical plate provides a solution for this long-standing problem.

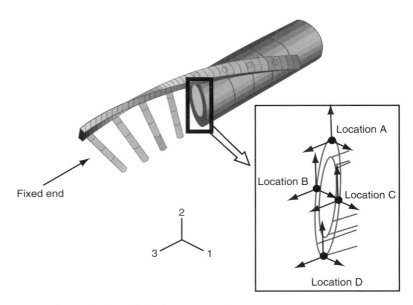

FIGURE 11.30 (See color insert following page 266.)
Different locations considered in the finite element analyses (FEA), for computing fracture gap movement. Location A is on the bone on the fracture gap and underneath the plate. Location D is on the bone at fracture gap but on the opposite side of the plate. Locations B and C are on the bone on the fracture gap and between locations A and D. *Note:* For better presentation of the fracture gap, the left bone fragment was made invisible.

11.5.3.3.2   Relative Fracture Gap Movement

Figure 11.30 illustrates the bone cross section at the fracture site. From the FEA, we have computed the relative movements (i.e., the combined movements of both the fracture fragments) at the fracture gap along the three axes (1, 2, and 3) at all the locations A. B, C, D (in Figure 11.30), for the bone–plate fracture fixation (with straight plate, 90° helical plate, and 180° helical plate) under axial, bending, and torsion loadings as depicted in Figures 11.31, 11.32, and 11.33, respectively.

It is observed that the fracture gap movement (or closure) along axis 3 (at locations A, B, C, and D) is maximum for 180° helical plate fixation, followed by 90° helical plate, and then by the straight plate in compression and bending (Figure 11.31). Hence, it is perceived that enhanced gap closure along the fracture site can be achieved by means of the helical plating, which also enhances the bone healing [8,17,18]. Further, it is noted that the gap closure becomes more uniform with the increase in the degree of contouring in helical plate, for all the loading conditions considered in this study. For torsional loading, it is noted from Figure 11.31c that only helical plate fixation provides gap closure, while the straight plate fixation opens up the gap at the fracture site.

Now, let us consider the lateral displacements at the fracture site, as shown in Figures 11.32 and 11.33. The lateral movement could eventually

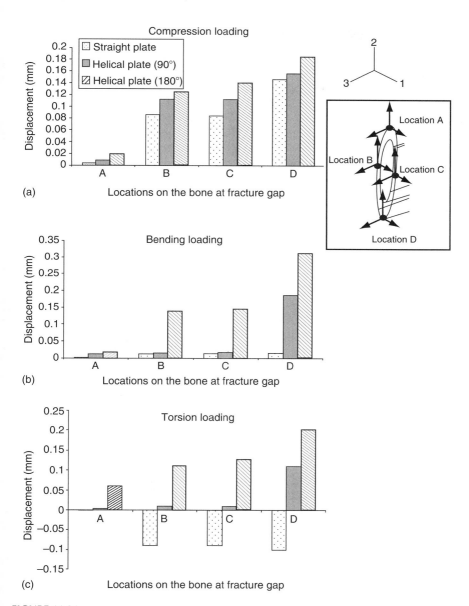

FIGURE 11.31

Relative movement along axis 3 at A, B, C, and D locations on the fracture gap (a) for compressive load, (b) for bending moment, and (c) for torsional load. Fracture gap closure is maximum in the 180° helical plate. It is seen that the fracture gap closes for all the plates in compression and bending loadings. However, in the case of torsion, the fracture gap closes for the helical plate only and opens up for the straight plate. "+" displacement means gap closure, "−" displacement means opening up.

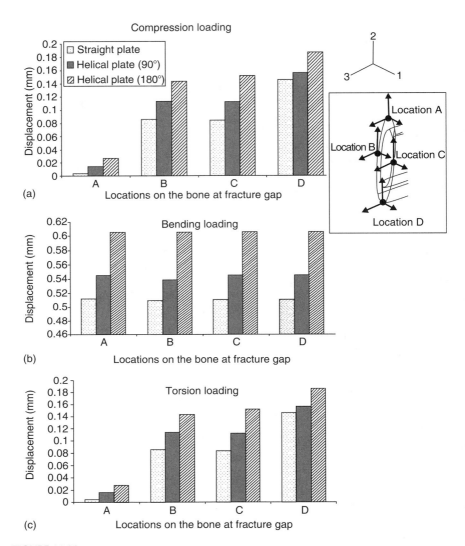

FIGURE 11.32
Relative movement along axis 2 at A, B, C, and D locations on the fracture gap (a) for compressive load, (b) for bending moment, and (c) for torsional load. The movements in helical plate along axis 2 are maximum in all the loading conditions indicating that the shear is maximum in helical plate along with the maximum gap closure.

lead to contact of the fractured surfaces, while causing compression and shear at the fracture gap. Whereas compression is definitely conducive to bone healing, studies on the role of shear at the fracture gap is somewhat controversial [19]. According to Russel [20], shear at the fracture gap delays healing, whereas Park et al. [19] report that shear at the fracture gap enhances the fracture healing. More recently, use of semirigid fixation

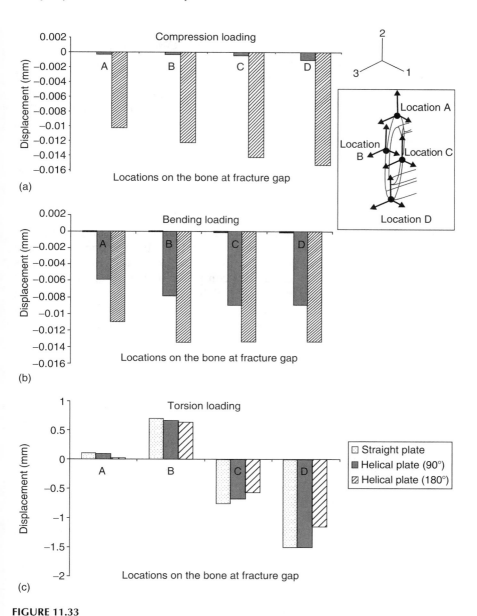

FIGURE 11.33
Relative movement (in shear) along axis 1 at A, B, C, and D locations on the fracture gap (a) for compressive load, (b) for bending moment, and (c) for torsional load. It is seen that the 180° helical plate has minimal shear displacement compared to the straight plate and 90° helical plate.

with increased shear interfragmentary movement has led to an increased amount of soft tissue formation and prolonged bone formation in the in vivo experiments conducted by Schell et al. [21]. As regards shear displacements along axes 1 and 2, the helical plate has more displacement compared to the straight plate, at all the four locations in uniaxial compression and bending loading conditions. Hence, the helical plate should not be used for transverse fractures and for bones subjected to considerable compression and bending. The helical plate fixation has less shear displacement along axis 1 compared to the straight plate fixation, and hence helical plate fixation should be preferably employed for oblique fractures in bones subjected to considerable torsional loading.

In conclusion (1) the axial compressive movement at the fracture gap (along axis 3, promoting gap closure) is greater for helical plate fixation than for straight plate fixation, and it increases with the degree of contouring, (2) the lateral (shear) displacements at the fracture gap (along axes 1 and 2) are generally greater for helical plate fixation than for straight plate fixation, and they are in the "smallness range" that is conducive to healing [19].

Hence, based on fracture gap movements (for oblique fractures of the bone), it can be concluded that helical plate fixation yields gap movements that are more conducive to healing (for compression, bending, and torsional loadings), when compared to straight plate fixation. This means that the callus formed during the healing phase will be subjected to beneficial compressive and shear stresses (at all locations shown in Figure 11.30) for helical plate fixation, which will induce consolidation of callus into bone and later aid remodeling of the bone.

11.5.3.3.3   *Stresses on the Plate and Screws*

The finite element contours of Von Mises stress in the plate reveal that in all the loading conditions, the plate is highly stressed near the screw hole at the fracture site (stressed approximately eight times more than the stresses at the extreme ends of the plate; typical results are summarized in Table 11.7). Hence, the plate failure (if it initiates) can be expected to take place at the screw hole near the fracture site.

For all the fixations, the Von Mises effective stress in the screw follows a typical trend: under compression and torsional loadings, the screws nearest to the fracture site are stressed more than the furthest screws, while in bending, the screws furthest from the fracture site are stressed more than the nearest screws. The obtained maximum stresses in the screws are tabulated in Table 11.8. It is to be noted that in the case of helical plate fixation, the screws are more stressed in torsion compared to the straight plate fixation; this may be taken to represent the price for higher holding capacity of the helical plate.

11.5.3.3.4   *Stress Shielding*

As regards stress shielding of the bone, the location of the NA for bending moment loading indicates the stress shielding offered by the plate on the bone. As discussed earlier, the optimal fracture fixation will not allow the fracture site to be in tension, i.e., the NA should at most be at the

TABLE 11.7

Stresses on Plate at the Center of the Plate (at the Fracture Site) and Extreme End of the Plate (i.e., Furthest from the Fracture Site)

Loading →	Straight Plate			Helical Plate (90°)			Helical Plate (180°)		
	Compression (MPa)	Bending (MPa)	Torsion (MPa)	Compression (MPa)	Bending (MPa)	Torsion (MPa)	Compression (MPa)	Bending (MPa)	Torsion (MPa)
Center of the plate	192.41	381.29	364.37	189.49	350.96	369.74	180.31	340.56	372.43
Furthest from the fracture	27.82	46.46	48.52	23.04	49.29	43.41	23.07	43.03	52.39

TABLE 11.8

Von Mises Effective Stresses on Furthest and Closest Screws from Fracture Fixation Configurations

Loading Type →	Straight Plate			Helical Plate (90°)			Helical Plate (180°)		
	Compression (MPa)	Bending (MPa)	Torsion (MPa)	Compression (MPa)	Bending (MPa)	Torsion (MPa)	Compression (MPa)	Bending (MPa)	Torsion (MPa)
Furthest screw from fracture	59.12	198.32	19.74	51.36	195.53	28.31	49.94	193.42	33.21
Closest screw from fracture	176.36	65.16	281.24	205.44	60.17	291.83	150.83	60.45	302.18

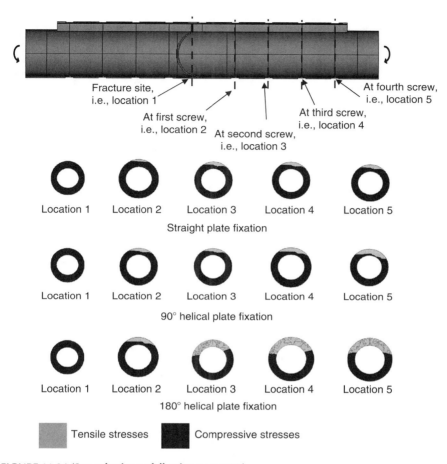

FIGURE 11.34 (See color insert following page 266.)
Locations of the neutral axis (NA) on the bone cross section at different places along the length of the bone for straight plate, 90° helical plate, and 180° helical plate fixations subjected to bending moment. The change of color from grey to black represents the NA. (a) Locations considered for the NA along the length of the bone; (b) cross sections of bone showing NA at different locations along the length of the bone fixed by the straight plate; (c) cross sections of bone showing NA at different locations along the length of the bone fixed by the 90° helical plate; and (d) cross sections of bone showing NA at different locations along the length of the bone fixed by the 180° helical plate.

plate–bone interface. On the other hand, away from the fracture interface, the NA should be located into the bone, so that the bone also bears tensile stress. Figure 11.34 depicts the location of the NA on the bone cross section at the fracture site, and at the sites of the first, second, third, and fourth screws for straight plate, 90° helical plate, and 180° helical plate fixations. The location of NA is also tabulated in Table 11.9, from which it is noted that the NA shift (away from fracture interface) into the bone for helical plate fixation is more than that for the straight plate. Further, the amount of the

TABLE 11.9

Location of the Neutral Axis at Different Sites along the Length of the Bone

Fixation Type	Computed Values (for the Bone) of the NA Location	At Fracture Site	At Cross Section at First Screw	At Cross Section at Second Screw	At Cross Section at Third Screw	At Cross Section at Fourth Screw
Straight plate	Neutral axis from top of the bone (mm)	Located inside the plate	1.5	2	2	2.5
90° helical plate	Neutral axis from top of the bone (mm)	Located inside the plate	2	2.6	3.2	4
180° helical plate	Neutral axis from top of the bone (mm)	Located inside the plate	2.8	4.3	5.5	6.8

shift in NA axis is also a function of the degree of contouring of helical plate. Thus, the helical plate offers less (and hence beneficial) stress shielding compared to the straight plate.

11.5.3.3.5   *Remarks on Helical Plate Fixation*

The advantages offered by helical plates are that:

1. Nearly uniform fracture gap closure, beneficial to healing, is obtained at all locations of the fracture interface for all the loading conditions and the uniformity of fracture gap closure improves with the increased contouring of the plate, i.e., increasing the pitch.

2. In bending moment loading, the NA is located inside the plate at the fracture site (similar to that of straight plate). Away from the fracture site, due to the helical shape of the plate, the NA shifts into the bone (as shown in the Figure 11.34) and can hence allow the bone to take on both normal tensile and compressive stresses; also the location of NA is dependent on the degree of contouring of the plate.

3. In torsional loading, the bone elements are subjected to tensile stresses on the long diagonal planes. Because the axis of helical plate will be parallel to the orientation of the tensile stress (as schematically shown in Figure 11.35), the helical plate will absorb the tensile stresses caused by the torsional loading; this is not the case with straight plate fixation, where the tensile stresses open up oblique fracture gaps. Thus, the helical plate can be optimally employed for treating spiral fractures.

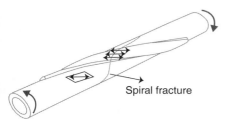

Spiral fracture

FIGURE 11.35
Schematic of a fracture-fixed bone by a 180°
helical plate under torsion.

4. From the clinical point of view, helical plates have freedom at the entry point, during minimally invasive surgery according to Fernandez [11].

5. Screw-holding power for helical plate fixation is higher (compared to that of the straight plate), as the screws are inclined at different orientations (as indicated in Table 11.6), thus avoiding sequential screw loosening.

Apart from the advantages that helical plate fixation offers, helical plates are not commercially available. Thus, currently, we need to contour a straight plate into a helical plate as a temporary solution. As discussed earlier, the procedure for contouring a straight plate to a helical plate is as follows: the straight plate is bent to form a semicircular sector of a ring, and then its ends are twisted in opposite directions to form a helical plate.

However, the repeatability of contouring the straight plate into a helical plate is questionable. During contouring, excessive screw holes deformation could take place and hence hamper easy insertion of the screws during surgery. The situation is even worse if we contour the locking compression plate (LCP), because the locking mechanism will be damaged due to excessive deformation at screw hole during contouring. Residual stresses can also develop in the helical plate while contouring, and this can have an impact on the fatigue property of the helical plate. Similarly, excessively deformed screw hole acts as a stress raiser, and the plate will tend to fail at the screw hole during contouring itself. Thus, contouring the plates should be replaced by manufacturing helical plates to near net shape.

While this study demonstrates the advantages offered by helical plate fixations, more preferred anatomical locations for helical plating need to be explored. Thus, our experimental and FEA have opened a new arena of study in modern fracture fixation through helical plates.

11.6 Conclusion

In this chapter, we have carried out the analyses of plate fixation of the fractured bone under axial and bending moment loadings.

In the case of the plate screwed to the bone and compression loading of the bone–plate assembly, the inner screw is found to transfer the majority of the force from one half of the fractured bone to the other half (as depicted in Figures 11.2 and 11.3).

Next, the bone–plate assembly under pure bending is analyzed. Initially the plate is assumed to be bonded to the bone, and we have determined the normal stresses acting on bone and plate, as influenced by their moments of inertia and moduli. For these stresses, the forces and moments shared by the bone and the plate (as depicted in Figure 11.4) are calculated. These forces and moments are then applied to the bone–plate assembly (Figure 11.5), in which the plate is screwed to the bone by means of two screws on either side of the fracture site. In order to analyze this case (1) the flexure-bending moment equations for the bone and the plate are worked out separately; (2) the boundary condition of zero deflection and slope at the extreme ends held by the two outer screws are applied to determine the expressions for the bone and the plate deflections; and (3) the bone and the plate deflections at the site of the inner screw locations are equated, in order to determine the value of the force W in the screws.

A similar procedure was adopted for the case of three screws on either side of the fracture site. In both the cases of two- and three-screw fixation modes, it was found that for the NA to be located at the plate–bone interface, the plate should have an elastic modulus of 210 GPa and thickness of 3 mm.

Further investigation to reduce the stress shielding has also been studied by means of stiffness-graded plates, in which the stiffness was varied both across the plate thickness and plate length. Analyses were carried out for the cases of (1) the plate screwed to the bone by means of one, two, and three screws on either side of the fracture site, as depicted in Figure 11.10; (2) for the callus modulus to be equal to 1% of the bone modulus (21 GPa) and 100% of the bone modulus at the fracture site, in order to simulate initial and total fracture healing phases; and (3) the modulus of plate varying from 210 to 21 GPa both longitudinally and transversely in the plate.

It is seen that the screw locations from the fracture site (as depicted in Figure 11.10) have a greater bearing on the stress shielding than the stiffness variation of the plate. From Table 11.4, it is seen that fixation modality S4-2 (with two screws on either side of the fracture site, one being close to the edge of the plate) is optimal, in terms of (1) yielding minimal deflection at the fracture site, (2) providing adequately high maximum compressive stress at the callus, and (3) providing adequately low stress shielding away from the fracture site.

Finally, we have demonstrated the efficacy of using helical plate fixation for improved osteosynthesis of fractured bone, especially for oblique and spiral fractures, and particularly for torsional loading. The advantages offered by helical plate fixation (relative to straight plate fixation) are in terms of greater holding strength, greater energy to fracture, and higher screw pullout strength. Further, based on our FEA of plate–bone assembly (under compression, bending, and torsional loadings), for oblique fractures of the bone, the helical plate fixation is shown to yield (1) greater axial displacement and fracture gap closure (conducive to callus consolidation into bone) as well as (2) greater shear displacements (in the magnitude range beneficial to fractured bone healing) compared to straight plate fixation. In particular, for torsional loading and for oblique (and spiral) fractures of the bone, it is more beneficial to use helical plate fixation for all of the above-mentioned reasons.

References

1. Ramakrishna, K., Sridhar, I., Sivashankar, S., Khong, K.S., and Ghista, D.N. 2004. Design of fracture fixation plate for necessary and sufficient bone stress shielding, *Japanese Society of Mechanical Engineers*, 47c(4):1086–1094.
2. Ramakrishna, K., Sridhar, I., Sivashanker, S., Ganesh, V.K., and Ghista, D.N. 2005. Analysis of internal fixation of a long bone, *Journal of Mechanics in Medicine and Biology*, 5(1):89–103.
3. Albright, J.A., Johnson, T.R., and Saha, S. 1978. Principles of internal fixation, in *Orthopaedic Mechanics: Procedures and Devices*, eds. D.N. Ghista and R. Roaf, Academic Press, New York.
4. David, M.N. and Perren, S.M. 1976. Force measurements in screw fixation, *Journal of Biomechanics*, 9(11):669–675.
5. Cordey, J., Borgeaud, M., and Perren, S.M. 2000. Force transfer between the plate and the bone: Relative importance of the bending stiffness of the screws and the friction between plate and bone, *Injury, International Journal of Care Injured*, 31 (S-C):21–28.
6. Deluca, P.A., Lindsey, R.W., and Ruwe, P.A. 1988. Refracture of bones of the forearm after the removal of compression plates, *Journal of Bone and Joint Surgery*, 70(9):1372–1376.
7. Panagiotopoulos, E., Fortis, A.P., Millis, Z., Lambiris, E., Kostopoulos, V., and Vellios, L. 1994. Pattern of screw loosening in fractures fixed with conventional and functional plates, *Injury, International Journal of Care Injured*, 25(8):515–517.
8. Perren, S.M. 2002. Evolution of the internal fixation of long bone fractures: The scientific basis of biological internal fixation; choosing a new balance between stability and biology, *The Journal of Bone and Joint Surgery*, 84B(8): 1093–1110.
9. Ganesh, V.K., Ramakrishna, K., and Ghista, D.N. 2005. Biomechanics of bone-fracture fixation by stiffness-graded plates in comparison with stainless-steel plates, *Biomedical Engineering Online*, 4:46.
10. Fujihara, K., Haung, Z., Ramakrishna, S., Satknanantham, K., and Hamada, H. 2003. Performance study of braided carbon/PEEK composite compression bone plates, *Journal of Biomaterials*, 24(15):2661–2667.
11. Fernandez, A.A.D. 2002. The principle of helical implants: Unusual ideas worth considering, *Injury, International Journal of Care Injured*, 33(S-A):1–27.
12. Gardner, M.J., Griffith, M.H., and Lorich, D.G. 2005. Helical plating of the proximal humerus, *Injury, International Journal of Care Injured*, 6(10):1197–1200.
13. Apivatthakakul, T., Arpornchayanon, O., and Bavornratanavech, B. 2005. Minimally invasive plate osteosynthesis (MIPO) of the humeral shaft fracture.

Is it possible? A cadaveric study and preliminary report, *Injury, International Journal of Care Injured*, 36(4):530–538.

14. Yang, K. 2005. Helical plate fixation for treatment of comminuted fractures of the proximal and middle one-third of the humerus, *Injury, International Journal of Care Injured*, 36(1):75–80.

15. Ramakrishna, K., Sivashanker, S., Sridhar, I., Khong, K.S., and Ghista, D.N. Biomechanics of Helical Plate. *Proceedings of 3rd Conference on Biomechanics Organized by The International Association of Science and Technology for Development (IASTED)*, Spain, September 2005, pp. 220–223.

16. Ferguson, S.J., Wyss, U.P., and Pichora, D.R. 1996. Finite element stress analysis of a hybrid fracture fixation plate, *Medical Engineering and Physics*, 18(3):241–250.

17. Tencer, A.F. and Johnson, K.D. 1994. *Biomechanics in Orthopaedic Trauma, Bone Fracture and Fixation*, Martin Dunitz, London.

18. Lutz, E.C., Christa, A.H., Cornelia, N.W., Daniela, K., Walter, S., Kristen, J.M., and Augat, P. 1998. Effects of mechanical factors on the fracture healing process, *Clinical Orthopaedics and Related Research*, 355S:132–147.

19. Park, S.H., Kim, C., Harry, M., and Augusto, S. 1998. The influence of active shear or compressive motion on fracture-healing, *The Journal of Bone and Joint Surgery*, 80A:868–878.

20. Russel, T.A. 1992. General principles of fracture fixation, in *Chambells Operative Treatments*, ed. A.H. Crenshaw, Vol. 2, pp. 725–784.

21. Schell, H., Epari, D.R., Kassi, J.P., Bragulla, H., Bail, H.J., and Duda, G.N. 2005. The course of bone healing is influenced by the initial shear fixation stability, *Journal of Orthopaedic Research*, 23(5):1022–1028.

12

Human Lumbar Vertebral Body: Analysis of Its Functionally Optimal Structure

Dhanjoo N. Ghista, Sridhar Idapalapati, and Ramakrishna Kotlanka

CONTENTS

12.1 Introduction and Scope

Spine gives the body structure, support, and allows the body to bend with flexibility. It is also designed to protect the spinal cord. The spine is made up of 24 small bones (vertebrae) that are stacked on top of each other to create the spinal column. Between each vertebra, there is an intervertebral disk that helps to cushion and transmit the load between the vertebrae and keeps the vertebrae from rubbing against each other [1]. The flexibility of the spine is primarily due to the intervertebral disks [2]. Each vertebra is held to the others by groups of ligaments. There are also tendons that fasten muscles to the vertebrae. The normal spine has an "S"-like curve when looking at it from the lateral side. The "S" curve must have evolved to help a healthy spine to perform its role in providing stability, strength, and flexibility [3,4]. Natural structures usually evolve with larger cross section, where stresses are maximum, and leaner cross section, where stresses are minimum, thereby attaining minimum weight. Spinal biomechanical efficacy is to a large extent based on the optimal intrinsic designs of the spinal vertebral body (VB).

In the VB, the load-carrying and transmitting function is primarily done by the cortical VB, whose shape resembles a hyperboloid (HP) shell. We have hence modeled the cortical VB as an HP shell, whose geometry and composition is made up of its generators. This chapter analyzes the forces in the VB generators due to compression, bending, and torsional loadings. The unique feature of the HP geometry is that all the loadings are transmitted as axial forces in the generators of VB HP shell. This makes the VB a high-strength structure. Further, because the cortical VB material is intrinsically made up of its generators (through which all the loadings are transmitted axially), it also makes the VB a natural lightweight structure.

We then analyze for the optimal HP shape and geometry by minimizing the sum of the forces in the HP VB generators (due to its loadings) with respect to the HP shape parameter (angle β between pairs of generators). The value of β is determined to be 26.5°, which closely matches with the in vivo geometry of the VB based on the magnetic resonance imaging (MRI). In other words, for the HP shape parameter $\beta = 26.5°$, the VB generators' forces (under the combined loadings acting on the VB) are minimal, so as to then enable it to bear maximal amounts of loadings. In this way, we have demonstrated that the VB is an intrinsically functionally optimal structure. This chapter (along with Figures 12.2–12.10) is based on our paper on the Human lumbar VB as an intrinsic functionally optimal structure [5].*

12.2 Introduction: Concept of the Spinal Vertebral Body Being an Intrinsically Optimal Structure

In nature, anatomical structures are customized to be functionally optimal [6,7]. If it is a load-bearing structure, then it is adroitly designed to be a lightweight and high-strength structure. For example, a long bone is modeled

---

* With permission from the publisher WIT Press, Southampton, U.K.

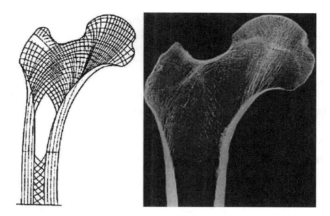

FIGURE 12.1
Wolff's drawing of trabecular orientation in the proximal part of the femur and the cross section of the femur. It is noted that the bone material distribution corresponds to the orientation of the stress trajectories. (Adapted from Wolff, J., *Das Gesetz der Transformation der Knochen*, Hirschwald, 1892.)

such that it can sustain maximum loading with least amount of material. Consider the case of the femur. Its shape and material density correspond to its stress trajectories under its functional loading (see Figure 12.1) as per Wolff's law [8]. In other words, there needs to be less density of bone where the stress trajectories are apart (such as in trabecular bone) and more density of bone where the stress trajectories are closer (as in cortical bone).

12.2.1 Optimal Dimensions of the Femur Cortical Bone

The diaphysial part of long bone (e.g., a femur) carries most of the bending stress. Using the Euler–Bernoulli flexural equation, its bending strength (BS), normalized with respect to its bending-moment bearing capacity, can be defined as

$$\text{BS} \equiv \frac{4M}{\pi\sigma_b r_e^3} = 1 - \left(\frac{r_i}{r_e}\right)^4 \tag{12.1}$$

where
  M is the moment on the bone
  σ_b is the maximum stress on the bone induced due to M
  r_e is the external radius of the bone
  r_i is the internal radius of the bone (considering long bone to be a hollow cylinder)

Also, its normalized weight factor (WF, i.e., normalized weight per unit length) can be represented as

$$\text{WF} \equiv \frac{\pi(r_e^2 - r_i^2)}{\pi r_e^2} = 1 - \left(\frac{r_i}{r_e}\right)^2 \tag{12.2}$$

Maximizing the function (BS − WF) with respect to (r_i/r_e), the optimal value of (r_i/r_e) is found to be $1/\sqrt{2}(= 0.707)$, at which long bone has minimum weight and maximum bending strength; the corresponding area ratio (inner to outer cross-sectional area) is 0.5. Based on our measurements of femur diaphysial cross sections, this area ratio is in the range 0.5 ± 0.2. This analysis shows that at $r_i/r_e = 0.707$, the long diaphysial bone segment has maximum bending stiffness for minimal weight.

12.2.2 Spinal Vertebral Body as an Optimal Structure

In this section, we will analyze the VB structural design by carrying out the stress analysis of how the cortical VB can bear uniaxial compression, bending, and torsional loadings. Then, the relationship between the dimensions of VB (based on physiological loading conditions) that makes it to be a functionally optimal (lightweight and high-strength) structure is analyzed. In other words, we will provide the relationship between the geometrical parameters of the VB that makes it an intrinsically optimal structure. Finally, the optimal design parameters obtained from the analysis are compared with published MRI scans of VB.

12.3   Vertebral Body Shape and Membrane Stresses

12.3.1   Hyperboloid Geometry of the Vertebral Body

The HP anatomy and geometry of the cortical VB is formed by two families of generators [9–11], as shown in Figure 12.2a. We will analyze, using shell membrane theory, how this HP VB geometry enables the VB to efficiently sustain (1) compressive loading "C" on the VB, to cause axial compression in both sets of generators, (2) bending moment "M," to result in compressive forces in one set of generators (i.e., on the compression side of the neutral axis) and tensile forces in other set of generators, and (3) torsional loading "T," to result in compressive forces (per unit length) in one family of generators and tensile forces in the other family of generators oriented in the other direction.

Figure 12.2b illustrates the HP geometry of the spinal VB. If we intersect the HP shell surface with a vertical plane parallel to the yz plane but at $x = -a$, then the intersecting curves are given by

$$\frac{a^2 + y^2}{a^2} - \frac{z^2}{b^2} = 1 \quad \text{or} \quad z = \pm\left(\frac{b}{a}\right)y \tag{12.3}$$

which have the same slope as the asymptotes. Based on the HP geometry, the HP surface can be generated by a pair of intersecting lines inclined at

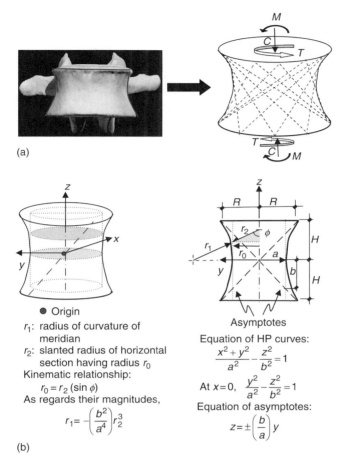

(a)

(b)

- ● Origin
- r_1: radius of curvature of meridian
- r_2: slanted radius of horizontal section having radius r_0

Kinematic relationship:

$$r_0 = r_2 (\sin \phi)$$

As regards their magnitudes,

$$r_1 = -\left(\frac{b^2}{a^4}\right) r_2^3$$

Asymptotes

Equation of HP curves:

$$\frac{x^2 + y^2}{a^2} - \frac{z^2}{b^2} = 1$$

At $x = 0$, $\quad \dfrac{y^2}{a^2} - \dfrac{z^2}{b^2} = 1$

Equation of asymptotes:

$$z = \pm \left(\frac{b}{a}\right) y$$

FIGURE 12.2 (See color insert following page 266.)
(a) Shows a typical vertebral body (VB) where the cortical VB is shaped as a hyperboloid (HP) shell formed of two sets of generators. The height of the HP can be expanded or reduced by the inclination of the generators. (b) Depicts geometry of HP shells. (Adopted from Ghista, D.N., Fan, S.C., Ramakrishna, K., and Sridhar, I., *Int. J. Des. Nat.*, 1, 34, 2006.)

an angle $\beta = \tan^{-1}(a/b)$ in the vertical plane tangent to the waist circle $(r_0 = a)$ [12].

The construction of the cortical VB HP, by a set of generators [13], is illustrated in Figure 12.3, where the end-plate radius AN is R, radius of the waist circle is "a," and height of the VB is $2H$. Based on it, we define the HP VB geometrical parameter β as

$$\tan \beta = \frac{\sqrt{(R^2 - a^2)}}{H} = \frac{a}{b} \tag{12.4}$$

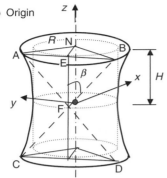

ABCD constitutes a plane tangent to the waist circle at F.

$$\tan \beta = \frac{EB}{FE} = \frac{\sqrt{R^2 - a^2}}{H} \qquad (I)$$

Coordinates of point A are $[-a, \sqrt{R^2 - a^2}, H]$

Substituting into the HP equation: $\dfrac{x^2 + y^2}{a^2} - \dfrac{z^2}{b^2} = 1$

We get $\dfrac{a^2 + R^2 - a^2}{a^2} - \dfrac{H^2}{b^2} = 1$

i.e., $\dfrac{R^2}{a^2} - \dfrac{H^2}{b^2} = 1$ or $b = \dfrac{aH}{\sqrt{R^2 - a^2}}$ (II)

From (I and II) we get: $\tan \beta = \dfrac{\sqrt{R^2 - a^2}}{H} = \dfrac{a}{b}$ (III)

FIGURE 12.3 (See color insert following page 266.)
Geometry of cortical vertebral body (VB) hyperboloid (HP) shell: The generators AD and BC form the basis of the construction of the HP shell. (Adopted from Ghista, D.N., Fan, S.C., Ramakrishna, K., and Sridhar, I., *Int. J. Des. Nat.*, 1, 34, 2006.)

The primary dimensional parameters of the VB HP are hence (R, a, and H), and tan β provides a relationship among them.

12.3.2   Membrane Stresses in the Vertebral Body Cortex

We adopt the membrane theory of shells to analyze the stresses in the cortical VB [14]. Membrane stresses (meridian stress σ_ϕ and hoop stress σ_θ) have a relationship with the normal pressure p_r, as depicted in Figure 12.4. The equilibrium of forces in the radial (r) direction yields

$$-2\sigma_\phi t(r_2 d\theta) \sin\left(\frac{d\phi}{2}\right) + 2\sigma_\theta t(r_1 d\phi) \sin\left(\frac{d\theta}{2}\right)$$
$$+ p_r \left[2r_1 \sin\left(\frac{d\phi}{2}\right) \cdot 2r_2 \sin\left(\frac{d\theta}{2}\right)\right] = 0 \qquad (12.5)$$

where, in the case of an HP, r_2 is considered to be positive and r_1 is considered to be negative, and their magnitudes in terms of a, b, and ϕ are

$$r_1 = \frac{a^2 b^2}{(a^2 \sin^2 \phi - b^2 \cos^2 \phi)^{3/2}} \qquad (12.6)$$

$$r_2 = \frac{a^2}{(a^2 \sin^2 \phi - b^2 \cos^2 \phi)^{1/2}} \qquad (12.7)$$

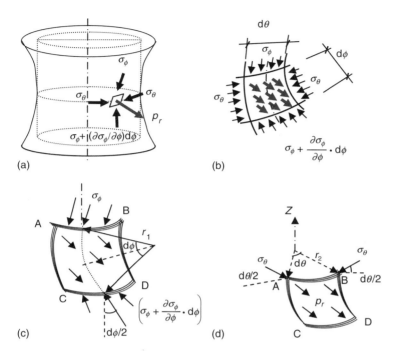

FIGURE 12.4 (See color insert following page 266.)
Stresses acting on an element of the vertebral body (VB) hyperboloid (HP) shell: (a) HP shell element in equilibrium, (b) σ_ϕ and σ_θ equilibrating internal pressure p_r, (c) orientation of σ_ϕ, and (d) orientation of σ_θ. (Adopted from Ghista, D.N., Fan, S.C., Ramakrishna, K., and Sridhar, I., *Int. J. Des. Nat.*, 1, 34, 2006.)

For small angle θ, $\sin \theta \approx \theta$, which leads to

$$-\sigma_\phi t r_2(d\theta)(d\phi) + \sigma_\theta t r_1(d\phi)(d\theta) = -p_r[r_1(d\phi)r_2(d\theta)]$$

or

$$\frac{\sigma_\phi t}{r_1} - \frac{\sigma_\theta t}{r_2} = p_r \qquad (12.8)$$

Denoting $N_\phi = \sigma_\phi t$ and $N_\theta = \sigma_\theta t$ as stresses per unit wall thickness (or stress resultants), with t being the VB wall thickness, we obtain

$$\frac{N_\phi}{r_1} - \frac{N_\theta}{r_2} = p_r \qquad (12.9)$$

which is the "membrane equation" for the HP VB shell. This is because for an HP shell, r_1 is negative and r_2 is positive. Now p_r, the internal pressure in the cancellous bone within the VB cortical shell, is negligible. Hence, by putting $p_r = 0$ (i.e., for an internally nonpressurized cortical VB HP shell), Equation 12.9 results in

$$N_\phi = \left(\frac{r_1}{r_2}\right)N_\theta \qquad (12.10)$$

Substituting $r_1 = (b^2/a^4)r_2^3$, from Figure 12.2, into Equation 12.10, the following relationship between the stress resultants is obtained:

$$N_\phi = \left(\frac{b^2}{a^4}r_2^2\right)N_\theta \qquad (12.11)$$

12.4 Analysis for Forces in the Vertebral Body Generators under Different Loadings

12.4.1 Stress Analysis of the Vertebral Body under Axial Compression

We will now analyze for stresses in the HP shell (generators) due to a uniaxial compressive force, as shown in Figure 12.5. Assume that there are

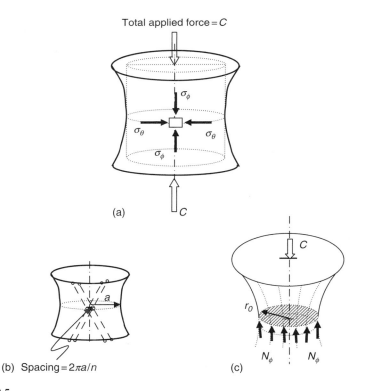

FIGURE 12.5
Stresses at waist section of the hyperboloid (HP) shell: (a) stress components, (b) equivalent straight bars (aligned with the generators) placed at equal spacing to take up the stresses, and (c) equilibrium of forces on a shell segment. (Adopted from Ghista, D.N., Fan, S.C., Ramakrishna, K., and Sridhar, I., *Int. J. Des. Nat.*, 1, 34, 2006.)

two sets of "*n*" number of straight bars placed at equal spacing of $2\pi a/n$ measured at the waist circle, to constitute the HP surface, as shown in Figure 12.5b. Due to the axisymmetric nature of the vertical load, no shear stresses are incurred in the shell, i.e., $\sigma_{\phi\theta} = 0$, as in Figure 12.5a. We then delineate a segment of the HP shell, and consider its force equilibrium (as illustrated in Figure 12.5c). Force equilibrium at any horizontal section gives

$$(2\pi r_0)N_\phi(\sin\phi) = C \qquad (12.12)$$

Now, considering the segment at the waist circle where $\phi = 90°$ and $r_2 = r_0 = a$ (throat radius), we get

$$(2\pi a)N_{\phi(\phi=90°)} = C \quad \text{or} \quad N_{\phi(\phi=90°)} = \frac{C}{2\pi a}, \quad \text{compressive} \qquad (12.13)$$

At the waist circle where $r_2 = a$, Equation 12.11 yields

$$N_{\theta(\phi=90°)} = \left(\frac{a^4}{b^2}\frac{1}{r_2^2}\right)N_{\phi(\phi=90°)} = \left(\frac{a^2}{b^2}\right)N_{\phi(\phi=90°)}$$

which, combining with Equation 12.13, leads to

$$N_{\theta(\phi=90°)} = \left(\frac{a^2}{b^2}\right)\frac{C}{2\pi a} = \frac{C}{2\pi a}(\tan\beta)^2 \qquad (12.14)$$

which is compressive in nature.

From Figure 12.6, the equivalent resultant compressive force "F_c" in a fiber generator of the VB HP shell is given by

$$F_c^2 = \left[N_\phi\left(\frac{\pi a}{n}\right)\right]^2 + \left[N_\theta\left(\frac{\pi b}{n}\right)\right]^2 \qquad (12.15)$$

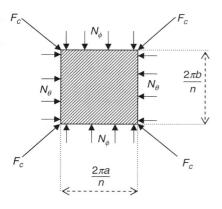

FIGURE 12.6
Equivalent diagonal forces in the intersecting bars to take up the stresses around a shell element. (Adopted from Ghista, D.N., Fan, S.C., Ramakrishna, K., and Sridhar, I., *Int. J. Des. Nat.*, 1, 34, 2006.)

Substituting Equations 12.13 and 12.14 into Equation 12.15, we obtain

$$F_c = \frac{C}{2n \cos \beta} = \frac{C\sqrt{H^2 + R^2 - a^2}}{2nH} \tag{12.16}$$

Thus, the total axial loading is transmitted into the HP shell's straight generators as compressive forces.

12.4.2 Vertebral Body Stress Analysis under Bending Moment

When the VB is subjected to a bending moment (M), normal stresses (σ_y) are developed at the waist circle ($r_0 = a$) cross section, as shown in Figure 12.7. The bending moment sustained at the waist circle is given by

$$M = 2 \int_0^a \sigma_y \left[2\frac{t}{\cos \alpha} dy \right] y \tag{12.17}$$

where σ_y is the compressive stress normal to the cross section (due to the bending moment M) acting on the two rectangular elements of the length $2(t/\cos \alpha)$ and width dy.

Also

$$\sigma_y = \frac{y}{a}(\sigma_a) \tag{12.18}$$

where σ_a is the stress at $y = a$ (at the waist circle).

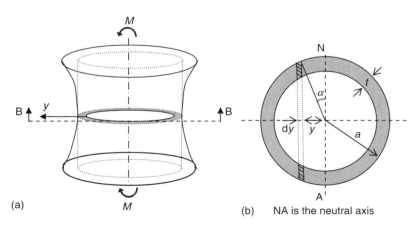

(a)   (b)   NA is the neutral axis

FIGURE 12.7
(a) Bending moment on the vertebral body (VB) and (b) plan view at the section BB (i.e., at waist circle). (Adopted from Ghista, D.N., Fan, S.C., Ramakrishna, K., and Sridhar, I., *Int. J. Des. Nat.*, 1, 34, 2006.)

Upon combining Equations 12.17 and 12.18 we get

$$M = 4t(\sigma_a) \int_0^a \frac{y^2}{a \cos \alpha} dy \tag{12.19}$$

After substituting for $y = a \sin\alpha$ and $dy = a \cos\alpha \, d\alpha$, Equation 12.19 can be rewritten as

$$M = 4t(\sigma_a) \int_0^{\pi/2} \frac{a^2 \sin^2 \alpha}{a \cos \alpha} a \cos \alpha \, d\alpha \tag{12.20}$$

Upon integrating Equation 12.20, we get

$$M = \pi a^2 t(\sigma_a) \tag{12.21}$$

The normal stress at the waist circle in terms of bending moment (M) can be written as

$$(\sigma_a) = \frac{M}{\pi a^2 t} \tag{12.22}$$

Then the stress resultant, N_ϕ, on the waist-circle element at "a" distance from the neutral axis, is given by

$$(N_\phi)_a = (\sigma_a)t$$

Thus, from Equation 12.22, we obtain

$$(N_\phi)_a = \frac{M}{\pi a^2} \tag{12.23}$$

According to Figure 12.8c, the force (F_m) in the VB HP generator is given by

$$F_m^2 = \left[\left(\frac{\pi a}{n} (N_\phi)_a \right)^2 + \left(\frac{\pi b}{n} (N_\theta)_a \right)^2 \right] \tag{12.24}$$

Upon substituting N_θ value from Equation 12.11, we get

$$F_m^2 = \left[\left(\frac{\pi a}{n} (N_\phi)_a \right)^2 + \left(\frac{\pi b}{n} \left(\frac{a^2}{b^2} (N_\phi)_a \right) \right)^2 \right] \tag{12.25}$$

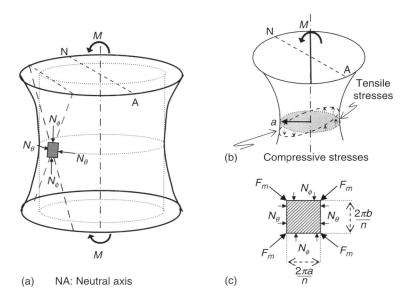

FIGURE 12.8 (See color insert following page 266.)
(a) Stress resultants at the waist section of vertebral body (VB) hyperboloid (HP) shell under bending, (b) equilibrium of forces on a shell segment, and (c) equivalent diagonal forces in the intersecting bars to take up the stresses around a shell element on the compression side of the VB HP shell. (Adopted from Ghista, D.N., Fan, S.C., Ramakrishna, K., and Sridhar, I., *Int. J. Des. Nat.*, 1, 34, 2006.)

Since $\tan \beta = a/b$, Equation 12.25 reduces to

$$F_m^2 = \left(\frac{M}{\pi a^2}\right)^2 \left(\frac{\pi a}{n}\right)^2 [1 + \tan^2 \beta]$$

or

$$F_m = \frac{M}{na \cos \beta} \tag{12.26}$$

where "F_m" can either be compressive or tensile force, based on the location of the generators relative to the plane about which the bending moment is applied.

12.4.3 Vertebral Body Stress Analysis under Torsional Loading

Next, we analyze the compressive and tensile forces in the HP shell generators when the VB is subjected to pure torsion (T). In this case (refer to Figure 12.9a), the normal stress resultants are zero, and we only have the shear stress resultant, as given by

$$N_\phi = N_\theta = 0 \quad \text{and} \quad N_{\phi\theta} = \tau t \tag{12.27}$$

The equilibrium of a segment of the shell at a horizontal section (as in Figure 12.9b) gives

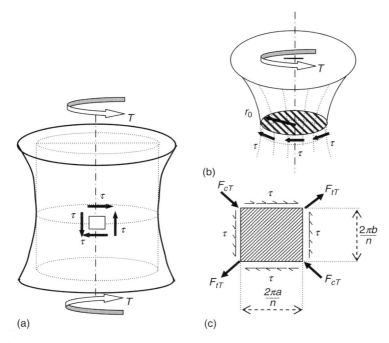

FIGURE 12.9
(a) Stress resultants in the hyperboloid (HP) shell element ($N_\phi = N_\theta = 0$ and $N_{\phi\theta} = \tau\, t$) due to torsion T acting on the vertebral body (VB); (b) equilibrium of a shell segment under torsion (T) and shear stresses (τ) (or shear stress resultant $N_{\phi\theta}$); (c) equivalent diagonal forces in the intersecting generators to take up the stresses around the shell element. (Adopted from Ghista, D.N., Fan, S.C., Ramakrishna, K., and Sridhar, I., *Int. J. Des. Nat.*, 1, 34, 2006.)

$$[(2\pi r_0)N_{\phi\theta}]r_0 = T \quad \text{or} \quad (2\pi r_0^2)N_{\phi\theta} = T \tag{12.28}$$

At the waist circle, $r_2 = r_0 = a$ (throat radius), we obtain for the shear stress resultant

$$[(\tau \cdot t)(2\pi a)]a = T, \quad \text{i.e.,} \quad N_{\phi\theta} = \frac{T}{2\pi a^2} \tag{12.29}$$

Now, consider an element at the waist circle as shown in Figure 12.9c. The equivalent compressive force (F_{cT}) and tensile force (F_{tT}), in the directions aligned to their respective set of shell generators, are given by

$$F_{cT}^2 = F_{tT}^2 = \left(N_{\phi\theta}\frac{\pi a}{n}\right)^2 + \left(N_{\phi\theta}\frac{\pi b}{n}\right)^2$$

or

$$|F_{cT}| = |F_{tT}| = \frac{T}{2na \sin\beta} \tag{12.30}$$

Thus, a torsional loading on the VB HP shell is taken up by one set of generators being in compression and the other set of generators being in tension.

12.5 Optimal Design

12.5.1 Structural Analogy of the Vertebral Body to the Cane Stool

The above analyses illustrate how the intrinsic HP shape design of the VB enables the loadings to be transmitted as axial (compressive/tensile) forces through the generators of the HP shell. In this regard, the VB can be compared to an HP cane stool (shown in Figure 12.10), which is an ideal high-strength and lightweight structure. This is because all the loading exerted on it (by a person sitting on it) is transmitted (to the ground) as axial forces in the cane generators. Now a structure (such as a cane stool) is strongest in compression (provided its length is less than the buckling length). This makes the cane stool a high-strength and high load-bearing structure.

If the two sets of canes (at $\pm\beta$) of the cane stool are encircled at the waist circle by a band, it increases the load-carrying capacity of the cane stool. If additionally, these two sets of canes ($\pm\beta$) are tied at all their intersecting points (as shown in Figure 12.10), their functional lengths are reduced, and this further enhances the strength and load-carrying capacity of the cane stool. Further, the cane stool is very light as it is just made up of discrete canes (as generators of the HP structure). This structural configuration makes the cane stool a very simple but effective load-bearing, high-strength, and lightweight structure. Incidentally, such a cane stool of radius 200 mm and height 175 mm has a nominal weight of 2.5 kg (or 25 N), but can easily bear a load of at least 5000 N, which is 200 times its weight.

FIGURE 12.10
The humble hyperboloid (HP)-shaped cane stool (with permission from www.exbali.com) weighing 2.5 kg but capable of bearing 5000 N in compression. (Adopted from Ghista, D.N., Fan, S.C., Ramakrishna, K., and Sridhar, I., *Int. J. Des. Nat.*, 1, 34, 2006.)

Now, the spinal VB cortex has similar structural configuration and properties as that of the cane stool, to make it an efficient load-bearing and load-transmitting, high-strength, lightweight structure. The VB wall can be deemed to be primarily comprised of the two sets of generators. Just as in the case of cane stool (Figure 12.10), the VB wall transmits all the loading as axial forces through its generators. This is the basis for a high-strength and lightweight VB design.

12.5.2 Optimization of the Hyperboloid Shape of the Vertebral Body

Now the spinal VB has a definitive value range of the HP-shape geometrical parameter β, and hence of its HP shape. In order to determine the structural basis of this β value, we will calculate that value of β that makes the combined axial force in its generators (due to combined compression, bending, and torsional loadings) to be minimum. In that case, this optimized VB structure will be able to sustain maximal loading before the ultimate failure load of its generators is reached.

The VB is subjected to the combined compression, bending moment, and torsional loadings. Under this combined loading, the forces in the generators given by Equations 12.16, 12.26, and 12.30 can be combined using the principle of linear superposition. Hence, for its optimal intrinsic design with respect to the HP shape parameter β, to sustain the combined loadings (i.e., compression from Equation 12.16, bending moment from Equation 12.26, and torsion from Equation 12.30), we need to have

$$\frac{d}{d\beta}[\text{Combined forces in the generators}] = 0 \tag{12.31}$$

Thus, from Equations 12.16, 12.26, and 12.30, we obtain

$$\frac{d}{d\beta}\left(\frac{C}{2n\cos\beta} + \frac{M}{na\cos\beta} + \frac{T}{2na\sin\beta}\right) = 0$$

or

$$\left(\frac{\sqrt{R^2 - a^2}}{H}\right)^3\left(\frac{C}{2} + \frac{M}{a}\right) = \frac{T}{2a} \tag{12.32}$$

Equation 12.32 gives the relationship between the applied loading and the geometry of the VB. For a specific set of values of R and H and functionally occurring ratios of the loading values C, M, and T, the value of the HP shape parameter "a" can be calculated from Equation 12.32 for the intrinsic design of the VB. In Equation 12.32, considering the representative values of $C = 1000$ N, $M = T = 3$ Nm, along with $R = 21.6$ mm and $H = 14.75$ mm, based on Guo et al. [15] and Zhou et al. [11], we obtain $a \approx 20.3$ mm; hence, from Equation 12.4, the value of the parameter $\beta = 26.5°$.

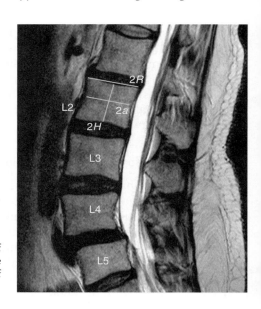

FIGURE 12.11
Magnetic resonance imaging (MRI) of lumbar vertebrae: $H/R = 0.7$ (average of L2 to L5) and a/R is 0.91 (average of L2 to L5).

Hence, the optimal lightweight high-strength spinal VB geometry is given by $\beta = 26.5°$ with $a/R = 0.939$ (for $H = 14.75$ mm). The a/R value of 0.91 measured from the lumbar vertebrae MRI scan, shown in Figure 12.11, confirms our analysis. Thus, the intrinsic design of the VB HP geometry is such that it bears the combined loadings of compression as well as bending and torsion, by minimizing the axial forces in the generators. In other words, it can sustain and transmit maximal values of the loadings with minimal amounts of material (because the entire loading is transmitted as axial forces through the HP generators).

In summary, herein an internal stress analysis of the HP VB under compression, bending moment, and torsion loadings is carried out. The analysis shows that all the loading states are transmitted by the VB HB generators as axial forces, thereby making the VB a high-strength structure with a high load-bearing capacity. Explicit expressions for these axial forces in the VB generators under compression, bending moment, and torsion loading conditions are obtained (in terms of VB geometrical parameters).

Minimization of the total axial force for the combined loadings acting on the VB gives the value of the HP shape parameter $\beta = 26.5°$, for which $a/R = 0.939$, which closely matches the measured value of $a/R = 0.91$ from the VB MRI. Thus, for this value of β, the spinal VB can maximize its load-bearing capacity. We have thereby demonstrated that the VB shape and material distribution are modeled (by the loading sustained by it) to be an optimal high-strength and lightweight structure, in the same way as a femur shape and material distribution are based on the stress trajectories (Figure 12.1) due to the loading sustained by it.

12.6 Overview of Vertebral Body Fixators and Impact of the Intrinsic Design on Better Anterior Fixation

Spinal VB fails if the load exceeds the sustainable limits. This generally happens during impact loading conditions such as car crash or fall from a height. Failure of the VB is very painful to the person, as the fractured VB impinges on the nerve roots and the spinal cord and disrupts the stability* of the spine. VB injury can either cause burst fractures or dislocation of the VB. Burst fractures are more frequent to T12, L1, L2, and L3 of spine VB, and cover up to 66.16% of all the spinal injuries [16]. Burst fractures cause loss of sensory and neural stimulation below the level of the injury, often resulting in paraplegia.

The characteristics of burst fracture (Figure 12.12) are that the VB is partially or completely commuted, with the fragments of the posterior wall retropulsed into the spinal canal causing neural injury. However, the posterior ligamentous complex will still be intact. These fractures are unstable in flexion–compression. As a result, the VB height reduces, and the spinal canal is often extremely narrowed by the protruding posterior wall fragments [16–18].

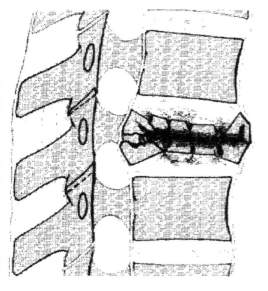

FIGURE 12.12
An axial burst fracture of the vertebral body (VB). (From Aebi, M., Thalgott, J.S., and Webb, J.K., *AO ASIF Principles in Spine Surgery*, Springer, Heidelberg, 1998. With permission.)

---

* A stable spine should be one that can withstand axial forces posteriorly, and rotational forces, thus being able to function to hold the body erect without progressive kyphosis and to protect the spinal cord contents from further injury.

The goals of surgical treatment in burst fracture are to (1) achieve pain free and stable spine, (2) enable neurological recovery, (3) restore the capacity of the spine to withstand physiological loads, (4) cause minimal resection of injured fragments, and (5) employ small implants. The methods available are posterior technique (with posterior fixators), anterior technique (with anterior fixators), or a combination of posterior and anterior techniques. A comparison of the biomechanical efficacies of fixation techniques has been reported by Rezaian and Ghista [17,18], and clinically by Verlaan et al. [19].

Posterior fixators as illustrated in Figure 12.13 stabilize the spine by reinforcing the spine posteriorly and in tension due to bending under torso weight; thereby they increase the fractured spine stiffness, and protect the fractured VB from being overstressed. Distraction applied by the posterior systems generates an anteriorly directed force on the retropulsed fragments (by the posterior ligament complex) to retract the fragments and decompress the spinal cord.

According to Rezaian and Ghista [17,18], as the body weight acts anterior to the spinal column, the spine is subjected to a flexion moment such that the fractured VB is subjected to a compressive force. In the absence of the posterior fixator, the tensile force is not adequately resisted by the posterior spinal column. Hence, the burst VB is subjected to high compressive stresses causing posterior displacement of the fractured fragment, and impingement

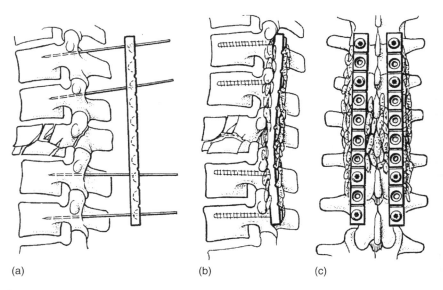

(a)   (b)   (c)

FIGURE 12.13
Posterior fixation technique as per AO ASIF recommendation: (a) the pedicles are entered and Kirschner wires are inserted in each hole; (b) the pates are fitted in position over the Kirschner wires; and (c) later bone graft is used between the vertebrae. (From Aebi, M., Thalgott, J.S., and Webb, J.K., *AO ASIF Principles in Spine Surgery*, Springer, Heidelberg, 1998. With permission.)

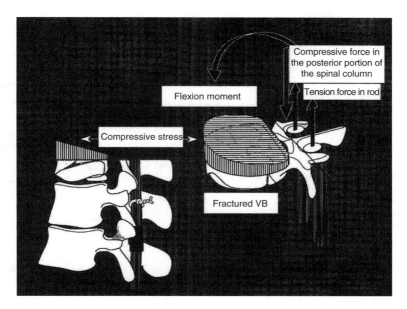

FIGURE 12.14
Forces exerted on a lumbar spinal cross section due to the anteriorly acting weight load. Note that under the flexion bending moment acting on the spinal column the anterior portion of the spinal column is in compression, so that the fractured vertebral body (VB) will be in compression. However, the tensile force in the posterior portion of the spinal column will be reduced by the compression force applied to it by the posterior devices. This will, in turn, reduce the compressive force and stresses on the fractured VB. (From Rezaian, S.M. and Ghista, D.N., *IEEE Eng. Med. Biol.*, 13, 525, 1994. Copyright 1994.)

of the neural structures. The role of the posterior fixator is to bear this tensile force by putting compression on the posterior column. This is because the posterior fixator and fractured VB act as a composite structure (when subjected to flexion bending) with the neutral axis located posteriorly in the spinal column (Figure 12.14).

Even then, the fractured VB is subjected to some compressive stresses and can undergo kyphosis or the VB may not be able to sustain such compressive force. Also, as the posterior system is offset from the spinal column, the amount of tensile stress and strain on this posterior fixator can cause it to fail [20,21]. Thus, a brace may need to be worn in conjunction with posterior fixator. Also, the disadvantage of using posterior fixator (generally spans 5–6 VB segments) is the reduction in the flexibility of the spinal system. These factors make a supporting case for anterior fixation technique.

Anterior fixation technique is used to support the anterior column when instability persists, particularly with the loss of height of VB. A dynamic compression plate (DCP) can be placed anteriorly as shown in Figure 12.15. However, because such a plate will be subjected to bending, loosening of the screws due to poor fixation in the cancellous bone and backing out of screws can be a problem that can cause erosion of aorta and vena cava.

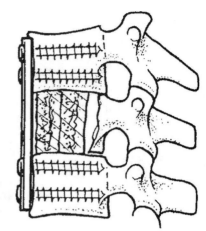

FIGURE 12.15
Anterior placement of the dynamic compression plate
according to AO ASIF. (From Aebi, M., Thalgott, J.S.,
and Webb, J.K., *AO ASIF Principles in Spine Surgery*,
Springer, Heidelberg, 1998. With permission.)

Thus, the anterior fixator placed along the loading axis of the spine seems
to be an appropriate treatment option for burst fractures, like the Rezaian
fixator shown in Figure 12.16. This technique requires the removal of
some of the fractured VB, and two disks (upper and lower portions) of the

FIGURE 12.16 (See color insert following page 266.)
Rezaian spinal fixator is placed along the loading axis of the spine. (From Rezaian, S.M. and
Ghista, D.N., *IEEE Eng. Med. Biol.*, 13, 525, 1994. Copyright 1994.)

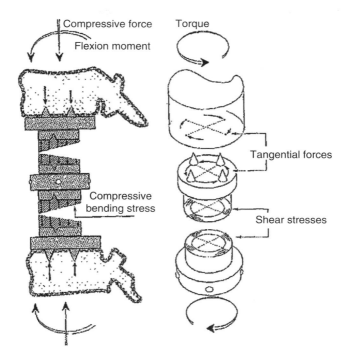

FIGURE 12.17
Rezaian spinal fixator biomechanics. (From Rezaian, S.M. and Ghista, D.N., *IEEE Eng. Med. Biol.*, 13, 525, 1994. Copyright 1994.)

fractured VB. The Rezaian fixator is embedded into the end plates of the VB and allows adjustment of the height by the turnbuckle technique [22]. Figure 12.17 demonstrates the manner in which the Rezaian fixator is simple and at the same time an efficient anterior fixator, which bears and transmits compression, bending moment, and torsional loadings on the spinal column. As illustrated in Figures 12.15 and 12.17, the novelty of this fixator is that all the forces and stresses are transmitted directly through the body of the fixator [17,18], and there is no bending sustained by the fixator.

The Rezaian fixator is fixed to the top and bottom of the VBs by means of four spikes (which form part of the fixator). This makes the VB rest on the fixator and hence directly transmit the forces through the fixator. This is the reason for the high success rate of this fixator, enabling the paraplegic patient to become ambulatory after a few days of hospitalization. The disadvantages are that the spikes may not be sufficient to hold the fixator within the spinal column due to the distraction and rotational forces. There will be some movement at the bone fixator interface, which can progress to a dislocation of the fixator. This problem needs attention.

One of the recent developments in treating burst fractures is the combination of lateral and anterior fixator, where anterior fixator constitutes a

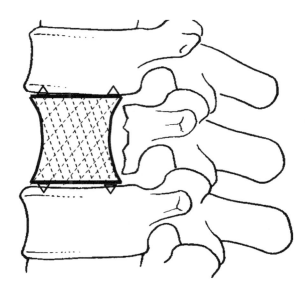

FIGURE 12.18

Schematic of our proposed (2-piece) Vertebral Body Cage (VBG) to hold the fractured vertebral body (VB) together in its original hyperboloid shape. The two components of the hyperboloid VBG will have spikes along their top and bottom edges, which will engage the top and bottom intact vertebral bodies. The two components of the VBG are to be placed separately, to enclose the fractured VB. It is expected that the VBG will help the fractured VB segments to heal and mould into the original hyperboloid shape.

titanium cage as shown in Figure 12.18, and the lateral fixator is a short rod bridging only two VBs [23]. The titanium cage is aligned along the axis of the VB, and due to its hollowness it gives more room for grafting. However, the fixation needs to be secured by additional lateral fixators, as the cage alone is unstable in torsion.

What is conceptually proposed is that a fixator be designed to simulate the cortical VB. It would be made up of two end-plate rings with spikes to fix them into adjacent disks. The rings could then be connected by criss-crossed generators. The fractured VB pieces could be enclosed within this fixator, and even solidified by introducing hydroxyapatite (HA) based polymethylmethacrylate (PMMA) cement.

References

1. Adams, M.A. and Dolan, P. 2005. Spine biomechanics, *Journal of Biomechanics*, 38(10), 1972–1983.
2. Goel, V.K., Grosland, N.M., and Scifert, J. 1997. Biomechanics of the lumbar disc, *Journal of Musculoskeletal Research*, 1(2), 81–94.

3. White, A.A. and Panjabi, M.M. 1990. *Clinical Biomechanics of Spine*, Lippincott: Philadelphia.

4. Levine, A.M. 1992. Lumbar and sacral spine trauma, in *Skeletal Trauma: Fractures, Dislocations, Ligamentous Injuries*, eds. B.D. Browner, Saunders: Philadephia, pp. 805–848.

5. Ghista, D.N., Fan, S.C., Ramakrishna, K., and Sridhar, I. 2006. Human lumbar vertebral body as an intrinsic functionally-optimal structure, *International Journal of Design and Nature*, 1(1), 34–47.

6. Collins, M.W., Hunt, D.G., and Atherton, M.A., 2004. *Optimisation Mechanics in Nature*, WIT Press: Southampton.

7. Collins, M.W., Atherton, M.A., and Bryant, J.A., 2005. *Nature and Design*, WIT Press: Southampton.

8. Wolff, J., *Das Gesetz der Transformation der Knochen*, Hirschwald, 1892. (Translated by P. Maquet and R. Furlong (1986) as *The Law of Bone Remodeling*, Springer: Berlin, 1986).

9. Nissan, M. and Gilad, I. 1986. Dimensions of human lumbar vertebrae in the sagittal plane, *Journal of Biomechanics*, 19(9), 753–758.

10. Punjabi, M., Goel, V.K., Oxland, T., Takata, K., Duranceau, J., Krag, M., and Price, M. 1992. Human lumbar vertebrae: Quantitative three-dimensional anatomy, *Spine*, 17(3), 298–306.

11. Zhou, S.H., McCarthy, I.D., McGregor, A.H., Coombs, R.R.H., and Hughes, S.P.F. 2000. Geometrical dimensions of the lower lumbar vertebrae—Analysis of data from digitized CT images, *European Spine Journal*, 9(3), 242–248.

12. Harris, J.W. and Stocker, H. 1998. *Hyperboloid of Revolution in Handbook of Mathematics and Computational Science*, Springer-Verlag: New York, pp. 112.

13. Steinhaus, H. 1999. *Mathematical Snapshots*, Dover: New York.

14. Raamachandran, J. 1993. *Thin Shells: Theory & Problems*, Universities Press (India) Limited.

15. Guo, L.X., Teo, E.C., and Qiu, T.Z. 2004. Prediction of biomechanical characteristics of intact and injured lower thoracic spine segment under different loads, *Journal of Musculoskeletal Research*, 8(2&3), 87–99.

16. Aebi, M., Thalgott, J.S., and Webb, J.K. 1998. *AO ASIF Principles in Spine Surgery*, Springer: Heidelberg.

17. Rezaian, S.M. and Ghista, D.N. 1994. Clinical biomechanics of spinal fixation—Anterior, posterior and lateral, *IEEE Engineering in Medicine and Biology*, 13(4), 525–531.

18. Rezaian, S.M. and Ghista, D.N. 1998. *Modern Management of Spinal Injury: Anterior Spinal Fixation in Comparison with Posterior and Lateral*, Dorrance: Pittsburgh.

19. Verlaan, J.J., Diekerhof, C.H., Buskens, E., van der Tweel, I., Verbout, A.J., Dhert, W.J.A, and Oner, F.C. 2004. Surgical treatment of traumatic fractures of the thoracic and lumbar spine: A systematic review of the literature on techniques, complications and outcome, *Spine*, 29(7), 803–814.

20. Brunski, J.B., Hill, D.C., and Moskowitz, A. 1983. Stresses in a Harrington distaction rod: Their origin and relationship to fatigue fractures in vivo, *Journal of Biomechanical Engineering Transactions of ASME*, 105(2), 101–107.

21. Gaines, R.W., Carson, W., Satterlee, C.C., and Groh, G. 2004. Experimental evaluation of seven different spinal fracture internal fixation devices using non failure stability testing: The load sharing and unstable-mechanism concepts, *Spine*, 16(8), 902–909.

22. Rezaian, S.M., 1983. US patent 4401112.
23. Hasegawa, K. and Hara, T. 2004. Titanium mesh cage in spinal reconstruction surgery: Biomechanics and clinical application, in *Advances in Spinal Fusion Molecular Science, Biomechanics, and Clinical Management*, eds. K. Lewandrowski, L.W. Donald, D.J. Trantolo, M.J. Yaszemski, and A.A. White, Marcel Dekker: New York.

13

Human Spinal Intervertebral Disc: Optimal Structural Design Characteristics

Dhanjoo N. Ghista, Sridhar Idapalapati, and Ramakrishna Kotlanka

CONTENTS

13.1 Scope

What are the structural features of the spinal intervertebral disc that make it an intrinsically optimal structure? This is because it effectively contains its lateral and axial deformations, while providing the necessary flexibility to the spine. How this is achieved forms the basis of this chapter. The intervertebral disc (IVD), as illustrated in Figure 13.1, consists of the annulus

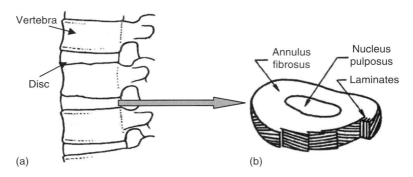

FIGURE 13.1
(a) The location of intervertebral disc within the spinal column. (b) Schematic of the disc structure. The NP is surrounded by annulus fibrosus. This outer layer has lamellar structure with highly ordered collagen structure.

fibrosus (AF) enclosing the nucleus pulposus (NP). When the IVD is loaded in axial compression, the NP gets pressurized and transmits radial stress to the AF, which in turn gets stressed. However, the annulus fibrosus' is a stress-stiffening solid resembling that of a hyper-elastic material. This increase in elastic modulus under loading in turn prevents the annulus fibrosus from deforming in proportion to the applied loading. In other words, as the IVD gets loaded, its deformation does not increase in the same proportion as the loading to which it is subjected. This is what makes the IVD an intrinsically optimal structure.

This chapter analytically models how this is made possible in an intact IVD under uniaxial compression. It also demonstrates that if a ruptured disc is denucleated surgically as a treatment for back pain (to prevent irritation of the spinal nerve structures by the nucleus pulposus, as it is squeezed out through the ruptured disc under compression loading), then the absence of nucleus pulposus no longer stresses the annulus fibrosus as effectively as in the case of an intact IVD.

Hence, the denucleated disc in fact deforms more than the intact disc under compression loading, and hence loses its intrinsic capacity to contain its deformation under increasing loading. This result serves as a contraindication for nucleotomy, and emphasizes that the nucleus pulposus needs to be substituted by a biocompatible gel-filled balloon, to simulate the beneficial effects of the NP.

This chapter (along with the figures) is based on our paper Ghista et al. [1], published in the *International Journal of Design and Nature.*

* With permission from the Publisher WIT Press, Southampton, U.K.

13.2 Introduction: Concept of the Intervertebral Disc as an Optimal Structure

The IVD, as the principal component of the intervertebral joint (shown in Figure 13.1a), sustains and transmits axial compression, bending, and torsional loadings. It is centrally pressurized by the NP and surrounded by the annulus (Figure 13.1b). The annulus fibers are oriented helically, at almost 30°–50° [2]. Under torsion, the torsional shear stresses on a disc element will result in diagonally oriented tensile and compressive stresses. It is revealing that these stresses due to torsion of the disc can thus be directly absorbed by the angled fibers of the annulus. Thus, the IVD is ideally designed for compression and bending as well as for torsion [3–11]. In this chapter, we will concentrate on the compression loading of the IVD, in demonstrating its optimal design characteristics. The IVD also functions as the shock-absorbing component of the spinal unit, comprising of two adjacent vertebral bodies on either side of the IVD [12].

Figure 13.2 illustrates the spinal disc model. As indicated earlier, what makes the IVD an optimal structure is the role of the NP to contain the disc axial and radial deformations. The causative mechanism is that when the disc is loaded in axial compression (or bending or torsion), the NP fluid gets pressurized and stresses the surrounding annulus. The annulus is a

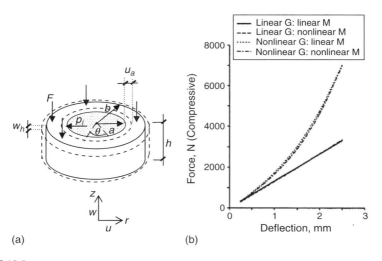

(a)   (b)

FIGURE 13.2
(a) Geometry and deformation variables of the spinal disc, loaded in compressive force *F*. Note that *u* is depicted as expansive radial deformation, while *w* is depicted as shortening axial deformation. (Adopted from Ghista, D.N., Fan, S.C., Sridhar, I., and Ramakrishna, K., *Int. J. Des. Nat.*, 1, 146, 2007.) (b) Comparison of the effects of including linear and nonlinear material (M) and geometry (G) solution options on compressive behavior of the disc. (From Fagan, M.J., Julian, S., Siddall, D.J., and Mohsen, A.M., *Proc. Inst. Mech. Eng., Part H: J. Eng. Med.*, 216, 299, 2002a. With permission.)

stress-stiffening solid, such that its elastic modulus (E) increases with the increase in stress (its stress–strain property is shown in Figure 13.2b) [13,14]. Hence under increased loadings, its elastic modulus value also increases, so that the deformations are thereby contained.

In this chapter, an elasticity model of the disc, as a closed thick-walled fluid-filled cylinder, is employed to determine its stress and deformations under uniaxial compressive loading, and demonstrate the role of NP in containing the disc deformations. It is also demonstrated that the nucleotomized disc will undergo larger deformations than the normal disc, for the same levels of loading, thereby drawing attention to the efficacy of nucleotomy to treat a ruptured disc and associated back pain.

This is the constitution of the spinal disc wherein the stress-dependent Young's modulus of the disc annulus can be represented as [13]:

$$E = E_0 + 375.3\sigma^{0.473} \tag{13.1}$$

where
  E_0 (the residual Young's modulus) $= 4.2$ MPa
  the stress σ is expressed in MPa

Further, for the simplicity of analysis we assume that the disc annulus is isotropic, so that $E_z = E_r = E_\theta = E$. As the disc gets compressed (by increasing the applied compressive force F), the annulus stresses (σ_z, σ_r, σ_θ) keeps increasing. For each updated value of E for the enhanced stress state of the disc (in response to increasing values of the compression force F on the disc), the σ (in Equation 13.1) is taken to be equal to the maximum value of the principal stress (which happens to be the axial stress σ_z). For this relationship, as the disc is loaded, the annulus stress state $\sigma = (\sigma_z)$ increases. Correspondingly, its E increases, so as to thereby contain the disc deformations.

In this chapter, the mechanism of disc deformation containment for vertical loading is delineated. Compressive loading (F) on the disc causes compressive axial stress (σ_z) in the annulus and also pressurizes the NP fluid, which then exerts hydrostatic pressure (p_i), and hence compressive radial stress σ_r on the annulus. This internal pressure in turn causes circumferential hoop tensile stress (σ_θ) in the annulus. These stresses in turn influence the strain state in the disc through its elastic modulus, and hence the axial and radial deformations of the disc by virtue of Equation 13.1.

13.3 Disc Model Analysis: Disc Stresses, Displacements, and Deformed Geometry

The disc is considered to be a thick-walled isotropic cylinder, whose geometry and deformations are depicted in Figure 13.2a. In this analysis, linear stress–strain constitutive relations have been employed. However, the IVD does undergo large deformations. According to Fagan et al. [5,15], under compressive loading of the order of 2000 N, the deformations are of the order of 1 mm (see Figure 13.2b). Hence, in order to compute the disc deformations under compressive loading, small incremental loadings are adopted so that the resulting strains are infinitesimal. Thus, for each incremental load state, (1) the NP pressure is determined, (2) the incremental stresses are computed, and the total stress state is computed, (3) the disc material modulus value is revised as per Equation 13.1, (4) the disc deformations are determined, and (5) its geometry is updated.

Equilibrium equations: Because of the axial symmetry of the disc geometry and loading conditions, there are no shear stresses, and the stress-equilibrium equations are given by

$$\text{in the radial direction,} \quad \frac{d\sigma_r}{dr} + \frac{\sigma_r - \sigma_\theta}{r} = 0 \tag{13.2}$$

$$\text{in the axial direction,} \quad \frac{d\sigma_z}{dz} = 0 \tag{13.3}$$

The strain–displacement relations: Let u be the radial displacement and w be the axial displacement, as shown in Figure 13.2.

$$\text{radial strain,} \quad \varepsilon_r = \frac{\sigma_r}{E} - \frac{\nu(\sigma_\theta + \sigma_z)}{E} = \frac{du}{dr} \tag{13.4a}$$

$$\text{circumferential strain,} \quad \varepsilon_\theta = \frac{\sigma_\theta}{E} - \frac{\nu(\sigma_z + \sigma_r)}{E} = \frac{u}{r} \tag{13.4b}$$

$$\text{axial strain,} \quad \varepsilon_z = \frac{\sigma_Z}{E} - \frac{\nu(\sigma_r + \sigma_\theta)}{E} = \frac{dw}{dz} \tag{13.4c}$$

where
E is the isotropic Young's modulus
ν is the Poisson's ratio of the disc material

The stresses in the disc in terms of displacements are obtained by matrix inversion of Equation 13.4 and are given by

$$\text{radial stress,} \quad \sigma_r = \frac{E}{1+\nu} \left[\frac{\nu}{(1-2\nu)} \left(\frac{du}{dr} + \frac{u}{r} + \frac{dw}{dz} \right) + \frac{du}{dr} \right] \tag{13.5a}$$

$$\text{circumferential (hoop) stress,} \quad \sigma_\theta = \frac{E}{1+\nu} \left[\frac{\nu}{(1-2\nu)} \left(\frac{du}{dr} + \frac{u}{r} + \frac{dw}{dz} \right) + \frac{u}{r} \right] \tag{13.5b}$$

$$\text{axial stress,} \quad \sigma_z = \frac{E}{1+\nu} \left[\frac{\nu}{(1-2\nu)} \left(\frac{du}{dr} + \frac{u}{r} + \frac{dw}{dz} \right) + \frac{dw}{dz} \right] \tag{13.5c}$$

Note that σ_θ, σ_r, σ_z are adopted to be positive for tensile stress.

Now, by substituting the constitutive relations (Equation 13.5) into the equilibrium equations (Equations 13.2 and 13.3), two partial differential equations in displacements u and w are obtained, as follows:

$$\frac{d}{dr}\left[\frac{\nu}{1-2\nu}\left(\frac{du}{dr}+\frac{u}{r}+\frac{dw}{dz}\right)+\frac{du}{dr}\right]+\frac{1}{r}\left(\frac{du}{dr}-\frac{u}{r}\right)=0 \tag{13.6a}$$

$$\frac{d}{dz}\left[\frac{\nu}{1-2\nu}\left(\frac{du}{dr}+\frac{u}{r}+\frac{dw}{dz}\right)+\frac{dw}{dz}\right]=0 \tag{13.6b}$$

The solutions of Equations 13.6a and 13.6b can be expressed as

$$u=\frac{A}{r}+Br \tag{13.7}$$

$$w=Cz+D \tag{13.8}$$

where, A, B, C, and D are the constants of integrations. These constants can be determined by applying appropriate boundary conditions, as will be shown later.

We will first show that the IVD deformations u_a and w_h are interrelated. As the NP is incompressible [2], its volume after deformation is unchanged, so that

$$\pi a^2 h = \pi(a+u_a)^2(h-w_h)$$

This can be simplified, by neglecting higher-order terms ($u_a w_h$ and $u_a^2 w_h$), to yield

$$2\pi a h u_a - \pi a^2 w_h = 0$$

or

$$u_a = \left(\frac{a}{2h}\right)w_h \tag{13.9}$$

It is to be noted that according to deformation defined in Figure 13.2a, w_h is the shortening deformation at $z=h$, while u_a is the radial expansion deformation at $r=a$.

Now, the integration constants A, B, C, and D are evaluated subjected to the boundary conditions

$$u_{r=a}=u_a=\frac{A}{a}+Ba \tag{13.10a}$$

$$\sigma_r = 0 \text{ at } r = b \tag{13.10b}$$

$$w = 0 \text{ at } z = 0 \tag{13.10c}$$

$$w = -w_h \text{ at } z = h \tag{13.10d}$$

Using the above boundary conditions and utilizing Equations 13.5a, 13.7, and 13.8, the constants in Equations 13.7 and 13.8 are obtained as

$$A = \frac{(1 - 2\nu)u_a a b^2}{b^2 + a^2(1 - 2\nu)} \tag{13.11a}$$

$$B = u_a \left[\frac{a^2(1 - 2\nu) + 2\nu b^2}{ab^2 + a^3(1 - 2\nu)} \right] \tag{13.11b}$$

$$C = -\frac{w_h}{h} = -\frac{2u_a}{a} \tag{13.11c}$$

$$D = 0 \tag{13.11d}$$

Using these constants A, B, C, and D in Equations 13.7 and 13.8, the stresses in Equation 13.5 can be obtained, by means of Equation 13.6, as follows:

$$\sigma_r = \frac{E}{1 + \nu} \left[\frac{\nu(2B + C)}{(1 - 2\nu)} + B - \frac{A}{r^2} \right]$$
$$= -\frac{E}{1 + \nu} \left[\frac{u_a a(1 - 2\nu)}{a^2(1 - 2\nu) + b^2} \right] \left(\frac{b^2}{r^2} - 1 \right) \tag{13.12a}$$

$$\sigma_\theta = \frac{E}{1 + \nu} \left[\frac{\nu(2B + C)}{(1 - 2\nu)} + B + \frac{A}{r^2} \right]$$
$$= \frac{E}{1 + \nu} \left[\frac{u_a a(1 - 2\nu)}{a^2(1 - 2\nu) + b^2} \right] \left(\frac{b^2}{r^2} + 1 \right) \tag{13.12b}$$

$$\sigma_z = \frac{E}{(1 + \nu)} \left[\frac{\nu(2B + C)}{(1 - 2\nu)} + C \right] = -\frac{2u_a E}{a(1 + \nu)} \left[\frac{a^2(1 - 2\nu) + b^2(1 + \nu)}{a^2(1 - 2\nu) + b^2} \right]$$
$$= -\frac{w_h E}{(1 + \nu)} \left[\frac{a^2(1 - 2\nu) + b^2(1 + \nu)}{a^2(1 - 2\nu) + b^2} \right] \tag{13.12c}$$

Then from Equations 13.7, 13.11a, and 13.11b, the radial displacement is given by

$$u_r = \frac{A}{r} + Br = \frac{u_a}{ar} \left[\frac{a^2 b^2(1 - 2\nu) + r^2[a^2(1 - 2\nu) + 2\nu b^2]}{a^2(1 - 2\nu) + b^2} \right] \tag{13.13a}$$

and hence u_b (at $r = b$) is given by

$$u_b = \frac{2bu_a}{a}\left(\frac{a^2(1-2\nu)+\nu b^2}{a^2(1-2\nu)+b^2}\right) \tag{13.13b}$$

It is to be noted (from Equation 13.12c) that the σ_z is uniform throughout the disc, and the minus sign implies that σ_z is compressive.

13.4 Stress Analysis of the Healthy Disc under Compression Loading (*F*): Determining Disc Deformations and Stresses in Terms of *F*

For an axially applied force *F* (as illustrated in Figure 13.3), the equilibrium equation is

$$F = \pi a^2 \sigma_f - \pi(b^2 - a^2)\sigma_z \tag{13.14}$$

where
  σ_f is the hydrostatic pressure in the fluid
  σ_z is the axial stress in the annulus (as shown in Figure 13.3). Its sign is
    taken to be negative in Equation 13.14, because positive σ_z is considered as tensile stress

Because the disc height (*h*) is small, therefore σ_f is approximately constant, and hence

$$\sigma_f = -\sigma_r\big|_{r=a} = p_i \text{ (the pressure in NP)} \tag{13.15}$$

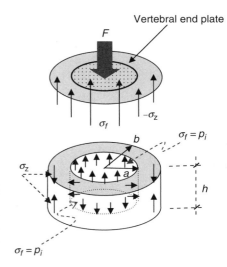

FIGURE 13.3 (See color insert following page 266.)
Normal stresses σ_f and σ_z under the applied force compressive *F*. (Adopted from Ghista, D.N., Fan, S.C., Sridhar, I., and Ramakrishna, K., *Int. J. Des. Nat.*, 1, 146, 2007.)

Combining Equations 13.15 and 13.12a, we get

$$p_i = \frac{E(1-2\nu)}{(1+\nu)} \left(\frac{u_a}{a}\right) \left(\frac{b^2 - a^2}{a^2(1-2\nu) + b^2}\right) \tag{13.16}$$

Then, substituting for u_a from Equation 13.16 into Equation 13.12c, we obtain

$$p_i = -\frac{1-2\nu}{2} \left(\frac{b^2 - a^2}{a^2(1-2\nu) + b^2(1+\nu)}\right) \sigma_z \tag{13.17}$$

The axial stress (σ_z) in the annulus is then obtained in terms of the applied force F, by substituting the expression for p_i from Equations 13.15 and 13.17 into Equation 13.14 as

$$\sigma_z = -\frac{2}{\pi} \left(\frac{F}{b^2 - a^2}\right) \left(\frac{a^2(1-2\nu) + b^2(1+\nu)}{2b^2(1+\nu) + a^2(3-6\nu)}\right) \tag{13.18}$$

Then, from Equations 13.17 and 13.18, the NP pressure in terms of the applied compressive force F is expressed as

$$p_i = \frac{(1-2\nu)}{\pi} \left[\frac{F}{3a^2(1-2\nu) + 2b^2(1+\nu)}\right] \tag{13.19}$$

From Equations 13.18 and 13.12c, we get the radial deformation of the IVD at the inner surface as

$$u_a = \frac{1}{\pi} \frac{1}{E} \left(\frac{F}{b^2 - a^2}\right) \left[\frac{a^3(1-2\nu) + ab^2}{2b^2(1+\nu) + a^2(3-6\nu)}\right] \tag{13.20}$$

Then, from Equations 13.20 and 13.9, we get the axial deformation as

$$w_h = \frac{2}{\pi} \frac{1}{E} \left(\frac{F}{b^2 - a^2}\right) \left[\frac{h(a^2(1-2\nu) + b^2)}{2b^2(1+\nu) + a^2(3-6\nu)}\right] \tag{13.21}$$

By substituting Equation 13.20 into Equation 13.13b, the radial deformation at the outer surface of the annulus, u_b, is expressed in terms of the applied force F as

$$u_b = \frac{2}{\pi} \frac{1}{E} \left(\frac{F}{b^2 - a^2}\right) \left[\frac{a^2 b(1-2\nu) + b^3}{2b^2(1+\nu) + a^2(3-6\nu)}\right] \tag{13.22}$$

Finally, from Equations 13.12a, 13.12b, and 13.20, we obtain the expressions for σ_r and σ_θ, in terms of applied load F, as

$$\sigma_r = \frac{1 - 2\nu}{\pi} \left(\frac{F}{b^2 - a^2} \right) \left[\frac{a^2}{2b^2(1 + \nu) + a^2(3 - 6\nu)} \right] \left(1 - \frac{b^2}{r^2} \right) \tag{13.23}$$

$$\sigma_\theta = \frac{1 - 2\nu}{\pi} \left(\frac{F}{b^2 - a^2} \right) \left[\frac{a^2}{2b^2(1 + \nu) + a^2(3 - 6\nu)} \right] \left(1 + \frac{b^2}{r^2} \right) \tag{13.24}$$

So now, we have obtained the expressions for the disc stresses and deformation in terms of the applied compressive force F.

It is seen that as the disc gets loaded in compression (by the force F), (1) both σ_z and p_i increase, by virtue of Equations 13.18 and 13.19, (2) the increased p_i (which is a function of F) causes both σ_r and σ_θ and u_b to increase as per Equations 13.22 through 13.24, (3) the axial (shortening) deformation w_h increases according to Equation 13.21. Finally, the stresses $(\sigma_r, \sigma_\theta, \sigma_z)$ are expressed in terms of F by Equations 13.18, 13.23, and 13.24, while the deformations $(u_a, u_b,$ and $w_h)$ are expressed in terms of F by means of Equations 13.20 through 13.22.

13.5 Mechanism and Computation of Disc Deformation

The nucleus pulposus gets pressurized when the load F acts on it, as per Equation 13.19. All the stresses increase with loading according to Equations 13.18, 13.23, and 13.24, and so does the annulus modulus E by Equation 13.1. Now E (the elastic modulus corresponding to the deformed state of the disc under load F) will be greater than its value in the unloaded state of the disc, as per Equation 13.1. Hence, as per Equations 13.21 and 13.22, both the axial and radial deformations will be contained, because the term E is in the denominator of these expressions.

This is attributed to the disc design, wherein the annulus contains the nucleus pulposus. This dependency of E on p_i is also reported by Shirazi-Adl et al. [13] and Ranu et al. [16,17] based on the experimental and finite-element analysis of the annulus. The following procedure is adopted to determine the disc deformation in response to compressive load:

Step 1

We start from the unloaded state, $\sigma_{z0} = 0$, for which $E = E_0$ as per Equation 13.1.

1. Initially an incremental compressive force of $\Delta F_1 = 1$ N is applied on the unstressed disc of dimensions $(a_0, b_0,$ and $h_0)$, and the resulting incremental stresses $(\Delta \sigma_{r1}, \Delta \sigma_{\theta 1}, \Delta \sigma_{z1})$ are computed using Equations 13.23, 13.24, and 13.18.
2. Next, the maximum value of these three stresses $(\Delta \sigma_{z1}, \Delta \sigma_{r1}, \Delta \sigma_{\theta 1})$, which happens to be $\Delta \sigma_{z1}$, is noted. Then, based on $\Delta \sigma_{z1}, E = E_1$ is

computed according to the relation (based on Equation 13.1): $E_1 = E_0 \, (= \, 4.2) + 373.3 \, [|\Delta\sigma_{z1}|]^{0.473}$.

3. Disc deformations (w_{h1}, u_{a1}, and u_{b1}), corresponding to the incremental stresses are also computed from Equations 13.20 through 13.22, based on the above calculated value of $E = E_1$.

4. Disc geometry is now updated to: $h_1 = h_0 - w_{h1}$, $a_1 = a_0 + u_{a1}$, $b_1 = b_0 + u_{b1}$.

Step 2

1. Again, an incremental $\Delta F_2 = 1$ N is applied on the deformed geometry of the disc (a_1, b_1, and h_1), and the incremental stresses ($\Delta\sigma_{r2}$, $\Delta\sigma_{\theta2}$, $\Delta\sigma_{z2}$) are determined.

2. Next, the maximum value of these three stresses ($\Delta\sigma_{z2}$, $\Delta\sigma_{r2}$, $\Delta\sigma_{\theta2}$) is noted, which happens to be $\Delta\sigma_{z2}$.

3. Stress state is upgraded to its current value σ_{z2} (by adding $\Delta\sigma_{z2}$ to $\Delta\sigma_{z1}$), and E_2 is computed based on Equation 13.1, as

$$E_2 = E_0(=4.2) + 373.3[|(\Delta\sigma_{z1} + \Delta\sigma_{z2})|]^{0.473} = E_0 + 373.3(\sigma_{z2})^{0.473}$$

4. Then the incremental disc deformations (w_{h2}, u_{a2}, and u_{b2}) are determined corresponding to the current values of (σ_{z2}, σ_{r2}, $\sigma_{\theta2}$), with E_2 as the updated annulus modulus. The total disc deformation is now: $w_{h1} + w_{h2}$, $u_{a1} + u_{a2}$, $u_{b1} + u_{b2}$.

5. Deformed disc geometry is now updated to $h_2 = h_1 - (w_{h1} + w_{h2})$, $a_2 = a_1 + (u_{a1} + u_{a2})$, $b_2 = b_1 + (u_{b1} + u_{b2})$.

Step 3

Step 2 is repeated until the total compressive force reaches 2000 N, in order to obtain the final deformed geometry at the desired applied load.

The resulting graphs of disc deformations w_h, u_a, and u_b versus force (F) are depicted in Figure 13.4, which depict the "disc-hardening" effect whereby the disc deformations do not increase linearly with F.

13.6 Disc Herniation, Back Pain, and Nucleotomy

If the load F becomes very large, σ_θ can exceed the annulus material rupture value, and cause the annulus to develop radial cracks. Then the NP breaks through the annulus. A herniated disc occurs most often in the lumbar region of the spine, especially at the lumbar L4–L5. This is because the lumbar spine carries most of the body's weight. People between the ages of 30 and 50 appear to be more vulnerable, because the elasticity and water content of the nucleus decrease with age. The pain resulting from herniation

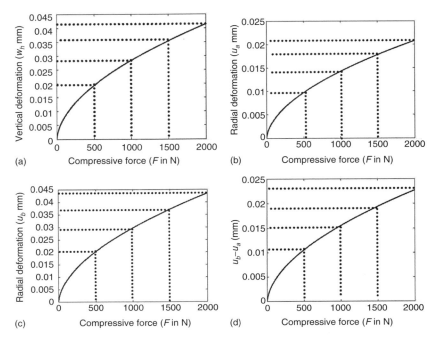

FIGURE 13.4

(a) Disc vertical deformation versus compressive force on the annulus. (b) Radial bulge at $r = a$ versus compressive force on the annulus. (c) Disc radial bulge at $r = b$ versus compressive force. (d) Disc $u_b - u_a$ versus F. The initial disc geometrical parameters adopted are $a = 11$ mm, $b = 25$ mm, and $h = 11$ mm, and the annulus residual modulus E_0 is taken to be 4.2 MPa. (Adopted from Ghista, D.N., Fan, S.C., Sridhar, I., and Ramakrishna, K., *Int. J. Des. Nat.*, 1, 146, 2007.)

may be combined with radiculopathy (neurological deficit). The deficit may include numbness, weakness, and reflex loss. These changes are caused by compression of the nerve structures, created by pressure from the NP material. Percutaneous nucleotomy is carried out, in order to remove the NP from the sequestered disc, and thereby alleviate the back pain [18]. A probe is inserted into the centre of herniated disc under fluoroscope monitoring, and the NP is removed through the probe. The analysis for (1) volume aspiration of the NP fluid with respect to the time for different external suction pressures and (2) the pressure drop in NP fluid with respect to the time is reported by Ghista et al. [12].

13.7 Nucleotomized Disc Model Analysis: Geometry, Stresses, and Displacements

For the nucleotomized disc, only the axial equilibrium needs to be satisfied (Equation 13.3), as there is no internal pressure; and the radial and circumferential hoop stresses are identically equal to zero. The boundary conditions $w = -w_{h,\text{nu}}$ at $z = h$ and $w = 0$ at $z = 0$, yield

$$w = w_{\text{nu}} = -w_{h,\text{nu}} \frac{z}{h} \tag{13.25}$$

The circumferential strain is related to the axial strain by (the Poisson's ratio) as

$$\varepsilon_\theta = \frac{u_{\text{nu}}}{r} = -v\varepsilon_z = -v\frac{dw_{\text{nu}}}{dz} = v\frac{w_{h,\text{nu}}}{h} \tag{13.26}$$

Hence, the radial displacements at $r = a$ and $r = b$ for the nucleotomized disc are given by

$$u_{a,\text{nu}} = v\frac{w_{h,\text{nu}}}{h}a \quad \text{and} \quad u_{b,\text{nu}} = v\frac{w_{h,\text{nu}}}{h}b \tag{13.27}$$

13.7.1 Stress Analysis for a Vertical Loading on the Nucleotomized Disc

For a vertically applied force F, the equilibrium of the disc is shown in Figure 13.5; the minus sign is employed because the axial stress $\sigma_{z,\text{nu}}$ (assumed to be tensile) acts on the vertebral end plate, and the axial stress $\sigma_{z,\text{nu}}$ in the annulus is hence given by

$$\sigma_{Z,\text{nu}} = -\frac{1}{\pi}\left(\frac{F}{b^2 - a^2}\right) \tag{13.28}$$

Using Hook's law (3), the axial deformation is related to $\sigma_{z,\text{nu}}$ and hence to the applied force F, so that the decrease in disc height is given by

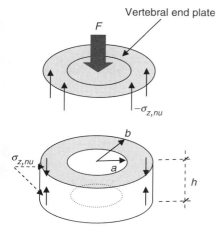

FIGURE 13.5 (See color insert following page 266.) Normal stress $\sigma_{z,\text{nu}}$ equilibrating the applied force F in a nucleotomized disc. (Adopted from Ghista, D.N., Fan, S.C., Sridhar, I., and Ramakrishna, K., *Int. J. Des. Nat.*, 1, 146, 2007.)

$$w_{h,\text{nu}} = \frac{1}{\pi} \frac{1}{E} \left(\frac{Fh}{b^2 - a^2} \right) \tag{13.29}$$

Then, from Equations 13.27 and 13.29, the radial expansion of the disc at $r = a$ is given by

$$u_{a,\text{nu}} = \frac{1}{\pi} \frac{\nu}{E} \left(\frac{Fa}{b^2 - a^2} \right) \tag{13.30}$$

Similarly, the radial expansion of the disc at $r = b$ is given by

$$u_{b,\text{nu}} = \frac{1}{\pi} \frac{\nu}{E} \left(\frac{Fb}{b^2 - a^2} \right)$$

or

$$u_{b,\text{nu}} = \frac{b u_{b,\text{nu}}}{a} \tag{13.31}$$

13.7.2 Determination of Disc Deformation in Nucleotomized Disc

The same procedure as outlined in Section 13.5 is used to determine the incremental and final deformations of the nucleotomized disc under uniaxial compressive load of 2000 N. The resulting graphs of disc deformations $w_{h,\text{nu}}$, $u_{a,\text{nu}}$, $u_{b,\text{nu}}$ versus F and $(u_{b,\text{nu}} - u_{a,\text{nu}})$ versus F are plotted in Figure 13.6, along side the deformations of the normal disc in order to provide a comparison.

It is seen that the nucleotomized disc has considerable greater deformations than the normal disc. These deformations can result in compression of the spinal cord nerve structures as well as the facet joints. Thus the removal of NP has adverse effects like disc collapse and excessive radial bulging. This trend has also been experimentally demonstrated by Meakin and Huikins [19] and Judith et al. [20].

13.8 Conclusion: For the IVD to Retain Its Optimal Structural Feature

Based on these results, in order to retain the stress-stiffening characteristic of the disc and mimic the normal disc load-deformation behavior, it is not advisable to carry out nucleotomy on herniated discs. Instead, it is advisable to replace NP with a gel-filled balloon in the case of disc herniation [21].

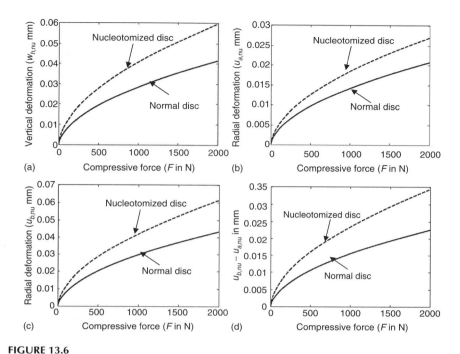

FIGURE 13.6
(a) Disc vertical deformation versus compressive force on the annulus with and without NP. (b) Disc $u_{a,nu}$ versus F with and without NP. (c) Disc $u_{b,nu}$ versus F with and without NP. (d) Disc $(u_{b,nu} - u_{a,nu})$ versus F with and without NP. The initial disc geometric parameters adopted are $a = 11$ mm, $b = 25$ mm, and $h = 11$ mm and the residual modulus E_0 is 4.2 MPa. (Adopted from Ghista, D.N., Fan, S.C., Sridhar, I., and Ramakrishna, K., *Int. J. Des. Nat.*, 1, 146, 2007.)

This chapter illustrates the natural anatomical-physiological design of the intervertebral disc as an optimal load-bearing and deformation-containing structure. This is because of the composite design of the IVD, in which the NP is enclosed by the annulus. Thus, when the IVD is loaded, the NP gets pressurized, its annulus stress increases, the annulus (stress-dependent) modulus increases, and hence the annulus deformation are contained. This is the salient feature of the IVD as an optimal structure, namely its ability to contain its axial and radial deformations under increased loading.

References

1. Ghista, D.N., Fan, S.C., Sridhar, I., and Ramakrishna, K. 2007. The optimal structural design of the human spinal intervertebral disc, *International Journal of Design and Nature*, 1(2): 146–160.
2. Humzah, M.D. and Soames, R.W. 1988. Human intervertebral disc: Structure and function, *The Anatomical Record*, 220(4): 337–356.

3. Shirazi-Adl, A. 1989. On the fibre composite material models of disc annulus-comparison of predicted stresses, *Journal of Biomechanics*, 22(4): 357–365.
4. Shirazi-Adl, A. 1990. Finite-element simulation of changes in the fluid content of human lumbar discs: Mechanical and clinical implications, *Spine*, 17(2): 206–212.
5. Fagan, M.J., Julian, S., Siddall, D.J., and Mohsen, A.M. 2002a. Patient-specific spine models. Part 1: Finite element analysis of the lumbar intervertebral disc—A material sensitivity study, *Proceedings of the Institution of Mechanical Engineers, Part H: Journal of Engineering in Medicine*, 216(5): 299–314.
6. Lin, H.S., Liu, Y.K., Ray, G., and Nikravesh, P. 1978. Systems identification for material properties of the intervertebral joint, *Journal of Biomechanics*, 11(2): 1–14.
7. Belytschko, T., Kulak, R.F., Schultz, A.B., and Galante, J. O. 1974. Finite element stress analysis of an intervertebral disc, *Journal of Biomechanics*, 7(3): 277–285.
8. Wu, H.C. and Yao, R.F. 1976. Mechanical behavior of the human annulus fibrosus, *Journal of Biomechanics*, 9(1): 1–7.
9. Kulak, R.F., Belytschko, T.B., and Schultz, A.B. 1976. Nonlinear behavior of human intervertebral disc under axial load, *Journal of Biomechanics*, 9(6): 377–386.
10. Nick, D., Pope, M.H., and Krag, M.H. 1987. Mechanical model for the human intervertebral disc, *Journal of Biomechanics*, 20(9): 839–850.
11. Martinez, J.B., Oloyede, V.O.A., and Broom, N.D. 1997. Biomechanics of load-bearing of the intervertebral disc: an experimental and finite element model, *Medical Engineering and Physics*, 19(2): 145–156.
12. Ghista, D.N., Subbaraj, S., Mazumdar, J., and Rezaian, S.M. 1998. The biomechanics of back pain, *IEEE Engineering in Medicine and Biology Magazine*, 17(3): 36–41.
13. Shirazi-adl, A., Srivastava, S.C., and Ahmed, A.M. 1984. Stress analysis in the lumbar disc body unit in compression, *Spine*, 9:120–134.
14. Yamada, H. *Strength of Biological Materials*, Williams and Wilkins, Baltimore, 1970.
15. Fagan, M.J., Julian, S., and Mohsen, A.M. 2002b. Finite element analysis in spine research, *Proceedings of the Institution of Mechanical Engineers, Part H: Journal of Engineering in Medicine*, 216: 281–298.
16. Ranu, H.S. 1995. Pressure–volume studies in human lumbar spinal discs, *Proceedings of the Fourteenth Southern Biomedical Engineering Conference*, April 1995, *LA, USA*, pp. 263–265.
17. Ranu, H.S., Denton, R.A., and King, A.I. 1979. Pressure distribution under an intervertebral disc-an experimental study, *Journal of Biomechanics*, 12(10): 807–812.
18. Amoretti, N., Huchot, F., Flory, P., Brunner, P., Chevallier, P., and Bruneton, J.N. 2005. Percutaneous nucleotomy: Preliminary communication on a decompressor probe in percutaneous discectomy. Ten case reports, *Clinical Imaging*, 29: 98–101.
19. Meakin, J.R. and Hukins, D.W.L. 2000. Effect of removing the nucleus pulposus on the deformation of the annulus fibrosus during compression of the intervertebral disc, *Journal of Biomechanics*, 33(5): 575–580.
20. Judith, R.M., Thomas, W.R., and David, W.L.H. 2001. The effect of partial removal of the nucleus pulposus from the intervertebral disc on the response of the human annulus fibrosus to compression, *Clinical Biomechanics*, 16: 121–128.
21. Goel, V., Dooris, A.P., Dennis, M., and Rengachary, S. Biomechanics of artificial disc in *Advances in Spinal Fusion Molecular Science, Biomechanics, and Clinical Management*, Lewandrowski, K., Donald. L.W., Trantolo, D.J., Yaszemski, M.J., and White, A.A., (Eds.), Marcel Dekker, New York, 2004.

Section V

Fitness and Sports Engineering Mechanics

14

Biomechanics of Fitness Index: Optimal Walking and Jogging Modes, and Hip Joint Assessment

Dhanjoo N. Ghista, Jor Huat Ong, and Geok Hian Lim

CONTENTS

14.1 Scope

This chapter defines a fitness index based on a patient's heart rate (HR) response to physical exertion. This is useful in cardiac clinics to assess a patient's cardiac fitness status, as a preventive care measure or to rule out cardiac dysfunction before subjecting the patient to further tests. Next, we have developed the analysis for the optimal walking modality of a person as a determinant of the stride length and frequency of the individual. In the case of jogging, the optimal jogging frequency is postulated to correspond with the natural oscillation frequency of the leg. Lastly, an assessment of the hip joint status is carried out, based on a simple-compound leg pendulum analysis.

14.2 Development of a Cardiac Fitness Index (Based On Exercise and Postexercise Heart Rate Data)

Everyone knows that exercise is good for health. So, you decide to start walking or even jogging. Before doing so, you want to assess your current fitness level. In this regard, in a cardiac clinic, the HR response to physical exertion (usually on a treadmill) is a common test to assess a patient's cardiac fitness status, as a preventive care measure or to rule out cardiac dysfunction before subjecting the patient to further tests. This test entails assessment of the patient's capability to continue exertion up to a certain HR value without having angina pain symptoms. Herein, we propose a means of quantifying this HR response to treadmill during and after exercising, in the form of a differential equation (DEq) model. The parameters of this DEq are then formulated into one nondimensional number or index, which can be employed as a cardiac fitness index (CFI) [1].

14.2.1 Fitness Assessment Model

The fitness assessment model consists of a first-order differential equation model describing the HR response (y) to exertion (exercise, jogging, etc.) monitored in terms of the workload or power (W, measured in watts), where y is defined as follows:

$$y = \frac{\text{HR}(t) - \text{HR(rest)}}{\text{HR(rest)}} \tag{14.1}$$

The subject is exercised on the treadmill for a period of time t_e (minutes) at a constant workload or power exerted (W), while the HR(t) (and hence y) is monitored. Now, we develop a model to simulate (1) the $y(t)$ response to the power expended during exercise, i.e., during $t < t_e$, and (2) thereafter for $y(t)$ decay, after the termination of exercise. In a way, t_e represents the exercise endurance of the subject. In order to compare the fitness capacity of subjects, they are exercised at the same workload.

The model DEq for y response to exercise on the treadmill is given by

$$\frac{dy}{dt} + k_1 y = C_0 W \tag{14.2}$$

where (1) k_1 is a parameter and (2) C_0 is a conversion factor to express W in the same units as the other terms of the equation. The y solution to Equation 14.2 is represented by

$$y = \frac{C_0 W}{k_1} (1 - e^{-k_1 t})$$

$$= \frac{y_e (1 - e^{-k_1 t})}{(1 - e^{-k_1 t_e})} \quad \text{for} \quad t \leq t_e \text{ during exercise} \tag{14.3}$$

$$y = y_e e^{-k_2 (t - t_e)} \quad \text{for} \quad t \geq t_e \text{ during the recovery period when } W = 0 \tag{14.4}$$

where $y_e = y \, (t = t_e)$, and k_1 and k_2 are the model parameters which can serve as cardiac fitness parameters (in units of min^{-1}).

A sample $y(t)$ response is illustrated in Figure 14.1 for the purpose of our discussion. Now,

$$\frac{dy}{dt} = \frac{k_1 y_e e^{-k_1 t}}{(1 - e^{-k_1 t_e})} \quad \text{for} \quad t \leq t_e$$

$$= -k_2 y_e e^{-k_2 (t - t_e)} \quad \text{for} \quad t \geq t_e \tag{14.5}$$

For $t < t_e$, as k_1 increases, y at t_e decreases and dy/dt also decreases (i.e., the rate of rise of HR is also lesser). So, for a healthier subject, k_1 is greater. For $t \geq t_e$, as k_2 increases, the HR recovery is faster. Hence, a higher value of k_2 is also associated with a healthier subject. Based on this observation, we can develop a fitness measure combining the parameters k_1 and k_2 into a single nondimensional index which is higher for a fit person and lower for an unfit person. This Cardiac Fitness Index (CFI) is given by

$$\text{CFI} = k_1 \, k_2 \, t_e^2 \tag{14.6}$$

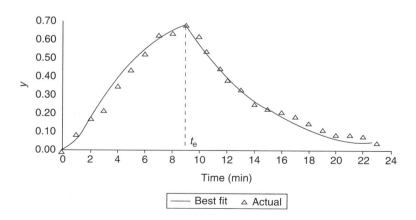

FIGURE 14.1
Sample subject y versus t data. (Adopted from figure 2 of Lim, G.H., Ghista, D.N., Koo, T.Y., Tan, J.C.C., Eng, P.C.T., and Loo, C.M., *Int. J. Comput. Appl. Technol.*, 21, 38, 2004.)

According to this formulation of CFI, for subjects exercised at identical workloads, a healthier subject would have (1) greater k_1 (i.e., slower rate of increase of HR during exercise); (2) greater k_2 (i.e., faster rate of decrease of HR following exercise); (3) greater t_e (i.e., exercise endurance); and hence (4) higher value of CFI.

14.2.2 Application to Fitness Evaluation

A small study was conducted with five subjects of varying fitness levels to test out the relevance of the CFI. Each subject was exercised on the treadmill for a period of time t_e (10 min) at a constant speed (7.5 km/h), while their HR was monitored. The HR was further monitored for 10 min after the subject got off the treadmill. There were five subjects for the test:

> Subject 1 (male) played tennis regularly at least three times a week for at least 2 h each time.
>
> Subject 2 (female) exercised a couple of times a week, running, playing tennis, etc.
>
> Subject 3 (female) exercised about three times a week by going to the gym or playing tennis.
>
> Subject 4 (male) swam once a week.
>
> Subject 5 (male) was a competitive sprinter who trained almost every day for at least an hour.

The results of the experiment are shown in Tables 14.1 and 14.2. The values of k_1 and k_2 were obtained (by parameter estimation) by simulating this data by means of Equations 14.3 and 14.4.

The computed values of k_1, k_2, and CFI are displayed in Table 14.3. The general trend for the fitness can be estimated from the CFI. Subjects 1 and 5 who exercised regularly have significantly higher values of CFI than subject 4 who exercised very little compared to the other four subjects.

Hence, from this experiment, we conclude that the CFI can provide a reliable quantification of a person's fitness for exercise. We will nevertheless need to evaluate CFI for a large spectrum of people (normal volunteers and cardiac patients), and then compute its distribution curve, to determine the efficacy of this index, in order to yield distinct separation of CFI ranges for healthy subjects and unfit patients.

14.2.3 Affirmation of Our Indices

Recent work by Falcone et al. [2] has identified that marked HR increase with exercise could serve as a novel and easily computed parameter that could be clinically useful as an independent predictor of adverse cardiac events, especially among patients with documented coronary artery disease. As can be noted from Table 14.1, both subjects 3 and 4 have enhanced HR increase (than the other three subjects) with exercise.

TABLE 14.1

Heart Rate Responses of the Subjects

Time (min)	Heart Rate of Subjects (HR)				
	1	2	3	4	5
0	72	85	90	85	65
1	126	141	155	132	111
2	137	151	159	147	108
3	135	156	168	159	113
4	133	155	165	165	110
5	134	152	171	165	113
6	134	152	176	172	111
7	134	154	173	172	108
8	133	153	172	173	116
9	134	155	173	171	118
10	133	155	173	173	118
11	87	125	137	145	84
12	78	105	115	119	82
13	78	100	107	118	78
14	76	93	105	112	73
15	80	91	103	107	79
16	79	94	100	106	72
17	77	88	102	105	72
18	76	87	99	107	74
19	73	83	101	100	70
20	72	86	100	100	70

Also, Cole et al. [3] found that a delayed decrease in the HR during the first minute after graded exercise was a powerful predictor of the overall mortality, independent of workload, the presence or absence of myocardial perfusion defects, and changes in HR during exercise. Hence, this affirms our hypothesis that for a healthier subject, the rate of decrease (k_2) of the HR from its value at t_e back to the resting HR value was a good indicator of fitness.

14.3 Optimal Walking Modality

Now that we have been declared fit to exercise, let us start walking daily. For this purpose, we will provide the basis of optimal (i.e., least tiring) walking mode, in terms of the stride length and frequency customized for an individual (based on her/his leg mass and length). Hence, the problem consists of analysis of (1) the optimal mode of leg swing, to determine the optimal stride frequency (SF) that minimizes the muscle involvement or work and (2) the optimal stride length, based on the concept of body adopting the most comfortable stride configuration.

TABLE 14.2

Computed y versus Time for the Five Subjects

Time (min)	y Values of Subjects				
	1	**2**	**3**	**4**	**5**
0	0	0.000	0.000	0.000	0.000
1	0.750	0.659	0.722	0.553	0.708
2	0.902	0.776	0.767	0.729	0.662
3	0.875	0.835	0.867	0.871	0.738
4	0.847	0.824	0.833	0.941	0.692
5	0.861	0.788	0.900	0.941	0.738
6	0.861	0.788	0.956	1.024	0.708
7	0.861	0.812	0.922	1.024	0.662
8	0.847	0.800	0.911	1.035	0.785
9	0.861	0.824	0.922	1.012	0.815
10	0.847	0.824	0.922	1.035	0.815
11	0.208	0.471	0.522	0.706	0.292
12	0.083	0.235	0.278	0.400	0.262
13	0.083	0.176	0.189	0.388	0.200
14	0.056	0.094	0.167	0.318	0.123
15	0.111	0.071	0.144	0.259	0.215
16	0.097	0.106	0.111	0.247	0.108
17	0.069	0.035	0.133	0.235	0.108
18	0.0556	0.024	0.100	0.259	0.138
19	0.014	−0.024	0.122	0.176	0.077
20	0	0.012	0.111	0.176	0.077

14.3.1 Stride Frequency (SF) Analysis

For optimal walking mode simulation, we can employ a simple-compound pendulum model to simulate the limb motion. It is deemed that the energy of walking can be minimized if the leg pacing or stride frequency (SF) is adjusted to that resulting from the undamped oscillatory motion of the free swinging of the leg, modeled as a simple–compound pendulum under the effect of gravity. This is because in this mode of leg motion, the influence of the leg muscles (to accelerate and decelerate the leg during one stride) is minimal.

The equation of (free undamped oscillatory) motion of the cylindrical simple–compound pendulum model of the free-swinging leg (Figure 14.2) is [11]

TABLE 14.3

Values of k_1, k_2, and CFI

Subject	k_1	k_2	CFI
1	2.15	1.26	270.90
2	1.67	0.54	90.18
3	1.38	0.47	64.86
4	0.67	0.28	18.76
5	3.04	0.51	155.04

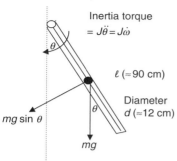

FIGURE 14.2
Model of the free-swinging leg motion simulation as a cylindrical simple-compound pendulum.

$$J \left(\text{polar moment of inertia about the hip joint (O)} = \frac{m\ell^2}{3} \right)$$

$$mg \left(\frac{\ell}{2} \right) \sin \theta + J\ddot{\theta} = 0$$

or

$$k \sin \theta + J\ddot{\theta} = 0 \qquad (14.7)$$

where $k = mg\left(\frac{\ell}{2}\right)$, J about O (for a cylinder) $= m\ell^2/3$, and θ is the angle of rotation.

From Equation 14.7, we obtain

$$\ddot{\theta}(t) + \left(\frac{3g}{2l} \right) \sin \theta(t) = 0 \qquad (14.8)$$

For small θ, we can approximate $\sin \theta = \theta$ (for instance, $\sin 23° = 0.39$, and $23° = 0.4$ rad).

Hence, Equation 14.8 becomes

$$\ddot{\theta}(t) + \left(\frac{3g}{2l} \right) \theta = 0 \qquad (14.9)$$

or

$$\ddot{\theta}(t) + \omega_n^2 \, \theta(t) = 0 \qquad (14.10)$$

where

$$\omega_n = (3g/2l)^{1/2} \qquad (14.11)$$

is the natural frequency of oscillation.

We adopt the solution (for angular displacement θ) of Equation 14.10 (as depicted in Figure 14.3) to be

$$\theta(t) = \theta_0 \cos (\omega_n t) \qquad (14.12)$$

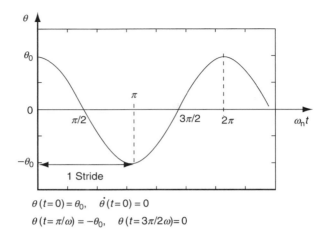

$\theta(t=0) = \theta_0, \quad \dot{\theta}(t=0) = 0$

$\theta(t=\pi/\omega) = -\theta_0, \quad \theta(t=3\pi/2\omega) = 0$

FIGURE 14.3
Variation of the angle $\theta(t)$ during the leg pendulum swing, at its natural frequency, ω_n. One stride is from $\theta = \theta_0$ to $-\theta_0$.

Based on Equation 14.11, for a leg length $\ell = 0.9$ m,

$$\omega_n \text{ (undamped natural angular frequency of oscillation)} = \left(\frac{3g}{2\ell}\right)^{1/2} \approx 4 \text{ rad/s}$$

$$(14.13)$$

and the natural frequency of oscillation $(f_n) = \omega_n/2\pi = 4/2\pi$ cycles/s

$$= 0.65 \text{ cycles/s} \qquad (14.14)$$

Now, since 1 cycle $= 2$ strides, we have the optimal stride frequency (SF):

$$f_s = \frac{\omega_n}{\pi} = \frac{1}{\pi}\left(\frac{3g}{2l}\right)^{1/2} = \frac{1.22}{\sqrt{\ell}}$$

$$= 1.285 \text{ strides/s (for } \ell = 0.9 \text{ m)} \qquad (14.15)$$

14.3.2   Stride Length (SL) Analysis

The left and right leg gait features (illustrated in Figure 14.4) are as follows:

1. At t_1, when the right leg (RL) is about to make heel strike, the left leg foot is flat on the ground. At t_2, when the right leg is at heel strike, the left leg is in toe-off mode.

2. When the left leg (LL) is swinging past the right leg (RL) at $\theta = 0$, its foot is just clearing the ground and its center of mass (CM) is at a height of $\ell/2$ above the ground; from instant t_2 to t_3, its CM_2^L is raised to CM_3^L by an amount of $\ell(1 - \cos\theta)/2$.

3. From instant t_2 to t_3, the CM of RL is raised from CM_2^R to CM_3^R and its hip joint is raised from H_2 to H_3.

4. Then, from t_2 to t_3, as H_2 is raised to H_3, CM_2^B is raised to CM_3^B; hence, the CM^B level increases by $\ell(1 - \cos\theta)$.

Let us now assess what the optimal stride length ought to be. When the left leg is swinging from instant t_2 to t_4, the right leg is in the stance phase and the right foot is rolling from heel to toe on the ground. The orientations of the legs and the positions of the body center of mass (CM^B) are illustrated in Figure 14.4.

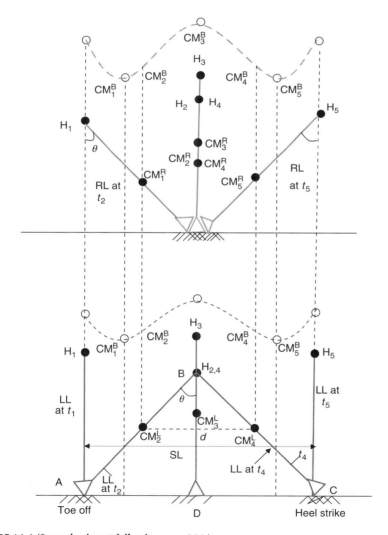

FIGURE 14.4 (See color insert following page 266.)
Depiction of the orientation of left and right legs, and their centers of mass (CM_i^L, CM_i^R) at times t_1 to t_5; H_1: hip joint locations. The right leg is in stance phase from t_2 to t_4; the left leg is in swing phase from t_2 to t_4. Note, that SL (stride length) $= 2d = 2l \sin\theta$; H_2 is at $l \cos\theta$ above the ground, while H_3 is at l above the ground; hence CM^B is raised by $l(1 - \cos\theta)$ from t_2 to t_3.

At point B in Figure 14.4, the orientations of the left leg at t_2 and t_4 and the stride length form a triangle (ABC) of base length AC = SL (the stride length) and height $(h) = BD = \ell \cos \theta$. Now, we want to assess what SL should be in terms of the leg length ℓ, i.e., what should be the shape of this triangle ABC. For this purpose, let us invoke the concept of "symmetry for optimality;" in other words, the body adopts an optimal comfortable stride configuration that provides symmetry. In this context, we propose that the triangle $AH_{2,4}C$ (or ABC) in Figure 14.4 be an equilateral triangle, for which $\theta = 30°$, and hence SL = AC = l.

Now, let us examine the rationality of this solution. While deriving the expression for $\omega_n = (3g/2l)^{1/2}$ in Equation 14.15, we made the assumption of "small θ" for which $\sin \theta = \theta$. Now, the highest value of θ up to which we can have $\sin \theta = \theta$ is $\theta = \pi/6$ or $30°$ (for which $\theta = 0.52$ radians and $\sin \theta = 0.5$). Hence, from the SL analysis, we can also have maximum θ for walking stride-frequency model analysis (in Section 14.3.1) to be $30°$, for which SL = AC $\approx l$ (the leg length).

However, the lengths of sides $AH_{2,4}$ and $CH_{2,4}$ are not precisely the leg length l, but a bit greater. This is because at A, the leg is in the toe-off phase, while at B the leg is in the heel-strike phase. Hence, we can approximate $AH_{2,4} = CH_{2,4} = 1.1l$, so the optimal SL is $1.1l$. Further, in our limited testing of student volunteers, we also found that the most comfortable or preferred stride length (PSL) corresponds to PSL = $1.1l$ (the leg length), for which $\theta = 30°$. For this value of θ, the optimal walking speed is

$$V_W^o = (PSL)\ (natural\ SF)$$

$$= (PSL)\ (\omega_n/\pi) = (1.1l\ \omega_n)/\pi$$

$$= \frac{(1.1l)}{\pi} \left(\frac{3g}{2l}\right)^{\frac{1}{2}} = 1.35 l^{\frac{1}{2}} \tag{14.16}$$

For a person whose leg length is 1.2 m, the value of $V_W^o = 1.48$ m/s or hence this person will take 18 min to walk 1 mile (or 1600 m).

14.4 Optimal Jogging Frequency

14.4.1 Introduction

Now that we have been walking for some time, let us decide to go one step further, and start jogging. In this regard, we have earlier postulated that the leg motion frequency, corresponding to the natural-frequency of oscillation of the leg about the hip joint, would be least tiring and most comfortable for the jogger. It is again based on the rationale that if the leg motion frequency were to be either lesser or greater than the fundamental frequency of oscillation of the leg, the leg muscles would have to exert more power to either decelerate or accelerate the limb from its natural oscillation frequency. Hence, for the least tiring–jogging mode, it would be advantageous to jog at the natural oscillation frequency of the leg.

Kaneko et al. [4] in their study of optimum stride frequency (SF) in constant speed running measured the energy cost as well as mechanical power and mechanical efficiency under various combinations of step frequency and stride length. They observed that the most efficient SF is indeed consistent with the SF chosen or preferred by the subjects themselves. Similarly, Cavanagh and Kram [5] showed that while individuals have different preferred stride frequency (PSF), they appeared to maintain their PSF across a range of running speeds. It was observed that over the range of running velocities between 3.15 and 4.12 m/s the PSF increased only 4%, but the PSL increased 28% (from 2.27 to 2.98 m) in the same velocity range. It can be inferred that (1) the running economy is the criterion for the adoption of a PSF over the running speed range of 3.15 to 4.12 m/s, while (2) a specific combination of SF and SL is physiologically (and automatically) employed by runners/joggers to yield the most comfortable (least tiring) jogging speed of 3.15 to 4.12 m/s.

Indeed observation of marathon runners can reveal that runners with relatively smoother and more comfortable running gait, associated with their PSF invariably end up as front-finishing runners. All of this suggests that for distance running, the freely selected (or preferred) stride frequency (PSF) can be related to economy in running. Herein, we will assume that the PSF is naturally selected to correspond to the natural frequency of oscillation of the swinging lower limb, for which the muscle force and hence the work done are a minimum. This assumption is similar to the proposal by Holts et al. [6,7] and Schot and Decker [8], where the resonance of the freely swinging lower limb (modeled as a simple pendulum) has been proposed as a mechanism responsible for lowest metabolic cost in walking.

14.4.2 Methodology

In our study, 10 healthy subjects (3 females and 7 males) with a mixture of fitness levels were chosen. Six of these subjects were well-conditioned athletes, who exercise regularly and represented the varsity in volleyball. The level of fitness for each of the participants was gauged on the basis of the Bruce incremental treadmill protocol, namely VO_2 max test running on a treadmill [5]. The daily resting HR average over a period of 30 days and the body mass index (BMI) of each participant were used to reinforce the classification of fitness derived from the Bruce incremental treadmill protocol—VO_2 max test. In the study, the volunteers were asked to jog at their least tiring pace with the aim of maintaining endurance over a long distance. While running, the PSL and PSF of the subjects were recorded.

In order to validate our hypothesis analytically, the leg is analyzed as a double-compound pendulum [9]. Expressions for the SFs are derived in terms of the limb segments' masses, as well as mass moments of inertia, lengths, and locations of centers of mass. Based on the anthropometric parameters of these subjects, their respective model-based SFs are then computed by substituting the anthropometric parameters' values in the derived expression for the natural oscillation frequency. These are compared with the experimentally determined SF.

14.4.3 Analysis of Natural Frequency of the Lower Limb (Based On Its Double-Compound Pendulum Model)

The natural frequencies of the lower limb are determined by modeling the whole leg as a double-compound pendulum, pivoted at the hip joint [9]. In this model, the whole leg is divided into two segments hinged at the knee joint. The foot is considered to be an integral part of the lower leg, on the assumption that the ankle joint rotation of the foot is small compared to the rotations of the joints at the hip and knee positions.

Figure 14.5 shows the equivalence between the anatomical and the double-compound pendulum model for the lower limb. The hip and the knee joints are represented by hinge joints (A and B). Segment AB represents the thigh, while segment BC represents the shank and the foot. The segment masses (m_1 and m_2) and their mass moments of inertia (I_1 and I_2) are located at the centers of mass CG_1 and CG_2 of the upper and lower limb segments (AB and BC), respectively.

In Figure 14.6, R denotes the horizontal component of reaction at B (knee joint), while θ and φ denote the angles of inclination to the vertical by limb segments AB and BC, respectively. Assuming small oscillations, the equation of motion for the thigh segment (AB) is obtained by taking moments about the hip joint (A), resulting in [9]:

$$I_A\ddot{\theta} = -m_1 g a L_1 \sin\theta - m_2 g L_1 \sin\theta + R L_1 \cos\theta \qquad (14.17)$$

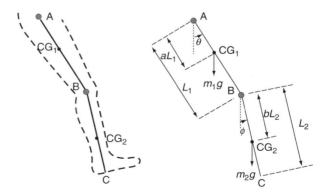

FIGURE 14.5
Double-compound pendulum model of the swinging leg while jogging. (Adopted from figure 1 of Huat, O.J., Ghista, D.N., Beng, N.K., and John, T.C.C., *Int. J. Comput. Appl. Technol.*, 21, 46, 2004.)

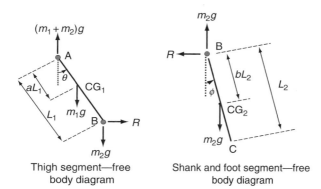

Thigh segment—free body diagram   Shank and foot segment—free body diagram

FIGURE 14.6
Modeling the free-swinging leg (while jogging), as a double-compound pendulum, along with the free body diagrams of the limb segment. (Adopted from figure 1 of Huat, O.J., Ghista, D.N., Beng, N.K., and John, T.C.C., *Int. J. Comput. Appl. Technol.*, 21, 46, 2004.)

Rearranging Equation 14.17 gives

$$R = \left(\frac{I_A}{L_1}\right)\ddot{\theta} + (am_1 + m_2)g\theta \qquad (14.18)$$

From Figure 14.5, it can be seen that the horizontal displacement (x_2) at the center of gravity (CG_2) of the lower leg (shank + foot segment) is given as ($bL_2\phi + L_1\theta$). Hence, from consideration of equilibrium of the horizontal forces, we get

$$R = -m_2\ddot{x}_2 = -m_2(bL_2\ddot{\phi} + L_1\ddot{\theta}) \qquad (14.19)$$

Equations 14.18 and 14.19 are combined, to obtain

$$\left(\frac{I_A}{L_1} + m_2L_1\right)\ddot{\theta} + (bm_2L_2)\ddot{\phi} + (am_1 + m_2)g\theta = 0 \qquad (14.20)$$

where

$$I_A = I_1 + m_1(aL_1)^2 \qquad (14.21)$$

Now, consider the rotation of the lower segment (BC) about its center of gravity (CG_2), in Figure 14.6. By taking moments about CG_2, the following equation is obtained:

$$I_2\ddot{\phi} = RbL_2 \cos\phi - m_2gbL_2 \sin\phi \qquad (14.22)$$

Rearranging Equation 14.22 gives

$$R = \left(\frac{I_B}{bL_2}\right)\ddot{\phi} - bm_2L_2\ddot{\phi} + m_2g\phi \tag{14.23}$$

where $I_B = I_2 + m_2(bL_2)^2$.

Equations 14.19 and 14.23 are combined to obtain the following differential equation:

$$(m_2L_1)\ddot{\theta} + \left(\frac{I_B}{bL_2}\right)\ddot{\phi} + (m_2g)\phi = 0 \tag{14.24}$$

Now, by assuming sinusoidal oscillations, we have

$$\theta = \theta_{max}\sin(\omega t + \alpha), \quad \text{and} \quad \phi = \phi_{max}\sin(\omega t + \alpha) \tag{14.25}$$

Then, by substituting θ and φ from Equation 14.25 into Equations 14.20 and 14.24, the resulting equations of motion (rewritten in matrix form) are given [9] as

$$\begin{bmatrix} \left(\frac{I_A}{L_1} + m_2L_1\right)\omega^2 - (am_1 + m_2)g & bm_2L_2\omega^2 \\ m_2L_1\omega^2 & \left(\frac{I_B}{bL_2}\right)\omega^2 - m_2g \end{bmatrix} \left\{\begin{array}{c} \theta_{max} \\ \varphi_{max} \end{array}\right\} = \left\{\begin{array}{c} 0 \\ 0 \end{array}\right\} \tag{14.26}$$

The frequency equation for this double-compound pendulum model (of the leg) is obtained by setting the determinant of the (2×2) square matrix of Equation 14.26 to be zero. Hence, the two natural frequencies of the lower limb are determined by finding the roots of ω^2 from the resulting frequency equation.

The calculations of the natural frequencies require data regarding the weights, centers of mass, lengths, and mass moments of inertia of the two limb segments (AB and BC). The length of the thigh (L_1) and the length of the shank plus foot (L_2) can be measured directly. The segments mass, moments of inertia, and center of mass are estimated from the data compiled by Winter [10], where the mass of each segment is expressed as a percentage of the total body mass. Based on the data compiled by Winter [10], if the total body mass is denoted as M kg, then:

- Mass of the thigh, $m_1 = 0.1M$
- Mass of the shank and foot, $m_2 = 0.061M$
- Moment of inertia about the hip joint for the thigh segment, $I_A = m_1(0.54L_1)^2$
- Moment of inertia about the knee joint for shank and foot segment, $I_B = m_2(0.735L_2)^2$
- Parameters related to the centers of mass are $a = 0.433$ and $b = 0.606$

Note that SF is defined as the number of successive contacts of the same foot in 1 s. Hence, the relationship between the computed natural frequency (ω_n radians per second) and the computed stride frequency (CSF) is

$$\text{CSF} = 2\left(\frac{\omega_n}{2\pi}\right)\text{Hz} = \frac{\omega_n}{\pi}\text{Hz} \qquad (14.27)$$

For calculating the CSF, the lowest natural frequency of the model is employed, corresponding to the oscillation of the two segments (AB and BC) being in phase, as occurs during jogging or endurance running.

14.4.4 Results and Remarks

The heights, weights, thin lengths, and shank plus foot lengths for all the volunteer subjects were measured. Measurements were taken on both the legs and then averaged. Based on these measurements, the estimates of the mass and inertia properties of the two segments (AB and BC) were calculated from the data compiled by Winters [10].

It has been found from our studies, that the first six subjects are well-conditioned athletes who exercise regularly and have represented the varsity in volleyball. Their resting HRs are less than 65 beats per minute, and their BMI is in the normal range, except for subject 2 who is slightly underweight. Subject 7 has a fitness indicator index "superior" according to the $\dot{V}O_2$ max test. The rest of the volunteers (subjects 8–10) do not exercise regularly, and this is reflected in their $\dot{V}O_2$ max classification as well as in their resting HRs. It is further noted that subject 8 is "underweight," and subjects 9 and 10 are "overweight."

Figure 14.7 shows a plot of the CSF versus the PSF for all the subjects. In this figure, it can be clearly seen that the values for the first six subjects are along the 45° line and the other four data points are for less-fit volunteers (classified as either "good" or "fair" under the $\dot{V}O_2$ max test), who have

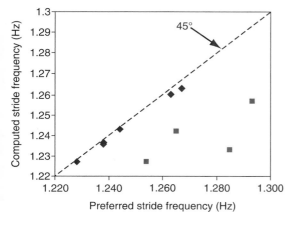

FIGURE 14.7
Computed stride frequency (CSF) versus preferred stride frequency (PSF).

average normal HRs and are either underweight or overweight as compared to the norm. It does suggest that our modeling the whole leg as a double-compound pendulum model (pivoted at the hip) accurately predicts the PSFs of the volunteers.

Table 14.4 indicates a good match between PSFs and CSFs, and also provides the values of their jogging speeds based on their PSLs. The results suggest that as the fitness level of an individual increases, the PSF becomes closer to the natural frequency of the lower limb (modeled as a double-compound pendulum), and hence will result in the least muscular effort (or least metabolic cost) required for jogging. This is in line with our suggestion that SF is the critical factor that determines the muscular effort required to generate the most economical jogging speed.

14.5 Evaluation of the Hip Joint Characteristics

Now, while we have been jogging for quite some time, we might be beginning to feel some pain in the hip joint. So, how do we evaluate the hip joint status? For this purpose, we incorporate a viscous damping factor, b, in our earlier walking model (as depicted in Figure 14.8), so that the damping moment about the hip joint is represented by $b\dot{\theta}$.

14.5.1 Analysis of Free Damped Oscillatory Motion of the Free-Swinging Leg

By referring to Figure 14.8, the differential equation of free-damped oscillatory motion of the simple–compound pendulum model of the free-swinging leg motion (incorporating the moment $b\dot{\theta}$ across the hip joint) is [11]

$$J\ddot{\theta}(t) + b\dot{\theta}(t) + k\theta = 0 \tag{14.28}$$

where (1) $k = mg\ell/2$ (refer Figure 14.8); (2) the moment of inertia J (about O) for a cylinder model of the leg $= m\ell^2/3$; (3) θ is the angle of rotation.

We can now rewrite this equation as follows:

$$\ddot{\theta} + \left(\frac{b}{J}\right)\dot{\theta}(t) + \left(\frac{k}{J}\right)\theta(t) = 0$$

or, as

$$\ddot{\theta} + 2n\dot{\theta}(t) + \omega_n^2\theta(t) = 0 \tag{14.29}$$

where the damping parameter $2n = b/J$, the undamped natural oscillatory angular frequency is ω_n ($\omega_n^2 = k/J$, $J = ml^2/3$), and hence

$$\omega_n^2 = 3g/2l \tag{14.30}$$

TABLE 14.4

Measured Parameters and Computed Stride Frequency

Subject	Mass (kg)	Height (m)	Thigh (m)	Shank+ Foot (m)	VO₂ max (mL/kg/min)	Average Resting Heart Rate (bpm)	BMI (kg/m²)	Preferred Stride Frequency (Hz)	Computed Stride Frequency (Hz)	Preferred Stride Length (m)	Jogging Speed (m/s)
F1	53	1.63	0.38	0.450	56.024	65	19.948	1.267	1.263	2.167	2.775
F2	45	1.65	0.380	0.440	41.597	65	16.529	1.263	1.260	2.070	2.642
M3	63	1.71	0.405	0.465	47.555	64	21.545	1.238	1.237	2.026	2.533
M4	67	1.75	0.4	0.465	53.390	60	21.878	1.244	1.243	2.449	3.083
M5	65	1.75	0.41	0.475	49.905	63	21.224	1.228	1.227	2.043	2.534
M6	66	1.71	0.403	0.473	57.374	60	22.571	1.238	1.236	2.423	3.037
F7	46	1.67	0.395	0.455	38.140	72	16.494	1.265	1.242	1.983	2.534
M8	54	1.78	0.41	0.475	44.261	78	18.043	1.254	1.227	2.252	2.856
M9	78	1.72	0.385	0.450	41.846	75	26.366	1.293	1.257	1.950	2.546
M10	78	1.76	0.405	0.470	38.018	76	25.181	1.285	1.233	2.238	2.907

Note: The symbols F and M represent female and male, respectively.

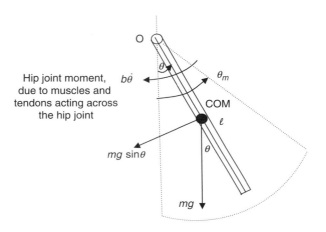

FIGURE 14.8
Simple-compound leg pendulum model, with: hip joint resistive moment (due to damping) $= b\dot{\theta}$; $J\ddot{\theta} =$ inertia torque about O = moment of inertia force at CM about O + inertia torque about COM. The moment of the weight component ($mg \sin\theta$) about O is given by $mg \sin\theta \cdot \ell/2 = mg\ell\theta/2 = k\theta$, for small θ.

We assume a solution of Equation 14.28 as [12]

$$\theta = Ce^{rt} \tag{14.31}$$

Upon substituting Equation 14.31 into Equation 14.28 we get

$$r^2 + 2nr + \omega_n^2 = 0$$

so that

$$r = -n \pm \left(n^2 - \omega_n^2\right)^{1/2} \tag{14.32}$$

For the case of small damping (i.e., $n^2 < \omega_n^2$), we have (for $\omega_d^2 = \omega_n^2 - n^2$)

$$r_1 = -n + i\omega_d \quad \text{and} \quad r_2 = -n - i\omega_d \tag{14.33}$$

where ω_d is the damped angular frequency.

Upon substituting these roots into Equation 14.31, we get two solutions of Equation 14.28. The sum or the difference of these two solutions multiplied by a constant will also be a solution. Hence, the solutions for Equation 14.28 can be given by

$$\theta_1 = \frac{C_1}{2}\left(e^{r_1 t} + e^{r_2 t}\right) = C_1 e^{-nt} \cos \omega_d t \tag{14.34a}$$

$$\theta_2 = \frac{C_2}{2}\left(e^{r_1 t} - e^{-r_2 t}\right) = C_2 e^{-nt}\sin\omega_d t \tag{14.34b}$$

Hence, by adding these two solutions, we obtain the general solution (for angular displacement θ) of Equation 14.28 as [12]

$$\theta = e^{-nt}(C_1\cos\omega_d t + C_2\sin\omega_d t) \tag{14.35}$$

for underdamped free oscillation with viscous damping.
   Therefore,

$$\dot{\theta} = e^{-nt}\left[(C_2\omega_d - nC_1)\cos\omega_d t - (C_1\omega_d + nC_2)\sin\omega_d t\right] \tag{14.36}$$

where

$$\omega_d = \text{damped angular frequency (of damped oscillations)}$$
$$= \left(\omega_n^2 - n^2\right)^{1/2} = \omega_n\left[1 - \left(n^2/\omega_n^2\right)\right]^{1/2}, \quad n < \omega_n \tag{14.37}$$

and C_1 and C_2 are to be determined from the initial conditions.
   Then, the period of damped oscillations is given by [12]

$$\tau_d = \frac{2\pi}{\omega_d} = \frac{2\pi}{\omega_n\sqrt{1 - \left(n^2/\omega_n^2\right)}} \tag{14.38}$$

so that

$$(\omega_d/\omega_n) = \sqrt{1 - \left(n^2/\omega_n^2\right)} \tag{14.39}$$

as depicted in Figure 14.9.
   For initial conditions:

$$\theta_0(t = 0) = \theta_0 \quad \text{and} \quad \dot{\theta}(t = 0) = 0 \tag{14.40}$$

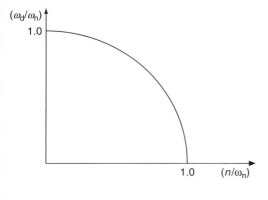

FIGURE 14.9
Graphical representation of the equation: $(\omega_d/\omega_n)^2 + (n^2/\omega_n^2) = 1$.

we get the values of the constants in Equation 14.35, as

$$C_1 = \theta_0 \quad \text{and} \quad C_2 = \frac{C_1 n}{\omega_d} = \frac{n\theta_0}{\omega_d} \tag{14.41}$$

Then, the solution (for angular displacement θ) to Equation 14.35 can be put down as

$$\theta = e^{-nt}[\theta_0 \cos \omega_d t + (n\theta_0/\omega_d) \sin \omega_d t]$$
$$= Ae^{-nt} \cos(\omega_d t - \alpha_d) \tag{14.42}$$

where A is the amplitude of oscillation and α is the phase angle.
Then, from the above equation, we have

$$A(\cos \omega_d t \cos \alpha_d + \sin \omega_d t \sin \alpha_d t) = \theta_0 \cos \omega_d t + \left(\frac{n\theta_0}{\omega_d}\right) \sin \omega_d t \tag{14.43}$$

so that

$$A \cos \alpha_d = \theta_0 \quad \text{and} \quad A \sin \alpha_d = n\theta_0/\omega_d \tag{14.44}$$

Based on Equations 14.35, 14.41, and 14.44, the amplitude of oscillation (A) is given by

$$A = (C_1^2 + C_2^2)^{1/2} = \left(\theta_0^2 + \frac{n^2\theta_0^2}{\omega_d^2}\right)^{1/2} \tag{14.45}$$

This response is depicted in Figure 14.10. Now, from Equation 14.42, we have

$$\text{for } (\omega_d t - \alpha_d) = 0, \quad t \text{ (for } \theta_{m1}) = \frac{\alpha_d}{\omega_d} \tag{14.46a}$$

$$\text{for } (\omega_d t - \alpha_d) = \pi, \quad t \text{ (for } \theta_{m2}) = \frac{\pi + \alpha_d}{\omega_d} \tag{14.46b}$$

$$\text{for } (\omega_d t - \alpha_d) = 2\pi, \quad t \text{ (for } \theta_{m3}) = \frac{2\pi + \alpha_d}{\omega_d} \tag{14.46c}$$

Also, from Equations 14.35, 14.41, and 14.44, the phase angle,

$$\alpha_d = \tan^{-1}\left(\frac{C_2}{C_1}\right) = \tan^{-1}\left(\frac{n}{\omega_d}\right) \tag{14.46d}$$

and the oscillation period (of damped oscillation)

$$\tau_d = \frac{2\pi}{\omega_d} \tag{14.46e}$$

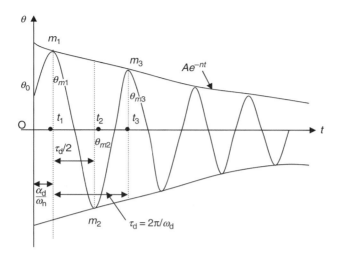

FIGURE 14.10
Equation 14.42 "θ versus t" damped oscillatory response of the leg model, depicting amplitude θ_{mi} at t_i.

We may regard the response solution for θ in Equation 14.43 to represent an underdamped harmonic motion having a decreasing amplitude Ae^{-nt} (due to damping parameter n), a phase angle α_d, and an oscillatory period $\tau_d = 2\pi/\omega_d$, as represented in Figure 14.10 [12].

From Equation 14.42 for the oscillation (angular) displacement θ, we note that for θ_{m1}, θ_{m2}, and θ_{m3}:

1. $\cos(\omega_d t - \alpha_d) = 1$ at $\omega_d t - \alpha_d = 0$ and 2π for θ_{m1} and θ_{m3}, so that

$$t_1 \text{(for oscillating amplitude, } \theta_{m1}) = \frac{\alpha_d}{\omega_d} \tag{14.47a}$$

and $t_3 \text{(for } \theta_{m3}) = \frac{(\alpha_d + 2\pi)}{\omega_d}$, so that

$$t_3 - t_1 = \tau_d = 2\pi/\omega_d \tag{14.47b}$$

as indicated in Figure 14.10

2. $\cos(\omega_d t - \alpha_d) = -1$ at $\omega_d t - \alpha_d = \pi$ for θ_{m2}, so that

$$t_2 = \frac{\alpha_d + \pi}{\omega_d} = \frac{\alpha_d}{\omega_d} + \frac{\tau_d}{2} \tag{14.47c}$$

3. $\theta_{m1} = \theta(t_1) = Ae^{-nt_1}$, $\theta_{m2} = \theta(t_2) = - Ae^{-nt_2}$

$$\theta_{m3} = \theta(t_3) = Ae^{-nt_3} \tag{14.47d}$$

Now, from Equation 14.43 we can put down

$$\dot{\theta} = -Ae^{-nt}\omega_d \sin(\omega_d t - \alpha_d) - nAe^{-nt}\omega_d \cos(\omega_d t - \alpha_d) = 0$$

which, in turn, gives

$$\tan(\omega_d t - \alpha_d) = -\frac{n}{\omega_d} \quad \text{for } \dot{\theta} = 0 \tag{14.48}$$

In Figure 14.10, we note that

1. Points of extreme θ (of $\dot{\theta} = 0$) are separated by equal time intervals $\Delta t_i = \pi/\omega_d = \tau_d/2$, as depicted in the figure, so that (as indicated by Equation 14.46e)

$$\omega_d = \pi/\Delta t_i \quad \text{and} \quad \tau_d = 2\Delta t_i = 2\pi/\omega_d \tag{14.49}$$

2. Maximum amplitudes θ_{mi} occur at times t_i, such that [12]

$$t_{i+2} - t_i = t_3 - t_1 = \Delta t_{13} = \frac{2\pi}{\omega_d} = \tau_d \tag{14.50}$$

and

$$\frac{\theta_{mi}}{\theta_{m(i+2)}} = \frac{\theta_{m1}}{\theta_{m3}} = \frac{Ae^{-nt_i}}{Ae^{-n(t_i+\tau_d)}} = e^{n\tau_d} = e^{\delta} \tag{14.51}$$

where δ, the logarithmic decrement, is given from Equations 14.50 and 14.51 as

$$\delta = l_n\left(\frac{\theta_{mi}}{\theta_{m(i+2)}}\right) = \ln\left(\frac{\theta_{m1}}{\theta_{m3}}\right) = n\tau_d = \frac{2\pi n}{\omega_d} \tag{14.52}$$

We can hence determine δ by merely measuring two successive extremes $\theta(s)$, θ_{m1} and θ_{m3}, of oscillation of the swinging "leg pendulum".

Now, in Equation 14.52, the damping parameter (n), responsible for δ, is a measure of hip joint viscosity and pathology. In order to evaluate it, we can measure two successive amplitudes θ_{m1} and θ_{m3} and τ_d or Δt_{13} (the time between θ_{m1} and θ_{m3}). Then, based on Equation 14.52, we can determine both n and ω_d as follows:

$$\left.\begin{array}{l} n = \dfrac{\ln(\theta_{m1}/\theta_{m3})}{\tau_d} = \dfrac{\ln(\theta_{m1}/\theta_{m3})}{\Delta t_{13}} \\[3mm] \omega_d = \dfrac{2\pi}{\tau_d} = \dfrac{2\pi}{\Delta t_{13}} \end{array}\right\} \tag{14.53}$$

Hence, from Equations 14.52 and 14.53, we can evaluate both ω_d (damped frequency) and n (measure of damping). Then, from Equations 14.37 and 14.53, we can also evaluate ω_n (natural oscillatory frequency) as

$$\omega_n = \left(\omega_d^2 + n^2\right)^{1/2} = \left[\frac{4\pi^2}{\Delta t_{13}^2} + \frac{\ln^2(\theta_{m1}/\theta_{m3})}{\Delta t_{13}^2}\right]^{1/2}$$

$$= \frac{\left[4\pi^2 + \ln^2(\theta_{m1}/\theta_{m3})\right]^{1/2}}{\Delta t_{13}} \tag{14.54}$$

Upon evaluating both n and ω_n, from Equations 14.53 and 14.54, we can determine the values of the hip joint parameters b and k in Equation 14.30 as

$$b = 2nJ, \quad k = \omega_n^2 J, \quad \text{where } J = ml^2/3 \tag{14.55}$$

These constitute the damping and stiffness parameters of the hip joint, for assessment purposes. Therefore, we can compute the hip joint parameters (b and k), illustrated in Figure 14.8, as follows:

$$b = 2nJ = 2nml^2/3 \tag{14.56}$$

$$k = \omega_n^2 J = \omega_n^2 ml^2/3 \tag{14.57}$$

14.5.2 Application

In order to implement this analysis, we can ask the subject to stand on a platform and swing her/his leg, so as to measure the oscillatory period (τ_d) and the oscillating amplitude decrement (δ) from the successive values of θ_1 and θ_3 and the time interval (Δt_{13}) between them, based on Equations 14.50 and 14.52. We can also compute ω_d and the value of the hip joint damping parameter n from Equation 14.53, $\omega_n = (n^2 + \omega_d^2)^{1/2}$ from Equation 14.54, the value of the joint damping b from Equation 14.56, and the value of the joint stiffness parameter k from Equation 14.57.

Now, suppose it is not feasible to experimentally measure δ. In that case, we can still obtain representative values of the joint damping parameter ($b = 2nJ$) from the preferred stride frequency (PSF, f_p) by approximating PSF (f_p) $= 2f$ (leg oscillation frequency). For the leg oscillation frequency (f), we have from the following equation:

$$f(\text{cycles/s or Hz}) = \frac{\omega_d}{2\pi} = \frac{\left(\omega_n^2 - n^2\right)^{1/2}}{2\pi} \tag{14.58}$$

and the oscillation period

$$\tau_d = 2\pi/\omega_d = 1/f \tag{14.59}$$

Then, from measuring PSF (f_p), we can determine

(leg oscillation frequency) $f = f_p/2$ and $\omega_d = 2\pi f = \pi f_p$ (14.60)

By measuring the subject's leg length (ℓ), we can evaluate (employing Equation 14.30) $\omega_n = (k/J)^{1/2} = (3g/2\ell)^{1/2}$, and hence compute the damping parameters (from Equations 14.37 and 14.30)

$$n = \left(\omega_n^2 - \omega_d^2\right)^{1/2} \quad \text{and} \quad b = 2nJ = 2nml^2/3$$ (14.61)

Now, let us take the case of a subject whose $\ell = 1$ m, leg mass $m = 10$ kg, and preferred walking stride frequency (PSF, f_p) $= 1.1$ strides/s. We can then put down (from Equation 14.58)

$$f_p(\text{PSF}) = 2(\text{oscillating frequency}, f) = 2f = \omega_d/\pi = 1.1 \text{ Hz}$$ (14.62)

Hence, from Equation 14.62, we can compute the damped frequency oscillation frequency

$$\omega_d = 1.1\pi = 3.5 \text{ Hz, from PSF}$$ (14.63)

On the other hand, based on Equation 14.30, we have

$$\omega_n = \left(\frac{k}{J}\right) = \left(\frac{3g}{2l}\right)^{1/2} = 3.84 \text{ Hz}$$ (14.64)

Then, since based on Equation 14.38, $\omega_d^2 = \omega_n^2 - n^2$, we have

$$n = \left(\omega_n^2 - \omega_d^2\right)^{1/2} = \left(3.84^2 - 3.5^2\right)^{1/2} = 1.58 \text{ s}^{-1}$$ (14.65)

The corresponding value of b is then given (from Equation 14.56) by

$$b = 2nJ = 2(1.58 \text{ s}^{-1})\left(\frac{ml^2}{3}\right) = 10.55 \text{ kgm}^2\text{s}^{-1} \quad \text{for } m = 10 \text{ kg}$$ (14.66)

and the joint stiffness (k) can also be obtained (from Equations 14.30 and 14.57) as

$$k = \omega_n^2 \frac{ml^2}{3} = \left(\frac{3g}{2l}\right)\left(\frac{ml^2}{3}\right) = \frac{mgl}{2} = 49 \text{ Nm}$$ (14.67)

Overdamped oscillatory motion of the free-swinging leg: Now, it is possible that the joint damping can be significant and $n > \omega_n$. In that case, both the roots of

$$r = -n \pm \left(n^2 - \omega_n^2\right)^{1/2}$$ (14.68)

will be real and negative, and the system is overdamped. The motion will not be oscillatory but will be aperiodic.

We can put down the general solution as

$$\theta = C_1 e^{r_1 t} + C_2 e^{r_2 t} \tag{14.69}$$

Now, we put the initial conditions

$$\theta_0(t = 0) = \theta_0 \quad \text{and} \quad \dot{\theta}(t = 0) = 0 \tag{14.70}$$

Therefore, from Equations 14.69 and 14.70, $C_1 + C_2 = \theta_0$ and $r_1 C_1 + r_2 C_2 = 0$, giving

$$C_1 = \frac{-r_2 \theta_0}{(r_1 - r_2)} \quad \text{and} \quad C_2 = \frac{r_1 \theta_0}{(r_1 - r_2)}$$

and Equation 14.69 becomes

$$\theta = \frac{-r_2 \theta_0}{(r_1 - r_2)} e^{r_1 t} + \frac{r_1 \theta_0}{(r_1 - r_2)} e^{r_2 t} \tag{14.71}$$

and the solution will look like that shown in Figure 14.11.

Now,

$$\left. \begin{array}{l} r_1 = -n + \left(n^2 - \omega_n^2\right)^{1/2} \\ r_2 = -n - \left(n^2 - \omega_n^2\right)^{1/2} \end{array} \right\} \tag{14.72}$$

It can be seen, from Equation 14.72, that both r_1 and r_2 will be negative.

At two time intervals, $t = t_1$ and t_2, we have

$$\left. \begin{array}{l} \theta_1 = \frac{-r_2 \theta_0}{(r_1 - r_2)} e^{r_1 t_1} + \frac{r_1 \theta_0}{(r_1 - r_2)} e^{r_2 t_1} \\ \theta_2 = \frac{-r_2 \theta_0}{(r_1 - r_2)} e^{r_1 t_2} + \frac{r_1 \theta_0}{(r_1 - r_2)} e^{r_2 t_2} \end{array} \right\} \tag{14.73}$$

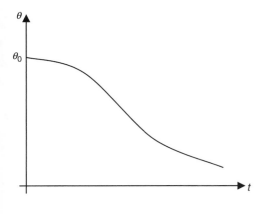

FIGURE 14.11
Graphical representation of Equation 14.71.

Herein, the unknowns are r_1 and r_2. Upon monitoring θ_1 and θ_2 at t_1 and t_2 as the leg swings down from θ_0, we can solve for r_1 and r_2 from Equation 14.73 and then solve for ω_n and n from Equation 14.72, to characterize the patient's hip status in terms of $b = 2\ nml^2/3$ and $k = \omega_n^2\ ml^2/3$.

References

1. Lim, G.H., Ghista, D.N., Koo, T.Y., Tan, J.C.C., Eng, P.C.T., and Loo, C.M., 2004, Cardiac Fitness biomathematical model of HR response to VO_2 during and after exercise stress-testing, *International Journal of Computer Applications in Technology*, 21(1/2), 38–45.
2. Falcone, C., Buzzi, M.P., Klersy, C., and Schwartz, P.J., 2005, Rapid Heart-rate increase at onset of exercise predicts adverse cardiac events in patients with coronary artery disease, *Circulation*, 112, 1959–1964.
3. Cole, C.R., Blackstone, E.H., Pashkow, F.J., Snader, C.E., and Lauer, M.S., 1999, Heart-rate recovery immediately after exercise as a predictor of mortality, *New England Journal of Medicine*, 341, 1351–1357.
4. Kaneko, K., Matsumo, M., Ito, A., and Fuchimoto, T., 1987, Optimum step frequency in constant running speed, In: Jonsson, B.(ed.), *Biomechanics, X-B*, Vol 6B, Human Kinetic Publishers, Inc, Champion, IL,USA, pp. 803–807.
5. Cavanagh, P.R. and Kram, R., 1990, Stride length in distance running: Velocity, body dimensions, and added mass effects, *Medicine and Science in Sports and Exercise*, 21, 467–479.
6. Holt, K.G., Hamill, J., and Andres, R.O., 1990, The force-driven harmonic oscillator as a model for human locomotion, *Human Movement Science*, 9, 55–68.
7. Holt, K.G., Hamill, J., and Andres, R.O., 1991, Prediction minimal energy costs of human walking, *Medicine and Science in Sports and Exercise*, 23(1/2), 491–498.
8. Schot, P.K. and Decker, M.J., 1998, The force driven harmonic oscillator model accurately predicts the preferred stride frequency for backward walking, *Human Movement Science*, 17, 67–76.
9. Huat, O.J., Ghista, D.N., Beng, N.K., and John, T.C.C., 2004, Optimal stride frequency computed from the double-compound pendulum model of the leg, and verified experimentally as the preferred SF of jogging, *International Journal of Computer Applications in Technology*, 21, 46–51.
10. Winter, D.A., 1990, *Biomechanics and Motor Control of Human Movement*, 2nd ed., John Wiley & Sons, New York.
11. Milsum, J.H., 1966, *Biological Control Systems Analyses*, McGraw-Hill Book Company, New York, pp. 180–185.
12. Timoshenko, S., Young, D.H., and Weaver, Jr., W., 1974, *Vibration Problems in Engineering*, John Wiley & Sons, New York, pp. 63–71.

15

Analysis of Spinning Ball Trajectories of Soccer Kicks and Basketball Throws

Dhanjoo N. Ghista and Xiang Liu

CONTENTS

15.1 Scope

It is a common observation that putting spin to a pitched baseball or to a kicked soccer-ball imparts an out-of-plane curve to the ball trajectory. This mechanism has been successfully used in soccer to deceive goalkeepers, in basketball throws to obtain a better entering angle into the net, and in baseball to induce the batter to swing and miss. So then, what is the mechanism of this curving soccer kick or a backspinning basketball throw? In addressing this issue, this chapter deals with the mechanics of a spinning ball trajectory, as to how deviations in the ball's original trajectory are governed by different amounts of angular velocity and translational velocity. It is shown as to how the amount and direction of spin plays an important role in delineating the trajectory and the position of the ball in its trajectory. Then, this analysis is employed to simulate a number of interesting situations in soccer and basketball, including the famous Ronaldinho's goal against England in the 2002 World-cup semifinals.

15.2 Theory: Lateral Force on the Spinning Ball of a Soccer Kick

When a soccer ball is kicked with a spin about the vertical axis of the ball, it will swerve laterally from its vertical–planar trajectory because of the lateral force exerted by the air on the ball, as illustrated in Figure 15.1a through c.

In Figure 15.1, the ball is kicked from right to left at velocity u. The air is flowing in the opposite direction from left to right; u_∞ is the horizontal velocity of air, ω is the counter-clockwise angular velocity imparted to the ball, Γ is the Circulation, and u is the horizontal velocity of kick. In Figure 15.1, it is noted that the ball has a higher velocity on one side (Figure 15.1b), and a higher pressure intensity on the opposite side (Figure 15.1c), which results in a transverse force as shown in Figure 15.1c. This lateral or transverse force makes the ball curve sideways, as it travels from right to left in the air.

Also, if the ball is kicked and spun in a vertical plane, it will have a lift force (upward or downward) resulting from a combination of angular and translational velocities caused by the air. As seen in Figure 15.1, the airflow on one side of the ball is retarded relative to that on the other side, and the pressure becomes greater than that on the other side, so the ball is pushed laterally. The greater the velocity of the spin, the larger is this lateral

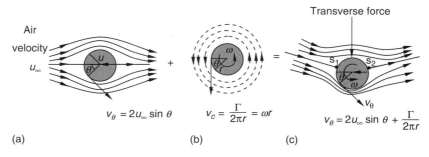

FIGURE 15.1
Effect of circulatory flow superimposed on translatory flow. Here, the ball is kicked to the left (as shown in (a)) and the spin imparted to it is counter-clockwise (as shown in (b)). The resultant airflow pattern (c) causes a transverse force on the ball, causing the ball to curve sideways as it travels from right to left.

force or lift force. This phenomenon of lateral or lift motion produced by imposing circulation over a uniform fluid stream, known as the Magnus effect, can be used to explain the deviation of spinning balls from their normal trajectories.

The corresponding lift force or lateral force (Figure 15.1) can be calculated by Equation 15.1 derived from the Kutta–Joukowski Law, being first noted by the German physicist, H.G. Magnus (1802–1870) and in honor of the German and Russian fluid dynamists M.W. Kutta (1867–1944) and N.E. Joukowski (1847–1927). They independently showed that, for a body of any shape, the transverse force per unit length is $\rho u_\infty \, \Gamma$, and is perpendicular to the direction of the air velocity u_∞. The formula for this lateral or transverse force (acting on the spinning ball) is given by

$$F_L = \tfrac{1}{2}\rho\pi R^3 \omega\, v_{0h} \tag{15.1}$$

where
  F_L is the lateral force
  ρ is the air density
  R is the radius of the ball
  ω is the angular velocity of the spin
  v_{0h} is the initial horizontal velocity

A brief derivation of this expression is provided in the Appendix, while some interesting applications of this phenomenon are given in the references [1–5].

Using this simplified equation for lateral or lift force, we have studied different cases of balls spinning with different velocities. Our results show that the angular velocity of the ball during its motion in the air causes that ball to deviate from its original trajectory, by different amounts for different angular velocities and initial translatory velocities. Hence, the amount and direction of spin play an important role in deciding the trajectory and final position of the ball.

In Figure 15.2, a case study of such a soccer kick is illustrated, to be analyzed later in Section 15.3.3. It is seen that the spin of the ball contributes about 6.10 m in the direction of y, which results from the acceleration in y direction caused by the lateral (or transverse) force due to spin or angular velocity (ω) imparted to the ball. In the feasible ranges of the angular velocity or spin imparted to the ball and the initial velocity with which the ball is kicked, the deviation (because of the spin) can vary from 2 to 5 m in the normal kicking range. It is no wonder that a corner kick, if properly taken with the right combination of initial velocity v_0 and ω, can make the ball swerve into the goal just under the bar and place the ball in the top-far corner of the goal (as illustrated later on in Figure 15.5).

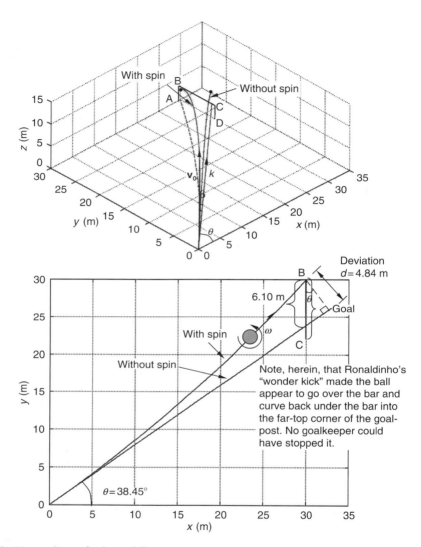

FIGURE 15.2 (See color insert following page 266.)
Analytical simulation of the trajectory of Ronaldinho's famous free kick in the quarter-final match against England in the 2002 World Cup (won by Brazil). The top figure shows the 3-D trajectories of the ball, with and without spin. The bottom figure shows the top view (or the horizontal projection) of the ball trajectory to its final location B into the goal. In doing so, to the goalkeeper Seaman, the ball must have actually appeared to be sailing over the bar, only to see it curve back to dip below the bar into the goal. In the figure, BC represents the goal bar, k is the unit vector making an angle θ with the x-axis, and β is the angle that the initial velocity vector ($\mathbf{v_0}$) makes with Ok (in the xOy plane). The initial velocity vector $\mathbf{v_0}$ lies in the zOk plane. The lateral deviation of the ball along the goal bar is 6.10 m, as shown in the figure.

15.3 Analysis of the Soccer Kick

15.3.1 Theory: Trajectory of a Spinning Soccer-Ball Kick

As shown in Figure 15.3, the ball is kicked with an initial velocity v_0 at angle β with the horizontal in the zOk plane, making an angle θ with the zOx plane. The conventional governing equations, of the ball's (x, y, z) displacement–time relations without spin, are

$$x = (v_0 \cos \beta) \cos \theta t \tag{15.2}$$

$$y = (v_0 \cos \beta) \sin \theta t \tag{15.3}$$

$$z = v_0 \sin \beta t - \tfrac{1}{2} g t^2 \tag{15.4}$$

where

$v_0 \cos \beta \; (= v_{ok})$ is the horizontal component of the initial velocity along Ok

$OE = (v_{ok}) t$ is the distance covered along the horizontal axis Ok (on the ground)

g is the gravity acceleration

Now, when the ball is kicked by imparting it spin, i.e., angular velocity ω (counter-clockwise, looking down on the ball) about the vertical axis through the ball (refer to Figure 15.4), then the ball displacement–time relations (in the $[k, j, z]$ coordinate frame) are given by (refer to Figure 15.4)

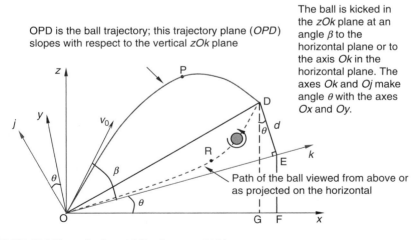

OPD is the ball trajectory; this trajectory plane (*OPD*) slopes with respect to the vertical *zOk* plane

The ball is kicked in the *zOk* plane at an angle β to the horizontal plane or to the axis *Ok* in the horizontal plane. The axes *Ok* and *Oj* make angle θ with the axes *Ox* and *Oy*.

Path of the ball viewed from above or as projected on the horizontal

FIGURE 15.3 (See color insert following page 266.)
Notations for soccer-ball kick-velocity and trajectory. The orthogonal lines (or axes) Ok and Oj are in the horizontal plane xOy, and make angles θ with the Ox and Oy axes, respectively. ED is the total horizontal deviation (d) of the ball when it lands on the ground at D. The curve ORD is the horizontal projection of the trajectory of OPD.

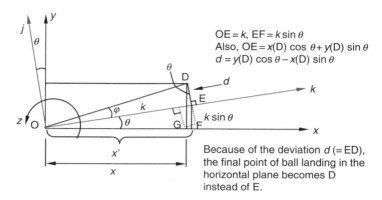

OE = k, EF = $k \sin\theta$
Also, OE = x(D) $\cos\theta$ + y(D) $\sin\theta$
$d = y$(D) $\cos\theta$ − x(D) $\sin\theta$

Because of the deviation d (= ED),
the final point of ball landing in the
horizontal plane becomes D
instead of E.

FIGURE 15.4 (See color insert following page 266.)
Ball displacements in the horizontal plane. The ball is kicked in the zOk vertical plane.
However, because of the counter-clockwise angular velocity (ω) imparted to it, it has deviated
by an amount "d" (= ED) perpendicular to Ok (i.e., parallel to Oj axis) when it lands on the
ground.

Horizontal displacement along Ok axis,

$$k(=\text{OE}) = (v_0 \cos\beta)t = v_{ok}t \tag{15.5}$$

Ball displacement in the "j" direction (normal to Ok),

$$j(=\text{ED}) = d \tag{15.6}$$

wherein d is given by the following Equations 15.8 and 15.10
    Vertical displacement along Oz,

$$z = (v_0 \sin\beta)t - \tfrac{1}{2}gt^2 \tag{15.7}$$

wherein

$$d = \tfrac{1}{2}a_{\text{L}}t^2 \tag{15.8}$$

with a_{L} (given by Equation 15.10) being the lateral acceleration due to the
lateral force (caused by ω).
    From Equation 15.5

$$t = k(\text{displacement})/v_{ok} = \frac{\text{OE}}{(v_0 \cos\beta)} \tag{15.9}$$

with β being the angle of the tangent to the trajectory with respect to the
horizontal Ok axis.

The lateral acceleration

$$a_L = \frac{F_L}{m}$$

$$= \frac{1}{2m} \rho \pi R^3 \, \omega v_{ok} = \frac{1}{2m} \rho \pi R^3 \omega (v_0 \cos \beta) \tag{15.10}$$

where
F_L is the lateral force (Equation 15.1)
m is the mass of the ball

Now, in the (x, y, z) coordinate frame, the ball displacement–time relations are given by (Figures 15.3 and 15.4)
Ball displacement along Ox axis,

$$x = [(v_0 \cos \beta) \cos \theta] t - d \sin \theta \tag{15.11}$$

Ball displacement in the Oy direction,

$$y = [(v_0 \cos \beta) \sin \theta] t + d \cos \theta \tag{15.12}$$

Vertical displacement of the ball,

$$z = (v_0 \sin \beta) t - \tfrac{1}{2} g t^2 \tag{15.13}$$

where d is given by Equation 15.8, and based on Equation 15.9

$$t = \frac{k(\text{displacement}) \text{ or } OE}{v_0 \cos \beta (= v_{ok})} = \frac{(x + d \sin \theta)/\cos \theta}{v_{ok}(= v_{ox}/\cos \theta)} = \frac{x + d \sin \theta}{v_{ox}}$$

$$= \frac{OF}{v_{ox}} = \frac{x'}{v_{ox}} \tag{15.14}$$

The resulting trajectory (OPD) of the ball is shown in Figure 15.3, along with the curved horizontal projection of the ball trajectory (ORD) in the horizontal plane. D is the point at which the ball lands on the ground and P is the highest point of the ball trajectory.

Let us take some reasonable data: ball radius $R = 10$ cm, $\omega = 30$ rad/s, distance traveled along $Ok(OE) = 40$ m, initial velocity $v_0 = 23$ m/s, angle β of the initial velocity vector with respect to the axis $Ok = 20°$, air density $\rho = 1.25$ kg/m^3, ball mass $m = 0.5$ kg. If we substitute these data into Equations 15.5 through 15.14, then

- From Equation 15.14, the time taken for the ball to land, $t = 40/23$ $\cos \beta = 1.85$ s
- From Equations 15.8 and 15.10, the ball deviation ($d = $ DE) perpendicular to Ok in the kOj plane, when it lands ($= $ "j" displacement of the ball)

$$= \frac{\rho \pi R^3 \omega (v_0 \cos \beta) t^2}{4m} = 4.35 \text{ m} \tag{15.15}$$

- Distance OD traveled by the ball in the horizontal plane (Equations 15.5, 15.6, and 15.15)

$$= (k^2 + j^2)^{1/2} = (\text{OE}^2 + \text{ED}^2)^{1/2} = 40.24 \text{ m} \tag{15.16}$$

- The z coordinate of the highest point P of the trajectory is obtained by putting the z (or vertical) velocity of the ball ($v_0 \sin \beta - gt) = 0$, and substituting t (time taken for the ball to go from O to P) $= v_0 \sin \beta / g = 0.8$ s into Equation 15.7, to obtain $z(P) = 3.16$ m
- The (k, j) coordinates (in the coordinate plane kOj) of the point R (the projection of the ball-trajectory point P on the horizontal plane) $= (17.29, 0.81)$

15.3.2 Exemplification of the Theory: Computation of the Spinning Ball Trajectories

We will now solve some realistic soccer situations.

Example 15.1
Let us analyze how a spinning kick from the goal line can make the ball swerve into net, as illustrated in Figure 15.5.

Here, we are given the final coordinates of the ball to be at C (41, 0, 2.4), the top-far corner of the goal. The ball is kicked by a right footer in the vertical plane zOk at $v_0 = 28$ m/s, β (angle made by V_0 with Ok) $= 19°$, ball mass $m = 0.5$ kg, ball radius $R = 0.1$ m, air density $\rho = 1.25$ kg/m^3. We need to determine the values of θ (illustrated in Figure 15.5), t and ω, such that the ball will land in the top-far corner of the goalpost.

The governing relations are (as adopted from Equations 15.5 through 15.14): displacement along Ok, distance k (Equation 15.5)

$$= (v_0 \cos \beta) t = v_{0k} t = \text{OE} = \text{OD} \cos \theta \tag{15.17}$$

deviation, d (Equations 15.8 and 15.10)

$$= \frac{\rho \pi R^3 \omega v_{0k} t^2}{4m} = \frac{\rho \pi R^3 \omega (v_0 \cos \beta) t^2}{4m} = \text{ED} = \text{OD} \sin \theta \tag{15.18}$$

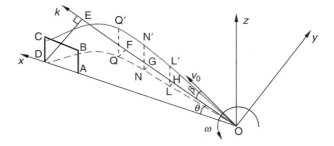

FIGURE 15.5 (See color insert following page 266.)
Corner kick by a right footer, straight into the goal. The player kicks the ball in the zOk plane, with a counter-clockwise angular velocity ω. The ball curves along OL'N'Q' to C into the far-top corner. The deviations of the ball trajectory projected on the horizontal plane are HL, GN, FQ, and ED.

final vertical displacement z (Equation 15.13) $= (v_0 \sin \beta)t - \frac{1}{2}gt^2 = DC$

$$(15.19)$$

$$t = \frac{k(\text{distance})}{v_{0k}} = \frac{OE}{v_{0k} \ (\text{horizontal velocity in the } k \text{ direction})} \qquad (15.20)$$

These relations are rewritten as

$$OE = (v_0 \cos \beta)t = OD \cos \theta = x(D) \cos \theta, \text{ where OD}$$
$$= 41 \text{ m (as per data)} \qquad (15.21)$$

$$ED(=d) = \left(\frac{\rho \pi R^3 \omega v_0 \cos \beta}{4m} \right) t^2 = OD \sin \theta \qquad (15.22)$$

$$DC = (v_0 \sin \beta)t - \frac{1}{2}gt^2 = 2.4 \text{ m} \qquad (15.23)$$

These 3 equations have to be solved for t, θ, and ω.
From Equation 15.23, we get $t = 1.54$ s.
Then, from Equation 15.21 and 15.22, we get

$$(v_0 \cos \beta)^2 t^2 + \left(\frac{1}{4m} \rho \pi R^3 v_0 \cos \beta \right)^2 \omega^2 t^4 = OD^2 \qquad (15.24)$$

Putting $t = 1.54$ s, $v_0 = 28$ m/s, $\beta = 19°$, $m = 0.5$ kg, $R = 0.1$ m, $\rho = 1.25$ kg/m³, and $OD = 41$ m into Equation 15.24, we obtain $\omega = 35.18$ rad/s.
Finally from Equation 15.21, we compute $\theta = 6.06°$. So we have shown that if the ball is kicked with an initial velocity of 28 m/s and angular velocity of 35 rad/s at $\theta = 6°$ and $\beta = 19°$, the ball will swerve into the goal.

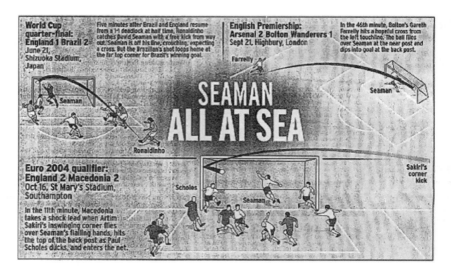

FIGURE 15.6
Illustration of Sakiri's corner kick into the far-top corner of the goalpost, as simulated in
Example 15.1.

On October 16, in the 11th min of the Euro 2004 qualifier match between
England and Macedonia, Macedonia's Artim Sakiri carried out this feat, as
illustrated in Figure 15.6.

Example 15.2
To plan a freekick from top of the box, by making the ball bend around the
players' wall (Figure 15.7).

Case 1: Right-footer kick, in the zOk vertical plane with $\boldsymbol{\omega} = \omega i_z$ (Figure 15.7).

We adopt coordinates of the final location of the ball top-near corner of
the goal-net point $C = (16.5, 5.5, 2.4)$, initial velocity $v_0 = 14$ m/s, angle of the

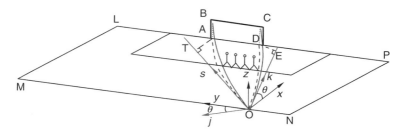

FIGURE 15.7 (See color insert following page 266.)
Left-footer kick (OB) and right-footer kick (OC) around the players' wall into the goalpost. The
ball is kicked in the vertical planes zOs and zOk.

initial velocity vector with the horizontal $\beta = 46°$, air density $\rho = 1.25 \text{ kg/m}^3$, ball radius $R = 0.1$ m, ball mass $m = 0.5$ kg.

We want to determine the angular velocity (or spin ω) to be imparted to the ball and the angle θ of the vertical plane zOk with the plane zOx, such that the ball ends up at C into the top corner of the goal. For this purpose, we first solve Equations 15.11 through 15.13, in which there are 3 unknowns, namely d, θ, and t.

In Equation 15.13, by substituting $z = 2.4$ m and the data values of $v_0 = 14 \text{ m/s}$ and $\beta = 46°$, we get $t = 1.78$ s.

From Equations 15.11 and 15.12,

$$x(D) \cos \theta + y(D) \sin \theta = OE \text{ (in Figures 15.7 and 15.4)} = (v_0 \cos \beta)t \quad (15.25)$$

By substituting $x(D) = 16.5$ m, $y(D) = 5.5$ m, $t = 1.78$ s, $v_0 = 14$ m/s, and $\beta = 46°$, we get $\theta = 12.92°$.

Also, from Equations 15.11 and 15.12, we can put down

$$y \cos \theta - x \sin \theta = d, \quad (15.26)$$

from which we get $d = 1.67$ m.

Then, from Equations 15.8 and 15.10, we get (for $d = 1.67$ m and $t = 1.78$ s) $\omega = 27.69$ rad/s.

Case 2: Left-footer kick, in the zOs vertical plane, with $\boldsymbol{\omega} = -\omega i_z$ (Figures 15.7 and 15.8).

We need the coordinates of (the far-top corner of the goal-net) point B ($= 16.5$, 12.8, 2.4) in Figure 15.7 and point A ($= 16.5$, 12.8, 0) in Figure 15.8. We take $v_0 = 16$ m/s, $\beta = 61°$, $\rho = 1.25 \text{ kg/m}^3$, $R = 0.1$ m.

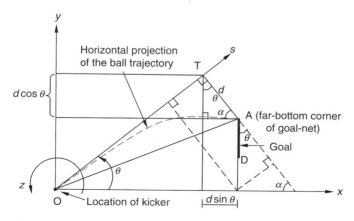

FIGURE 15.8 (See color insert following page 266.)
Geometry of the left-footer kick in the horizontal xOy plane: zOs is the vertical plane in which the ball is kicked, TA (d) is the horizontal deviation of the ball trajectory, A is the horizontal projection of the ball-location B (into the net).

The geometry of the kick and ball trajectory is illustrated in Figure 15.8. The (x, y) coordinates of the ball are now given by the following equations (instead of Equations 15.11 and 15.12):

$$x = [(v_0 \cos \beta) \cos \theta]t + d \sin \theta \qquad (15.27)$$

$$y = [(v_0 \cos \beta) \sin \theta]t - d \cos \theta \qquad (15.28)$$

For the vertical displacement of the ball, we employ Equation 15.7

$$z = (v_0 \sin \beta)t - \tfrac{1}{2}gt^2 \qquad (15.29)$$

The deviation of the ball trajectory is given by Equations 15.8 and 15.10, as

$$d = \frac{1}{2}a_L t^2 = \frac{\rho \pi R^3 \omega (v_0 \cos \beta)t^2}{4m} \qquad (15.30)$$

From Equation 15.29, by substituting $z = 2.4$ m and $\beta = 61°$, we get $t = 2.67$ s. From Equations 15.27 and 15.28,

$$x(A) \cos \theta + y(A) \sin \theta = OT = (v_0 \cos \beta)t = 20.73 \qquad (15.31)$$

By substituting $x(A) = 16.5$ m and $y(A) = 12.8$ m, we get $\theta = 44.70°$. Similarly, from Equations 15.27 and 15.28,

$$x(A) \sin \theta - y(A) \cos \theta = d(=TA) = \frac{\rho \pi R^3 \omega (v_0 \cos \beta)t^2}{4m} \qquad (15.32)$$

By substituting, in Equation 15.32, the value of $x(A)$ and $y(A)$, ρ, R, v_0, β, m (as provided earlier), we obtain $\omega = 23.10$ rad/s, $d = 2.51$ m. It would appear that the goalkeeper would find it more difficult to judge a spot kick taken by a left footer.

15.3.3 Case Study: Analysis of the Famous Ronaldinho Goal against England in the Quarter-Final of the 2002 World Cup

In the 2002 World Cup, Brazil won a free kick 30 m out on the right flank. Ronaldinho kicked the ball with his right foot and made the ball spin anti-clockwise (view from top) (Figure 15.9). The shot was aimed at the far corner of the goal and the ball just dipped under the bar. This wonder goal won the game for Brazil. So let us simulate his goal, by doing an inverse analysis. Obtaining the initial shot angles β and θ by reviewing the video tape, we carried out an inverse analysis, by computing the values of the angular velocity (ω), the initial velocity (v_0), and the in-flight time (t) of the ball (Figure 15.2).

From the video, we estimated, $\theta = 38.45°$, $\beta = 43°$, and the coordinates of the ball landing into the goal (at the top right corner B) to be: $x(B) = 30$,

Brazil won a free kick when Scholes tackled Kleberson from behind, 30 m out on the right flank. Five Brazilians lined up across the edge of the penalty area, seemingly ready for the ball to be crossed towards the far-post.

| 50th minute |
| ENGLAND 1 |
| BRAZIL 2 |

Goalkeeper David Seaman obviously expected this too. He was only 3 m off his line and took a small step forward when Ronaldinho struck the free kick.

But Ronaldinho's shot was aimed at the far-top corner and dipped just under the bar with Seaman flapping helplessly. In a few short minutes, Ronaldinho had turned the game on its head.

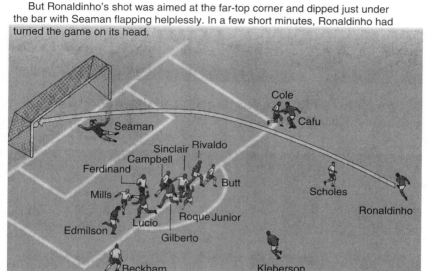

FIGURE 15.9 (See color insert following page 266.)
Ronaldinho's wonder goal, the famous right-foot free kick, that made the ball curve into the far-top corner of the goalpost and won the game for Brazil in the 2002 World-cup quarter-finals.

$y(B) = 30$, $z(B) = 2.4$. We now solve Equations 15.9 through 15.13, for the following unknown variables v_0, $\omega(= \omega i_z)$, and t. On the basis of Figure 15.2, the relevant equations are

$$x(B) = 30 = [(v_0 \cos \beta) \cos \theta]t - d \sin \theta \tag{15.33}$$

$$y(B) = 30 = [(v_0 \cos \beta) \sin \theta]t + d \cos \theta \tag{15.34}$$

$$z(B) = 2.4 = (v_0 \sin \beta)t - \tfrac{1}{2}gt^2 \tag{15.35}$$

From Equations 15.33 and 15.34,

$$x(B) \cos \theta + y(B) \sin \theta = (v_0 \cos \beta)t = 42.08 \tag{15.36}$$

where $\theta = 38.45°$ and $\beta = 43°$.
From Equation 15.35,

$$2.4 = v_0 \times (\sin 43°)t - \tfrac{1}{2}gt^2 \tag{15.37}$$

We then solve Equations 15.36 and 15.37, to obtain the values of $v_0 = 21$ m/s and $t = 2.74$ s. Then based on Equations 15.8 and 15.10, the deviation from the vertical plane in which the ball was kicked (with angular velocity ω), is computed from either Equation 15.33 or 15.34, as:

$$d = \left(\frac{\rho \pi R^3 v_0 \cos \beta t^2}{4m}\right)\omega = 4.84 \text{ m,}$$

for geometrical construction of the trajectory (as illustrated in Figure 15.2), giving:

$$\omega = 21.35 \text{ rad/s.} \tag{15.38}$$

The deviation along the goal line is $d/\cos\theta = 6.10$ m. If the ball had not spun, it would have sailed over the top of the goal. Even if the ball had been kicked with a lesser initial velocity and a smaller angle β, it would not have curved that much, and it would have been easy for David Seaman (England's goalkeeper) to grab the ball because at that time he stood just in the middle of the goal, as seen in Figure 15.9.

However, the tremendous angular velocity (of 21.35 rad/s) imparted to the ball made it swerve and sail over Seaman, making it impossible to reach it, despite of his height (as shown in Figure 15.10). From the analysis

FIGURE 15.10 (See color insert following page 266.)
Seaman tried to reach the ball but failed, and the ball just dipped below the bar into the top corner of the goal-net.

and computation, it can be seen that this wonder kick had the adroit combination of $\theta = 38.45°$, $\beta = 43°$, $v_0 = 21$ m/s, $\omega = 21.35$ rad/s, making the ball appear to come from outside and above and across the bar into the net.

15.4 Basketball Foul Throw Analysis

One can often see (in an NBA game) that when a player shoots a ball into the basket, the ball spins backwards when it leaves the hand of the player. Does this technique help to improve the accuracy (and stability) of the throw? The answer is yes. The backward spin produces an upward lift force, that enables the ball to have a higher trajectory and a bigger entering angle into the basket than a throw without spin. The advantage of a bigger entering angle (in Figure 15.11) is that it reduces the possibility of the ball hitting the rim and rebounding out, and makes the ball enter the basket squarely. Here, we have studied the case of a throw, taken from the foul line.

The following conventional equations of trajectory are used for describing the ball motion without spin (Figure 15.11):

$$x = v_0 \cos \theta t \tag{15.39}$$

$$z = v_0 \sin \theta t - \tfrac{1}{2} g t^2 \tag{15.40}$$

where θ is the angle between the initial velocity and the horizontal direction (the attacking angle or the shooting angle).

However, if a backspin (ω) is imparted to the ball, the corresponding equations for the spinning ball motion trajectory become altered to

$$x = v_0 \cos \theta t \tag{15.41}$$

$$z = v_0 \sin \theta t + \tfrac{1}{2}(a_L - g)t^2 \tag{15.42}$$

where a_L the lift acceleration (as a result of ω) is given by Equation 15.10, as before in the case of a soccer ball.

Suppose in making a foul shot, as the ball leaves the player A's hands, it is about 2.1 m above the ground, i.e., $z_0 = 2.1$ m, in Figure 15.12. Then as the ball travels in a vertical plane to enter the basket, it covers 0.95 m in the vertical direction (z direction) and 3.97 m in the horizontal direction (x direction). If a backspin is imparted to the ball when throwing, then because of a_L (in Equation 15.42), the ball has a smaller acceleration towards the ground, due to the lift force generated by the spin.

In our simulation of player A's throw, we found out computationally that shots with initial velocity of 7–8 m/s, would need to have angular velocity of backspin ranging from 1 to 10 rad/s (for different initial attacking angles to the horizontal plane) in order to be able to enter the basket. The two cases

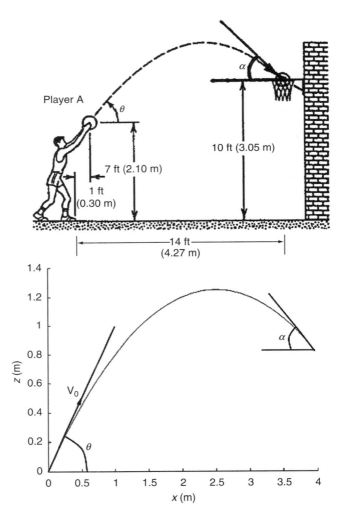

FIGURE 15.11
The definitions of θ and α for basketball trajectory.

studied (and illustrated in Figure 15.12) show that the ball with spin is lifted by the lift force compared to the ball without spin. The deviations in these two cases of $v_0 = 7$ and 8 m/s are 0.09 and 0.05 m, respectively; the corresponding lift accelerations a_L (computed from Equation 15.10) are 0.296 m/s^2 and 0.35 m/s^2, respectively. It also means that if a player wants the ball to enter the basket without spin, he has to throw the ball at a higher angle θ and a greater velocity v_0 to make the basket.

Our study also suggests that it is optimal for player A to throw the ball with an initial velocity of 7–8 m/s, with an angular velocity ω of 1–10 rad/s, and an initial shooting angle (θ) varying from 30° to 65°. By throwing in this way, he can make the ball pass directly through the middle of the

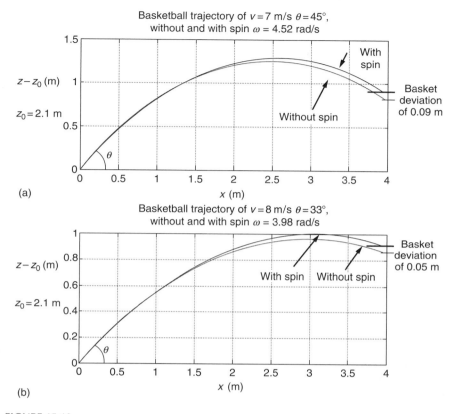

FIGURE 15.12
Basketball trajectories for different throw parameters (v, θ, ω). The origin corresponds to the point at which the ball leaves the hands at $z_0 = 2.1$ m above the ground.

basket. If he now wants to achieve the same result with a lower initial velocity, he would have to make the ball spin in the air at more than 100 rad/s, which is impossible in practice (if other parameters remain unchanged). On the other hand, if he would have to make a steeper throw with a higher velocity of throw, this is far more difficult to control. Hence for player A, the adroit combination of throw parameters to make a foul throw at an initial velocity v_0 of 7–8 m/s, is with a backspin angular velocity of $\omega = 1$–10 rad/s, at $\theta = 30°$–$65°$.

15.5 Concluding Remarks

In a soccer kick or a basketball throw, there is bound to be some spin imparted to the ball because of the orientation of the hand and foot at the time of releasing and making ball contact, respectively. This spin will alter the trajectory of the ball in an unpredictable way. Hence, in order to shoot accurately, it is better to impart a deliberate spin to the ball.

Herein, we have simulated some soccer situations involving corner kick and free kick taken just outside the box, and demonstrated how a spin-imparting kick can make the ball deceptively swerve into the goal. The type of kick, making the swerve, can also be employed by a halfback to pass to a forward or to a striker out of reach of the defenders. As regards corner kicks, it needs to be recognized that a right footer taking the kick from the left corner will make the ball swerve towards the goal, while a left footer taking the kick from the left flank will make the ball swerve away from the goal. The reverse holds good when the kick is taken from the right corner.

As regards basketball, it is often seen that tall forwards are not so prolific as shorter guards in foul shooting. Part of the reason could be that their greater height requires a flatter ball trajectory, which is more difficult. However, if they were to impart backspin to the ball, it would make the ball arch more, so as to easily enter the net.

Appendix

Equation 15.1 is derived by Watts and Bahill [5]. They identified three dimensionless parameters to describe the results of their lift-force experiments:

$$C_L = \frac{2F_L}{\rho A v^2}, \quad SP = \frac{R\omega}{v}, \quad Re = \frac{2vR}{\nu} \tag{15.A1}$$

where
  C_L is the lift coefficient
  SP is the spin coefficient
  Re is the Reynolds number
  A is the cross-sectional area of the ball
  R is the radius of the ball
  v is the horizontal velocity
  ω is the angular velocity of the spin
  ν is the kinematic viscosity of the fluid
  ρ is air density

Watts and Bahill [5] plotted C_L versus SP for different types of balls, as depicted in Figure 15.A1. Typical values of SP for baseballs range between 0.1 and 0.2. In this range, C_L varies almost linearly with SP. Infact, for SP less

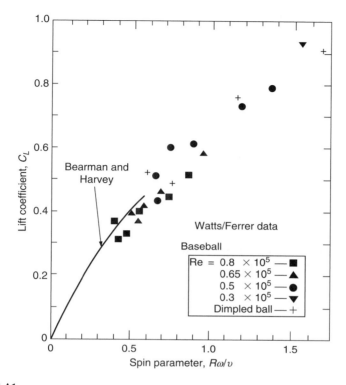

FIGURE 15.A1

Plot of experimentally derived values of C_L and SP. (Adapted from Watts, R.G. and Bahill, A.T., *Keep Your Eye on the Ball: Curve Balls, Knuckleballs, and Fallacies of Baseball*, Freeman, New York, 2000. With permission.)

than 0.4, the relation between C_L and SP is almost a straight line of slope unity. Hence, from $C_L \approx$ SP, we get

$$\frac{2F_L}{\rho A v^2} = \frac{R\omega}{v} \qquad (15.A2)$$

Substituting $A = \pi R^2$, and rearranging the equation yields:

$$F_L = \tfrac{1}{2}\rho\pi R^3 \omega v \qquad (15.A3)$$

Out of interest, it may be mentioned that the transverse force (F_L), on a cylindrical body (of length L and radius R) is given by

$$F_L^C = \pi\rho_a(2LR)R\omega v_\infty = \pi\rho_a A^c R\omega \vartheta_{oh} = 2\rho_a V^c \omega \vartheta_{oh} \qquad (15.A4)$$

where

v_∞ (the air-flow velocity away from the body) corresponds to the ϑ_{oh} the initial horizontal velocity imparted to the body

$2LR$ is the vertical area (A^c) of the projection of the surface exposed to the flow

R is the radius of body cross-section

$\pi R^2 L$ represent the body volume (V^c)

On the other hand, for airflow past a spherical body, we have (from Equation 15.A3):

$$F_L^S = \frac{1}{2\pi^{1/2}} \rho_a A_s^{3/2} \omega \vartheta_{oh} = \frac{3}{8} \rho_a V^S \omega \vartheta_{oh} \tag{15.A5}$$

where

A_s is the projected area exposed to airflow

V^S is the volume of the spherical body

A comparison of expressions Equations 15.A4 and 15.A5 shows that F_L^C is about 5–10 times bigger than F_L^S, as can be expected.

References

1. Watts, R.G. and Ferrer, R. 1987, The lateral force on a spinning sphere, *American Journal of Physics*, 55(1), 40–44.
2. Miller, S. and Bartlett, R. 1996, The relationship between basketball shooting kinematics, distance and playing position, *Journal of Sports Sciences*, 4, 243–253.
3. Mehta, R.D. and Wood, D.H. 1980, Aerodynamics of the cricket ball, *New Scientist*, 87(1213), 442–447.
4. Mehta, R.D. 1985, Aerodynamics of sports balls, *Annual review of fluid Mechanics*, 17, 151–189.
5. Watts, R.G. and Bahill, A.T. 2000, *Keep Your Eye on the Ball: Curve Balls, Knuckleballs, and Fallacies of Baseball*, Freeman, New York.

16

Mechanics of Baseball Pitching and Batting

A. Terry Bahill and David G. Baldwin

CONTENTS

This chapter discusses the pitch, the bat–ball collision, and the swing of the bat. Section 16.1, based on Bahill and Baldwin [1], describes the pitch in terms of the forces on the ball and the ball's movement. Section 16.2, based on Bahill [2] and Bahill and Baldwin [3], discusses bat–ball collisions in terms of the sweet spot of the bat and the coefficient of restitution (CoR). Section 16.3 based on Bahill and Baldwin [3], presents a model for bat–ball collisions, a new performance criterion, and the resulting vertical sweet spot of the bat. Section 16.4, based on Bahill [2] and Bahill and Karnavas [4], presents experimental data describing the swing of a bat and suggests ways of choosing the best bat for individual batters.

This chapter is about the mechanics of baseball. To understand the whole baseball enterprise, read Bahill et al. [5]. They populate a Zachman framework with nearly 100 models of nearly all aspects of baseball.

16.1 Pitch

Batters say that the ball hops, drops, curves, breaks, rises, sails, or tails away. The pitcher might tell you that he or she throws a fastball, screwball, curveball, drop curve, flat curve, slider, changeup, split-fingered fastball, splitter, forkball, sinker, cutter, two-seam fastball, or four-seam fastball. This sounds like a lot of variation. However, no matter how the pitcher grips or throws the ball, once it is in the air its motion depends only on gravity, its velocity, and its spin.* In engineering notation, these pitch characteristics are described respectively by a "linear velocity vector" and an "angular velocity vector," each with magnitude and direction. The magnitude of the linear velocity vector is called "pitch speed" and the magnitude of the angular velocity vector is called the "spin rate." These vectors produce a force acting on the ball that causes a deflection of the ball's trajectory.

In 1671, Isaac Newton [6] noted that spinning tennis balls experienced a lateral deflection mutually perpendicular to the direction of flight and of spin. Later in 1742, Benjamin Robins [7] bent the barrel of a musket to produce spinning musket balls and also noted that the spinning balls experienced a lateral deflection perpendicular to the direction of flight and to the direction of spin. In 1853, Gustav Magnus (see Refs. [8 and 9]) studied spinning shells fired from rifled artillery pieces and found that the range depended on crosswinds. A crosswind from the right lifted the shell and gave it a longer range: a crosswind from the left made it drop short. Kutta and Joukowski studied cylinders spinning in an airflow. They were the first to model this force with an equation, in 1906. Although these four experiments sound quite different (and they did not know about each other's

---

* This statement is true even for the knuckleball, because it is the shifting position of the seams during its slow spin en route to the plate that gives the ball its erratic behavior. Equations 16.1 through 16.7 give specific details about the forces acting on the ball.

papers), they were all investigating the same underlying force. This force, commonly called the Magnus force, operates when a spinning object (like a baseball) moves through a fluid (like air) which results in it being pushed sideways. Two models explain the basis of this Magnus force: one is based on conservation of momentum and the other is based on Bernoulli's principle [10–12]. We will now apply the right-hand rules to the linear velocity vector and the angular velocity vector in order to describe the direction of the spin-induced deflection of the pitch.

16.1.1 Right-Hand Rules and the Cross Product

In vector analysis, the right-hand rules specify the orientation of the cross product of two vectors. Figure 16.1a shows that the cross (or vector) product, written as **u** × **v**, of nonparallel vectors **u** and **v** is perpendicular to the plane of **u** and **v**: the symbol × represents the cross product. The angular right-hand rule, illustrated in Figure 16.1b, is used to specify the orientation of a cross product **u** × **v**. If the fingers of the right hand are curled in the direction from **u** to **v**, the thumb will point in the direction of the vector **u** × **v**. The coordinate right-hand rule is illustrated in Figure 16.1c. The index finger, middle finger, and thumb point in the directions of **u**, **v**, and **u** × **v**, respectively, in this local coordinate system. The vectors of Figure 16.1d represent the angular velocity vector (spin), the linear velocity vector (direction), and the spin-induced deflection force of a spinning pitch.

16.1.2 Right-Hand Rules Applied to a Spinning Ball

The spin axis of the pitch can be found by using the angular right-hand rule. As shown in Figure 16.2, if you curl the fingers of your right hand in the direction of spin, your extended thumb will point in the direction of the spin axis.

The direction of the spin-induced deflection force can be described using the coordinate right-hand rule. Point the thumb of your right hand in the

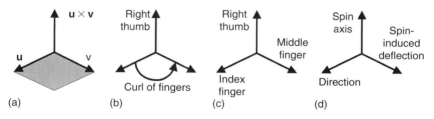

FIGURE 16.1
(a) The vector (or cross) product of vectors **u** and **v** is perpendicular to the plane of **u** and **v**. (b) The angular right-hand rule: If the fingers of the right hand are curled in the direction from **u** to **v**, the thumb will point in the direction of the vector **u** × **v**, which is pronounced *yōō cross vē*. (c) The coordinate right-hand rule: The index finger, the middle finger, and the thumb point in the directions of **u**, **v**, and **u** × **v**, respectively. (d) For a baseball, the cross product of the spin axis and the direction of motion gives the direction of the spin-induced deflection. (From Bahill, A.T., http://www.sie.arizona.edu/sysengr/slides. With permission. Copyright 2004.)

FIGURE 16.2
The angular right-hand rule: For a rotating object, if the fingers are curled in the direction of rotation, the thumb points in the direction of the spin axis. (Photograph courtesy of Zach Bahill. From Bahill, A.T., http://www.sie.arizona.edu/sysengr/slides. With permission. Copyright 2004.)

direction of the spin axis (as determined from the angular right-hand rule), and point your index finger in the direction of forward motion of the pitch (Figure 16.3). Bend your middle finger so that it is perpendicular to your index finger. Your middle finger will be pointing in the direction of the spin-induced deflection (of course, the ball also drops due to gravity). The spin-induced deflection force will be in a direction represented by the cross product of the angular and the linear velocity vectors of the ball: angular velocity × linear velocity = spin-induced deflection force. Or mnemonically, Spin axis × Direction = Spin-induced deflection (SaD Sid). This acronym only gives the direction of deflection. The equation yielding the magnitude of the spin-induced deflection force is more complicated and is discussed in Section 16.1.4.

16.1.3 Deflection of Specific Kinds of Pitches

Figures 16.4 and 16.5 show the directions of spin (circular arrows) and spin axes* (straight arrows) of some common pitches from the perspective of the pitcher (Figure 16.4 represents a right-hander's view and Figure 16.5 a

---

* These could be labeled spin vectors, because they suggest both magnitude and direction.

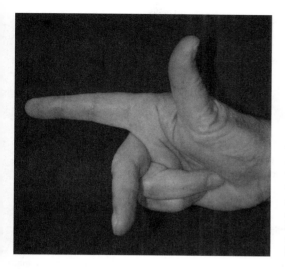

FIGURE 16.3
The coordinate right-hand rule: For a baseball, if the thumb points in the direction of the spin axis and the index finger points in the direction of forward motion of the pitch, then the middle finger will point in the direction of the spin-induced deflection. (Photograph courtesy of Zach Bahill. From Bahill, A.T., http://www.sie.arizona.edu/sysengr/slides. With permission. Copyright 2004.)

left-hander's view). We will now consider the direction of deflection of each of these pitches.

Figure 16.4 illustrates the fastball, curveball, and slider, distinguished by the direction of the spin axis. When a layperson throws a ball, the fingers are the last part of the hand to touch the ball. If the ball is thrown with an overhand motion, the fingers touch the ball on the bottom and thus impart backspin to the ball. Most pitchers throw the fastball with a three-quarter arm delivery, which means the arm does not come straight over the top, but rather it is in between over the top and sidearm. This delivery rotates the spin axis from the horizontal as shown in Figure 16.4. The curveball is also thrown with a three-quarter arm delivery, but this time the pitcher rolls his or her wrist and causes the fingers to sweep in front of the ball. This produces a spin axis as shown for the curveball of Figure 16.4. This pitch

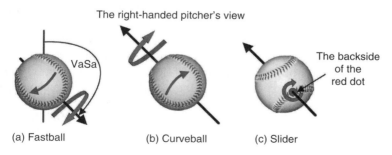

The right-handed pitcher's view

(a) Fastball   (b) Curveball   (c) Slider

FIGURE 16.4 (See color insert following page 266.)
The direction of spin (circular arrows) and the spin axes (straight arrows) of a three-quarter arm (a) fastball, (b) curveball, and (c) slider from the perspective of a right-handed pitcher, meaning the ball is moving into the page. VaSa is the angle between the vertical axis and the spin axis. (From Bahill, A.T., http://www.sie.arizona.edu/sysengr/slides. With permission. Copyright 2005.)

The left-handed pitcher's view

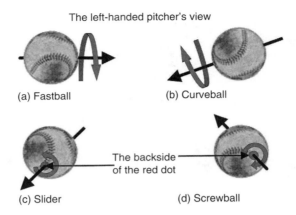

(a) Fastball

(b) Curveball

(c) Slider

The backside
of the red dot

(d) Screwball

FIGURE 16.5 (See color insert following page 266.)
The direction of spin (circular arrows) and the spin axes (straight arrows) of an overhand (a) fastball, (b) curveball, (c) slider, and (d) screwball from the perspective of a left-handed pitcher, meaning the ball is moving into the page. (From Bahill, A.T., http://www.sie.arizona.edu/sysengr/slides. With permission. Copyright 2004.)

will curve at an angle from upper right to lower left as seen by a right-handed pitcher. Thus, the ball curves diagonally. The advantage of the drop in a pitch is that the sweet area of the bat is about 2 in. long (5 cm) [2] but only one-third of an inch (8 mm) high [3,13]. Thus, when the bat is swung in a horizontal plane, a vertical drop is more effective than a horizontal curve at taking the ball away from the bat's sweet area.

The overhand fastball shown in Figure 16.5 has a predominate backspin, which gives it lift, thereby decreasing its fall due to gravity. But when the fastball is thrown with a three-quarter arm delivery (as in Figure 16.4), the lift is reduced, but it introduces lateral deflection (to the right for a right-handed pitcher). A sidearm fastball (from a lefty or a righty) tends to have some topspin, because the fingers put pressure on the top half of the ball during the pitcher's release. The resulting deflection augments the effects of gravity and the pitch "sinks."

The slider is thrown somewhat like a football. Unlike the fastball and curveball, the spin axis of the slider is not perpendicular to the direction of forward motion (although the direction of deflection is still perpendicular to the cross product of the spin axis and the direction of motion). As the angle between the spin axis and the direction of motion decreases, the magnitude of deflection decreases, but the direction of deflection remains the same. If the spin axis is coincident with the direction of motion, as for the backup slider, the ball spins like a bullet and undergoes no deflection. Therefore, a right-handed pitcher usually throws the slider so that he or she sees the axis of rotation pointed up and to the left. This causes the ball to drop and curve from the right to the left. Rotation about this axis allows some batters to see a red dot at the spin axis on the top right side of the ball (see Figure 16.6). Baldwin et al. [14] and Bahill et al. [15] show pictures of this spinning red dot. Seeing this red dot is important, because if the batter can see this red

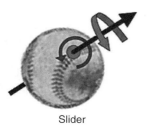

FIGURE 16.6
The batter's view of a slider thrown by a right-handed pitcher: the ball is coming out of the page. The red dot reveals that the pitch is a slider. (From Bahill, A.T., http://www.sie.arizona. edu/sysengr/slides. With permission. Copyright 2004.)

Slider

dot, then he or she will know the pitch is a slider and he or she can better predict its trajectory. We questioned 15 former major league hitters; 8 remembered seeing this dot, but 2 said it was black or dark gray rather than red. For the backup slider, the spin causes no horizontal deflection and the batter might see a red dot in the middle of the ball.

16.1.4 Forces Acting on a Ball in Flight

A ball in flight is influenced by three forces as shown in Figure 16.7: gravity pulling downward, air resistance or drag operating in the opposite direction of the ball's motion, and, if it is spinning, a force perpendicular to the direction of motion. The force of gravity is downward, $F_{gravity} = mg$, where m is the mass of the ball and g is the gravitation constant: its magnitude is the ball's weight. The magnitude of the force opposite to the direction of flight is

$$F_{drag} = 0.5 \; \rho \pi r_{ball}^2 C_d v_{ball}^2 \qquad (16.1)$$

where
ρ is air mass density
v_{ball} is the ball speed
r_{ball} is the radius of the ball [10, p. 161]

Typical values for these parameters are given in Table 16.1. Of course SI units can be used in this equation, but if English units are to be used in Equations 16.1 through 16.7, then ρ is measured in lb-s^2/ft^4, v_{ball} is measured in feet per second (ft/s), r_{ball} is measured in feet (ft), F_{drag} is measured in

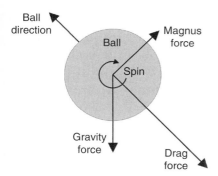

FIGURE 16.7
The forces acting on a spinning ball moving in a fluid. (From Bahill, A.T., http://www.sie.arizona. edu/sysengr/slides. With permission. Copyright 2007.)

TABLE 16.1

Typical Baseball and Softball Parameters for Line Drives

	Major League Baseball	Little League	NCAA Softball
Ball	Baseball	Baseball	Softball
Ball weight (oz)	5.125	5.125	6.75
Ball weight, $F_{gravity}$ (lb)	0.32	0.32	0.42
Ball radius (in.)	1.45	1.45	1.9
Ball radius, r_{ball} (ft)	0.12	0.12	0.16
Pitch speed (mph)	85	50	65
Pitch speed, v_{ball} (ft/s)	125	73	95
Distance from front of rubber to tip of plate (ft)	60.5	46	43
Pitcher's release point: (distance from tip of plate, height) (ft)	(55.5, 6)	(42.5, 5)	(40.5, 2.5)
Bat–ball collision point: (distance from tip of plate, height) (ft)	(3, 3)	(3, 3)	(3, 3)
Bat type	Wooden C243	Aluminum	Aluminum
Typical bat weight (oz)	32	23	25
Maximum bat radius (in.)	1.375	1.125	1.125
Speed of sweet spot (mph)	60	45	50
Coefficient of restitution (CoR)	0.54	0.53	0.52
Backspin of batted ball (rps)	10–70	10–70	10–70
Backspin of batted ball, ω (rad/s)	63–440	63–440	63–440
Desired ground contact point from the plate (ft)	120–240	80–140	80–150
Air weight density (lb_m/ft^3)	0.075	0.075	0.075
Air mass density ρ ($lb\text{-}s^2/ft^4$)	0.0023	0.0023	0.0023

Note: Air density is inversely related to temperature, altitude, and humidity.

pounds (lb), and in later equations ω is measured in radians per second (rad/s). For the drag coefficient, C_d, we use a value of 0.5. This drag coefficient is discussed in Section 16.1.7.

Table 16.2 shows typical parameters for major league pitches. We estimate that 90% of major league pitches fall into these ranges, except for a few pitchers that have consistently slower fastballs. The pitch speed is the speed at the release point: the ball will be going 10% slower when it crosses the

TABLE 16.2

Typical Values for Major League Pitches

Type of Pitch	Initial Speed (mph)	Initial Speed (m/s)	Spin Rate (rpm)	Spin Rate (rps)	Rotations between Pitcher's Release and the Point of Bat–Ball Contact
Fastball	85–95	38–42	1200	20	8
Slider	80–85	36–38	1400	23	10
Curveball	70–80	31–36	2000	33	17
Changeup	60–70	27–31	400	7	4
Knuckleball	60–70	27–31	30	½	¼

plate. In this chapter, the equations are general and should apply to many types of spinning balls. However, whenever we give specific numerical values they are (unless otherwise stated) for major league baseball.

The earliest empirical equation for the transverse force on a spinning object moving in a fluid is the Kutta–Joukowski lift theorem

$$\mathbf{L} = \rho \mathbf{U} \times \mathbf{\Gamma} \tag{16.2}$$

where
  $\mathbf{L}$ is the lift force per unit length of cylinder
  ρ is the fluid density
  $\mathbf{U}$ is the fluid velocity
  $\mathbf{\Gamma}$ is the circulation around the cylinder
  $\mathbf{L}$, $\mathbf{U}$, and $\mathbf{\Gamma}$ are vectors

When this equation is tailored for a baseball [10, pp. 77–81], we get the magnitude of the spin-induced force acting perpendicular to the direction of flight

$$F_{\text{perpendicular}} = F_{\text{Magnus}} = 0.5\,\rho\pi r_{\text{ball}}^3 \omega\, v_{\text{ball}} \tag{16.3}$$

where ω is the spin rate. This is usually called the Magnus force. This force can be decomposed into a force lifting the ball up and a lateral force pushing it sideways.

$$F_{\text{upward}} = 0.5\,\rho\pi r_{\text{ball}}^3 \omega\, v_{\text{ball}} \sin \text{VaSa} \tag{16.4}$$

where VaSa is the angle between the vertical axis and the spin axis (Figures 16.4 and 16.8). The magnitude of the lateral force is

$$F_{\text{sideways}} = 0.5\,\rho\pi r_{\text{ball}}^3 \omega\, v_{\text{ball}} \cos \text{VaSa} \tag{16.5}$$

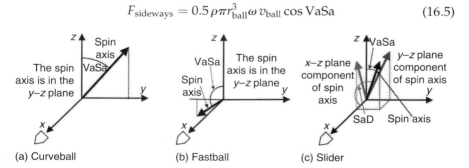

(a) Curveball   (b) Fastball   (c) Slider

FIGURE 16.8 (See color insert following page 266.)
Rectangular coordinate system and illustration of the angles VaSa and SaD for (a) curveball, (b) three-quarter arm fastball, and (c) slider all thrown by a right-handed pitcher. The origin is the pitcher's release point. For the curveball, the spin axis is in the y–z plane. For the fastball, the spin axis is also in the y–z plane, but it is below the y-axis. For the slider, the spin axis has components in both the y–z and x–z planes. (From Bahill, A.T., http://www.sie.arizona.edu/sysengr/slides. With permission. Copyright 2006.)

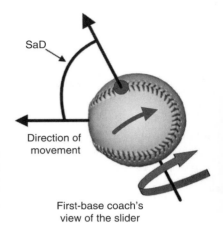

FIGURE 16.9 (See color insert following page 266.) The first-base coach's view of a slider thrown by a right-handed pitcher. This illustrates the definition of the angle SaD. (From Bahill, A.T., http://www.sie.arizona.edu/sysengr/slides. With permission. Copyright 2007.)

SaD

Direction of movement

First-base coach's view of the slider

Finally, if the spin axis is not perpendicular to the direction of motion (as in the case of the slider), the magnitude of the cross product of these two vectors will depend on the angle between the spin axis and direction of motion; this angle is called SaD (Figures 16.8 and 16.9). In aeronautics, it is called the angle of attack.

$$F_{\text{lift}} = 0.5\,\rho\pi r_{\text{ball}}^3\,\omega\,v_{\text{ball}}\,\sin\text{VaSa}\,\sin\text{SaD} \qquad (16.6)$$

$$F_{\text{lateral}} = 0.5\,\rho\pi r_{\text{ball}}^3\,\omega\,v_{\text{ball}}\,\cos\text{VaSa}\,\sin\text{SaD} \qquad (16.7)$$

During the pitch, gravity is continuously pulling the ball downward, which changes the direction of motion of the ball by 5° to 10° during its flight. However, the ball acts like a gyroscope, so the spin axis does not change. This means that, for a slider, the angle SaD increases and partially compensates for the drop in ball speed in Equations 16.6 and 16.7.

16.1.5 Comparison of the Slider and Curveball

Let us now compare the magnitude of this lateral spin-induced deflection force (Equation 16.7) for two specific pitches, namely the slider and the curveball. The magnitude of the lateral spin-induced deflection of the slider is less than that of a curveball for the following four reasons:

1. For the curveball, the angle between the spin axis and the direction of motion (SaD) is around 85°. For the slider, it is around 60°. The magnitude of the cross product is proportional to the sine of this angle. Therefore, the slider's deflection force is less than the curveball's by the ratio $\frac{\sin 60}{\sin 85}$: the slider force equals 0.87 times the curveball force. The angle between the vertical axis and the spin axis (VaSa) has no effect because it is about the same for the slider and the curveball.

2. Curveball spins at up to 33 revolutions per second (rps) and the slider probably spins around 23 rps [16], and hence the slider's deflection force is smaller because of its slower rotation. Thus, the slider force equals 0.7 times the curveball force.

3. Deflection force also depends on the speed of the pitch. Assume a 75 mph (34 m/s) curveball and an 85 mph (38 m/s) slider: the slider force equals 1.13 times the curveball force.

    Therefore, for the three effects of this example, the total slider force equals 0.69 times the curveball force.

4. Furthermore, the curveball is slower, so it is in the air longer. Therefore, the deflection force has longer to operate and the total deflection due to this effect is greater. An 85 mph (38 m/s) slider travels from the pitcher's release point, 5 ft (1.5 m) in front of the rubber, to the point of bat–ball collision, 1.5 ft (0.5 m) in front of the plate, in 453 ms, whereas a 75 mph (34 m/s) curveball is in the air for 513 ms: squaring these durations gives a ratio of 0.78. The total deflection is proportional to total force times duration squared: therefore, the ratio deflection of the slider with respect to the curveball is

$$(\text{ratio-force}_{\text{spin axis}})(\text{ratio-force}_{\text{spin rate}})(\text{ratio-force}_{\text{speed}})$$
$$(\text{ratio-durations squared}) = (0.87)(0.7)(1.13)(0.78) = 0.54$$

In summary, the magnitude of the lateral spin-induced deflection of the slider is about half that of the curveball.

The screwball (sometimes called a "fadeaway" or "in-shoot") was made popular in the early 1900s by Christy Mathewson and Mordecai "Three Fingered" Brown and was repopularized by the left-hander Carl Hubbell in the 1930s. Therefore, we show it from the left-hander's perspective in Figure 16.5. Of the pitches shown in Figures 16.4 and 16.5, it is the least used, in part, because the required extended pronation of the hand strains the forearm and elbow. At release, the fingers are on the inside and top of the ball. The deflection of the left-hander's screwball is the same as the deflection of a right-hander's slider. The spin of the screwball is basically like that of a slider, so its deflection will be less than that of a curveball for the reasons given above.

The direction of deflection of these pitches is variable depending on the direction of the spin axis. The direction of this axis varies with the angle of the arm during delivery and the position of the fingers on the ball at the time of release. By controlling his or her arm angle and finger positions, the pitcher controls the direction of deflection.

16.1.6 Vertical Deflection

Tables 16.3 and 16.4 show the magnitude of the spin-induced drop for three kinds of pitches at various speeds, as determined by our simulations. Our

TABLE 16.3

Gravity- and Spin-Induced Drop (with English Units)

Pitch Speed and Type	Spin Rate (rpm)	Duration of Flight (ms)	Drop due to Gravity (ft)	Spin-Induced Vertical Drop (ft)	Total Drop (ft)
95 mph fastball	−1200	404	2.63	−0.91	1.72
90 mph fastball	−1200	426	2.92	−0.98	1.94
85 mph slider	+1400	452	3.29	+0.74	4.03
80 mph curveball	+2000	480	3.71	+1.40	5.11
75 mph curveball	+2000	513	4.24	+1.46	5.70

baseball trajectory simulator includes the effects of lift and drag due to spin on the ball [10,11,17,18]. Looking at one particular row, a 90 mph (40.2 m/s) fastball is in the air for 426 ms, so it drops 2.92 ft (0.89 m) due to gravity ($\frac{1}{2} g t^2$, where the gravitational constant g is 32.2 ft/s² (9.8 m/s²) and t is the time from release until the point of bat–ball collision). But the backspin lifts this pitch 0.98 ft (0.3 m), producing a total drop of 1.94 ft (0.59 m) as shown in Tables 16.3 and 16.4. In the spin rate column, negative numbers are backspin and positive numbers are topspin. In the spin-induced vertical drop column, negative numbers mean the ball is being lifted up by the Magnus force. All of the pitches in Tables 16.3 and 16.4 were launched horizontally—that is, with a launch angle of zero. The angle VaSa was also set to zero (simulating an overhand delivery): therefore, pitches thrown with a three-quarter arm delivery would have smaller spin-induced deflections than given in Tables 16.3 and 16.4.

Vertical misjudgment of the potential bat–ball impact point is a common cause of batters' failure to hit safely [3,13]. The vertical differences between the curveballs and fastballs in Tables 16.3 and 16.4 are greater than 3 ft (1 m), whereas the difference between the two speeds of fastballs is around 3 in. (7 cm) and the difference between the two speeds of curveballs is around 7 in. (18 cm). However, the batter is more likely to make a vertical error because speed has been misjudged than because the kind of pitch has been misjudged [3,13]. A vertical error of as little as one-third of an inch (8 mm) in the batter's swing will generally result in a failure to hit safely [3,13], as is shown in Section 16.3.

TABLE 16.4

Gravity- and Spin-Induced Drop (with SI Units)

Pitch Speed and Type	Spin Rate (rad/s)	Duration of Flight (ms)	Drop due to Gravity (m)	Spin-Induced Vertical Drop (m)	Total Drop (m)
42.5 m/s fastball	−126	404	0.80	−0.28	0.52
40.2 m/s fastball	−126	426	0.89	−0.30	0.59
38.0 m/s slider	+147	452	0.95	+0.23	1.23
35.8 m/s curveball	+209	480	1.13	+0.43	1.56
33.5 m/s curveball	+209	513	1.29	+0.45	1.74

The spin on the pitch also causes a horizontal deflection of the ball. In "deciding" whether to swing, the horizontal deflection is more important than the vertical, because the umpire's judgment with respect to the corners of the plate has more precision than his or her judgment regarding the top and bottom of the strike zone. However, after the batter has decided to swing and is trying to "track and hit" the ball, the vertical deflection becomes more important.

The right-hand rules for the lateral deflection of a spinning ball also apply to the batted ball, except it is harder to make predictions about the magnitude of deflection because we have no data about the spin on a batted ball. The right-hand rules can be applied to tennis, where deflections are similar to baseball, but not to American football, because spin-induced deflections of a football are small [19]. A professional quarterback throws a pass at around 80 mph with 12 rps spin.

16.1.7 Modeling Philosophy

Although our equations and discussion might imply great confidence and precision in our numbers, it is important to note that our equations are only models. The Kutta–Joukowski equation and subsequent derivations are not theoretical equations, they are only approximations fit to experimental data. There are more complicated equations for the forces on a baseball (e.g., see [20–25]). Furthermore, there is much that we did not include in our model. We ignored the possibility that air flowing around certain areas of the ball might change from turbulent to laminar flow en route to the plate. Our equations did not include effects of shifting the wake of turbulent air behind the ball. En route to the plate, the ball loses 10% of its linear velocity and 2% of its angular velocity: we did not include this reduction in angular velocity. We ignored the difference between the center of gravity and the geometrical center of the baseball [9]. We ignored possible differences in the moments of inertia of different balls. We ignored the precession of the spin axis. In computing velocities due to bat–ball collisions, we ignored deformation of the ball and energy dissipated when the ball slips across the bat surface. Finally, as we have already stated, we treated the drag coefficient as a constant.

We used a value of 0.5 for the drag coefficient, C_d. However, for speeds over 80 mph this drag coefficient may be smaller [10, p. 157; 20,23,24]. There are no wind-tunnel data showing the drag coefficient of a spinning baseball over the range of velocities and spin rates that characterize a major league pitch. Sawicki et al. [22] summarize data from a half-dozen studies of spinning baseballs, nonspinning baseballs, and other balls and showed C_d between 0.15 and 0.5. In most of these studies, the value of C_d depended on the speed of the airflow. In the data of Ref. [25], the drag coefficient can be fit with a straight line of $C_d = 0.45$, although there is considerable scatter in these data. The drag force causes the ball to lose about 10% of its speed en route to the plate. The simulations of Ref. [26] also studied this loss in speed.

Data shown in Figure 9 of Ref. [26] for the speed lost en route to the plate can be nicely fitted with *PercentSpeedLost* $= 20$ C_d, which implies $C_d = 0.5$.

It is somewhat surprising that given the multitude of modern computer camera pitch-tracking devices such as the QuesTec system, the best-published experimental data for the spin rate of different pitched baseballs come from Selin's cinematic measurements of baseball pitches [16]. And we have no experimental data for the spin on the batted ball. Table 16.2 summarizes our best estimates of speed and spin rates for most popular major league pitches.

There is uncertainty in the numerical values used for the parameters in our equations. However, the predictions of the equations match baseball trajectories quite well. When better experimental data become available for parameters such as C_d and spin rate, then values of other parameters will have to be adjusted to maintain the match between the equations and actual baseball trajectories.

The value of this present study lies in comparisons rather than absolute numbers. Our model emphasizes that the right-hand rules show the direction of the spin-induced deflections of a pitch. The model provides predictive power and comparative evaluations relative to the behavior of all kinds of pitches.

Stark [27] explained that models are ephemeral: they are created, they explain a phenomenon, they stimulate discussion, they foment alternatives, and then they are replaced by new models. When there are better wind-tunnel data for the forces on a spinning baseball, then our equations for the lift and drag forces on a baseball will be supplanted by newer parameters and equations. But we think our models, based on the right-hand rules showing the direction of the spin-induced deflections, will have permanence: they are not likely to be superseded.

16.1.8   Somatic Metaphors of Pitchers

A pitcher uses his or her hand as a metaphor for the ball when asked to demonstrate the trajectory of a particular kind of pitch (such as a screwball). But he or she derives a mental model of a specific pitch from the feelings of arm angle and his or her fingers on the ball as the pitch is being released. By imagining slight shifts in these sensations, the pitcher can create subtly differing models that can provide pitch variability to his or her repertoire. For example, he or she might model the screwball with fingers on top of the ball when it is released (resulting in a downward deflection) or with fingers on the side of the ball (resulting in a flatter deflection).

The batter finds it hard to distinguish subtle differences in the spin direction of a specific kind of pitch. For example, a 95 mph (42.5 m/s) fastball thrown directly overhand looks much like a 95 mph fastball thrown with the arm angle lowered by 20°. The vertical difference in the potential bat–ball contact point, however, is significant. For the 95 mph fastball with a 1200 rpm backspin shown in Tables 16.3 and 16.4, the pitch thrown with the lower arm angle would drop about three-quarters of an inch (2 cm) farther than the overhand pitch. Three-quarters of an inch is bigger than the vertical sweet spot. Mental models of pitch differences allow the pitcher to take advantage of the batter's difficulty in recognizing a wide variety of spin directions and detecting small shifts in arm angle.

16.1.9 Summary

Somatic metaphors are pervasive in everyday life, so it is not surprising to find that baseball pitchers make use of these modeling devices in their work. We have shown how a pair of widely used engineering metaphors, the right-hand rules, provides a formalized approach to describing the pitchers' mental models, allowing prediction of the deflection direction of each pitch. Besides describing the behavior of the pitched ball, these rules can also be used in describing the deflection direction of the batted ball. To determine the direction of deflection of the pitched or the batted ball, point the thumb of your right hand in the direction of the Spin axis and your index finger in the Direction of motion of the ball; your middle finger will indicate the direction of the Spin-induced deflection (SaD Sid).

16.2 Bat–Ball Collisions

16.2.1 Sweet Spot of the Bat

For skilled batters, we assume that most bat–ball collisions occur near the sweet spot of the bat, which is, however, difficult to define precisely. The horizontal sweet spot has been defined as the center of percussion (CoP), the node of the fundamental bending vibrational mode, the antinode of the hoop mode, the maximum energy transfer area, the maximum batted-ball speed area, the maximum CoR area, the minimum energy loss area, the minimum sensation area, and the joy spot [2,28]. Let us now examine each of these definitions.

1. **Center of percussion.** For most collision points, when the ball hits the bat it produces a translation of the bat and a rotation of the bat. However, if the ball hits the bat at the center of mass there will be a translation but no rotation. Whereas, if the bat is fixed at a pivot point and the ball hits the bat at the CoP for that pivot point, then there will be a rotation about that pivot point but no translation (and therefore no sting on the hands). The pivot point and the CoP for that pivot point are conjugate points, because if instead the bat is fixed at the CoP and the ball hits the pivot point then there will be a pure rotation about the CoP. The CoP and its pivot point are related by the following equation derived by Sears et al. [29], where the variables are defined in Figure 16.10:

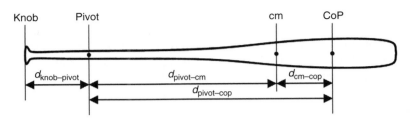

FIGURE 16.10
Definition of distances on a bat. (From Bahill, A.T., http://www.sie.arizona.edu/sysengr/slides. With permission. Copyright 2001.)

$$d_{\text{pivot-cop}} = \frac{I_{\text{pivot}}}{m_{\text{bat}}d_{\text{pivot-cm}}} \qquad (16.8)$$

The CoP is not one fixed point on the bat. There is a different CoP for every pivot point. If the batter chokes up on the bat, the pivot point (and consequently the CoP) will change. In fact, the pivot point might even change during an individual swing. In this chapter, we assume that the pivot point is 6 in. (15 cm) from the knob.

There are three common experimental methods for determining the CoP.

Method 1: Pendular motion: Hang a bat at a point 6 in. (15 cm) from the knob with 2 or 3 ft (1 m) of string. Hit the bat with an impact hammer. Hitting it off the CoP will make it flop like a fish out of water, because there is a translational force and a rotational force at the pivot point. Hitting it near the CoP will make it swing like a pendulum (as shown in Figures 12 and 13 of Ref. [29]).

Method 2: Toothpick pivot: Alternatively, you can pivot the bat on a toothpick through a hole at the pivot point 6 in. from the knob and strike the bat at various places. When struck near the CoP for that pivot point the toothpick will not break. At other places, the translational forces will break the toothpick.

Method 3: Equivalent pendulum: A third method for measuring the distance between the pivot point and the CoP is to make a pendulum by putting a mass equal to the bat's mass on a string and adjusting its length until the pendulum's period and the bat's period are the same. This method has the smallest variability.

2. **Node of the fundamental mode**. The node of the fundamental bending vibrational mode is the area where this vibrational mode (roughly between 150 and 200 Hz for a wooden bat) of the bat has a null point [20,30–33]. To find this node, grip a bat about 6 in. from the knob with your fingers and thumb. Lightly tap the barrel at various points with an impact hammer. The area where you feel no vibration and hear almost nothing (except the secondary

vibration crack or ping at 500 to 800 Hz) is the node. A rubber mallet could be used in place of an impact hammer: the point is, the hammer itself should not produce any noise. The antinode of the third bending vibrational mode may also be important [34].

3. **Antinode of the hoop mode.** For hollow metal and composite baseball and softball bats, there is another type of vibration, called a hoop vibration. The walls of a hollow bat deform during a bat–ball collision. The walls are crushed in and then bounce back out. This vibration can be modeled as a hoop or a ring around the bat; this ring deforms like the vertical cross-sectional area of a water drop falling from a faucet; first the water drop is tall and skinny, in free fall it is round, and when it hits the ground it becomes short and fat. The location of the antinode of the first hoop mode is another definition of the sweet spot [34,35].

4. **Maximum energy transfer area.** A collision at the maximum energy transfer area transfers the most energy to the ball [36]. This derivation is reproduced in Ref. [10]. This definition says that the best contact area on the bat is that which loses the least amount of energy to bat translation, rotation, vibration, etc. This would be a more useful definition if it specified maximum "useful" energy transfer—the useful energy is that which moves the ball in the same direction as the trajectory of the bat. In this definition, energy stored in the spin of the ball is not useful.

5. **Maximum batted-ball speed area.** There is an area of the bat that produces the maximum batted-ball speed [32,33,37,38]. This area is about 5 or 6 in. from the end of the barrel for wooden bats and about 7 in. from the end of the barrel for aluminum bats [32,33]. This would be a more useful definition if it specified ball velocity rather than ball speed (since the bat is a three-dimensional object).

6. **Maximum coefficient of restitution area.** The CoR is commonly defined as the ratio of the relative speed after a collision to the relative speed before the collision. In our studies, the CoR is used to model the energy transferred to the ball in a collision with a bat. If the CoR were 1, then all the original energy would be recovered in the motion of the system after impact. But if there were losses due to energy dissipation or energy storage, then the CoR would be less than 1. For example, in a bat–ball collision there is energy dissipation: both the bat and the ball increase slightly in temperature. Duris and Smith [46] said in their presentation that 100 bat–ball collisions in rapid succession raised the temperature of a softball by 10°F. Also both the bat and the ball store energy in vibrations. Not all of this energy will be transferred to the ball. (For now, we ignore the kinetic energy stored in the ball's spin.) The maximum CoR area is the area that produces the maximum CoR for a bat–ball collision [32,36].

7. **Minimum energy loss area.** There is an area that minimizes the total (translation plus rotation plus vibration) energy lost in the handle. This area depends on the fundamental bending mode, the second mode, and the CoP [39].

8. **Minimum sensation area.** For most humans, the sense of touch is most sensitive to vibrations between 200 and 400 Hz. For each person there is a collision area on the bat that would minimize these sensations in the hands [40].

9. **Joy spot.** Finally, Williams and Underwood [41] stated that hitting the ball at the joy spot makes you the happiest. The joy spot was centered 5 in. (13 cm) from the end of the barrel.

These nine areas are different, but they are close together. We group them together and refer to this region as the sweet spot. We measured a large number of bats (youth, adult, wood, aluminum, ceramic, titanium, etc.) and found that the sweet spot was 15%–20% of the bat length from the barrel end of the bat. This finding is in accord with Refs. [20,30–32,39–42] as well as Worth Sports Co. (personal communication) and Easton Aluminum Inc. (personal communication). In our ideal bat weight experiments [4,43] and our variable moment of inertia experiments [2] for adult bats, the center of the sweet spot was defined to be 5 in. (13 cm) from the barrel end of the bat.

It does not make sense to try getting greater precision in the definition of the sweet spot, because the concept of a sweet spot is a human concept, and it probably changes from human to human. For one example, in calculating the CoP, the pivot point of the bat must be known and this changes from batter to batter, and it may even change during the swing of an individual batter.

Table 16.5 shows general properties for a standard Hillerich and Bradsbury Louisville Slugger wooden C243 pro stock 34 in. (86 cm) bat

TABLE 16.5

Parameters for a C243 Wooden Bat

Stated Length (in.)	34
Period (s)	1.634
Mass (kg)	0.905
I_{knob} (kg m^2)	0.342
I_{pivot} (kg m^2)	0.208
I_{cm} (kg m^2)	0.048
Measured $d_{knob-cm}$ (cm)	57
Measured $d_{knob-cop}$ (cm)	69
Calculated $d_{knob-cop}$ (cm)	69
Measured $d_{pivot-cop}$ (cm)	55
Calculated $d_{pivot-cop}$ (cm)	54
Measured $d_{knob-firstNode}$ (cm)	67

TABLE 16.6

Distance in Centimeters from the Barrel End to the Center of the Sweet Spot
for a 34 in. Wooden Bat

Definition of Sweet Spot	This Study of a C243 Wooden Bat	References
Center of percussion for a 15 cm pivot point[a]	16 calculated	16.5 [38][b]
	18 experimental method 1:	More than 15 [47]
	15 experimental method 2:	17 [36]
	14 experimental method 3:	
Maximum energy transfer area		20 [36]
Maximum batted-ball speed area		14 [32,33]
		17 [38]
Maximum coefficient of restitution area		15 [32]
Node of fundamental vibration mode[c]	18 measured	17 [33]
		17 [38]
		17 [39]
Minimum sensation area		17 [40]
Minimum energy area		15 [39]
		15–18 [31]
Joy spot		13 [41]

[a] The center of percussion for a uniform rod would be 15 cm from the end [48]. This is a lower limit for a bat.
[b] Ref. [38] used a 33 in. bat and their CoP was $16/84 = 19\%$ from the barrel end: scaling for a 34 in. bat yields 16.5 cm.
[c] The node of the fundamental vibration mode of an open-ended pipe is 0.224 times the length. For a 34 in. pipe, it would be 19 cm from the end. This is an upper limit for a bat.

with the barrel end cupped out to reduce weight. Table 16.6 shows sweet spot parameters for this and similar 34 in. wooden bats. These modern scientific methods of calculating the center of the sweet spot of the bat are all only a few centimeters above the true value given by Williams a quarter century ago. Table 16.7 shows several other parameters for a variety of commercially available bats.

There is no sweet spot of the bat: however, there is a sweet area and for a 34 in. wooden bat, it is 5 to 7 in. (13 to 18 cm) from the barrel end of the bat. We presented nine definitions for the sweet spot of the bat. Some of these definitions had a small range of experimentally measured values (e.g., 1 cm for the node of the fundamental vibration mode), whereas others had a large range of experimentally measured values (e.g., 10 cm for the maximum batted-ball speed area). But of course, none of these definitions have square sides. They are all bowl shaped. So the width depends on how far you allow the parameter to decline before you say that you are out of the sweet area. In general, the sweet area is about 2 in. wide. Our survey of retired major league batters confirmed that the sweet spot of the bat is about 2 in. (5 cm) wide. Therefore, most of the sweet spot definitions of this chapter fall within this region. In summary, recent scientific analyses have validated William's statement that the sweet spot of the bat is an area 5 to 7 in. from the end of the barrel.

TABLE 16.7

Properties of Typical Commercially Available Bats

League	Stated Weight (oz)	Length (in.)	Period (s)	Mass (kg)	Distance from the Knob to Center of Mass, $d_{knob-cm}$ (m)	Moment of Inertia with Respect to the Knob, I_{knob} (kg m^2)	Moment of Inertia with Respect to the Center of Mass, I_{cm} (kg m^2)
Tee ball	17	25	1.420	0.478	0.346	0.083	0.026
Little League	22	31	1.570	0.634	0.448	0.174	0.047
High school	26	32	1.669	0.764	0.510	0.269	0.070
Softball	23	33	1.584	0.651	0.477	0.193	0.045
Softball, end loaded	26	34	1.667	0.731	0.505	0.255	0.069
Softball, end loaded	29	34	1.674	0.810	0.506	0.285	0.078
Major league, R161 (wood)	32	34	1.654	0.920	0.571	0.356	0.056
Major league, C243 (wood)	32	34	1.634	0.905	0.570	0.342	0.048

16.2.2 Coefficient of Restitution

The CoR is commonly defined as the ratio of the relative speed after a collision to the relative speed before the collision [10,29,32]. In our studies, the CoR is used to model the energy transferred to the ball in a collision with a bat. If the CoR were 1.0, then all the original energy would be recovered in the motion of the system after impact. But if there were losses due to energy dissipation or energy storage, then the CoR would be less than 1.0. For example, in a bat–ball collision there is energy dissipation: both the bat and the ball increase slightly in temperature. Also both the bat and the ball store energy in vibrations. This energy is not available to be transferred to the ball and therefore the ball velocity is smaller. (We ignore the kinetic energy stored in the ball's spin.)

The CoR depends on many things including the shape of the object that is colliding with the ball. When a baseball is shot out of an air cannon onto a flat wooden wall, most of the ball's deformation is restricted to the outer layers: the cowhide cover and the four yarn shells. However, in a high-speed collision between a baseball and a cylindrical bat, the deformation penetrates into the cushioned cork center. This allows more energy to be stored and released in the ball and the CoR is higher. In our model, the CoR for a baseball–bat collision is 1.17 times the CoR of a baseball–wall collision. The CoR also depends on the speed of the collision. Our computer programs use the following equations for the CoR: for an aluminum bat and a softball:

$$CoR = 1.17\,(0.56 - 0.001\ CollisionSpeed) \tag{16.9}$$

for a wooden bat and a baseball

$$CoR = 1.17\,(0.61 - 0.001\ CollisionSpeed) \qquad (16.10)$$

where *CollisionSpeed* (the sum of the magnitudes of the pitch speed and the bat speed) is in miles per hour. These equations come from unpublished data provided by Jess Heald of Worth Sports Co. and they assume a collision at the sweet spot. Our baseball CoR equation is in concordance with data from six studies summarized in a report to the NCAA [44]: $CoR = 1.17\,(0.57 - 0.0013\ CollisionSpeed)$.

The CoR also depends on where the ball hits the bat, because different locations produce different vibrations in the bat [20,30,32,33]. Increasing the humidity of the ball from 10% to 90% decreases the CoR by roughly 15%. Ball temperature affects the CoR [20,45]. Bat temperature also affects the CoR: so bat warmers in the dugout would increase the CoR. But we will not consider these complexities in this chapter.

In the past, the CoR of a baseball–bat collision was mostly a property of the ball, because a wooden bat does not deform during a bat–ball collision. But hollow metal and composite baseball and softball bats do deform during the collision; thus, they play an important part in determining the CoR. During a collision, energy is stored in the ball and in the bat. Most of the energy stored in the ball is lost. This energy loss is modeled with the CoR. If the CoR is half, then three-fourth of the energy is lost (because kinetic energy is proportional to velocity squared). Most of the energy stored in the bat is not lost, but is transferred to the ball. This increases the batted-ball speed. This matching of the bat to the ball to increase batted-ball speed is called the trampoline effect [34]. Because most of the energy stored in the ball is lost and most of the energy stored in the bat is returned, the batter would prefer to have energy stored in the bat rather than in the ball. A hard (or stiff) ball will deform the bat more and therefore store more energy in the bat, which, by the above argument, will increase batted-ball speed. Therefore, the hardness (or stiffness) of the ball becomes another regulated parameter. Today, softballs are typically marked with a CoR number and a stiffness number. The stiffness is the amount of slowly applied force that is required to deform a softball by ¼ inch (0.64 cm) [46].

16.2.3 Performance Criterion

In most engineering studies, the most important decision is choosing the performance criterion. For a batter hitting a ball, what is the most important performance criterion? Kinetic energy imparted to the ball? Momentum imparted to the ball? Batted-ball speed? Accuracy? Launch angle? Batted-ball spin rate? Batted-ball spin axis? Efficiency of energy transfer? or Distance from the plate where the ball hits the ground? For most studies in the baseball literature, the performance criterion was maximizing batted-ball speed. Where in baseball would other performance criteria be more appropriate?

In calculating knockdown power, kinetic energy would be appropriate. The Colt 0.45 automatic pistol was designed for battles in the Philippines in the early years of the twentieth century, with the performance criterion of "Knock down the charging warrior before he or she can chop off your head with a machete." The existing 0.38 would kill him, but he or she would chop off your head before he or she would die. A solution for this problem was the 0.45 caliber munition with a muzzle kinetic energy of 370 ft-lb$_m$ (502 J). (The kinetic energy of bullets is given in units of foot-pounds, but the pounds are not pounds-force, rather they are pounds-mass.) So 1 ft-lb$_m$ = 1.36 J. In contrast, a baseball traveling at 97 mph (43 m/s) has 100 ft-lb$_m$ (136 J) of kinetic energy. This explains why a hit-batter can be hurt, but not knocked down by a pitch.

As an aside, the energy stored in the spin of a baseball is $KE_{spin} = \frac{I\omega^2}{2} = \frac{mr_{ball}^2\omega^2}{5}$. Substituting in nominal values for a baseball spinning at 1200 rpm yields $KE_{spin} = \frac{0.145 \times 0.0014 \times 15.791}{5} = 0.6$ J (0.5 ft-lb$_m$), which is much lesser than the translational energy.

Here are some potential performance criteria for a pitcher: (1) minimize the number of pitches per inning, by getting the hitter to hit an early pitch for a grounder (this would reduce the batter's opportunities to learn the pitches and lessen pitcher fatigue), (2) minimize the number of runs, (3) maximize batter intimidation, and (4) generate impressive statistics (e.g., strikeouts, wins, ERA, saves) that would generate high salaries.

16.2.4   Vertical Size of the Sweet Spot

We need a model for batting success that shows the relative importance of bat weight, bat speed, launch angle, bat shape, and coefficient of friction. These are all under the batter's control. We [3,13] developed a new performance criterion: the probability of getting a hit. The old performance criterion of maximizing batted-ball speed works well for home runs, but only 4% of batted balls in play are home runs.

We now introduce a new criterion for the batted ball, the distance from the plate where the ball first hits the ground. Assume that the batter wants to hit a line drive. He or she wants the ball to clear the infielders without bouncing, and to hit the grass in front of the outfielders. Thus, a major league baseball player wants the ball to hit the ground between 120 and 240 ft (37 to 73 m) from the plate. These numbers were given in Table 16.1.

We now make the following assumptions. The batter is using a Louisville Slugger C243 wooden bat and is hitting a regulation baseball. The pitch speed is 85 mph (38 m/s). The speed of the sweet spot of the bat is 60 mph (27 m/s): this is the average value for the San Francisco Giants measured by Bahill and Karnavas [43]. These speeds would produce a CoR of 0.54. The bat weighs 32 oz (0.91 kg) and the ball weighs 5.125 oz

(0.145 kg). We can put these data into the following equation from Ref. [2] to get a batted-ball speed of 106 mph (47 m/s), which is a reasonable value:

$$v_{\text{ball-after}} = v_{\text{ball-before}} + \frac{(1 + \text{CoR})(v_{\text{bat-before}} - v_{\text{ball-before}})}{1 + \dfrac{m_{\text{ball}}}{m_{\text{bat}}} + \dfrac{m_{\text{ball}}d^2_{\text{cm-ss}}}{I_{\text{cm}}}} \tag{16.11}$$

This performance criterion is used in the next section to define the vertical sweet spot of the bat.

16.3 Model for Bat–Ball Collisions

Baseball and softball batters swing a narrow cylinder with the axis more or less parallel with the ground. Thus, the transverse curvature of the bat's face (hitting surface) is a vertical curvature. In combination with the vertical offset of the bat and ball trajectories, this vertical curvature strongly influences the ball's vertical launch velocity, angle, and spin rate. These launch characteristics can be included in a vector describing a specific point on the bat's face; a vector field can specify the launch characteristics of all the points on the face. Each vector determines the batted ball's behavior—the distance it travels in the air until it first strikes the ground (range), how long it stays in the air (hang time), and, for ground balls, the time taken for the ball to reach the positional arc of infielders (ground time).

The set of success probabilities associated with a specific vertical arc on the bat's face is called the vertical sweetness gradient of that arc. The face's vector field represents sweetness gradients in both the longitudinal (horizontal) and transverse (vertical) dimensions of the bat. However, we restrict our current discussion to vertical collision considerations and the radial placement of the ball in play in fair territory.

We integrated many models as shown in Figure 16.11. One of the input parameters in the overall model is the offset between the bat and the ball. This offset is defined in Figure 16.12. The basic principle of this model is that we break up the bat and ball velocities into normal and tangential components. We apply conservation of energy. And then we apply conservation of linear and angular momentum. This technique is suggested in Figure 16.13.

Finally, Figure 16.14 shows the full model. It illustrates the initial vertical configuration of the bat and ball at the instant of collision. The initial parameters of the collision are:

1. Initial velocity vector of bat's contact point ($v_{\text{bat},0}$)
2. Initial normal component of the bat's velocity vector ($v_{\text{bat},0,n}$)
3. Initial tangential component of the bat's velocity vector ($v_{\text{bat},0,t}$)
4. Initial velocity vector of ball ($v_{\text{ball},0}$)

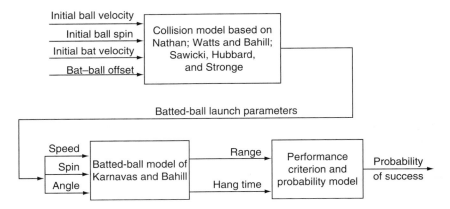

FIGURE 16.11
Our model used components from several other models. (From Bahill, A.T., http://www.sie.arizona.edu/sysengr/slides. With permission. Copyright 2004.)

5. Initial normal component of the ball's velocity vector ($v_{ball,0,n}$)
6. Initial tangential component of the ball's velocity vector ($v_{ball,0,t}$)
7. Bat–ball offset distance (D) from the ball's center perpendicular to the trajectory plane of the bat's transverse center
8. Vertical angle (θ) between the line connecting ball and bat centers (line of centers) and the horizontal plane ($z = 0$)
9. Vertical angle (γ) between the horizontal plane and the ball's trajectory plane

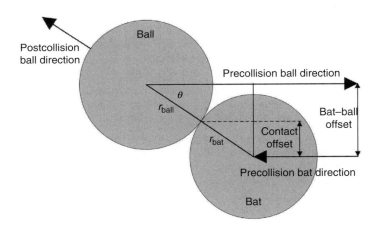

FIGURE 16.12
Definition of the bat–ball offset. (This figure does not show the effect that pitch spin has on the postcollision ball direction. For most collisions, the ball is going down at a 10° angle and the bat is going up at a 10° angle. These angles are not shown in this figure.) (From Bahill, A.T., http://www.sie.arizona.edu/sysengr/slides. With permission. Copyright 2004.)

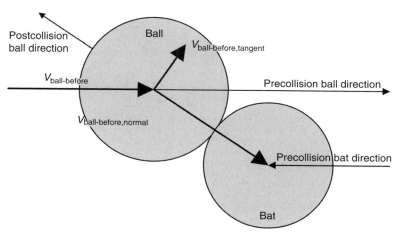

FIGURE 16.13
The bat and ball velocities are decomposed into normal and tangential components. (This figure ignores the spin of the ball.) (From Bahill, A.T., http://www.sie.arizona.edu/sysengr/slides. With permission. Copyright 2004.)

10. Vertical angle (ψ) between the horizontal plane and the bat's trajectory plane
11. Mass of bat (m_{bat})
12. Mass of ball (m_{ball})
13. Radius of bat at contact point (r_{bat})
14. Radius of ball (r_{ball})

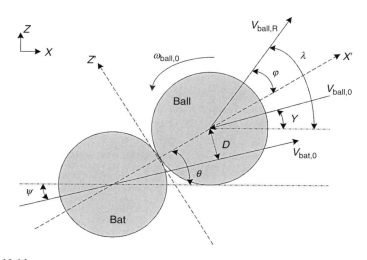

FIGURE 16.14
Initial vertical configuration of the bat–ball collision. (From Bahill, A.T., http://www.sie.arizona.edu/sysengr/slides. With permission. Copyright 2004.)

15. Coefficient of restitution (CoR) of the bat–ball impact
16. Coefficient of friction (μ) during bat–ball impact
17. Angular velocity ($\omega_{ball,0}$) of the pitch

The characteristics m_{ball}, r_{ball}, CoR, and μ will be considered constants. The values of CoR and μ must be derived empirically.

The resultant vectors and angles used to calculate launch velocity and angle are:

1. Resultant velocity vector of bat's contact point ($v_{bat,R}$)
2. Resultant normal component of the bat's velocity vector ($v_{bat,R,n}$)
3. Resultant tangential component of the bat's velocity vector ($v_{bat,R,t}$)
4. Resultant velocity vector of ball ($v_{ball,R}$). This is called the "launch velocity"
5. Resultant normal component of the ball's velocity vector ($v_{ball,R,n}$)
6. Resultant tangential component of the ball's velocity vector ($v_{ball,R,t}$)
7. Vertical angle (φ) between the line of centers and $v_{ball,R}$
8. Vertical angle (λ) between the horizontal plane ($z = 0$) and $v_{ball,R}$. This is called the "launch angle"

We also calculate the vertical launch spin rate ($\omega_{ball,R}$) and specify air density (ρ).

The model is set in the x–z plane of a coordinate system with origin at the contact point, 3 ft in front of the vertex of home plate and at a height of 3 ft. The positive z-axis points upward, positive x-axis points toward the pitcher, and positive y-axis points out of the plane [22]. In Figure 16.14, the x–z plane is reoriented so the x' axis lies along the bat–ball line of centers and the z' axis is tangential to the bat–ball contact point. Angular velocity is positive for pitch topspin and for batted-ball backspin. D is positive if the bat undercuts the ball.

The pitch does not fly horizontally. It is dropping downward at an angle between 4° and 12°, depending on the speed and type of pitch. The angle of descent (γ) of an average fastball is about 10° [10,20]. Batters generally uppercut the ball (ψ) with a 5° to 10° upward angle, which means the ball and bat are actually traveling in opposite directions, as shown in Figure 16.14. For Tables 16.8 through 16.10 we set $\gamma = \psi = 10°$.

16.3.1 Ball's Launch Velocity, Angle, and Spin Rate

The ball's launch parameters are calculated by decomposing the initial velocities of the bat and ball into their normal and tangential components at the point of contact. These velocities are used with the principles of conservation of momentum and conservation of energy to yield resultant normal and tangential velocities for the ball, which are then used to calculate the launch velocity of the ball and the angles φ and λ. The batted-ball

TABLE 16.8

Launch Parameters and Contact Offset for Various Bat–Ball Offsets

Bat–Ball Offset (in.)	Launch Velocity (mph)	Launch Angle (°)	Backspin Rate (rpm)	Contact Offset (in.)
1.50	82	58	4924	0.73
1.25	85	48	3991	0.61
1.00	88	39	3059	0.49
0.75	90	31	2127	0.37
0.50	91	23	1195	0.24
0.25	92	15	263	0.12
0.00	93	8	−669	0

angular velocity is calculated from the normal and tangential linear velocities of the ball and bat and the ball's initial angular velocity.

The collision of two partially elastic bodies with friction is described by numerous authors [48]. The first step in building the model is to calculate θ and, from this, the normal and tangential components of the initial velocity vectors. The vertical angle of the line of centers, $\theta = \psi + \sin^{-1}(D/(r_{bat} + r_{ball}))$. The initial velocity components of the ball are $v_{ball,0,n} = v_{ball,0} \cos\theta$ and $v_{ball,0,t} = v_{ball,0} \sin\theta$. The initial velocity components of the bat are $v_{bat,0,n} = v_{bat,0} \cos\theta$ and $v_{bat,0,t} = v_{bat,0} \sin\theta$.

The resultant normal velocity of the ball [10] is

$$v_{ball,R,n} = v_{ball,0,n} - (1 + CoR)[(m_{bat}v_{ball,0,n} - m_{bat}v_{bat,0,n})/(m_{ball} + m_{bat})]$$

$$(16.12)$$

Calculation of relative tangential velocity, resultant angular velocity of the ball, and final launch angle, λ, is described by Refs. [22,49]. In these models,

TABLE 16.9

Range and Hang Time for the Launch Parameters of Table 16.8

Bat–Ball Offset (in.)	Launch Velocity (mph)	Launch Angle (°)	Backspin Rate (rpm)	Range (ft)	Hang Time (s)
1.50	82	58	4924	129	6.4
1.25	85	48	3991	236	6.7
1.00	88	39	3059	306	6.1
0.75	90	31	2127	321	5.0
0.50	91	23	1195	285	3.6
0.25	92	15	263	213	2.2
0.00	93	8	−669	122	1.1

TABLE 16.10

Launch Parameters, Range, Hang Time, and the Probability of Batter's Success
for Nonnegative Offsets

Bat–Ball Offset (in.)	Launch Velocity (mph)	Launch Angle (°)	Backspin Rate (rpm)	Range (ft)	Hang Time (s)	Probability of Success
1.50	82	58	4924	129	6.4	0.00
1.25	85	48	3991	236	6.7	0.00
1.00	88	39	3059	306	6.1	0.00
0.75	90	31	2127	321	5.0	0.00
0.50	91	23	1195	285	3.6	0.09
0.25	92	15	263	213	2.2	1.00
0.00	93	8	−669	122	1.1	0.63

friction acts in the direction opposite to the slip of the ball. If friction is large
enough, it halts the relative tangential velocity (the combined velocities of
bat and ball surfaces relative to the contact point). When this occurs, slippage ceases, the ball sticks to the bat, and the ball begins to roll, contributing
to the launch angular velocity. These models account for bat recoil and
assume conservation of linear and angular momentum for tangential ball
and bat motions. Both models ignore deformation of the ball during collision (they assume it remains a perfect sphere).

The launch velocity, launch angle, and backspin rate for various bat–ball
offsets are shown in Table 16.8 and Figure 16.15.

Figure 16.15 indicates the launch angle and the center of the ball's area of
the contact with the bat. The distance of this contact point from the center
axis of the bat can be derived from Figure 16.12. $\sin \theta = \frac{\text{bat–ball offset}}{r_{\text{ball}} + r_{\text{bat}}} = \frac{\text{contact offset}}{r_{\text{bat}}}$ which gives

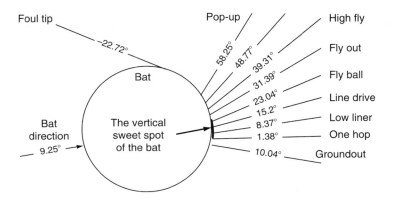

FIGURE 16.15

Common outcomes for some particular launch angles and bat–ball offsets from Table 16.8. The
indicated vertical sweet spot of the bat is about one-third of an inch (8 mm) high. (From Bahill,
A.T., http://www.sie.arizona.edu/sysengr/slides. With permission. Copyright 2004.)

$$\text{Contact offset} = \frac{\text{bat–ball offset} \times r_{bat}}{r_{ball} + r_{bat}} \tag{16.13}$$

This distance is inserted as an additional column in Table 16.8.

16.3.2 Range, Hang Time, and Ground Time

The launch velocity, launch angle, and spin rate are the input data into the equations of Ref. [10] to calculate the batted ball's range and hang time. The vertical distance traveled by the batted ball (without regard to lift or drag) is $z = v_{z0}\,t - 0.5\,gt^2$, where v_{z0} is the vertical velocity of the ball, t is the hang time, and g is the acceleration rate of gravity at the surface of the Earth (32.17 ft/s^2, 9.8 m/s^2). The horizontal distance traveled (again ignoring lift and drag) is $x = v_{x0}\,t$, where v_{x0} is the horizontal velocity component. However, the rotation of the ball creates a Magnus force acting vertically perpendicular to the trajectory. This force tends to lift the ball (if backspin) or depress the ball (if topspin). It is calculated as $F_{lift} = 0.5\,\rho\pi r^3_{ball}$ $\omega v_{ball}\,\sin VaSa$, where ρ is the air density. Friction of the ball passing through the air is a drag force acting directly counter to the trajectory. This force is calculated as $F_{drag} = 0.5\,\rho\pi r^2_{ball}\,C_d\,v^2_{ball}$. In our model, the drag and lift coefficients are constants. Table 16.8 shows the ranges and hang times that result from various offsets.

Ground time is not calculated for this chapter. It will be modeled by using the launch angle to find the angle of incidence on the first bounce. The incidental horizontal and vertical velocity components and launch spin rate will then be used to generate the bounce velocity, angle, and spin rate. An aerodynamics model will be used to find the flight characteristics between bounces, including the incidental angle on the subsequent bounce. Note that here CoR and μ will have values different from those for the bat–ball collision. As ω usually represents topspin on ground balls, angular velocity contributes to linear horizontal velocity and vice versa. If μ is large enough to overcome the combined angular and horizontal velocities, slippage stops and rolling begins.

16.3.3 Batting Success Probability Function

The characteristics of a batted ball can be associated with probability of success through a step function based on the potential of defensive players to prevent a base hit. Four kinds of batted-ball behavior are represented in the model:

1. Fly balls (range > 130 ft (40 m), hang time > 2 s)
2. Pop-ups (range < = 130 ft (40 m), hang time > 2 s)
3. Line drives (range > = 115 ft (35 m), hang time < = 2 s)
4. Grounders (range < 115 ft (35 m), hang time < = 1 s)

Our model incorporates several simplifying assumptions. Either an infielder or outfielder might catch a fly ball, depending on range and hang time. All pop-ups are caught by an infielder or a catcher so the probability of success is zero. Only infielders catch line drives and grounders.

Each batted ball is associated with defensive coverage formulated as a function of time. A defensive player prevents a base hit if he or she can reach the ball during hang time or ground time. To determine coverage, we positioned outfielders and infielders on two arcs—the outfield arc with a radius of 300 ft (91 m) and the infield arc with a radius of 115 ft (35 m). The outfield arc is divided into thirds and the infield arc into quarters, with a player positioned at the center of each arc segment. For example, the outfield arc has a length of 471.3 ft (300 × 1.571); thus, it is divided into three segments each of which is 157.1 ft long. The right fielder, then, is positioned 300 ft from home, 78.55 ft from the right field foul line, and 157.1 ft from the center fielder. The batted ball's range (from the range column of Table 16.10) yields a "range arc" with length equal to 1.571 times range (angle in radians times radius).

On fly balls, each player's position is the center of an ellipse representing defensive coverage by the player (a fly ball is illustrated in Figure 16.16). Hang time determines the dimensions of the ellipse for a specific batted ball. Probability of a base hit is the proportion of the range arc that is not overlapped by ellipses.

In Figure 16.16, the outfielders are positioned on the outfield arc. The dashed line shows the range arc for a low fly ball that is in the air for 3 s and travels 250 ft. Three-fourth of this range arc is overlapped by the 3 s fielder ellipses. Therefore, the probability of success is 0.25.

If a line drive or grounder passes the infield arc without encountering an infielder, it is considered a base hit. Therefore, only infielders' lateral

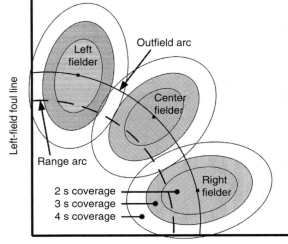

FIGURE 16.16
Range arc, outfield arc, and defensive coverage of each outfielder for batted balls that would be in the air for 2, 3, and 4 s. (From Bahill, A.T., http://www.sie.arizona.edu/sysengr/slides. With permission. Copyright 2004.)

movements provide coverage. Success is the proportion of range arc (on line drives) or infield arc (on grounders) not covered by infielders. In this model, batters do not beat out infield hits and pitchers do not catch line drives or grounders.

The model assumes the speed of outfielders and infielders is 23 ft/s (7 m/s). Outfielders' reaction delays are 8 ft (2.4 m) (0.35 s) forward, 12 ft (3.7 m) (0.52 s) sideward, and 15 ft (4.6 m) (0.65 s) backward. Infielders' reaction times are 12 ft sideward and 15 ft backward. These values were selected as "reasonable" and are not based on empirical data.

16.3.4 Example of Varying Offsets

For an example of collision evaluation, the model is solved at offset increments of 0.25 in. upward from zero offset. Pitch backspin is −1800 rpm and pitch speed is 85 mph (38 m/s). Contact occurs at the bat's area of maximum horizontal sweetness and the speed of the bat's contact point is 60 mph (27 m/s) the average value for the San Francisco Giants [43]. These speeds produce a CoR of 0.54. We measured the coefficient of friction, μ, to be 0.5 (see also Ref. [22]). The angles γ and ψ of ball and bat are both 10°. Other test values are $r_{bat} = 1.375$ in., $r_{ball} = 1.452$ in., $m_{bat} = 32.0$ oz (effective bat mass $= 20.0$ oz [33]), $m_{ball} = 5.125$ oz, and ρ at standard sea level conditions. These numbers are given in Tables 16.2 and 16.11. Ranges and hang times were found using a Pascal aerodynamics program. Launch

TABLE 16.11

Parameter Values Used to Compute the Vertical Size of the Sweet Spot SI Units

	Major League Baseball	Little League	NCAA Softball
Bat type	Wooden C243	Aluminum	Aluminum
Ball type	Baseball	Baseball	Softball
Pitch speed (m/s)	38	22	29
Speed of sweet spot (m/s)	27	20	22
CoR	0.54	0.53	0.52
Typical bat mass (kg)	0.9	0.6	0.7
Ball mass (kg)	0.145	0.145	0.191
Maximum bat radius (m)	0.035	0.029	0.029
Ball radius (m)	0.037	0.037	0.048
Distance from front of rubber to tip of plate (m)	18.4	14.0	13.1
Pitcher's release point: distance from tip of plate and height	17 m out 2 m up	13 m out 1.5 m up	12 m out 0.8 m up
Bat–ball collision point: distance from tip of plate and height	1 m out 1 m up	1 m out 1 m up	1 m out 1 m up
Backspin of batted ball (rad/s)	100–500	100–500	100–500
Desired ground contact point: distance from the plate (m)	37–73	24–43	24–46
Air density, ρ (kg/m^3)	1.04	1.04	1.04

Note: Air density is inversely related to temperature, altitude, and humidity.

values were computed using spreadsheets developed by A. Nathan (personal communication).

16.3.5   Results

Test results are given in Table 16.10. As hang time increases, probability of success decreases rapidly. Pop-ups are produced by offsets greater than 1.5 in. (3.8 cm). These are assigned a success probability of zero. Note the model assumes no outfield barriers. In most major league stadiums, long fly balls have a chance of clearing the wall (the average distances are 330 ft (100 m) down the foul lines and 400 ft (122 m) in center field). Thus, the model underestimates success for any range with a chance to be a home run.

The example shows how the model might be used to analyze collision parameters (e.g., offsets, bat velocity) or bat properties (e.g., bat radius). Relating initial conditions to sweetness provides a valuable criterion for these analyses.

16.3.6   Discussion

From this collision model, we get the launch velocity, the launch angle, and the backspin rate. We put these into our simulation for the batted ball that uses the following equations from Ref. [10, p. 80]:

$$F_{drag} = 0.25 \, \rho \pi r^2_{ball} v^2_{ball\text{-}after} \tag{16.14}$$

$$F_{Magnus} = 0.5 \, \rho \pi r^3_{ball} \omega v_{ball\text{-}after} \tag{16.15}$$

where
   ρ is air density
   $v_{ball\text{-}after}$ is the ball speed after its collision with the bat
   ω is the rotation rate
   r_{ball} is the radius of the ball

Values for these parameters are provided in Tables 16.1 and 16.11.

Some physicists (see Equation 1 in Ref. [25]) model the Magnus force with $F_L = \frac{1}{2}C_L \rho A v^2$, where A is the cross-sectional area of the ball and C_L is not a constant, but rather it is a nonlinear parameter that depends on the Reynolds number, the spin rate, the ball velocity, and, perhaps, C_D. However, we prefer the simpler formulation of Equation 16.15.

To show how Equations 16.14 and 16.15 work, let us now present a simple numerical example. Assume a 95 mph (42.5 m/s) fastball has 20 rps of pure backspin. Near the beginning of the pitch, the Magnus force will be straight up in the air, i.e., pure lift. Using English units and Table 16.1, we get

$$F_{drag} = 0.25 \, \rho \pi r^2_{ball} v^2_{ball}$$
$$= (0.25)(0.0023)(3.14)(0.12)^2 (139)^2 = 0.5 \, lb$$

and

$$F_{\text{Magnus}} = 0.5\,\rho\pi r_{\text{ball}}^3\,\omega v_{\text{ball-after}}$$
$$= (0.5)(0.0023)(3.14)(0.12)^3(126)(139) = 0.11\,\text{lb}$$

which is about one-third the force of gravity given in Table 16.1. This is consistent with Tables 16.3 and 16.4.

Using SI units and Table 16.11, we get

$$F_{\text{drag}} = 0.25\,\rho\pi r_{\text{ball}}^2\,v_{\text{ball}}^2$$
$$= (0.25)(1.2)(3.14)(0.037)^2(42.5)^2 = 2.3\,\text{n}$$

and

$$F_{\text{Magnus}} = 0.5\,\rho\pi r_{\text{ball}}^3\,\omega v_{\text{ball-after}}$$
$$= (0.5)(1.2)(3.14)(0.037)^3(126)(42.5) = 0.51\,\text{n}$$

which is about one-third the force of gravity, which is

$$F_{\text{gravity}} = mg = 0.145 \times 9.8 = 1.42\,\text{n} \tag{16.16}$$

This simulator allows us to calculate the trajectory of the batted ball. From the ball's trajectory we can compute where it will first hit the ground. Assume that the batter wants to hit a line drive that first hits the ground between 120 and 240 ft (37 to 73 m) from the plate. (The performance criterion is to maximize the probability that the batted ball will be a line drive that first hits the ground 120 to 240 ft from the plate.) From our simulations, the vertical offset between the ball and the bat should be between 0.15 and 0.45 in. (0.38 to 1.1 cm). Therefore, the vertical size of the sweet spot of the bat is one-third of an inch (8 mm). For the Little League the vertical size of the sweet spot is about the same. However, because the softball is bigger, for NCAA softball the vertical size of the sweet spot is a little less than half an inch.

This discussion is suggesting another performance criterion: efficiency. The batter wants to swing the bat so that as much energy as possible is transferred from the bat to the ball in a particular direction, namely 5° to 10° upward. Momentum in a perpendicular direction is not helpful (pop-ups and grounders). This performance criterion wants the batted-ball direction to be the same as the bat's direction before the collision, i.e., it wants a 5° to 10° uppercut and zero offset. A lot of previously used performance criteria were appropriate for home runs. This new performance criterion is designed for line drive singles or doubles.

At this point it is appropriate to caution young players; we are not advising that they ignore their coaches' advice to "swing level." Coaches and parents have difficulty differentiating between level horizontal swings and those with a 5° to 10° upward angle. The coach's admonition means do not swing with a 30° upward angle, because you do not want to launch the ball at 30°. In this context, swing level means swing with a 5° to 10° upward angle.

16.4   Swing of the Bat

Williams and Underwood [41] said that hitting a baseball is the hardest act in all of sports. This act is easier if the right bat is used, but it is difficult to determine the right bat for each individual. Therefore, we developed the Bat Chooser* to measure the swings of an individual, make a model for that person, and compute his or her Ideal Bat Weight [4,43]. The Bat Chooser uses individual swing speeds, CoR data, and the laws of conservation of momentum, and then it computes the ideal bat weight for each individual, trading off maximum batted-ball speed with accuracy. However, with the advent of lightweight aluminum bats, it is now possible for bat manufacturers to vary not only the weight but also the weight distribution. They can start with a lightweight aluminum shell, and add a weight inside the barrel to bring the bat up to its specified weight. This internal weight can be placed anywhere inside the barrel. When the weight is placed at the tip of the bat, the bat is said to be "end loaded." So now, there is a need to determine the best weight distribution in general, for certain classes of players and for individual players. These are the topics of this section.

16.4.1   Ideal Bat Weight and the Bat Chooser

Our instrument for measuring bat speeds, the Bat Chooser, has two vertical laser beams, each with an associated light detector. The subjects were positioned so that when they swung the bats, the sweet spot (which we defined to be an area on the bat that is centered 5 in. from the barrel end) of each bat passed through the laser beams. A computer (sampling once every 16 μs) recorded the time between interruptions of the laser beams. Knowing the distance between the laser beams (15 cm, 6 in.) and the time required for the bat to travel that distance, the computer calculated the horizontal speed of the bat's sweet spot for each swing. This is a simple model, because the motion of the bat is very complex, being comprised of a horizontal translation, a rotation about the batter's spine, a rotation about a point between the two hands (which may be moving), and a vertical motion.

In our variable moment of inertia experiments, to be described in the next section, and in our ideal bat weight experiments, each player was positioned so that bat speed was measured at the place where the subject's front foot hit the ground. We believe that this is the place where most players reach maximum bat speed. The batters were told to swing each of six bats as

---

* Bat Chooser and Ideal Bat Weight are trademarks of Bahill Intelligent Computer Systems.

fast as possible, while still maintaining control. They were told to "Pretend you are trying to hit a Randy Johnson fastball." In a 20 min interval of time, each subject swung each bat through the instrument five times. The order of presentation was randomized. A speech synthesizer announced the selected bat; for example, "Please swing bat Babe Ruth; that is bat B." For each swing, the name of the bat and the speed of the sweet spot were recorded.

To reduce bat swing variability we gave the batters a visual target to swing at. It was a knot on the end of a string hanging from the ceiling. Typically, this knot was 3 ft (1 m) off the floor. The height of this knot was very important for some batters. For one batter, bat speed increased 20% when the knot was lowered 1 ft (0.3 m).

16.4.2 Principles of Physics Applied to Bat Weight Selection

The speed of a baseball after its collision with a bat depends on many factors, not the least of which is the weight of the bat. In this section, we present data to help an individual player to decide if his or her preference is the most effective bat weight. Knowing the ideal bat weight can eliminate time-consuming and possibly misleading experimentation by ball players.

To find the best bat weight we must first examine the conservation of momentum equations for bat–ball collisions.

$$m_{bat}v_{bat\text{-}before} + m_{ball}v_{ball\text{-}before} = m_{bat}v_{bat\text{-}after} + m_{ball}v_{ball\text{-}after} \tag{16.17}$$

We want to solve for the ball's speed after its collision with the bat, called the "batted-ball speed," but first we should eliminate the bat's speed after the collision, because it is not easily measured. The CoR for a bat–ball collision can be modeled with

$$CoR = -\frac{v_{bat\text{-}after} - v_{ball\text{-}after}}{v_{bat\text{-}before} - v_{ball\text{-}before}} \tag{16.18}$$

The negative signs are there because $v_{ball\text{-}before}$ is in the direction from the pitching rubber to the plate, whereas the other three velocities go from the plate toward the rubber. Therefore, we define $v_{ball\text{-}before}$ to have a negative magnitude.

We can use the equation for the CoR to solve for $v_{ball\text{-}after}$, substitute the result into the equation for the conservation of momentum, and solve for the ball's speed after its collision with the bat. The result is

$$v_{ball\text{-}after} = \frac{-v_{ball\text{-}before}\left[CoR - \dfrac{m_{ball}}{m_{bat}}\right] + (1 + CoR)v_{bat\text{-}before}}{1 + \dfrac{m_{ball}}{m_{bat}}} \tag{16.19}$$

This means that the ball's speed after the collision will depend on the mass of the ball, the mass of the bat, the CoR, and the precollision speeds of the ball and bat.

16.4.3   Coupling Physics to Physiology

Physiologists have long known that muscle speed decreases with increasing load. This is why bicycles have gears. The rider can keep muscle speed in its optimal range while bicycle speed varies greatly. Therefore, to discover how muscle properties of individual ball players affect their best bat weights, we measured the bat speeds of many batters swinging bats of various weights. We plotted the data of bat speed versus bat weight, and used this to help calculate the best bat weight for each batter.

Over the last half century, physiologists have used three equations to describe the force–velocity relationship of muscles: that for the straight line ($y = Ax + B$), that for the rectangular hyperbola ($(x + A)\,(y + B) = C$), and that for the exponential ($y = Ae^{-Bx} + C$). Each of these equations has been best for some experimenters, under some conditions, with certain muscles, but usually the one for the hyperbola fits the data best. In our experiments, we fit all three and chose the equation that gave the best fit to the data of each subject's 30 swings. For example, for batters where the straight line fit was the best

$$v_{\text{bat-before}} = slope\ m_{\text{bat}} + intercept \tag{16.20}$$

where
  slope is the slope of the line
  intercept is the y-axis intercept.

Now to couple physiology to physics, we substituted this relationship into the previous equation to yield

$$v_{\text{ball-after}} = \frac{-v_{\text{ball-before}}\left[CoR - \dfrac{m_{\text{ball}}}{m_{\text{bat}}}\right] + (1 + CoR)(slope\ m_{\text{bat}} + intercept)}{1 + \dfrac{m_{\text{ball}}}{m_{\text{bat}}}}$$

$$\tag{16.21}$$

Next, you can either take the derivative with respect to the bat weight, set this equal to zero, and solve for the maximum batted-ball speed bat weight or you can get this result graphically, as suggested in Figure 16.17.

16.4.4   Ideal Bat Weight

The maximum batted-ball speed bat weight is probably not the best bat weight for any player. A lighter bat will give a player better control and more accuracy. Obviously, a trade-off must be made between maximum

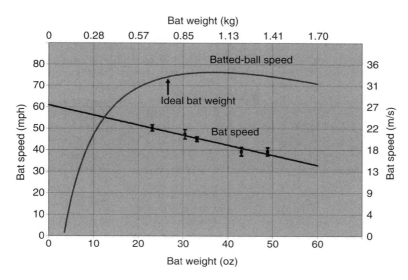

FIGURE 16.17
Bat speed (straight line) and batted-ball speed (curved line) for a typical member of the University of Arizona softball team. Her ideal bat weight is 28 oz. (From Bahill, A.T., http://www.sie.arizona.edu/sysengr/slides. With permission. Copyright 2007.)

batted-ball speed and controllability. Because the batted-ball speed curve is so flat around the point of the maximum batted-ball speed bat weight, we believe there is little advantage in using a bat as heavy as the maximum batted-ball speed bat weight. Therefore, we have defined the ideal bat weight to be the weight at which the ball speed curve drops 2% below the speed of the maximum batted-ball speed bat weight. We believe this gives a reasonable trade-off between distance and accuracy.* Of course, this is subjective and each player might want to weigh the two factors differently. It does, however, give a quantitative basis for comparison. For the player whose data are shown in Figure 16.17, the ideal bat weight was 28 oz (0.8 kg).

Not only is the ideal bat weight specific for each player, but it also depends on whether the player is swinging right or left handed. We measured two switch-hitters (one professional and one university ball player who later had a long professional career). One player's ideal bat weights were 1 oz (0.03 kg) different and the other's were 5 oz (0.14 kg) different. Switch-hitters were so different when hitting right and left handed that we treated them as different players.

It is difficult for most batters to determine the best bat for themselves. Therefore, we developed a system to measure the swings of an individual,

---

* We used 1% for major league baseball players and NCAA softball champions.

TABLE 16.12

Rules of Thumb for Recommending Bats

Group	Recommended Bat Weight
Baseball, major league	Height/3 + 7
Baseball, amateur	Height/3 + 6
Softball, fast-pitch	Height/7 + 16
Softball, slow-pitch	Weight/115 + 24
Junior league (13 and 15 years)	Height/3 + 1
Little League (11 and 12 years)	Weight/18 + 16
Little League (9 and 10 years)	Height/3 + 4
Little League (7 and 8 years)	2 × Age + 4

Note: Recommended bat weight is in ounces, age is in years, height is in inches, and body weight is in pounds.

make a model for that person, and recommend a specific bat weight for that person. However, this system is not conveniently available to most people. So we used our database of the 200 people who had been measured with our system and created simple equations that can be used to recommend a bat for an individual using common parameters such as age, height, and weight. These recommendations are given in Table 16.12. These rules of thumb were derived from our 200 subject database, with constraints of commercial availability and integer numbers, from Ref. [28].

16.4.5 Ideal Moment of Inertia

Bahill [2] presented the variable moment of inertia data that his group has gathered over the last two decades. In these studies, the subjects swung bats of the same weight, but different weight distribution (inertia). The bat speeds were measured and recorded. Then the data for each player were fit with a line of the form

$$v_{\text{bat-before}} = slope\ I_{\text{knob}} + intercept \tag{16.22}$$

where

slope is the slope of the line
I_{knob} is the moment of inertia of the bat with respect to the knob
intercept is the y-axis intercept

We model the swing of a bat as a translation and two rotations: one centered in the batter's body and the other between the batter's hands. Next, we compute the batted-ball speed (the speed of the ball after its collision with the bat). We use conservation of linear and angular momentum and the definition of the CoR to get the following equation, which has been previously derived [10,36]:

$$v_{\text{ball-after}} = \frac{-v_{\text{ball-before}}\left[CoR - \dfrac{m_{\text{ball}}}{m_{\text{bat}}} - \dfrac{m_{\text{ball}}\, d^2_{\text{cm-ss}}}{I_{\text{cm}}}\right] + (1 + CoR)v_{\text{bat-before}}}{1 + \dfrac{m_{\text{ball}}}{m_{\text{bat}}} + \dfrac{m_{\text{ball}}\, d^2_{\text{cm-ss}}}{I_{\text{cm}}}}$$

(16.23)

where

CoR is the coefficient of restitution of the bat–ball collision

$d_{\text{cm-ss}}$ is the distance between the center of mass and the sweet spot, which is assumed to be the point of collision

I_{cm} is the moment of inertia about the center of mass.

The term $v_{\text{bat-before}}$ is simply the velocity of the sweet spot. $v_{\text{ball-before}}$ is a negative number, because its direction is the opposite of $v_{\text{ball-after}}$.

The subjects swung bats composed of wooden bat handles with ¼ inch threaded rods attached to the end and brass disks fixed at various points on the rods. These bats had similar lengths and masses, but a wide range for moments of inertia. The moment of inertia of a bat is given with

$$I_{\text{knob}} = I_{\text{handle}} + m_{\text{disk}}d^2_{\text{knob-disk}}$$

(16.24)

where

I_{knob} is the inertia of the total bat with respect to the knob

I_{handle} is the inertia of the handle part of the bat with respect to the knob

m_{disk} is the mass of the disk on the end of the rod

$d_{\text{knob-disk}}$ is the distance from the knob to the disk

After a little bit of algebra, Bahill [2] derived the following equation for the batted-ball speed:

$v_{\text{ball-after}} = v_{\text{ball-before}}$

$$+ \frac{(1+CoR)[slope(I_{\text{handle}} + m_{\text{disk}}d^2_{\text{knob-disk}}) + intercept - v_{\text{ball-before}}]}{1 + \dfrac{m_{\text{ball}}}{m_{\text{bat}}} + \dfrac{m_{\text{ball}}\left(d_{\text{k-ss}} - \dfrac{m_{\text{disk}}d_{\text{knob-disk}}}{m_{\text{handle}} + m_{\text{disk}}} - \dfrac{m_{\text{handle}}d_{\text{knob-cm(handle)}}}{m_{\text{handle}} + m_{\text{disk}}}\right)^2}{I_{\text{handle}} + d^2_{\text{knob-disk}}\left(m_{\text{disk}} - \left[\dfrac{m_{\text{disk}}d_{\text{knob-disk}}}{m_{\text{handle}} + m_{\text{disk}}}\right]^2 m_{\text{bat}}\right) - \dfrac{2m_{\text{disk}}d_{\text{knob-disk}}}{m_{\text{handle}} + m_{\text{disk}}}\dfrac{m_{\text{handle}}d_{\text{knob-cm(handle)}}}{m_{\text{handle}} + m_{\text{disk}}} - \left(\dfrac{m_{\text{handle}}d_{\text{knob-cm(handle)}}}{m_{\text{handle}} + m_{\text{disk}}}\right)^2 m_{\text{bat}}}}$$

This equation is plotted in Figure 16.18 for a typical subject.

All of the batters in this study would profit (meaning would have higher batted-ball speeds) from using end-loaded bats.

At this point, it may be useful to reiterate that an end-loaded bat is not a normal bat with a weight attached to its end. Adding a weight to the end of a normal bat would increase both the weight and the moment of inertia. This is unlikely to help anyone. In the design and manufacture of an

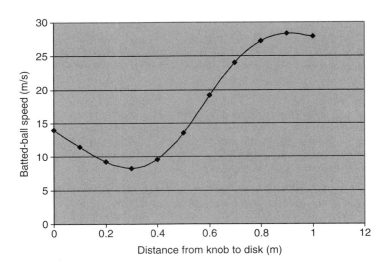

FIGURE 16.18
Batted-ball speed as a function of $d_{\text{knob-disk}}$ for one batter showing an optimal value at 0.9. (From Bahill, A.T., http://www.sie.arizona.edu/sysengr/slides. With permission. Copyright 2003.)

end-loaded bat, the weight is distributed so that the bat has a normal weight, but a larger than normal moment of inertia.

16.5 Summary

This chapter presented the right-hand rules that can be used to show the direction of spin-induced deflection for a spinning ball in any sport. They were summarized with the acronym SaD Sid. Then, we discussed the sweet spot of the bat. Nine different definitions were given for the horizontal sweet spot of a bat: most of them were in an area 5 to 7 in. (13 to 18 cm) from the end of the barrel. Next, this chapter presented a new model for bat–ball collisions and used it along with a new performance criterion, namely the probability of getting a hit. Previous models were designed for analyzing home runs, which constitute less than 4% of the batted balls in play. This new model was used to describe the vertical gradients of the sweet spot of the bat. The vertical size of the sweet spot is one-third of an inch (8 mm). Then the chapter showed that there is an ideal bat weight for each batter. A simple table gave rules of thumb for recommending bat weights. Finally, this chapter gave a recommendation that all batters would profit from using end-loaded bats. For nonmathematical aspects of baseball see Baldwin's autobiography, Snake Jazz [50].

List of Variables

CoP	Center of percussion of a bat
CoR	Coefficient of restitution of a bat–ball collision
$CollisionSpeed$	Sum of pitch speed and speed of the bat at the collision point
C_d	Coefficient of drag
d_{cm-cop}	Distance from the center of mass to the center of percussion
d_{cm-ss}	Distance from the center of mass to the sweet spot
$d_{knob-cm}$	Distance from the center of the knob to the center of mass
$d_{pivot-ss}$	Distance from the pivot point to the sweet spot
$d_{pivot-cm}$	Distance from the pivot point to the center of mass
$d_{pivot-cop}$	Distance from the pivot point to the center of percussion
g	Earth's gravitational constant
I_{cm}	Moment of inertia of the bat with respect to the center of mass
I_{knob}	Moment of inertia of the bat with respect to the knob
I_{pivot}	Moment of inertia of the bat with respect to the pivot point
m_{ball}	Mass of the ball
m_{bat}	Mass of the bat
r_{ball}	Radius of the ball
r_{bat}	Radius of the bat
$v_{ball-after}$	Speed of the ball after the bat–ball collision
$v_{ball-before}$	Speed of the ball before the bat–ball collision
$v_{bat-after}$	Speed of the bat after the bat–ball collision
$v_{bat-before}$	Speed of the bat before the bat–ball collision
ω	Ball rotation rate

References

1. Bahill, A.T. and Baldwin, D., Describing baseball pitch movement with right-hand rules, *Computers in Biology and Medicine*, 37:1001–1008, 2007.
2. Bahill, A.T., The ideal moment of inertia for a baseball or softball bat, *IEEE Transactions on Systems, Man, and Cybernetics—Part A Systems and Humans*, 34(2):197–204, 2004.
3. Bahill, A.T. and Baldwin, D., The vertical illusions of batters, *Baseball Research Journal*, 32:26–30, 2003.
4. Bahill, A.T. and Karnavas, W.J., The ideal baseball bat, *New Scientist*, 130(1763):26–31, 1991.
5. Bahill, A.T., Botta, R., and Daniels, J., The Zachman framework populated with baseball models, *Journal of Enterprise Architecture*, 2(4):50–68, 2006.
6. Newton, I., A letter of Mr. Isaac Newton, of the University of Cambridge, containing his new theory about light and colors, *Philosophical Transactions of the Royal Society*, 7:3075–3087, 1671–1672.
7. Robins, B., *New Principles of Gunnery: Containing, the Determination of the Force of Gun-Powder, and Investigation of the Difference in the Resisting Power of the Air to*

Swift and Slow Motions, J. Nourse, London, Available online through Early English Books Online, 1742.

8. Barkla, H. and Auchterlonie, L., The Magnus or Robins effect on rotating spheres, *Journal of Fluid Mechanics*, 47:437–447, 1971.

9. Briggs, L.J., Effect of spin and speed on the lateral deflection (curve) of a baseball; and the Magnus effect for smooth spheres, *American Journal of Physics*, 27(8):589–596, 1959.

10. Watts, R.G. and Bahill, A.T., *Keep Your Eye on the Ball: Curve Balls, Knuckleballs and Fallacies of Baseball*, New York: W.H. Freeman, 2000.

11. Bahill, A.T. and Karnavas, W.J., The perceptual illusion of baseball's rising fastball and breaking curve ball, *Journal of Experimental Psychology: Human Perception and Performance*, 19:3–14, 1993.

12. NASA, http://www.grc.nasa.gov/WWW/K-12/airplane/bga.html, retrieved July 2007.

13. Baldwin, D.G. and Bahill, A.T., A model of the bat's vertical sweetness gradient, *The Engineering of Sport 5*, M. Hubbard, R.D. Mehta, and J.M. Pallis (Eds.), *Proceedings of the 5th International Engineering of Sport Conference*, September 13–16, 2004, Davis, CA, International Sports Engineering Association (ISEA), Sheffield, UK, Vol. 2, pp. 305–311, 2004.

14. Baldwin, D.G., Bahill, A.T., and Nathan, A., Nickel and dime pitches, *Baseball Research Journal*, 35:25–29, 2007.

15. Bahill, A.T., Baldwin, D., and Venkateswaran, J., Predicting a baseball's path, *American Scientist*, 93(3):218–225, 2005.

16. Selin, C., An analysis of the aerodynamics of pitched baseballs, *The Research Quarterly*, 30(2):232–240, 1959.

17. Bahill, A.T. and Baldwin, D.G., The rising fastball and other perceptual illusions of batters. In: *Biomedical Engineering Principles in Sports*, G. Hung and J. Pallis (Eds.), Kluwer Academic, New York, pp. 257–287, 2004.

18. Bahill, A.T. and LaRitz, T., Why can't batters keep their eyes on the ball? *American Scientist*, 72:249–253, 1984.

19. Rae, W.J., Mechanics of the forward pass, In: *Biomedical Engineering Principles in Sports*, G. Hung and J. Pallis (Eds.), Kluwer Academic, New York, pp. 291–319, 2004.

20. Adair, R.K., *The Physics of Baseball*, HarperCollins, New York, 2002.

21. Adair, R.K., Comments on "How to hit home runs: Optimum baseball bat swing parameters for maximum range trajectories," by G.S. Sawicki, M. Hubbard, and W.J. Stronge, *American Journal of Physics*, 73(2):184–185, 2004.

22. Sawicki, G.S., Hubbard, M., and Stronge, W.J., How to hit home runs: Optimum baseball bat swing parameters for maximum range trajectories, *American Journal of Physics*, 71(11):1152–1162, 2003.

23. Sawicki, G.S., Hubbard, M., and Stronge, W.J., Reply to comments on "How to hit home runs: Optimum baseball bat swing parameters for maximum range trajectories," *American Journal of Physics*, 73(2):185–189, 2004.

24. Frohlich, C., Aerodynamic drag crisis and its possible effect on the flight of baseballs. *American Journal of Physics*, 52(4):325–334, 1984.

25. Nathan, A.M., Hopkins, J., Chong, L., and Kaczmarski, H., The effect of spin on the flight of a baseball, SABR conference, Seattle, June 2006.

26. Alaways, L.W., Mish, S.P., and Hubbard, M., Identification of release conditions and aerodynamic forces in pitched-baseball trajectories, *Journal of Applied Biomechanics*, 17:63–76, 2001.

27. Stark, L., *Neurological Control Systems, Studies in Bioengineering*, Plenum Press, New York, 1968.
28. Bahill, A.T. and Morna Freitas, M., Two methods for recommending bat weights, *Annals of Biomedical Engineering*, 23(4):436–444, 1995.
29. Sears, F.W., Zemansky, M.W., and Young, H.D., *University Physics*, Reading, MA: Addison-Wesley, 1976.
30. Van Zandt, L.L., The dynamical theory of the baseball bat, *American Journal of Physics*, 60(2):172–181, 1992.
31. Cross, R., The sweet spot of the baseball bat, *American Journal of Physics*, 66:772–779, 1998.
32. Nathan, A.M., Dynamics of the baseball–bat collision, *Amercian Journal of Physics*, 68:979–990, 2000.
33. Nathan, A.M., Characterizing the performance of baseball-bats, *American Journal of Physics*, 71:134–143, 2003.
34. Nathan, A.M., Russell, D.A., and Smith, L.V., The physics of the trampoline effect in baseball and softball bats, *Proceedings of the 5th Conference of Engineering of Sport*, M. Hubbard, R.D. Mehta, and J.M. Pallis (Eds.), International Sports Engineering Association (ISEA), Vol. 2, pp. 38–44, 2004.
35. Russell, D.A., Hoop frequency as a predictor of performance for softball bats, *Proceedings of the 5th Conference of Engineering of Sport*, M. Hubbard, R.D. Mehta, and J.M. Pallis (Eds.), International Sports Engineering Association (ISEA), Vol. 2, pp. 641–647, 2004. See also http://www.kettering.edu/ ~drussell/bats-new/sweetspot.html
36. Brancazio, P., Swinging for the Fences: The Physics of the Baseball Bat, paper presented at the New England section of the *American Physical Society meeting*, October 1987.
37. Crisco, J.J., Greenwald, R.M., and Penna, L.H., Baseball bat performance: A batting cage study, www.nisss.org/BBSPEED6a.html, 1999.
38. Vedula, G. and Sherwood, J.A., An experimental and finite element study of the relationship amongst the sweet spot, COP and vibration nodes in baseball bats, *Proceedings of the 5th Conference of Engineering of Sport*, M. Hubbard, R.D. Mehta, and J.M. Pallis (Eds.), International Sports Engineering Association (ISEA), Vol. 2, pp. 626–632, 2004.
39. Cross, R., Response to "Comment on 'The sweet spot of a baseball bat,'" *American Journal of Physics*, 69(2):231–232, 2001.
40. Adair, R.K., Comment on The sweet spot of a baseball bat, by Rod Cross, *American Journal of Physics*, 69(2):229–230, 2001.
41. Williams, T. and Underwood, J., *The Science of Hitting*, New York: Simon and Schuster, 1982.
42. Brancazio, P., *SportScience: Physical Laws and Optimum Performance*. New York: Simon and Schuster, 1984.
43. Bahill, A.T. and Karnavas, W.J., Determining ideal baseball bat weights using muscle force–velocity relationships, *Biological Cybernetics*, 62:89–97, 1989.
44. Crisco, J.J., NCAA Research Program on Bat and Ball Performance, final report, November 12, 1997.
45. Drane, P.J. and Sherwood, J.A., Characterization of the effect of temperature on baseball COR performance, *Proceedings of the 5th Conference of Engineering of Sport*, M. Hubbard, R.D. Mehta, and J.M. Pallis (Eds.), International Sports Engineering Association (ISEA), Vol. 2, pp. 59–65, 2004.

46. Duris, J. and Smith, L., Evaluation test methods used to characterize softballs, *Proceedings of the 5th Conference of Engineering of Sport*, M. Hubbard, R.D. Mehta, and J.M. Pallis (Eds.), International Sports Engineering Association (ISEA), Vol. 2, pp. 80–86, 2004.
47. Cross, R., Center of percussion of hand-held instruments, *American Journal of Physics*, 72(5):622–630, 2004.
48. Goldsmith, W., *Impact: The Theory and Physical Behavior of Colliding Solids*, London: Edward Arnold, 1960.
49. Watts, R.G. and Baroni, S., Baseball–bat collisions and the resulting trajectories of spinning balls, *American Journal of Physics*, 57:40–45, 1989.
50. Baldwin, D., *Snake Jazz*, Xlibris Corp, www.Xlibris.com, 2007.

17

Biodynamics Analysis of Women's Gymnastics: Yurchenko Layout Vault

Michael Koh and Leslie Jennings

CONTENTS

17.1 Introduction

Sports biomechanists typically provide the scientific basis that enable coaches to teach sports techniques and injury prevention by means of analyzing and reviewing movement performance data. The usual approach is to identify key performance variables by means of statistical methods or inductive analysis. Cinematographic studies have provided insights into the performance of a variety of gymnastics vaults. However, this understanding is limited only to the examples analyzed and cannot be generalized to understand the effects of technique variations beyond observed performances. In other words, they cannot answer the "what if?" questions. Although a well controlled experiment can provide an explanation for the observed outcome, the power of such experiments is also limited, as changes in one aspect of technique may inadvertently result in other changes in the movement that may influence the final outcome [1].

Computer simulation can be effectively employed by applying a validated mathematical (also known as biomechanical) model of a system being investigated, to evaluate its response to changes in the input parameters. The application of simulation models to sports movements ensure that the results are due to the interventions (changes in the input parameters) introduced [1]. Consequently, such applications yield valuable insights into how some of the training interventions may potentially alter athletic performance. These interventions may be investigated individually or in combination with several other training factors, and it is a matter of transcribing such factors into model parameters to be included in the simulation. While it may not be feasible to incorporate all movements' kinematic–kinetic relationships into the model, depending on the complexity of the model, the advantage in computer simulation studies is that the model used in the simulation is free from the physical limitations commonly associated with experimental research. In particular, the safety of the performer will not be compromised, as no further performance will be required once the data have been collected based on present performance capabilities. There will also be savings in time as many different simulations can be performed quickly and accurately by powerful computers.

However, computer simulation is not without its limitation. One of the most important of these is the trade-off between simplicity and accuracy. The construction of a computer model to represent a particular biomechanical system often incorporates simplifying assumptions. The complexity of a model can range from the simple two-segment model of Alexander [2], optimizing the plant angle and takeoff velocity of the high and the long jumps, to the very complex 17-segment model incorporating 46 muscle groups used by Hatze [3], to simulate the long jump takeoff. Therefore, the complexity of a model should depend on its objective [4]. Even though high-speed computers now enable complex mathematical expressions to be solved rapidly and accurately, thus affording greater flexibility in model construction, some researchers [4,5] recommend that the creation of a computer model should be as simple as possibly required for the simulation objective. In fact, some very compelling implications can be derived from quite simplistic models [6,7,8]. However, there are others [1] who hold the view that a complex model be created and subsequently simplified by adding constraints to the model.

Since the human body, and its motion, is far too complex to be perfectly duplicated on the computer [9], computer models should therefore be evaluated for their level of accuracy so as to give confidence in its predictions [10]. Typically, model evaluation involves determining if the performance data collected (e.g., linear and angular displacement histories) can be reproduced by matching the model output to the data. Such a comparison will provide some evidence for the appropriateness of the findings derived from the model and to lend support for the confidence in the use of the model, for that particular athlete.

17.2   Biomechanical Simulation Studies of Gymnastics Vaulting

Traditionally, studies of gymnastics vaulting have relied on the statistical analysis of data digitized from film, to understand the mechanics and techniques involved [11,12]. Description of mechanical variables derived from these analyses, such as the angle-of-takeoff and speed-of-rotation, frequently form the basis of instruction to other performers, regardless of their anthropometric structure, physiological capacity, and skill level. Kinematic data, while explicitly dependent on technique, are simultaneously implicitly dependent on the strength, flexibility, and somatotype of the particular athlete. Therefore, it would be most appropriate that definitive instruction to any participant should be based on an individual biomechanical assessment of the performance used in conjunction with computer simulation of a customized biomechanical model.

In surveying the literature on theoretical analysis of gymnastics vaulting, it is noted that the horse-impact phase and the kinematics of the preflight for the handspring vault have been studied [13]. For instance, the horse-impact phase has been modeled to comprise two distinct stages of compression and repulsion; the springboard takeoff velocity and the initial distance from the horse are seemingly the principal variables affecting the outcome of the handspring vault. However, the force exerted by the performer during the horse repulsion (push-off) phase is deemed to have only a minimal effect on the postflight characteristics of the vault as the gymnast has little control over the duration of the repulsion phase.

Other theoretical studies on the contribution of preflight characteristics to successful vaulting have concentrated on the counter rotational Hecht vault [6,14]. For instance, a two-segment simulation model has been used to show that the preflight requirements of the Hecht vault are different from those for the handspring somersault vaults [6]. A successful Hecht vault is largely due to a good preflight and a passive impact, where no shoulder torque is involved. As regards the role of preflight trajectory, in the reversal of total body rotation at horse impact in the Hecht vault, optimized simulations based on the mean preflight data of 27 competitive Hecht vaults indicate that more than 70% of the reversal of rotation could be produced from a suitable preflight trajectory [14]. The simulations also show that preflight impact angle contributes more significantly than horse elasticity and shoulder torques during the horse contact phase. Also, the takeoff speed is important to successful vaulting with the effect that any increase in horizontal takeoff velocity on the performance of the Hecht vault increases landing distance of the vault.

In summary, while it may appear that the variables used as input into computer simulation studies on gymnastics vaulting are typically the linear velocities of the center of mass just before horse impact, or the body angle of attack with the vaulting horse (Figure 17.1), in practice the variables

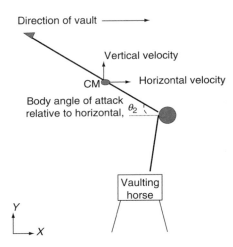

FIGURE 17.1
An example of typical variables used as input into computer simulation models of gymnastics vaulting. In this example, the linear velocities of the center of mass (CM) and the body angle of attack relative to the horizontal are illustrated.

measured and used as input into such models will depend on the research questions asked. An example using the Yurchenko layout vault will be detailed in a later section.

Optimization studies: The application of simulation models to sports movements has definitely provided a means of determining optimum techniques [15] to maximize a performance. Optimization involves the iterative use of a computer simulation to determine the parameters' values or control variables that minimize or maximize an objective function representing some performance criterion. If a critical factor in a problem can be isolated and characterized by an objective function that quantifies the performance, then optimizations may provide a suitable framework for analysis. Although, it is rarely possible to fully represent all the complexities of variable interactions, constraints and appropriate objectives for a complex problem, nevertheless, optimization formulation should be regarded as a good approximation [16]. For instance, the objective function for gymnastics vaulting can be put into a tractable analytical expression based on the primary objectives of vaulting, namely postflight height and distance [17]. Appendix 1 provides details of how the center of mass position or displacements of body segments can be expressed in a mathematical form. It has been shown that the performance of a male individual's handspring front somersault vault can be optimized, using postflight height and distance as the optimization criteria and angular momentum as a penalty function [17]. The approach involves a static optimization technique. Ideally, the formulation of the system dynamics and the use of dynamic optimization techniques, through an application of optimal control theory, should be more appropriate for the optimization of sports performance. The implications of such simulation studies are of significance for developing coaching methods that would bring about optimal performance. For instance, insights gained from optimized simulations of sports performances can offer coaches an understanding of joint torque contributions that result in optimal technique [18]. Coaches may then use these insights to introduce training interventions that would facilitate the optimal technique performance.

Questions: Simulation studies suggest that the success of a vault can be attributed primarily to the preflight characteristics, namely linear and angular velocities of the whole body center of mass and preflight trajectory. It appears that there exists an optimum combination of linear and angular velocities, along with individualized inertial characteristics, that will produce the best performance for each vault type. Clearly, the combination of mechanical factors would vary from one movement to another and between athletes performing the same skill due to their different physique and strengths. Consequently, the answer to the question of how a gymnast modifies technique to improve performance can best be provided by computer simulation.

Here, our vault of interest is known as the Yurchenko layout vault. The Yurchenko layout vault was pioneered by Natalia Yurchenko in the 1982 World Cup gymnastics competition. It comprises a forward running approach that finishes with a cartwheel half-turn (round-off skill) to orientate the body, so that the back faces the vaulting horse at the point of takeoff from the springboard. This is immediately followed by the gymnast taking off from the springboard using a back-flip action to impact the horse, and finally completing a one and a half somersault rotation with the body fully extended (layout) before landing. Figure 17.2 illustrates the vaulting sequence.

At present, there is a paucity of information (from an optimized simulation perspective) on the effect of initial conditions at horse contact on the postflight characteristics of continuous rotation vaults. Specifically, few theoretical studies have investigated the Yurchenko layout vault, to answer the following questions:

FIGURE 17.2
Vault sequence of the Yurchenko layout vault beginning with a backflip onto the horse during preflight; followed by a dynamic push-off (blocking action) during horse impact; and finally performing a $1\frac{1}{2}$ layout somersault rotation to land in the postflight.

- What is an optimal technique for the Yurchenko layout vault, for a given performance ability?
- Is there a "blocking" technique at contact with the horse, characterized by a decrease in horse impact time, and positive changes to the linear velocities at the end of impact?

17.3   Model Development

The optimization of a vault performance requires the development of a computer model that adequately represents the gymnast. As the validity of the simulation study will only be as good as the model itself, accurate measures of segment masses, center of mass, and the moment of inertia values are required to customize the model to the gymnast by using the elliptical zone modeling technique of Jensen [19]. The method yielded an error of less than 2% when the estimated total body weight was compared against the actual body weight of the subject. The data were used to customize the model.

We present, herein, a five segment rigid-linked model consisting of the hand, whole arm, upper and lower trunk and whole leg. Symmetry is assumed as the motion is essentially planar in nature. Figure 17.3 illustrates the equivalence between the rigid body model at the impact phase of the vault and the anatomical representation. The following assumptions are made in the model:

- Each segment is represented by a point mass located along the rigid body known as its center of mass (CM), the position of which is fixed throughout the motion.
- Segments are linked by revolute (hinge) joints.
- Inertial characteristics remain constant throughout the motion.
- Gravitational force acts downward through the each segment's CM, and is equal to the mass and gravitational acceleration (9.8 m/s^2).
- External force (ground reaction force, GRF) acting on the segments are considered to act at a point known as the center of pressure (COP).
- Net effect of muscle and ligament activity at any joint is represented in terms of net muscle moments at the joint.

From the original link-segment model, a free-body diagram of each segment is obtained, and this is typically done by "breaking" the model up at the joints. The internal forces that act across each joint are represented in the free-body diagram. Figure 17.4 depicts a free-body segment and the accompanying reaction forces and moments of force acting at the proximal and

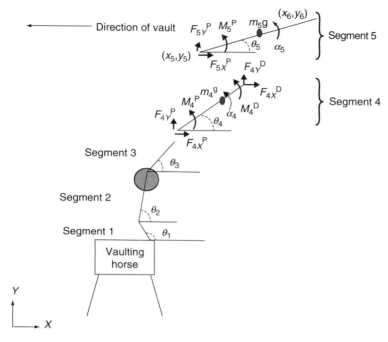

FIGURE 17.3
The multilinked rigid body model of the Yurchenko layout vault, depicted at horse impact. The model comprises of the hand (segment 1), whole arm (segment 2), upper trunk (segment 3), lower trunk (segment 4), and whole leg segments (segment 5). Symmetry is assumed due to the planar nature of the movement. Note that the segments 4 and 5 have been disjointed in order to illustrate the internal joint reaction forces at the proximal and distal ends of each segment and also the angular displacements, acceleration, and joint torques. The illustration holds true for all segments depicted. The coordinates (x_5, y_5) and (x_6, y_6) denote the digitized (x, y) coordinates of the whole leg segment at time, t.

distal ends of the segment in the generic case. Consider the planar movement of a single segment. Applying Newton's laws of motion yields the following equations:

$$\sum F_x = ma_x \Rightarrow F_x^P + F_x^D = ma_x$$

and

$$\sum F_y = ma_y \Rightarrow F_y^P + F_y^D - mg = ma_y \qquad (17.1)$$

where a_x, a_y, and g represent the horizontal and vertical acceleration of the segment CM and gravitational acceleration, respectively. The angular motion equivalent of Newton's law about the CM is given by

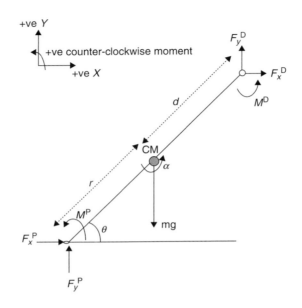

FIGURE 17.4
Free-body diagram of a single segment showing proximal and distal reaction forces in the horizontal and vertical directions represented by F_x^P, F_x^D, F_y^P, F_y^D, respectively. The mass of the segment is lumped at its center of mass (CM). The segment rotates with angular acceleration α about its CM and possesses horizontal and vertical linear acceleration a_x and a_y. The net muscle moments about the proximal and distal ends (joints) are denoted by M^P and M^D, respectively (also known as joint torques). The proximal distance of the segment CM is given by r, and its distal length, d. The segment subtends an angle θ with the horizontal.

$$\sum M = I_0\alpha \Rightarrow M^P + M^D + F_x^P r \sin\theta + F_y^D d \cos\theta$$
$$- F_y^P r \cos\theta - F_x^D d \sin\theta = I_0\alpha \tag{17.2}$$

where
  θ is the angle the segment makes with the horizontal
  α the angular acceleration of the segment about its CM
  I_0 is the transverse moment of inertia of the segment about its CM

Knowledge of the anthropometrics, kinematic data, and ground reaction forces will enable a systematic resolution of the equations to yield solutions for the net muscle moments at a joint (joint torques) and joint reaction forces of the free-body diagrams. Kinematic data are typically obtained from digitizing video images. Where ground reaction forces are not available, the equations are resolved systematically using kinematic data; beginning from the free end (most distal to contact point) of the multilinked rigid body system, and progressing toward the end in contact with the ground; each time substituting the results of the prior analysis of a more distal segment into the next set of equations of a more proximal segment to the point of contact. This approach, to obtain the ground reaction force, is known as the conventional inverse dynamics method [20].

For the example of the vault model described previously, the equations of motion to represent the vault can be formulated using a Newtonian approach [21] and are given by Equations 17.3 through 17.8. Note that there are alternative formulations which relates the segment CM's kinematics to the kinematics of the digitized points [10]. Refer to Figure 17.3 for a depiction of the variables used in these equations.

Segment 1 (hand),

$$F_{1x}^P + F_{1x}^D = m_1 a_{1x}^{CM}$$

and

$$F_{1y}^P + F_{1y}^D - m_1 g = m a_{1y}^{CM} \tag{17.3a}$$

$$M_1^P + M_1^D + F_{1x}^P r_1 \sin \theta_1 + F_{1y}^D d_1 \cos \theta_1 - F_{1y}^P r_1 \cos \theta_1 - F_{1x}^D d_1 \sin \theta_1 = I_1 \alpha_1 \tag{17.3b}$$

Segment 2 (upper limb),

$$F_{2x}^P + F_{2x}^D = m_2 a_{2x}^{CM}$$

and

$$F_{2y}^P + F_{2y}^D - m_2 g = m_2 a_{2y}^{CM} \tag{17.4a}$$

$$M_2^P + M_2^D + F_{2x}^P r_2 \sin \theta_2 + F_{2y}^D d_2 \cos \theta_2 - F_{2y}^P r_2 \cos \theta_2 - F_{2x}^D d_2 \sin \theta_2 = I_2 \alpha_2 \tag{17.4b}$$

Segment 3 (upper trunk),

$$F_{3x}^P + F_{3x}^D = m_3 a_{3x}^{CM}$$

and

$$F_{3y}^P + F_{3y}^D - m_3 g = m_3 a_{3y}^{CM} \tag{17.5a}$$

$$M_3^P + M_3^D + F_{3x}^P r_3 \sin \theta_3 + F_{3y}^D d_3 \cos \theta_3 - F_{3y}^P r_3 \cos \theta_3 - F_{3x}^D d_3 \sin \theta_3 = I_3 \alpha_3 \tag{17.5b}$$

Segment 4 (lower trunk),

$$F_{4x}^P + F_{4x}^D = m_4 a_{4x}^{CM}$$

and

$$F_{4y}^P + F_{4y}^D - m_4 g = m_4 a_{4y}^{CM} \tag{17.6a}$$

$$M_4^P + M_4^D + F_{4x}^P r_4 \sin \theta_4 + F_{4y}^D d_4 \cos \theta_4 - F_{4y}^P r_4 \cos \theta_4 - F_{4x}^D d_4 \sin \theta_4 = I_4 \alpha_4 \tag{17.6b}$$

Segment 5 (lower limb),

$$F_{5x}^P + F_{5x}^D = ma_{5x}^{CM}$$

and

$$F_{5y}^P + F_{5y}^D - m_5g = m_5a_{5y}^{CM} \tag{17.7a}$$

$$M_5^P + M_5^D + F_{5x}^P r_5 \sin\theta_5 + F_{5y}^D d_5 \cos\theta_5 - F_{5y}^P r_5 \cos\theta_5 - F_{5x}^D d_5 \sin\theta_5 = I_5\alpha_5 \tag{17.7b}$$

However, the distal end of the fifth segment is free and Equations 17.7a and 17.7b simplify to,

$$F_{5x}^P = m_5a_{5x}^{CM}$$

and

$$F_{5y}^P - m_5g = m_5a_{5y}^{CM} \tag{17.8a}$$

$$M_5^P + F_{5x}^P r_5 \sin\theta_5 - F_{5y}^P r_5 \cos\theta_5 = I_5\alpha_5 \tag{17.8b}$$

As an example, consider the following set of data which will be used as input to resolve the inverse dynamics problem at an instance of time.

The segment CM displacements are obtained by digitizing the ends of a segment and using appropriate scaling methods based on anthropometric models [19], to determine each segment's CM location. Consider for example, such a method applied to obtain the segment CM position for the whole lower limb (seen as segment 5 in Figure 17.3) at one instance in time, t. Let us suppose that the digitized coordinates of this segment be (x_5, y_5) and (x_6, y_6) for the proximal and distal ends of the segment, respectively, at this particular instant, t. Applying the ratio theorem, the x-coordinate of the whole lower limb segment CM at this one instance in time is simply:

$$x\text{-coordinate CM} = \frac{0.34}{0.78}x_5 + \left(\frac{0.78 - 0.34}{0.78}\right)x_6$$

where the anthropometric dimensions of 0.34 and 0.78 m are obtained from Table 17.1. By repeating this process iteratively to the (x, y) coordinates of the whole lower limb segment corresponding to every instant of time, t_i (where i is the number of data points), the whole lower limb segment CM x-coordinate trajectory can be obtained. The process is similar for the y-coordinate of the whole lower limb segment CM, namely, applying the ratio theorem throughout the displacement history with the same anthropometric dimensions of the whole lower limb taken from Table 17.1 for the purpose of this discussion. The data are then smoothed with quintic splines [22] whose derivatives yield the respective segments CM velocity and acceleration time histories.

TABLE 17.1

Subject Leg Segment Anthropometric Data and Kinematics at an Instance of Horse Impact

Segment	Mass (kg)	I_{CM} (kgm^2)	Proximal Radius of CM (m)	Length (m)	a_x^{CM}	a_y^{CM}	θ (rad)	Angular Acceleration (rad/s)
Whole lower limb	20.237	1.069	0.340	0.779	15.68	5.73	−0.560	0.908

The proximal forces F_{5x}^P, F_{5y}^P in Equation 17.9 and shown in Figure 17.3 are obtained by directly substituting the values of m_5, a_{5x}^{CM}, a_{5x}^{CM}, and the inertial parameters obtained from Table 17.1.

Solving for Equation 17.8a yields:

$$F_{5x}^P = 20.237 \times 15.68 = 317.32 \text{ N}$$

and

$$F_{5y}^P = 20.237 \times (9.81 + 5.73) = 314.48 \text{ N} \tag{17.9}$$

Solving for Equation 17.8b by substituting for F_{5x}^P and F_{5y}^P and using data from Table 17.1, yields the net muscle moment/joint torque (M_5^P) for the proximal joint of the lower limb segment (Segment 5):

$$M_5^P = 317.32 \times 0.34 \sin(-0.56) - 314.48 \times 0.34 \cos(-0.56) - 1.069 (0.908)$$
$$= 108.95 \text{ Nm} \tag{17.10}$$

The solutions for Equation 17.8, together with the corresponding parameter values for the lower trunk segment, are substituted into Equation 17.6, to resolve for the joint forces and net muscle moments accordingly for the lower trunk segment (Segment 4).

Note that at the joints where segment $k+1$ and k are linked (with segment $k+1$ assumed to be superior in position to segment k), the distal joint forces and net joint torque of segment k, are of the same magnitude but opposite in sign to the proximal joint forces and net torque of segment $k+1$ (see Figure 17.3). This means that

$$M_5^P = -M_4^D, \quad F_{5x}^P = -F_{4x}^D \quad \text{and} \quad F_{5y}^P = -F_{4y}^D$$

$$\text{In general, } M_{i+1}^P = -M_i^D, \quad F_{i+1_{x,y}}^P = -F_{i_{x,y}}^D \tag{17.11}$$

Subsequent backward substitution into Equations 17.5 through 17.3 will resolve the set of equations to obtain the set of joint forces and net muscle moments corresponding to the segments. These computations to resolve the sets of equations can be facilitated using matrix manipulation [23]. The joint torque histories are used as input into the simulation, together with the initial values comprising body-segment orientation, segment angular velocities and the CM displacements and linear velocities at impact with the horse. The computational steps to be carried out can be summarized as follows:

1. Establish link-segment model and equations of motions for model.
2. Substitute kinematic data, in conjunction with appropriate body-segment parameters (BSP) to solve for equations of motion for distal most segment of model.
3. Use these solutions, together with the BSP values and kinematic data, to solve the equations of motions of the preceding segment.
4. Repeat step (c), until all equations are resolved.

Many models have been created in gait studies [24] that are based on inverse dynamics approaches. The approach is computationally efficient and practical, albeit highly dependent on the accuracy of experimental data collection and processing; and, is limited by the assumptions of the model. In the case of normal gait analysis, research [24] has indicated that the inverse dynamics approach is an appropriate method to obtain gait kinetics.

17.4   Model Evaluation

To make use of the model for subsequent optimizations to determine, say, ideal states for sports performance, it is necessary to first evaluate the model. A common approach is to compare the angular displacement trajectories from the output data of the model against the kinematic data for the vault in question. It must be noted here that mathematical formulations of modeled situations are generally of standard methods, although some are more elegant than others. However, the different methods used should not influence the outcome of the results. Evaluation of a model is thus important, as it provides evidence for the appropriateness of the outcome of the model.

The present model is driven by the joint torques at the wrist, shoulder, mid-trunk, and hip. These are obtained from the conventional inverse dynamics approach [20] outlined previously. However, it is possible that the joint torques used as input to drive the model may not be able to reproduce the experimental motion precisely [25], even though its derivation is based on the same experimental data. As a consequence, a dynamic optimization technique is used to improve the precision of the joint torques [25,26]. This is not to suggest that the inverse dynamics approach is comprehensively inadequate. The question to ask is whether the computational expense of a dynamic optimization technique is justified for the analysis to be pursued. Dynamic optimization techniques are generally preferred and sometimes even necessary in the absence of accurate experimental data. In the event of predicting novel movements, it is certainly a desirable technique to be employed as it is able to integrate the influences of musculo-skeletal structures on biomechanical performance through the enforcement of the system dynamics into the solution process.

The MISER3 [27] constrained optimal control software, which computes constrained optimal control problems, is used in this experiment. In the algorithm (Table 17.2), we set up 14 states (five segment angles, whole body CM coordinates, and their first derivatives), 12 system parameters (initial segment orientations, angular velocities, vault duration and body lean), and 4 controls (joint torques).

The MISER3 algorithm integrates the ordinary differential equations to an accuracy of 10^{-9}, using a variable step, variable order, and Adam's method [5]. Readers are directed to the manual for greater detail in the setup of MISER3 [27]. The basic idea of the dynamic optimization method presented here is to alter the magnitude of the joint torque histories (wrist, shoulder, mid-trunk, and hip) continuously and systematically, until the displacement solutions of the differential equations governing the system closely matches the experimental data [25,26]. Once this is achieved, the set of segment angular accelerations are obtained; and, thence segment velocities and displacements. These are substituted into equations governing the geometry of the system, to produce the simulation.

TABLE 17.2

Model Parameters Used in MISER3

MISER3 States	Model Parameters	System Parameters	Initial Value	Controls	Joint Torques
X_1	(Hand) θ_1	Z_1	$\theta_1 (0)$	U_1	Wrist
X_2	(Arm) θ_2	Z_2	$\theta_2 (0)$	U_2	Shoulder
X_3	(Upper trunk) θ_3	Z_3	$\theta_3 (0)$	U_3	Mid-trunk
X_4	(Lower trunk) θ_4	Z_4	$\theta_4 (0)$	U_4	Hip
X_5	(Leg) θ_5	Z_5	$\theta_5 (0)$		
X_6	ω_1	Z_6	$\omega_1 (0)$		
X_7	ω_2	Z_7	$\omega_2 (0)$		
X_8	ω_3	Z_8	$\omega_3 (0)$		
X_9	ω_4	Z_9	$\omega_4 (0)$		
X_{10}	ω_5	Z_{10}	$\omega_5 (0)$		
X_{11}	^iCM_X or ^fx_E	Z_{11}	Duration of vault		
X_{12}	^iCM_y or ^fy_E	Z_{12}	Body lean		
X_{13}	^{i}velocity CM_x or ^{f}velocity x_E				
X_{14}	^{i}velocity CM_x or ^{f}velocity y_E				

Note: i denotes impact phase, f denotes postflight phase, (x_E, y_E) coordinates of the proximal end of segment 1, (CM_X, CM_Y) denotes coordinates of system CM; θ_i $(i = 1, \ldots, 5)$ denotes segment angle with respect to the horizontal axis, and ω_j $(j = 1, \ldots, 5)$ denotes segment angular velocity.

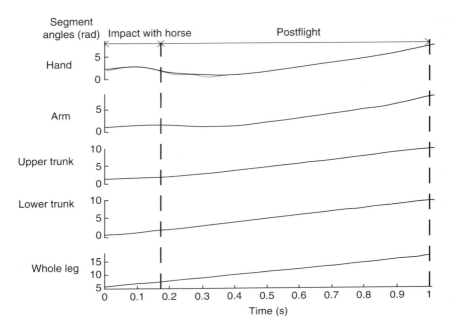

FIGURE 17.5

A common approach in model evaluation is to compare the angular displacement trajectories from the output data of the model against the kinematic data for the performance in question. Model evaluation of the Yurchenko layout vault is performed by comparing the closeness-of-fit between the segment angular displacement trajectories of the model (broken line) and the experimental data (solid line). The reproduced segment angular trajectories were a good fit to the experimental data. This closeness-of-fit enables the researcher to have greater confidence in the model. Evaluation of a model is thus important, as it provides evidence for the appropriateness of the outcome of the model.

Figure 17.5 depicts the results of the model evaluation for the Yurchenko layout vault model, based on the dynamic optimization technique used. The model is able to reproduce the movement closely throughout the vault duration, in terms of the segment angular trajectories. However, there exist some minor fluctuations of the reproduced hand segment angular displacement trajectory between 0.2–0.4 s, which are not deemed to be significant. Overall confidence in the use of the model for subsequent optimization studies is therefore good.

17.5 Defining an Optimization Criteria

To define an optimization criterion, it is important to first establish the outcomes or the research questions. To answer these questions, an objective function is formulated analytically. In general, the objective function of the optimal control problem consists of two parts [28]: (1) a terminal cost that exacts a penalty according to the final state of the system, which is defined by the position and alignment of the body segments at the end of postflight and (2) a cost that depends on the state of the system given by the dynamics equations and the controls (joint torques) used to arrive at the final solution of the problem. The optimization proceeds to find the set of joint torques that will produce the optimum technique (optimized parameters) at a minimal cost, based on the optimization criteria.

For this study, let us recall that the optimization problem is to determine an optimum technique for the Yurchenko layout vault, based on a given performance ability. For this purpose, performance ability refers to the level or standard achieved at the present time. The optimum technique may be characterized perhaps by changes at contact with the horse, such as a decrease in horse impact time, changes to the linear velocities at the end of impact, as well as horse impact body-segment configuration. The derivation of the objective function is thus based on the international judging criteria for the vault, in which a gymnast starts with a base score, from which deductions are made for performance faults such as a "break" in the form or poor dynamics of the vault (Figure 17.6).

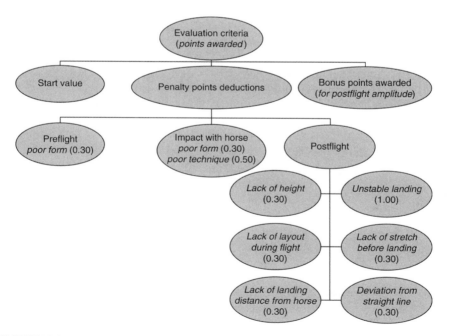

FIGURE 17.6
Evaluation scheme of vault summarized from the 1997 Judging Criteria of the Federation of International Gymnastics. Note that the greatest possibility of points deductions is in the postflight. This evaluation scheme has a direct bearing on the optimization criteria used in the simulation.

However, the performance may be awarded bonus points for good flight amplitude that is a high flight height after horse impact. Since the likelihood of points deduction is greater in the flight phase than during impact or preflight (flight phase prior to horse impact), the optimization criteria consist of an objective function that sufficiently represents the aim of maximizing the performance score during postflight while adhering to the movement constraints of the skill. Poor form (interpreted as flexed limbs) in the impact phase is excluded in the objective function, as the model uses rigid limb segments instead of two-segment limbs.

The objective function (G_0) given by Equation 17.12 consists of a terminal cost (ϕ_0, Equation 17.13) for meeting the 2 m postflight distance requirement (1st term of Equation 17.13), while attaining a landing angle of the legs (2nd term of Equation 17.13) represented by the system parameter (z_{12}) at the end of postflight. The landing angle (z_{12}) is defined by the angle of inclination of the legs, with respect to the horizontal.

$$G_0(\tau) = \underbrace{\phi_0[\theta(t_{term}), x_e(t_{term}), z]}_{\substack{\text{Terminal cost}(\phi_0), \\ \text{comprising body position} \\ \text{and alignment at terminal} \\ \text{time}(t_{term}), \text{also known as} \\ \text{end of flight phase after} \\ \text{horse impact. Elaborated} \\ \text{further in Equation 17.13.}}} + \underbrace{\int_{t_{imp}}^{t_{term}} \sum_{i=1}^{4} (\theta_i - \theta_{i+1})^2 dt}_{\substack{\text{Cost for maintenance} \\ \text{of straight body form} \\ \text{during flight, where} \\ \text{the respective} \\ \text{segment angles is} \\ \text{denoted by } (\theta_i)}}$$

$$\hspace{10cm} (17.12)$$

$$\phi_0(\theta(t_{term}), x_e(t_{term}), z) = \underbrace{[\min \{0, x_e(t_{term}) + \sum_{i=1}^{5} l_i \cos \theta_i(t_{term}) - 1.55\}]^2}_{\substack{\text{The point on the ground that meets the 2 m} \\ \text{distance requirement has an } x\text{-coordinate} \\ \text{of 1.55 m obtained from digitization.} \\ \text{This expression calculates, by geometry} \\ \text{(see Appendix), the } x\text{-coordinate of the} \\ \text{distal end of the leg segment at terminal} \\ \text{time } (t_{term}). \text{ It uses the } x\text{-coordinate of the} \\ \text{distal end of the hand segment } (x_e) \text{ and} \\ \text{the respective segment angles } (\theta_i). \text{ This} \\ \text{calculation of the distal end of the leg} \\ \text{segment is checked against the coordinate} \\ \text{of the point on the ground the cost value} \\ \text{is obtained when the distance} \\ \text{is attained.}}} + \underbrace{\sum_{i=3}^{5} (\theta_i(t_{term}) - z_{12})^2}_{\substack{\text{To keep the upper} \\ \text{body segments} \\ \text{aligned with the} \\ \text{legs, the parameter} \\ z_{12} \text{ which represents} \\ \text{the angle of} \\ \text{inclination the body} \\ \text{makes with the} \\ \text{ground at the end} \\ \text{of flight, is used.}}}$$

$$\hspace{10cm} (17.13)$$

where

θ_i represents the segment angles

t_{imp} and t_{term} are initial impact time and terminal time

ϕ_0 is a function of θ, x_e, and z at terminal time t_{term}

x_e is the x-coordinate of the distal end of the hand segment

The optimization proceeds to find the set of optimum parameters (z_i), in Table 17.2, that characterizes the optimum technique and the set of joint torque histories required to produce the prescribed motion at a minimum cost based on the objective function defined. For this analysis, the system parameters would be the segment angles, angular velocities, and the body lean at the end of postflight. Note that the inverse dynamics approach is used to compute a set of joint torques (M_i), in Equations 17.3b through 17.8b, that will serve as initial guesstimates to drive the optimization. The symbol τ (see Equation 17.12) is used in the optimization algorithm to distinguish from M_i. Preliminary work in testing the model indicated that the optimized vault meets the height requirements even without its inclusion in the objective function. Thus, to reduce computation time, height requirements are excluded from the objective function of the optimized simulations. However, the results of the optimizations were always checked graphically, to determine if the height requirements are satisfied.

Equal weightings for each variable in the objective function namely the terminal cost and the integrand (see Equation 17.12) are given, as it is equally important to meet the height and distance requirements spelt out in the evaluation scheme. A good height is one in which the hips are at least 1 m above the horse at the peak height of the vault. The distance requirement must be at least 2 m away from the far side of the horse at landing. Undoubtedly, the choice of weightings will affect the optimum solution. For instance, if a greater weight is placed on Equation 17.12 than on Equation 17.13, then the outcome of the optimization will be different to that where a greater weight is placed on Equation 17.13. It is a matter of balancing the priority of the layout in flight to achieving the horizontal distance at landing. It is possible to achieve a good layout at the expense of meeting the landing distance requirement. Thus, the experience of the researcher and knowledge of the skill demands of the movement contribute to making an accurate assessment of the appropriate weights in the objective function.

Subject-specific kinematic performance data and anthropometric measurements are input into the computer model. The optimization software MISER3 [27] is used to investigate the dynamics from impact with the horse until just before landing. The vault is optimized for: (1) the set of joint torques (τ); namely, wrist, shoulder, mid-trunk, and hip joint torques and (2) parameters (z), namely segment angles and angular velocities at impact, vault duration and landing angle (z_{12}). Details are provided in Table 17.2.

17.6 Technique of the Optimized Vault

The whole body CM trajectory from the optimized results of the simulation is illustrated in Figure 17.7. A comparison is made with the data CM trajectory. For the optimized vault, the CM trajectory demonstrates greater flight amplitude. In terms of judging, this should culminate in bonus points for the vault (Figure 17.6) because of the perceived increased flight time due to greater flight trajectory (also known as flight amplitude).

In addition to greater flight amplitude after horse impact, the simulation of the vault (dark model in Figure 17.7), also demonstrates a fully extended body (layout) throughout the flight right up to the instant before landing, compared with the actual vault performed (fair model Figure 17.7). This would translate to less point deduction under the penalty category of "lack of stretch before landing" and "insufficient layout." The simulation also depicts clearly that the height and distance requirements are satisfied. This means no points deductions under the penalty category of "insufficient height and distance." Overall, the optimized vault produces a superior technique compared with the actual performance in terms of flight

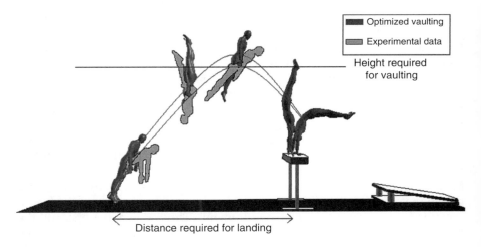

FIGURE 17.7
Comparison of whole body center of mass (CM) trajectory between the optimized vault (dark line) and actual vault performed. In terms of judging, this should culminate in bonus points for the vault because of the perceived increased flight time due to greater flight trajectory. Note that the CM is displaced higher and further for the optimized vault (dark model) compared to the physical performance (fair model). The vault is performed from a left to right direction with the vaulting horse at a height of 1.2 m. The actual performance demonstrates hip-flexion during the flight specifically near the instant before landing, compared with the optimized vault. On the other hand, the optimized simulation demonstrates a fully extended body (layout) through-out the flight right up to the instant before landing.

amplitude and body extension because of fewer penalty points deductions based on the judging criteria (Figure 17.6).

The optimization has produced a better flight trajectory than the data on two counts: (1) the criteria dictates a need to keep a good layout, as a result, the optimization sees to the attainment of this goal; (2) since a layout (fully extended body) is maintained during flight, it results in an increase in the transverse moment of inertia (I_{yy}). Therefore, to complete the same number of somersaults with an increased I_{yy}, it would be necessary to either increase the angular momentum or allow greater flight time to complete the motion. Figure 17.7 would suggest that the optimization results opted for greater flight time by increasing flight amplitude.

The impact technique also differs between the optimized solution and the data. The result of the optimization indicates that at initial impact, shoulder flexion increased by 9° and hip flexion by 8°. The shoulder angle ($\pi + \theta_2 - \theta_3$) is defined as the included angle between the trunk and the arm segments, while the hip angle ($\pi - \theta_4 + \theta_5$) is the included angle between the lower trunk and the leg segments. The increased shoulder flexion is consistent with the other experimental observations [29] that better Yurchenko layout vaults are performed by gymnasts who contact the horse with a larger shoulder joint flexion, as a result of reaching early for the horse, than those who did not. Table 17.3 compares the body configuration of the optimized vault at impact with the horse against the actual performance. A consequence of the optimized body configuration is that the body angle of attack [$\tan^{-1}(x_{11}/x_{12})$], determined by the angle subtended by the line joining the total body CM to the point of horse impact, and the horizontal axis, is increased by 9°.

The optimized simulation also produced a higher angular velocity ($z_6 - z_{10}$) at impact with the horse than the actual performance recorded. The implication is that to perform the optimal vault, the total angular momentum of the system must be increased during the flight phase prior to impact with the horse. This is consistent with the angular demands of the optimized vault. The larger total body moment of inertia, due to its fully extended body shape, would thus require a greater amount of angular momentum than the recorded vault.

TABLE 17.3

Comparison of Shoulder, Hip, and Body Angle between Actual Performance and the Optimized Vault

Parameter	Actual Performance (°)	Optimized Vault (°)
Shoulder angle	164	173
Hip angle	148	156
Body angle	32	41

TABLE 17.4

Comparison of Center of Mass (CM) Linear Velocities
between Actual Performance and the Optimized Vault
during the Horse Impact Phase of the Vault

Horizontal Velocity	Start of Impact (m/s)	End of Impact (m/s)
Actual vault	3.29	2.32
Optimized vault	As above	2.34
Vertical Velocity	**Start of Impact (m/s)**	**End of Impact (m/s)**
Actual vault	2.40	2.27
Optimized vault	As above	2.61

A "blocking" (push-off) technique is also observed. This is manifested by an increase in vertical takeoff velocity of the CM (x_{14}) and a reduction in horse contact time (z_{11}) by 0.003 s compared with the data. However, the reduction in impact duration (z_{11}) of the optimized vault is not regarded as significant as it fell within one frame of the capture rate of the video recorder. More importantly, there is a marked increase in the vertical horse takeoff velocity of the CM (x_{14}) for the optimized vault compared to the actual performance (Table 17.4). This result suggests that the vertical velocity component (x_{14}) is a key component to altering the flight trajectory of the vault. The finding is consistent with the mechanics of projectile motion. The reduction in the horizontal velocity (x_{13}) is similar between the two vaults. The result thus indicates that there is a trade-off between linear and horizontal velocities as a consequence of horse impact. However, the change in the magnitude of the horizontal velocity must not prevent the gymnast from meeting the distance requirements of the vault. It would appear, from the simulation, that the horizontal component velocity (x_{13}) of the actual vault is close to the optimal level and does not need further intervention. Table 17.4 illustrates the change in the CM linear velocities between the two vaults—actual and optimized.

17.7 Other Considerations

Sensitivity analyses show how sensitive an optimum solution is, with respect to the calculated parameters. It is a powerful tool that can be applied to simulation models in order to increase the level of confidence in the results obtained. The sensitivity analysis can be used on the parameters such as the inertial characteristics of the model, as well as on the variables such as input data. For instance, Alexander [2] found that the optimum techniques for high and long jumps were not sensitive to the muscle parameters used in his single torque generator model. As such, he was able to place more credence in his results.

The current vault model can be tested for sensitivity to small changes in the initial segment angles, initial segment angular velocities, and for changes in the inertial characteristics of the model. To test for sensitivity of the model output to small changes in the initial segment configuration, each segment's initial angle is reduced by $1°$ while maintaining the initial respective segmental angular velocities. The change in each of the parameters, as a result of the perturbation, is reported as a root-mean-square error from the original values. In this analysis, the model was found to be insensitive to small changes in the initial segment configuration, with a similar result for a small increase of $1°$. Likewise, to assess the sensitivity of the model output to the initial segment angular velocities, each of the latter is decreased by $1°s^{-1}$ while keeping the initial segmental configuration constant. This is repeated for a small increase in the parameters by the same magnitude. The results of the perturbation also indicated that the model output is insensitive to such changes except for the hand segment. Overall, the results imply that the model is generally robust to these measurement errors and this increases the confidence level in the optimized parameters.

17.8   Conclusion

Computer simulation entails the use of a validated mathematical model, to evaluate the response of the model to changes in the input parameters of a system being investigated. The application of simulation models to sports movements can ensure that the results are due to the interventions introduced. When used iteratively to determine parameter values or control variables against some performance criterion, it enables the incorporation of some optimum techniques. Mathematical formulations of modeled situations are generally of standard methods, although some are more elegant than others. However, the different methods used should not influence the outcome of the results. Evaluation of a model is thus important, as it provides evidence for the appropriateness of the outcome of the model. Evaluation is typically done by comparing the output of the model from some appropriate input data, against data collected experimentally or derived hypothetically.

Sensitivity analyses show how sensitive an optimum solution is to the calculated parameters. It is a powerful tool that can be applied to simulation models, in order to increase the level of confidence in the results obtained. It is done by perturbing the optimized results via introducing small changes to the values, and determining if the optimization process returns the parameters to the optimized value.

We have illustrated a dynamic optimization technique applied to determine an optimal Yurchenko layout vault technique for an individual gymnast, based on current performance capabilities [30,31]. The outcome of the optimization is a superior technique to the actual vault performed based on the judging criteria used in international competitions. The optimized vault has a greater flight height after horse impact as well as a more extended body than the actual performance. However, the technique requires a higher angular velocity at impact with the horse and thus greater angular momentum at impact. In addition, increased shoulder flexion is reported, and a "blocking" technique during horse impact has been identified that produces greater the vertical horse takeoff velocity than the data. The observations suggest that an effective blocking technique, a high flight angular velocity prior to horse impact, and an appropriate body-segment configuration at horse impact are key ingredients for a high scoring Yurchenko layout vault.

Appendix

With reference to Figure 17.3 and the conventions used, if the distal end of a segment's coordinate is known, say (x_e, y_e), as indicated in Equation 17.10 of the text, then the segment CM position can be expressed geometrically as follows:

For the x-coordinate,

$$CM_x = x_e + l_1 \cos \theta_1 + \cdots + l_{n-1} \cos \theta_{n-1} + r_n \cos \theta_n$$

Similarly, for the y-coordinate,

$$CM_y = y_e + l_1 \sin \theta_1 + \cdots + l_{n-1} \sin \theta_{n-1} + r_n \sin \theta_n$$

For the distal end of the leg segment, the expression simplifies to

$$[(x_e + l_1 \cos \theta_1 + \cdots + l_n \cos \theta_n, y_e + l_1 \sin \theta_1 + \cdots + l_n \sin \theta_n)]$$

References

1. Yeadon, M. and Challis, J.H. 1994. The future of performance-related sports biomechanics research. *Journal of Sports Sciences*, 16:349–356.
2. Alexander, R.M. 1990. Optimum takeoff techniques for high and long jumps. *Philosophical Transactions of the Royal Society*, B329:3–10.
3. Hatze, H. 1981. A comprehensive model for human motion simulation and its application to the takeoff phase of the long jump. *Journal of Biomechanics*, 14:13–18.

4. Barnes, S., Oggero, E., Pagnacco, G., and Berme, N. 1997. Simulation of human movement: goals, model formulation, solution techniques and considerations. In P. Allard, A. Cappozzo, A. Lundberg and C.L. Vaughan (Eds.), *Three-Dimensional Analysis of Human Locomotion*, pp. 281–306. Chichester: John Wiley & Sons.

5. Atkinson, K.E. 1989. *An Introduction to Numerical Analysis* (2nd ed.). Singapore: John Wiley & Sons.

6. King, M., Yeadon, M.R., and Kerwin, D.G. 1999. A two-segment simulation model of long horse vaulting. *Journal of Sports Sciences*, 17:313–324.

7. Koh, M.T.H. 2001. Optimal Performance of the Yurchenko layout vault (re-print) Eugene, Oregon: Microform Publications, University of Oregon.

8. Koh, M. and Jennings, L. 2007. Strategies in preflight for an optimal Yurchenko layout vault. *Journal of Biomechanics*, 40(6):1256–1261.

9. Alexander, R.M. 1992. Simple models of walking and jumping. *Human Movement Science*, 11:3–9.

10. Nigg, B.M. 1999. Modelling. In B.M. Nigg and W. Herzog (Eds.), *Biomechanics of the Musculo-Skeletal System* (2nd ed.), pp. 423–445. Chichester: John Wiley & Sons.

11. Elliott, B. and Mitchell, J. 1991. A biomechanical comparison of the yurchenko vault and two associated teaching drills. *International Journal of Sport Biomechanics*, 7:91–107.

12. Kwon, Y.H., Fortney, V.L., and Shin, I.S. 1990. 3-D analysis of yurchenko vaults performed by female gymnasts during the 1988 Seoul Olympic Games. *International Journal of Sport Biomechanics*, 6:157–176.

13. Dainis, A. 1981. A model for gymnastics vaulting. *Medicine and Science in Sports and Exercise*, 13:34–43.

14. Sprigings, E.J. and Yeadon, M. 1997. An insight into the reversal of rotation in the hecht vault. *Human Movement Science*, 16:517–532.

15. Yeadon, M. 1998. Computer simulation in sports biomechanics. In H.J. Riehle and M.M. Vieten (Eds.), *Proceedings of the XVI International Symposium on Biomechanics in Sports*, pp. 309–318. Germany, UVK: Universittsverlag Konstanz GmbH.

16. Luenberger, D.G. 1984. *Linear and Nonlinear Programming* (2nd ed.). London: Addison-Wesley.

17. Gervais, P. 1994. A prediction of an optimal performance of the handspring front salto longhorse vault. *Journal of Biomechanics*, 27:67–76.

18. Koh, M., Jennings, L.S. and Elliott, B. 2003. Role of joint torques generated in an optimised Yurchenko layout vault. *Sports Biomechanics*, 2(2):177–190.

19. Jensen, R.K. 1978. Estimation of the biomechanical properties of three body types using a photogrammetric method. *Journal of Biomechanics*, 11:349–358.

20. Winter, D.A. 1990. *Biomechanics and Motor Control of Human Movement* (2nd ed.). New York: Wiley Interscience.

21. Marshall, R., Jensen, R.K., and Wood, G. 1985. A general Newtonian simulation of an N-segment open chain method. *Journal of Biomechanics*, 18:359–367.

22. Wood, G.A. and Jennings, L.S. 1979. On the use of spline functions for data smoothing. *Journal of Biomechanics*, 12:477–479.

23. Koh, M. and Jennings, L.S. 2000. A formalism for 2-D segmented bodies (Tech. Rep. CADO 2000:21). http://www.cado.uwa.edu.au/Reports.php3. The University of Western Australia, Centre of Applied Dynamics and Optimisation.

24. Anderson, F.C. and Pandy, M.G. Static and dynamic optimization solutions for gait are practically equivalent. *Journal of Biomechanics*, 34:153–161.

25. Koh, M. and Jennings, L. 2002. Dynamic Optimisation: A solution to the inverse dynamics problem of biomechanics using MISER3. Dynamics of continuous, *Discrete and Impulsive Systems, Series B: Applications and Algorithms,* 9B(3):369–386.

26. Koh, M. and Jennings, L.S. 2003. Dynamic Optimization: Inverse analysis for the Yurchenko layout vault in women's artistic gymnastics. *Journal of Biomechanics,* 36(8):1177–1183.

27. Jennings, L.S., Fisher, M.E., Teo, K.L., and Goh, C.J. 2000. MISER3 Optimal control software (version 3): Theory and user manual. CADO, Department of Mathematics and Statistics, University of Western Australia.

28. Hocking, L.M. 1991. *Optimal Control: An Introduction to the Theory with Applications.* Oxford: Clarendon Press

29. Bohne, M., Mecham, C.J., Mitchell, K., and A-Smith, J. 2000. A biomechanical analysis of the Yurchenko layout vault. *Research Quarterly for Exercise and Sport,* 71(Suppl. 1):A-19.

30. Koh, M., Jennings, L., Elliott, B.C., and Lloyd, D. 2003. A predicted optimal performance of the Yurchenko layout vault in women's artistic gymnastics. *Journal of Applied Biomechanics,* 19(3):187–204.

31. Koh, M., Jennings, L., and Elliott, B. 2003. Role of joint torques generated in an optimized Yurchenko layout vault. *Sports Biomechanics,* 2(2):177–190.

Index

A

ABAQUS finite element package, 315
Actin filaments, 87
Actin–Myosin filaments, 73, 88, 107
Active elastance (E_a), of LV, 51–52
   as contractility index, 59–60
Airflow flow rate curve, 148
Airflow resistance (R_a), 205–206
Alveolar air composition, 179–180
Alveolar-wall displacement, 157
Alveoli, 153, 155, 174–177, 180,
   183, 202
Angiography, 52
Angle of attack, 454, 491–492, 507
Angular velocity, 425–427, 429–430,
   433, 435–436, 438–442,
   446–448, 457, 470–473,
   507, 510
Annulus fibrosus (AF), 381–382
Anterior fixation
   dynamic compression plate of,
     375–376
   requirements for, 376–378
Anthropometric measurements
   investigating dynamics of joint
     torque for, 505
   models for, 498
Antinodal points, 136–137, 139
Aorta
   blood flow and volume
     input during ejection phase,
     116–117
   diastolic and systolic pressure
     expression, 13
   elasticity, 12
   inflow rate $I(t)$, 116
   NDI, 12
   peripheral resistance, 12

   pressure response, 12
   resistance to flow in, 115
   schematic variation of pressure
     during cardiac cycle, 117
   volume elasticity of, 12
Aortic number, 14
Aortic pressure–time profile
   determination, 113–123
   application, 122
   coefficients a and b, determination of,
     119–120
   diastolic pressure $P_d(t)$ and systolic
     pressure $P_s(t)$ analysis,
     117–118
   model parameters m and λ,
     determination of, 119–121
   pressure P_2, determination of, 118
   validation of, 122–123
Aortic stiffness (arteriosclerosis)
   measurement
   clinical measure of, 124
   elasticmodulus (E) *vs.* wall
     stress (σ), 124
   pulse-wave velocity (PWV/V_p),
     124–125
Aortic valve biomechanical model,
   32–39
Arterial bifurcation, 130–133
Arterial branching, 132–133, 140
Arterial compliance, defined, 125
Arterial impedance, concept and
   implication of, 124–129
Arterial wall stress and elasticity, 11
Arteriosclerotic nondimensional
   index, 11
Asthma, 8, 153
Auscultation process, 29, 119
Automatic neuropathy, 223
Ayurvedic medicine, 114

513

H